Y0-CBF-144

Nonlinear Optics

Robert W. Boyd

The Institute of Optics
University of Rochester
Rochester, New York

Academic Press
San Diego New York Boston
London Sydney Tokyo Toronto

This book is printed on acid-free paper. ♾

Academic Press
A Division of Harcourt Brace & Company
525 B Street, Suite 1900, San Diego, California 92101-4495
http://www.apnet.com

United Kingdom Edition published by
ACADEMIC PRESS LIMITED
24–28 Oval Road, London NW1 7DX

Library of Congress Cataloging-in-Publication Data:

Boyd, Robert W., date.
Nonlinear optics / Robert W. Boyd.
p. cm.
Includes bibliographical references and index.
ISBN 0-12-121680-2 (alk. paper)
1. Nonlinear optics. I. Title.
QC446.2.B69 1992 91-9903
535.2—dc20 CIP

Printed In The United States Of America
99 00 01 02 SB 12 11 10 9 8 7

for Katherine, Jessica, John, and Brendan

Contents

Preface

Nonlinear optics is the study of the interaction of intense laser light with matter. This book is a textbook on nonlinear optics at the level of a beginning graduate student. The intent of the book is to provide an introduction to the field of nonlinear optics that stresses fundamental concepts and that enables the student to go on to perform independent research in this field. The author has successfully used a preliminary version of this book in his course at the University of Rochester, which is typically attended by students ranging from seniors to advanced PhD students from disciplines that include optics, physics, chemistry, electrical engineering, mechanical engineering, and chemical engineering. This book could be used in graduate courses in the areas of nonlinear optics, quantum optics, quantum electronics, laser physics, electrooptics, and modern optics. By deleting some of the more difficult sections, this book would also be suitable for use by advanced undergraduates. On the other hand, some of the material in the book is rather advanced and would be suitable for senior graduate students and research scientists.

The field of nonlinear optics is now thirty years old, if we take its beginnings to be the observation of second-harmonic generation by Franken and coworkers in 1961. Interest in this field has grown continuously since its beginnings, and the field of nonlinear optics now ranges from fundamental studies of the interaction of light with matter to applications such as laser frequency conversion and optical switching. In fact, the field of nonlinear optics has grown so enormously that it is not possible for one book to cover all of the topics of current interest. In addition, since I want this book to be accessible to beginning graduate students, I have attempted to treat the topics that are covered in a reasonably self-contained manner. This consideration also restricts the number of topics that can be treated. My strategy in deciding what topics to include has been to stress the fundamental aspects of nonlinear optics, and to include applications and experimental results only as necessary to illustrate these fundamental issues. Many of the specific topics that I have chosen to include are those of particular historical value.

Nonlinear optics is notationally very complicated, and unfortunately much of the notational complication is unavoidable. Because the notational aspects of nonlinear optics have historically been very confusing, considerable effort is made, especially in the early chapters, to explain the notational conventions. The book uses primarily the Gaussian system of units, both to establish a connection with the historical papers of nonlinear optics, most of which were written using the Gaussian system, and also because the author believes that the laws of electromagnetism are more physically transparent when written in this system. At several places in the text (see especially the appendices at the end of the book), tables are provided to facilitate conversion to other systems of units.

The book is organized as follows: Chapter 1 presents an introduction to the field of nonlinear optics from the perspective of the nonlinear susceptibility. The nonlinear susceptibility is a quantity that is used to determine the nonlinear polarization of a material medium in terms of the strength of an applied optical-frequency electric field. It thus provides a framework for describing nonlinear optical phenomena. Chapter 2 continues the description of nonlinear optics by describing the propagation of light waves through nonlinear optical media by means of the optical wave equation. This chapter introduces the important concept of phase matching and presents detailed descriptions of the important nonlinear optical phenomena of second-harmonic generation and sum- and difference-frequency generation. Chapter 3 concludes the introductory portion of the book by presenting a description of the quantum mechanical theory of the nonlinear optical susceptibility. Simplified expressions for the nonlinear susceptibility are first derived through use of the Schrödinger equation, and then more accurate expressions are derived through use of the density matrix equations of motion. The density matrix formalism is itself developed in considerable detail in this chapter in order to render this important discussion accessible to the beginning student.

Chapters 4 through 6 deal with properties and applications of the nonlinear refractive index. Chapter 4 introduces the topic of the nonlinear refractive index. Properties, including tensor properties, of the nonlinear refractive index are discussed in detail, and physical processes that lead to the nonlinear refractive index, such as nonresonant electronic polarization and molecular orientation, are described. Chapter 5 is devoted to a description of nonlinearities in the refractive index resulting from the response of two-level atoms. Related topics that are discussed in this chapter include saturation, power broadening, optical Stark shifts, Rabi oscillations, and dressed atomic states. Chapter 6 deals with applications of the nonlinear refractive index. Topics that are included are optical phase conjugation, self focusing, optical bistability, two-beam coupling, pulse propagation, and the formation of optical solitons.

Chapters 7 through 9 deal with spontaneous and stimulated light scattering and the related topic of acoustooptics. Chapter 7 introduces this area by presenting a description of theories of spontaneous light scattering and by describing the important practical topic of acousto-optics. Chapter 8 presents a description of stimulated Brillouin and stimulated Rayleigh scattering. These topics are related in that they both entail the scattering of light from material disturbances that can be described in terms of the standard thermodynamic variables of pressure and entropy. Also included in this chapter is a description of phase conjugation by stimulated Brillouin scattering and a theoretical description of stimulated Brillouin scattering in gases. Chapter 9 presents a description of stimulated Raman and stimulated Rayleigh-wing scattering. These processes are related in that they entail the scattering of light from disturbances associated with the positions of atoms within a molecule.

The book concludes with Chapter 10, which treats the electrooptic and photorefractive effects. The chapter begins with a description of the electrooptic effect and describes how this effect can be used to fabricate light modulators. The chapter then presents a description of the photorefractive effect, which is a nonlinear optical interaction that results from the electrooptic effect. The use of the photorefractive effect in two-beam coupling and in four-wave mixing is also described.

The author wishes to acknowledge his deep appreciation for discussions of the material in this book with his graduate students at the University of Rochester. He is sure that he has learned as much from them as they have from him. He also gratefully acknowledges discussions with numerous other professional colleagues, including N. Bloembergen, D. Chemla, R. Y. Chiao, J. H. Eberly, C. Flytzanis, J. Goldhar, G. Grynberg, J. H. Haus, R. W. Hellwarth, K. R. MacDonald, S. Mukamel, P. Narum, M. G. Raymer, J. E. Sipe, C. R. Stroud, Jr., C. H. Townes, H. Winful, and B. Ya. Zel'dovich. In addition, the assistance of J. J. Maki and A. Gamliel in the preparation of the figures is gratefully acknowledged.

Chapter 1

The Nonlinear Optical Susceptibility

1.1. Introduction to Nonlinear Optics

Nonlinear optics is the study of phenomena that occur as a consequence of the modification of the optical properties of a material system by the presence of light. Typically, only laser light is sufficiently intense to modify the optical properties of a material system. In fact, the beginning of the field of nonlinear optics is often taken to be the discovery of second-harmonic generation by Franken *et al.* in 1961, shortly after the demonstration of the first working laser by Maiman in 1960. Nonlinear optical phenomena are "nonlinear" in the sense that they occur when the response of a material system to an applied optical field depends in a nonlinear manner upon the strength of the optical field. For example, second-harmonic generation occurs as a result of the part of the atomic response that depends quadratically on the strength of the applied optical field. Consequently, the intensity of the light generated at the second-harmonic frequency tends to increase as the square of the intensity of the applied laser light.

In order to describe more precisely what we mean by an optical nonlinearity, let us consider how the dipole moment per unit volume, or polarization $\tilde{P}(t)$, of a material system depends upon the strength $\tilde{E}(t)$ of the applied optical field.* In the case of conventional (i.e., linear) optics, the induced

* Throughout the text, we use the tilde to denote a quantity that varies rapidly in time. Constant quantities, slowly varying quantities, and Fourier amplitudes are written without the tilde. See, for example, Eq. (1.2.1).

polarization depends linearly upon the electric field strength in a manner that can often be described by the relationship

$$\tilde{P}(t) = \chi^{(1)}\tilde{E}(t), \tag{1.1.1}$$

where the constant of proportionality $\chi^{(1)}$ is known as the linear susceptibility. In nonlinear optics, the nonlinear optical response can often be described by generalizing Eq. (1.1.1) by expressing the polarization $\tilde{P}(t)$ as a power series in the field strength $\tilde{E}(t)$ as

$$\begin{aligned} \tilde{P}(t) &= \chi^{(1)}\tilde{E}(t) + \chi^{(2)}\tilde{E}^2(t) + \chi^{(3)}\tilde{E}^3(t) + \cdots \\ &\equiv \tilde{P}^{(1)}(t) + \tilde{P}^{(2)}(t) + \tilde{P}^{(3)}(t) + \cdots. \end{aligned} \tag{1.1.2}$$

The quantities $\chi^{(2)}$ and $\chi^{(3)}$ are known as the second- and third-order nonlinear optical susceptibilities, respectively. For simplicity, we have taken the fields $\tilde{P}(t)$ and $\tilde{E}(t)$ to be scalar quantities in writing Eqs. (1.1.1) and (1.1.2). In Section 1.3 we show how to treat the vector nature of the fields; in such a case $\chi^{(1)}$ becomes a second-rank tensor, $\chi^{(2)}$ becomes a third-rank tensor, etc. In writing Eqs. (1.1.1) and (1.1.2) in the form shown, we have also assumed that the polarization at time t depends only on the instantaneous value of the electric field strength. The assumption that the medium responds instantaneously also implies (through the Kramers–Kronig relations)* that the medium must be lossless and dispersionless. We shall see in Section 1.3 how to generalize these equations for the case of a medium with dispersion and loss. In general, the nonlinear susceptibilities depend on the frequencies of the applied fields, but under our present assumption of instantaneous response we take them to be constants.

We shall refer to $\tilde{P}^{(2)}(t) = \chi^{(2)}\tilde{E}(t)^2$ as the second-order nonlinear polarization and to $\tilde{P}^{(3)}(t) = \chi^{(3)}\tilde{E}(t)^3$ as the third-order nonlinear polarization. We shall see later in this section that the physical processes that occur as a result of the second-order polarization $\tilde{P}^{(2)}$ are distinct from those that occur as a result of the third-order polarization $\tilde{P}^{(3)}$. In addition, we shall show in Section 1.5 that second-order nonlinear optical interactions can occur only in noncentrosymmetric crystals, that is, in crystals that do not display inversion symmetry. Since liquids, gases, amorphous solids (such as glass), and even many crystals do display inversion symmetry, $\chi^{(2)}$ vanishes identically for such media, and consequently they cannot produce second-order nonlinear optical interactions. On the other hand, third-order nonlinear optical interac-

* See, for example, Loudon (1983) for a discussion of the Kramers–Kronig relations.

tions (i.e., those described by a $\chi^{(3)}$ susceptibility) can occur both for centrosymmetric and noncentrosymmetric media.

We shall see in later sections of this book how to calculate the values of the nonlinear susceptibilities for various physical mechanisms that can lead to optical nonlinearities. For the present, we shall make a simple order-of-magnitude estimate of the size of these quantities for the common case in which the nonlinearity is electronic in origin. One might expect that the lowest-order correction term $\tilde{P}^{(2)}$ would be comparable to the linear response $\tilde{P}^{(1)}$ when the amplitude of the applied field strength $\tilde{E}$ was of the order of the characteristic atomic electric field strength $E_{\text{at}} = e/a_0^2$, where $-e$ is the charge of the electron and $a_0 = \hbar^2/me^2$ is the Bohr radius of the hydrogen atom (here $\hbar$ is Planck's constant divided by 2π, and m is the mass of the electron). Numerically, we find that $E_{\text{at}} = 2 \times 10^7$ esu.* We thus expect that under conditions of nonresonant excitation the second-order susceptibility $\chi^{(2)}$ will be of the order of $\chi^{(1)}/E_{\text{at}}$. For condensed matter $\chi^{(1)}$ is of the order of unity, and we hence expect that $\chi^{(2)}$ will be of the order of $1/E_{\text{at}}$, or that

$$\chi^{(2)} \simeq 5 \times 10^{-8} \text{ esu} = 5 \times 10^{-8} \frac{\text{cm}}{\text{statvolt}}. \tag{1.1.3}$$

Similarly, we expect $\chi^{(3)}$ to be of the order of $\chi^{(1)}/E_{\text{at}}^2$, which for condensed matter is of the order of

$$\chi^{(3)} \simeq 3 \times 10^{-15} \text{ esu} = 3 \times 10^{-15} \frac{\text{cm}^2}{\text{statvolt}^2}. \tag{1.1.4}$$

The most common procedure for describing nonlinear optical phenomena is based on expressing the polarization $\tilde{P}(t)$ in terms of the applied electric field strength $\tilde{E}(t)$, as we have done in Eq. (1.1.2). The reason why the polarization plays a key role in the description of nonlinear optical phenomena is that a time-varying polarization can act as the source of new components of the electromagnetic field. For example, we shall see in Section 2.1 that the wave equation in nonlinear optical media often has the form

$$\nabla^2 \tilde{E} - \frac{n^2}{c^2}\frac{\partial^2 \tilde{E}}{\partial t^2} = \frac{4\pi}{c^2}\frac{\partial^2 \tilde{P}}{\partial t^2}, \tag{1.1.5}$$

* Except where otherwise noted, we use the gaussian system of units in this book. Following standard conventions, we usually do not give the dimensions of physical quantities explicitly, but instead quote values simply in gaussian units or (equivalently for all cases treated herein) in electrostatic units, abbreviated esu. For the present case, the dimensions of E_{at} are statvolt/cm. See also the discussion in the appendix to this book on the conversion between systems of units.

where n is the refractive index and c is the speed of light in vacuum. We can interpret this expression as an inhomogeneous wave equation in which the polarization $\tilde{P}$ drives the electric field $\tilde{E}$. This equation expresses the fact that, whenever $\partial^2\tilde{P}/\partial t^2$ is nonzero, charges are being accelerated, and according to Larmor's theorem from electromagnetism, accelerated charges generate electromagnetic radiation.

1.2. Descriptions of Nonlinear Optical Interactions

In the present section, we present brief qualitative descriptions of a number of nonlinear optical interactions. In addition, for those processes that can occur in a lossless medium, we indicate how they can be described in terms of the nonlinear contributions to the polarization described by Eq. (1.1.2).* Our motivation is to provide the reader with an indication of the variety of nonlinear optical phenomena that can occur. These interactions are described in greater detail in later sections of this book. In this section we also introduce some notational conventions and some of the basic concepts of nonlinear optics.

Second-Harmonic Generation

As an example of a nonlinear optical interaction, let us consider the process of second-harmonic generation, which is illustrated schematically in Fig. 1.2.1.

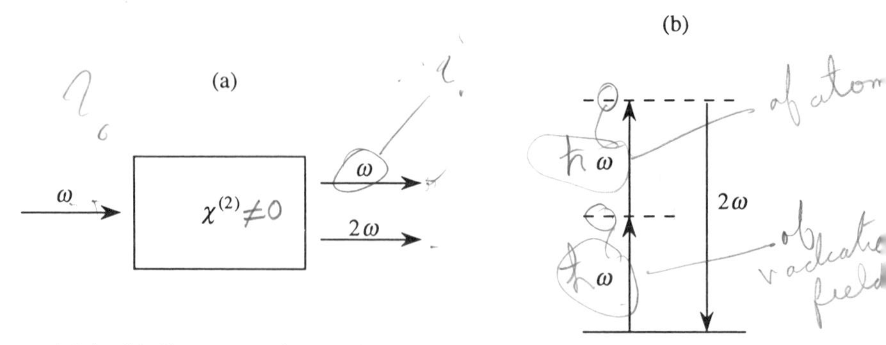

FIGURE 1.2.1 (a) Geometry of second-harmonic generation. (b) Energy-level diagram describing second-harmonic generation.

* Recall that Eq. (1.1.2) is valid only for a medium that is lossless and dispersionless.

Here a laser beam whose electric field strength is represented as

$$\tilde{E}(t) = Ee^{-i\omega t} + \text{c.c.} \tag{1.2.1}$$

is incident upon a crystal for which the second-order susceptibility $\chi^{(2)}$ is nonzero. The nonlinear polarization that is created in such a crystal is given according to Eq. (1.1.2) as $\tilde{P}^{(2)}(t) = \chi^{(2)}\tilde{E}^2(t)$ or as

$$\tilde{P}^{(2)}(t) = 2\chi^{(2)}EE^* + (\chi^{(2)}E^2e^{-2i\omega t} + \text{c.c.}). \tag{1.2.2}$$

We see that the second-order polarization consists of a contribution at zero frequency (the first term) and a contribution at frequency 2ω (the second term). According to the driven wave equation (1.1.5), this latter contribution can lead to the generation of radiation at the second-harmonic frequency. Note that the first contribution in Eq. (1.2.2) does not lead to the generation of electromagnetic radiation (because its second time derivative vanishes); it leads to a process known as optical rectification in which a static electric field is created within the nonlinear crystal.

Under proper experimental conditions, the process of second-harmonic generation can be so efficient that nearly all of the power in the incident radiation at frequency ω is converted to radiation at the second-harmonic frequency 2ω. One common use of second-harmonic generation is to convert the output of a fixed-frequency laser into a different spectral region. For example, the Nd:YAG laser operates in the near infrared at a wavelength of 1.06 μm. Second-harmonic generation is routinely used to convert the wavelength of the radiation to 0.53 μm, in the middle of the visible spectrum.

Second-harmonic generation can also be visualized by considering the interaction in terms of the exchange of photons between the various frequency components of the field. According to this picture, which is illustrated in part (b) of Fig. 1.2.1, two photons of frequency ω are destroyed and a photon of frequency 2ω is simultaneously created in a single quantum-mechanical process. The solid line in the figure represents the atomic ground state, and the dashed lines represent what are known as virtual levels. These levels are not energy eigenlevels of the free atom, but rather represent the combined energy of one of the energy eigenstates of the atom and of one or more photons of the radiation field.

The theory of second-harmonic generation is developed more fully in Section 2.6.

Sum- and Difference-Frequency Generation

Let us next consider the circumstance in which the optical field incident upon a nonlinear optical medium characterized by a nonlinear susceptibility $\chi^{(2)}$ consists of two distinct frequency components, which we represent in the form

$$\tilde{E}(t) = E_1 e^{-i\omega_1 t} + E_2 e^{-i\omega_2 t} + \text{c.c.} \tag{1.2.3}$$

Then, assuming as in Eq. (1.1.2) that the second-order contribution to the nonlinear polarization is of the form

$$\tilde{P}^{(2)}(t) = \chi^{(2)}\tilde{E}(t)^2, \tag{1.2.4}$$

we find that the nonlinear polarization is given by

$$\begin{aligned}\tilde{P}^{(2)}(t) = \chi^{(2)}[&E_1^2 e^{-2i\omega_1 t} + E_2^2 e^{-i2\omega_2 t} + 2E_1E_2 e^{-(\omega_1+\omega_2)t} \\ &+ 2E_1E_2^* e^{-i(\omega_1-\omega_2)t} + \text{c.c.}] + 2\chi^{(2)}[E_1E_1^* + E_2E_2^*].\end{aligned} \tag{1.2.5}$$

It is convenient to express this result using the notation

$$\tilde{P}^{(2)}(t) = \sum_n P(\omega_n)e^{-i\omega_n t}, \tag{1.2.6}$$

where the summation extends over positive and negative frequencies ω_n. The complex amplitudes of the various frequency components of the nonlinear polarization are hence given by

$$\begin{aligned}P(2\omega_1) &= \chi^{(2)}E_1^2 \quad \text{(SHG)},\\ P(2\omega_2) &= \chi^{(2)}E_2^2 \quad \text{(SHG)},\\ P(\omega_1+\omega_2) &= 2\chi^{(2)}E_1E_2 \quad \text{(SFG)},\\ P(\omega_1-\omega_2) &= 2\chi^{(2)}E_1E_2^* \quad \text{(DFG)},\\ P(0) &= 2\chi^{(2)}(E_1E_1^* + E_2E_2^*) \quad \text{(OR)}.\end{aligned} \tag{1.2.7}$$

Here we have labeled each expression by the name of the physical process that it describes, such as second-harmonic generation (SHG), sum-frequency generation (SFG), difference-frequency generation (DFG), and optical rectification (OR). Note that, in accordance with our complex notation, there is also a response at the negative of each of the nonzero frequencies given above:

$$\begin{aligned}P(-2\omega_1) &= \chi^{(2)}E_1^{*2}, & P(-2\omega_2) &= \chi^{(2)}E_2^{*2},\\ P(-\omega_1-\omega_2) &= 2\chi^{(2)}E_1^*E_2^* & P(\omega_2-\omega_1) &= 2\chi^{(2)}E_2E_1^*.\end{aligned} \tag{1.2.8}$$

However, since each of these quantities is simply the complex conjugate of one of the quantities given in Eq. (1.2.7), it is not necessary to take explicit account of both the positive and negative frequency components.*

We see from Eq. (1.2.7) that four different nonzero frequency components are present in the nonlinear polarization. However, typically no more than one of these frequency components will be present with any appreciable intensity in the radiation generated by the nonlinear optical interaction. The reason for this behavior is that the nonlinear polarization can efficiently produce an output signal only if a certain phase-matching condition (which is discussed in detail in Section 2.7) is satisfied, and usually this condition cannot be satisfied for more than one frequency component of the nonlinear polarization. Operationally, one often chooses which frequency component will be radiated by properly selecting the polarization of the input radiation and orientation of the nonlinear crystal.

Sum-Frequency Generation

Let us now consider the process of sum-frequency generation, which is illustrated in Fig. 1.2.2. According to Eq. (1.2.7), the complex amplitude of the nonlinear polarization describing this process is given by the expression

$$P(\omega_1 + \omega_2) = 2\chi^{(2)}E_1E_2. \tag{1.2.9}$$

In many ways the process of sum-frequency generation is analogous to that of second-harmonic generation, except that in sum-frequency generation the two input waves are at different frequencies. One application of sum-frequency generation is to produce tunable radiation in the ultraviolet spectral region by

* Not all workers in nonlinear optics use our convention that the fields and polarizations are given by Eqs. (1.2.3) and (1.2.6). Another common convention is to define the field amplitudes according to

$$\tilde{E}(t) = \tfrac{1}{2}(E_1' e^{-i\omega_1 t} + E_2' e^{-i\omega_2 t} + \text{c.c.}),$$
$$\tilde{P}(t) = \tfrac{1}{2}\sum_n P'(\omega_n) e^{-i\omega_n t},$$

where in the second expression the summation extends over all positive and negative frequencies. Using this convention, one finds that

$$P'(2\omega_1) = \tfrac{1}{2}\chi^{(2)}E_1'^2, \qquad P'(2\omega_2) = \tfrac{1}{2}\chi^{(2)}E_2'^2,$$
$$P'(\omega_1 + \omega_2) = \chi^{(2)}E_1'E_2', \qquad P'(\omega_1 - \omega_2) = \chi^{(2)}E_1'E_2'^*,$$
$$P'(0) = \chi^{(2)}(E_1'E_1'^* + E_2'E_2'^*).$$

Note that these expressions differ from Eqs. (1.2.7) by factors of $\tfrac{1}{2}$.

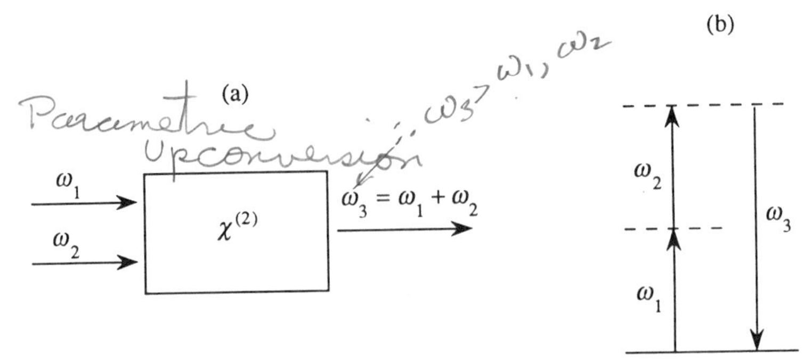

FIGURE 1.2.2 Sum-frequency generation. (a) Geometry of the interaction. (b) Energy-level description.

choosing one of the input waves to be the output of a fixed-frequency visible laser and the other to be the output of a frequency-tunable visible laser. The theory of sum-frequency generation is developed more fully in Sections 2.2 and 2.4.

Difference-Frequency Generation

The process of difference-frequency generation is described by a nonlinear polarization of the form

$$P(\omega_1 - \omega_2) = 2\chi^{(2)} E_1 E_2^* \tag{1.2.10}$$

and is illustrated in Fig. 1.2.3. Here the frequency of the generated wave is the difference of those of the applied fields. Difference-frequency generation can be used to produce tunable infrared radiation by mixing the output of a frequency-tunable visible laser with that of a fixed-frequency visible laser.

Superficially, difference-frequency generation and sum-frequency generation appear to be very similar processes. However, an important difference between the two processes can be deduced from the description of difference-frequency generation in terms of a photon energy-level diagram (part (b) of Fig. 1.2.3). We see that conservation of energy requires that for every photon that is created at the difference frequency $\omega_3 = \omega_1 - \omega_2$, a photon at the higher input frequency (ω_1) must be destroyed and a photon at the lower input frequency (ω_2) must be created. Thus, the lower-frequency input field is amplified by the process of difference-frequency generation. For this reason,

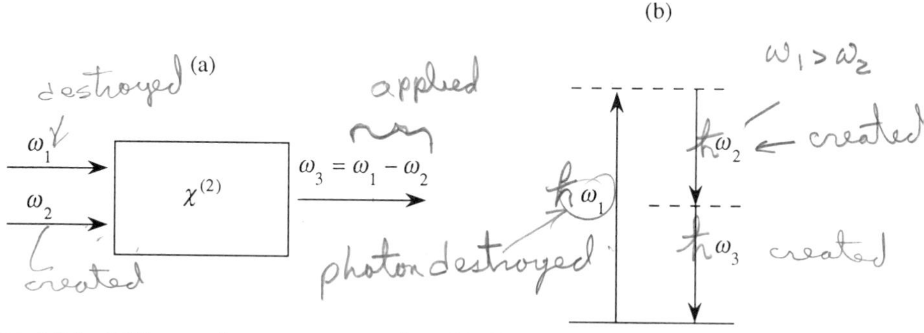

FIGURE 1.2.3 Difference-frequency generation. (a) Geometry of the interaction. (b) Energy-level description.

the process of difference-frequency generation is also known as optical parametric amplification.

According to the photon energy-level description of difference-frequency generation, the atom first absorbs a photon of frequency ω_1 and jumps to the highest virtual level. This level decays by a two-photon emission process that is stimulated by the presence of the ω_2 field, which is already present. Two-photon emission can occur even if the ω_2 field is not applied. The generated fields in such a case are very much weaker, since they are created by *spontaneous* two-photon emission from a virtual level. This process is known as parametric fluorescence and has been observed experimentally (Harris *et al.*, 1967; Byer and Harris, 1968).

The theory of difference-frequency generation is developed more fully in Section 2.5.

Optical Parametric Oscillation

We have just seen that in the process of difference-frequency generation the presence of radiation at frequency ω_2 or ω_3 can stimulate the emission of additional photons at these frequencies. If the nonlinear crystal used in this process is placed inside an optical resonator, as shown in Fig. 1.2.4, the ω_2 and/or ω_3 fields can build up to large values. Such a device is known as an optical parametric oscillator. Optical parametric oscillators are used primarily at infrared wavelengths, where other sources of tunable radiation are not readily available. Such a device is tunable because any frequency ω_2 (that is less than ω_1) can satisfy the condition $\omega_2 + \omega_3 = \omega_1$ for some frequency ω_3. In

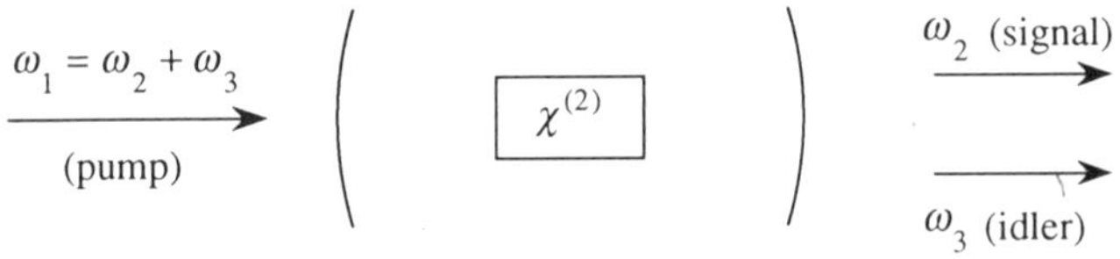

$\omega_1 = \omega_2 + \omega_3$ (pump) $\chi^{(2)}$ ω_2 (signal) ω_3 (idler)

FIGURE 1.2.4 The optical parametric oscillator. The cavity end mirrors have high reflectivity at frequencies ω_2 and/or ω_3.

practice, one controls the output frequency by adjusting the phase-matching condition, as discussed in Section 2.7. The applied field frequency ω_1 is often called the pump frequency, the desired output frequency is called the signal frequency, and the other, unwanted, output frequency is called the idler frequency.

Third-Order Polarization

We next consider the third-order contribution to the nonlinear polarization

$$\tilde{P}^{(3)}(t) = \chi^{(3)}\tilde{E}(t)^3. \tag{1.2.11}$$

For the general case in which the field $\tilde{E}(t)$ is made up of several different frequency components, the expression for $\tilde{P}^{(3)}(t)$ is very complicated. For this reason, we first consider the simple case in which the applied field is monochromatic and is given by

$$\tilde{E}(t) = \mathscr{E} \cos \omega t. \tag{1.2.12}$$

Then, since $\cos^3 \omega t = \frac{1}{4}\cos 3\omega t + \frac{3}{4}\cos \omega t$, the nonlinear polarization can be expressed as

$$\tilde{P}^{(3)}(t) = \tfrac{1}{4}\chi^{(3)}\mathscr{E}^3 \cos 3\omega t + \tfrac{3}{4}\chi^{(3)}\mathscr{E}^3 \cos \omega t. \tag{1.2.13}$$

The significance of each of the two terms in this expression is described briefly below.

Third-Harmonic Generation

The first term in Eq. (1.2.13) describes a response at frequency 3ω that is due to an applied field at frequency ω. This term leads to the process of third-harmonic generation, which is illustrated in Fig. 1.2.5. According to the photon description of this process, shown in part (b) of the figure, three photons of frequency ω are destroyed and one photon of frequency 3ω is created in each elementary event.

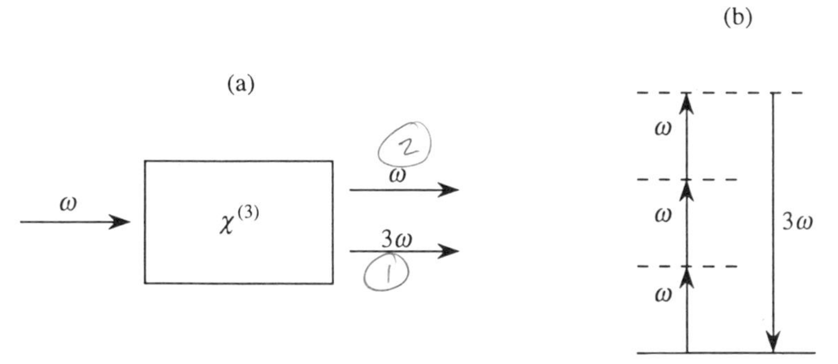

FIGURE 1.2.5 Third-harmonic generation. (a) Geometry of the interaction. (b) Energy-level description.

Intensity-Dependent Refractive Index

The second term in Eq. (1.2.13) describes a nonlinear contribution to the polarization at the frequency of the incident field; this term hence leads to a nonlinear contribution to the refractive index experienced by a wave at frequency ω. We shall see in Section 4.1 that the refractive index in the presence of this type of nonlinearity can be represented as

$$n = n_0 + n_2 I \tag{1.2.14a}$$

where n_0 is the usual (i.e., linear or low-intensity) refractive index, where

$$n_2 = \frac{12\pi^2}{n_0^2 c} \chi^{(3)} \tag{1.2.14b}$$

is an optical constant that characterizes the strength of the optical nonlinearity, and where $I = (n_0 c/8\pi)\mathcal{E}^2$ is the intensity of the incident wave.

Self-Focusing. One of the processes that can occur as a result of the intensity-dependent refractive index is self-focusing, which is illustrated in Fig. 1.2.6. This process can occur when a beam of light having a nonuniform transverse intensity distribution propagates through a material in which n_2 is positive. Under these conditions, the material effectively acts as a positive lens, which causes the rays to curve toward each other. This process is of great practical importance because the intensity at the focal spot of the self-focused beam is usually sufficiently large to lead to optical damage of the material. The process of self-focusing is described in greater detail in Section 6.2.

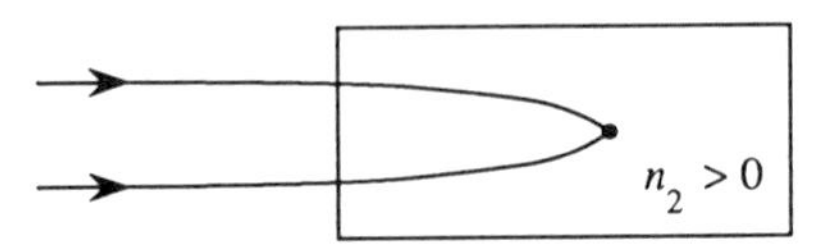

FIGURE 1.2.6 Self-focusing of light.

Third-Order Polarization (General Case)

Let us next examine the form of the nonlinear polarization

$$\tilde{P}(t) = \chi^{(3)}\tilde{E}(t)^3 \tag{1.2.15a}$$

induced by an applied field that consists of three frequency components:

$$\tilde{E}(t) = E_1 e^{-i\omega_1 t} + E_2 e^{-i\omega_2 t} + E_3 e^{-i\omega_3 t} + \text{c.c.} \tag{1.2.15b}$$

When we calculate $\tilde{E}^3(t)$, we find that the resulting expression contains 44 different frequency components, if we consider positive and negative frequencies to be distinct. Explicitly, these frequencies are

$$\omega_1, \omega_2, \omega_3, 3\omega_1, 3\omega_2, 3\omega_3, (\omega_1 + \omega_2 + \omega_3), (\omega_1 + \omega_2 - \omega_3),$$
$$(\omega_1 + \omega_3 - \omega_2), (\omega_2 + \omega_3 - \omega_1), (2\omega_1 \pm \omega_2), (2\omega_1 \pm \omega_3), (2\omega_2 \pm \omega_1),$$
$$(2\omega_2 \pm \omega_3), (2\omega_3 \pm \omega_1), (2\omega_3 \pm \omega_2),$$

and the negative of each. Again representing the nonlinear polarization as

$$\tilde{P}^{(3)}(t) = \sum_n P(\omega_n) e^{-i\omega_n t}, \tag{1.2.16}$$

we can write the complex amplitudes of the nonlinear polarization for the positive frequencies as

$$P(\omega_1) = \chi^{(3)}(3E_1E_1^* + 6E_2E_2^* + 6E_3E_3^*)E_1,$$
$$P(\omega_2) = \chi^{(3)}(6E_1E_1^* + 3E_2E_2^* + 6E_3E_3^*)E_2,$$
$$P(\omega_3) = \chi^{(3)}(6E_1E_1^* + 6E_2E_2^* + 3E_3E_3^*)E_3,$$
$$P(3\omega_1) = \chi^{(3)}E_1^3, \qquad P(3\omega_2) = \chi^{(3)}E_2^3, \qquad P(3\omega_3) = \chi^{(3)}E_3^3,$$
$$P(\omega_1 + \omega_2 + \omega_3) = 6\chi^{(3)}E_1E_2E_3, \qquad P(\omega_1 + \omega_2 - \omega_3) = 6\chi^{(3)}E_1E_2E_3^*,$$
$$P(\omega_1 + \omega_3 - \omega_2) = 6\chi^{(3)}E_1E_3E_2^*, \qquad P(\omega_2 + \omega_3 - \omega_1) = 6\chi^{(3)}E_2E_3E_1^*,$$
$$P(2\omega_1 + \omega_2) = 3\chi^{(3)}E_1^2E_2, \qquad P(2\omega_1 + \omega_3) = 3\chi^{(3)}E_1^2E_3,$$
$$P(2\omega_2 + \omega_1) = 3\chi^{(3)}E_2^2E_1, \qquad P(2\omega_2 + \omega_3) = 3\chi^{(3)}E_2^2E_3,$$

$$
\begin{aligned}
P(2\omega_3+\omega_1) &= 3\chi^{(3)}E_3^2E_1, & P(2\omega_3+\omega_2) &= 3\chi^{(3)}E_3^2E_2,\\
P(2\omega_1-\omega_2) &= 3\chi^{(3)}E_1^2E_2^*, & P(2\omega_1-\omega_3) &= 3\chi^{(3)}E_1^2E_3^*,\\
P(2\omega_2-\omega_1) &= 3\chi^{(3)}E_2^2E_1^*, & P(2\omega_2-\omega_3) &= 3\chi^{(3)}E_2^2E_3^*,\\
P(2\omega_3-\omega_1) &= 3\chi^{(3)}E_3^2E_1^*, & P(2\omega_3-\omega_2) &= 3\chi^{(3)}E_3^2E_2^*.
\end{aligned}
\tag{1.2.17}
$$

We have displayed these expressions in complete detail because it is very instructive to study their form. In each case the frequency argument of P is equal to the sum of the frequencies associated with the field amplitudes appearing on the right-hand side of the equation, if we adopt the convention that a negative frequency is to be associated with a field amplitude that appears as a complex conjugate. Also, the numerical factor (1, 3, or 6) that appears in each term on the right-hand side of each equation is equal to the number of distinct permutations of the field frequencies that contribute to that term.

Some of the nonlinear optical mixing processes described by Eq. (1.2.17) are illustrated in Fig. 1.2.7.

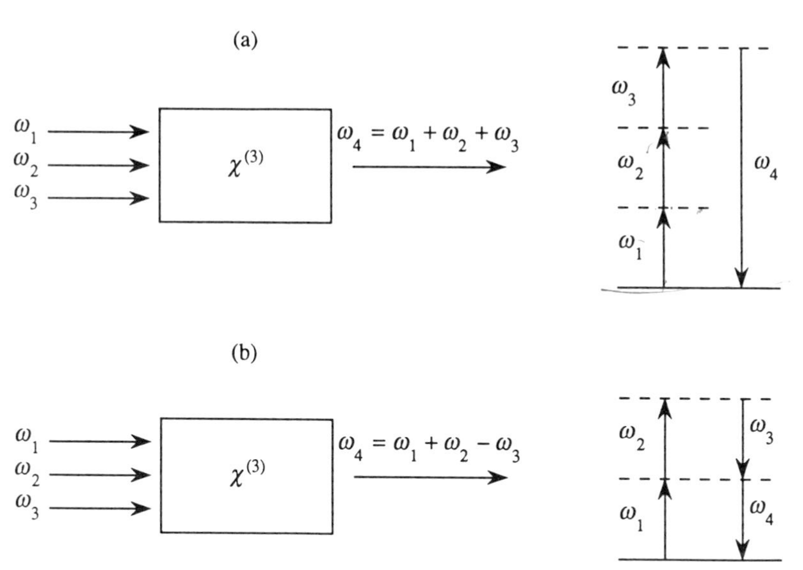

FIGURE 1.2.7 Two of the possible mixing processes described by Eq. (1.2.17) that can occur when three input waves interact in a medium characterized by a $\chi^{(3)}$ susceptibility.

Parametric versus Nonparametric Processess

All of the processes described thus far in this chapter are examples of what are known as parametric processes. The origin of this terminology is obscure, but the word parametric has come to denote a process in which the initial and final quantum-mechanical states of the system are identical. Consequently, in a parametric process population can be removed from the ground state only for those brief intervals of time when it resides in a virtual level. According to the uncertainty principle, population can reside in a virtual level for a time interval of the order of $\hbar/\Delta E$, where ΔE is the energy difference between the virtual level and the nearest real level. Conversely, processes that do involve the transfer of population from one real level to another are known as nonparametric processes. The processes that we describe in the remainder of the present section are all examples of nonparametric processes.

One difference between parametric and nonparametric processes is that parametric processes can always be described by a real susceptibility; conversely, nonparametric processes are described by a complex susceptibility by means of a procedure described in Section 1.2. Another difference is that photon energy is always conserved in a parametric process; photon energy need not be conserved in a nonparametric process, because energy can be transferred to or from the material medium. For this reason, photon energy level diagrams of the sort shown in Figs. 1.2.1, 1.2.2, 1.2.3, 1.2.5, and 1.2.7 to describe parametric processes play a less definitive role in describing nonparametric processes.

As a simple example of the distinction between parametric and nonparametric processes, we consider the case of the usual (linear) index of refraction. The real part of the refractive index is a consequence of parametric processes, whereas its imaginary part is a consequence of nonparametric processes, since the imaginary part of the refractive index describes the absorption of radiation, which results from the transfer of population from the atomic ground state to an excited state.

Saturable Absorption

One example of a nonparametric nonlinear optical process is saturable absorption. Many material systems have the property that their absorption coefficient decreases when measured using high laser intensity. Often the dependence of the measured absorption coefficient α on the intensity I of the incident laser radiation is given by the expression*

$$\alpha = \frac{\alpha_0}{1 + I/I_s}, \tag{1.2.18}$$

where α_0 is the low-intensity absorption coefficient, and I_s is a parameter known as the saturation intensity.

Optical Bistability. One consequence of saturable absorption is optical bistability. One way of forming a bistable optical device is to place a saturable absorber inside a Fabry–Perot resonator, as illustrated in Fig. 1.2.8. As the input intensity is increased, the field inside the cavity also increases, lowering the absorption that the field experiences and thus increasing the field intensity still further. If the intensity of the incident field is subsequently lowered, the field inside the cavity tends to remain large because the absorption of the material system has already been reduced. A plot of the input-versus-output characteristics thus looks qualitatively like that shown in Fig. 1.2.9. Note that

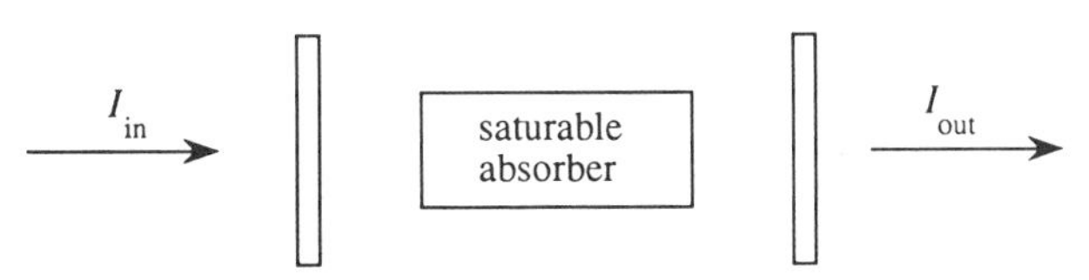

FIGURE 1.2.8 Bistable optical device.

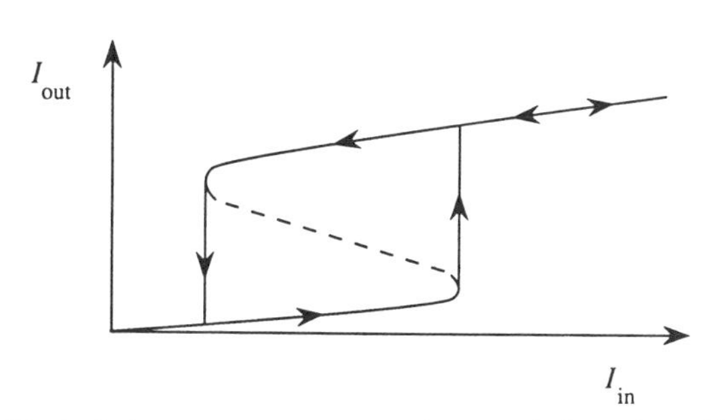

FIGURE 1.2.9 Typical input-versus-output characteristics of a bistable optical device.

* This form is valid for the case of homogeneous broadening of a simple atomic transition.

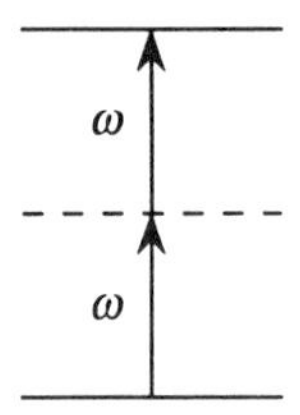

FIGURE 1.2.10 Two-photon absorption.

over an appreciable range of input intensities more than one output intensity is possible. The process of optical bistability is described in greater detail in Section 6.3.

Two-photon Absorption

In the process of two-photon absorption, which is illustrated in Fig. 1.2.10, an atom makes a transition from its ground state to an excited state by the simultaneous absorption of two laser photons. The absorption cross section σ describing this process increases linearly with laser intensity according to the relation

$$\sigma = \sigma^{(2)} I, \tag{1.2.19}$$

where $\sigma^{(2)}$ is a coefficient that describes two-photon absorption. (Recall that in conventional, linear optics the absorption cross section σ is a constant.) Consequently, the atomic transition rate R due to two-photon absorption scales as the square of the laser intensity, since $R = \sigma I/\hbar\omega$, or

$$R = \frac{\sigma^{(2)} I^2}{\hbar\omega}. \tag{1.2.20}$$

Two-photon absorption is a useful spectroscopic tool for determining the positions of energy levels that are not connected to the atomic ground state by a one-photon transition. Two-photon absorption was first observed experimentally by Kaiser and Garrett (1961).

Stimulated Raman Scattering

In stimulated Raman scattering, which is illustrated in Fig. 1.2.11, a photon of frequency ω is annihilated and a photon at the Stokes shifted frequency $\omega_S = \omega - \omega_v$ is created, leaving the molecule (or atom) in an excited state with energy $\hbar\omega_v$. The excitation energy is referred to as ω_v because stimulated Raman scattering was first studied in molecular systems, where $\hbar\omega_v$ corresponds to a vibrational energy. The efficiency of this process can be quite large,

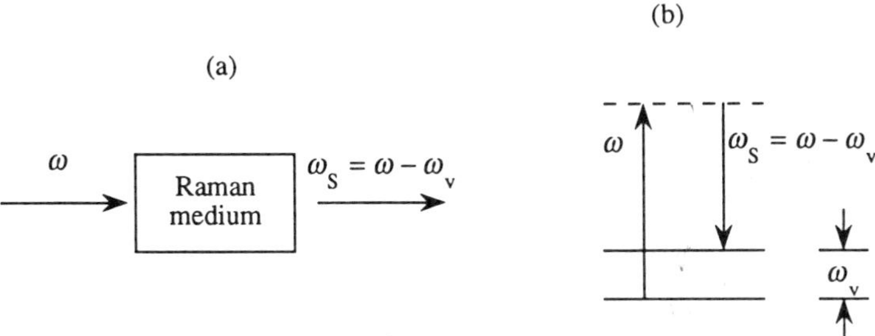

FIGURE 1.2.11 Stimulated Raman scattering.

with often 10% or more of the power of the incident light being converted to the Stokes frequency. In contrast, the efficiency of normal or spontaneous Raman scattering is typically many orders of magnitude smaller. Stimulated Raman scattering is described more fully in Chapter 9.

Other stimulated scattering processes such as stimulated Brillouin scattering and stimulated Rayleigh scattering also occur, and are described more fully in Chapter 8.

1.3. Formal Definition of the Nonlinear Susceptibility

Nonlinear optical interactions can be described in terms of the nonlinear polarization given by Eq. (1.1.2) only for a material system that is lossless and dispersionless. In the present section, we consider the more general case of a material with dispersion and/or loss. In this general case the nonlinear susceptibility becomes a complex quantity relating the complex amplitudes of the electric field and polarization.

We assume that we can represent the electric field vector of the optical wave as the discrete sum of a number of frequency components as

$$\tilde{\mathbf{E}}(\mathbf{r}, t) = \sum_n{}' \tilde{\mathbf{E}}_n(\mathbf{r}, t). \tag{1.3.1}$$

The prime on the summation sign of Eq. (1.3.1) indicates that the summation is to be taken over positive frequencies only. It is often convenient to represent $\tilde{\mathbf{E}}_n(\mathbf{r}, t)$ as the sum of its positive- and negative-frequency parts as

$$\tilde{\mathbf{E}}_n = \tilde{\mathbf{E}}_n^{(+)} + \tilde{\mathbf{E}}_n^{(-)}, \tag{1.3.2}$$

where

$$\tilde{\mathbf{E}}_n^{(+)} = \mathbf{E}_n e^{-i\omega_n t} \tag{1.3.3a}$$

and

$$\tilde{\mathbf{E}}_n^{(-)} = \tilde{\mathbf{E}}_n^{(+)*} = \mathbf{E}_n^* e^{i\omega_n t}. \tag{1.3.3b}$$

By requiring $\tilde{\mathbf{E}}_n^{(-)}$ to be the complex conjugate of $\tilde{\mathbf{E}}_n^{(+)}$ we are assured that the quantity $\tilde{\mathbf{E}}(\mathbf{r}, t)$ of Eq. (1.3.1) will be real, as it must be in order to represent a physical field. It is also convenient to define the spatially slowly varying field amplitude $\mathbf{A}_n$ by means of the relation

$$\mathbf{E}_n = \mathbf{A}_n e^{i\mathbf{k}_n \cdot \mathbf{r}}. \tag{1.3.4}$$

The total electric field of Eq. (1.3.1) can thus be represented in terms of these field amplitudes by either of the expressions

$$\begin{aligned} \tilde{\mathbf{E}}(\mathbf{r}, t) &= \sum_n{}' \mathbf{E}_n e^{-i\omega_n t} + \text{c.c.} \\ &= \sum_n{}' \mathbf{A}_n e^{i(\mathbf{k}_n \cdot \mathbf{r} - \omega_n t)} + \text{c.c.} \end{aligned} \tag{1.3.5}$$

On occasion, we shall express these field amplitudes using the alternative notation

$$\mathbf{E}_n = \mathbf{E}(\omega_n) \quad \text{and} \quad \mathbf{A}_n = \mathbf{A}(\omega_n). \tag{1.3.6}$$

This notation is very convenient, even though ω_n is actually a parameter rather than an argument showing a true functional dependence. In terms of this new notation, the reality condition of Eq. (1.3.3b) becomes

$$\mathbf{E}(-\omega_n) = \mathbf{E}(\omega_n)^* \quad \text{or} \quad \mathbf{A}(-\omega_n) = \mathbf{A}(\omega_n)^*. \tag{1.3.7}$$

Using this new notation, we can write the total field in the more compact form

$$\begin{aligned} \tilde{\mathbf{E}}(\mathbf{r}, t) &= \sum_n \mathbf{E}(\omega_n) e^{-i\omega_n t} \\ &= \sum_n \mathbf{A}(\omega_n) e^{i(\mathbf{k}_n \cdot \mathbf{r} - \omega_n t)}, \end{aligned} \tag{1.3.8}$$

where the unprimed summation symbol denotes a summation over all frequencies, both positive and negative.

Note that according to our definition of field amplitude, the field given by

$$\tilde{\mathbf{E}}(\mathbf{r}, t) = \mathscr{E} \cos(\mathbf{k} \cdot \mathbf{r} - \omega t) \tag{1.3.9}$$

is represented by the complex field amplitudes

$$\mathbf{E}(\omega) = \tfrac{1}{2}\mathscr{E} e^{i\mathbf{k} \cdot \mathbf{r}}, \qquad \mathbf{E}(-\omega) = \tfrac{1}{2}\mathscr{E} e^{-i\mathbf{k} \cdot \mathbf{r}}, \tag{1.3.10}$$

or alternatively by the slowly varying amplitudes

$$\mathbf{A}(\omega) = \tfrac{1}{2}\mathscr{E}, \qquad \mathbf{A}(-\omega) = \tfrac{1}{2}\mathscr{E}. \tag{1.3.11}$$

In either representation, factors of $\frac{1}{2}$ appear because the physical field amplitude $\mathscr{E}$ has been divided equally between the positive- and negative-frequency components.

Using a notation similar to that of Eq. (1.3.8), we can express the nonlinear polarization as

$$\tilde{\mathbf{P}}(\mathbf{r}, t) = \sum_n \mathbf{P}(\omega_n) e^{-i\omega_n t}, \tag{1.3.12}$$

where, as before, the summation extends over all positive- and negative-frequency components.

We now define the components of the second-order susceptibility tensor $\chi_{ijk}^{(2)}(\omega_n + \omega_m, \omega_n, \omega_m)$ as the constants of proportionality relating the amplitude of the nonlinear polarization to the product of field amplitudes according to

$$P_i(\omega_n + \omega_m) = \sum_{jk} \sum_{(nm)} \chi_{ijk}^{(2)}(\omega_n + \omega_m, \omega_n, \omega_m) E_j(\omega_n) E_k(\omega_m). \tag{1.3.13}$$

Here the indices ijk refer to the cartesian components of the fields. The notation (nm) indicates that, in performing the summation over n and m, the sum $\omega_n + \omega_m$ is to be held fixed, although ω_n and ω_m are each allowed to vary. Since the amplitude $E(\omega_n)$ is associated with the time dependence $\exp(-i\omega_n t)$, and the amplitude $E(\omega_m)$ is associated with the time dependence $\exp(-i\omega_m t)$, their product $E(\omega_n)E(\omega_m)$ is associated with the time dependence $\exp[-i(\omega_n + \omega_m)t]$. Hence the product $E(\omega_n)E(\omega_m)$ does in fact lead to a contribution to the nonlinear polarization oscillating at frequency $\omega_n + \omega_m$, as the notation of Eq. (1.3.13) suggests. Following convention, we have written $\chi^{(2)}$ as a function of three frequency arguments. This is technically unnecessary in that the first argument is always the sum of the other two. To emphasize this fact, the susceptibility $\chi^{(2)}(\omega_3, \omega_2, \omega_1)$ is sometimes written as $\chi^{(2)}(\omega_3; \omega_2, \omega_1)$ as a remainder that the first argument is different from the other two; or it may be written symbolically as $\chi^{(2)}(\omega_3 = \omega_2 + \omega_1)$.

Let us examine some of the consequences of the definition of the nonlinear susceptibility as given by Eq. (1.3.13) by considering two simple examples.

1. *Sum-frequency generation.* We let the input field frequencies be ω_1 and ω_2 and the sum frequency be ω_3, so that $\omega_3 = \omega_1 + \omega_2$. Then, by carrying out the summation over ω_n and ω_m in Eq. (1.3.13), we find that

$$\begin{aligned} P_i(\omega_3) = \sum_{jk} [&\chi_{ijk}^{(2)}(\omega_3, \omega_1, \omega_2) E_j(\omega_1) E_k(\omega_2) \\ &+ \chi_{ijk}^{(2)}(\omega_3, \omega_2, \omega_1) E_j(\omega_2) E_k(\omega_1)]. \end{aligned} \tag{1.3.14}$$

This expression can be simplifed by making use of the intrinsic permutation symmetry of the nonlinear susceptibility (this symmetry is discussed in more detail in Eq. (1.5.6) below), which requires that

$$\chi_{ijk}^{(2)}(\omega_m+\omega_n,\omega_m,\omega_n)=\chi_{ikj}^{(2)}(\omega_m+\omega_n,\omega_n,\omega_m). \tag{1.3.15}$$

Through use of this relation, the expression for the nonlinear polarization becomes

$$P_i(\omega_3)=2\sum_{jk}\chi_{ijk}^{(2)}(\omega_3,\omega_1,\omega_2)E_j(\omega_1)E_k(\omega_2), \tag{1.3.16}$$

and for the special case in which both input fields are polarized in the x direction the polarization becomes

$$P_i(\omega_3)=2\chi_{ixx}^{(2)}(\omega_3,\omega_1,\omega_2)E_x(\omega_1)E_x(\omega_2). \tag{1.3.17}$$

2. *Second-harmonic generation.* We take the input frequency as ω_1 and the generated frequency as $\omega_3=2\omega_1$. If we again perform the summation over field frequencies in Eq. (1.3.13), we obtain

$$P_i(\omega_3)=\sum_{jk}\chi_{ijk}^{(2)}(\omega_3,\omega_1,\omega_1)E_j(\omega_1)E_k(\omega_1). \tag{1.3.18}$$

Again assuming the special case of an input field polarized along the x direction, this result becomes

$$P_i(\omega_3)=\chi_{ixx}^{(2)}(\omega_3,\omega_1,\omega_1)E_x(\omega_1)^2. \tag{1.3.19}$$

Note that a factor of two appears in Eqs. (1.3.16) and (1.3.17), which describe sum-frequency generation, but not in Eqs. (1.3.18) and (1.3.19), which describe second-harmonic generation. The fact that these expressions remain different even as ω_2 approaches ω_1 is at first sight surprising, but is a consequence of our convention that $\chi_{ijk}^{(2)}(\omega_3,\omega_1,\omega_2)$ must approach $\chi_{ijk}^{(2)}(\omega_3,\omega_1,\omega_1)$ as ω_1 approaches ω_2. Note that the expressions for $P(2\omega_1)$ and $P(\omega_1+\omega_2)$ that apply for the case of a dispersionless nonlinear susceptibility [Eq. (1.2.7)] also differ by a factor of two. Moreover, one should expect the nonlinear polarization produced by two distinct fields to be larger than that produced by a single field (all of the same amplitude, say), because the total light intensity is larger in the former case.

In general, the summation over field frequencies ($\sum_{(nm)}$) in Eq. (1.3.13) can be performed formally to obtain the result

$$P_i(\omega_n+\omega_m)=D\sum_{jk}\chi_{ijk}^{(2)}(\omega_n+\omega_m,\omega_n,\omega_m)E_j(\omega_n)E_k(\omega_m), \tag{1.3.20}$$

where D is known as the degeneracy factor and is equal to the number of distinct permutations of the applied field frequencies ω_n and ω_m.

The expression (1.3.13) defining the second-order susceptibility can readily be generalized to higher-order interactions. In particular, the components of the third-order susceptibility are defined as the coefficients relating the amplitude of the nonlinear polarization to a product of three electric field amplitudes according to the expression

$$P_i(\omega_o + \omega_n + \omega_m) = \sum_{jkl} \sum_{(mno)} \chi^{(3)}_{ijkl}(\omega_0 + \omega_n + \omega_m, \omega_o, \omega_n, \omega_m) \times E_j(\omega_o) E_k(\omega_n) E_l(\omega_m). \tag{1.3.21}$$

We can again perform the summation over m, n, and o to obtain the result

$$P_i(\omega_o + \omega_n + \omega_m) = D \sum_{jkl} \chi^{(3)}_{ijkl}(\omega_0 + \omega_n + \omega_m, \omega_o, \omega_n, \omega_m) \times E_j(\omega_o) E_k(\omega_n) E_l(\omega_m), \tag{1.3.22}$$

where the degeneracy factor D represents the number of distinct permutations of the frequencies ω_m, ω_n, and ω_o.

1.4. Nonlinear Susceptibility of a Classical Anharmonic Oscillator

The Lorentz model of the atom, which treats the atom as a harmonic oscillator, is known to provide a very good description of the linear optical properties of atomic vapors and of nonmetallic solids. In the present section, we extend the Lorentz model by allowing the possibility of a nonlinearity in the restoring force exerted on the electron. The details of the analysis differ depending upon whether or not the medium possesses inversion symmetry.* We first treat the case of a noncentrosymmetric medium, and we find that such a medium can give rise to a second-order optical nonlinearity. We then treat the case of a medium that possesses a center of symmetry and find that the lowest-order nonlinearity that can occur in this case is a third-order nonlinear susceptibility. Our treatment is similar to that of Owyoung (1971).

The primary shortcoming of the classical model of optical nonlinearities presented here is that this model ascribes a single resonance frequency (ω_0) to each atom. In contrast, the quantum-mechanical theory of the nonlinear optical susceptibility, to be developed in Chapter 3, allows each atom to

* The role of symmetry in determining the nature of the nonlinear susceptibility is discussed from a more fundamental point of view in Section 1.5. See especially the treatment leading from Eq. (1.5.31) to (1.5.35).

possess many energy eigenvalues and hence more than one resonance frequency. Since the present model allows for only one resonance frequency, it cannot properly describe the complete resonance nature of the nonlinear susceptibility (such as, for example, the possibility of simultaneous one- and two-photon resonances). However, it provides a good description for those cases in which all of the optical frequencies are considerably smaller than the lowest electronic resonance frequency of the material system.

Noncentrosymmetric Media

For the case of noncentrosymmetric media, we take the equation of motion of the electron coordinate $\tilde{x}$ to be of the form

$$\ddot{\tilde{x}} + 2\gamma\dot{\tilde{x}} + \omega_0^2\tilde{x} + a\tilde{x}^2 = -e\tilde{E}(t)/m. \tag{1.4.1}$$

In this equation we have assumed that the applied electric field is given by $\tilde{E}(t)$, that the charge of the electron is $-e$, that there is a damping force of the form $-2m\gamma\dot{\tilde{x}}$,* and that the restoring force is given by

$$\tilde{F}_{\text{restoring}} = -m\omega_0^2\tilde{x} - ma\tilde{x}^2, \tag{1.4.2}$$

where a is a parameter that characterizes the nonlinearity of the response. We obtain this form by assuming that the restoring force is a nonlinear function of the displacement of the electron from its equilibrium position and retaining the linear and quadratic terms in the Taylor series expansion of the restoring force in the displacement $\tilde{x}$. We can understand the nature of this form of the restoring force by noting that it corresponds to a potential energy function of the form

$$U = -\int \tilde{F}_{\text{restoring}}\, d\tilde{x} = \tfrac{1}{2}m\omega_0^2\tilde{x}^2 + \tfrac{1}{3}ma\tilde{x}^3. \tag{1.4.3}$$

Here the first term corresponds to a harmonic potential and the second term corresponds to an anharmonic correction term, as illustrated in Fig. 1.4.1. This model corresponds to the physical situation of electrons in real materials, because the actual potential well that the atomic electron feels is not perfectly parabolic. The present model can describe only noncentrosymmetric media because we have assumed that the potential energy function U of Eq. (1.4.3)

* The factor of two is introduced to make γ the dipole damping rate. 2γ is therefore the full width at half maximum in angular frequency units of the atomic absorption profile in the limit of linear response.

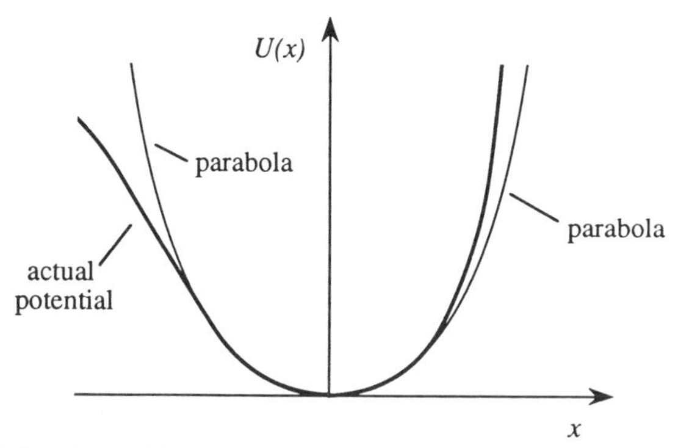

FIGURE 1.4.1 Potential energy function for a noncentrosymmetric medium.

contains both even and odd powers of x; for a centrosymmetric medium only even powers of $\tilde{x}$ could appear, because the potential function $U(\tilde{x})$ must possess the symmetry $U(\tilde{x}) = U(-\tilde{x})$. For simplicity, we have written Eq. (1.4.1) in the scalar-field approximation; the reason is that we cannot treat the tensor nature of the nonlinear susceptibility without making explicit assumptions regarding the symmetry properties of the material.

We assume that the applied optical field is of the form

$$\tilde{E}(t) = E_1 e^{-i\omega_1 t} + E_2 e^{-i\omega_2 t} + \text{c.c.} \tag{1.4.4}$$

No general solution to Eq. (1.4.1) for an applied field of the form (1.4.4) is known. However, if the applied field is sufficiently weak, the nonlinear term $a\tilde{x}^2$ will be much smaller than the linear term $\omega_0^2\tilde{x}$ for any displacement $\tilde{x}$ that can be induced by the field. Under this circumstance, Eq. (1.4.1) can be solved by means of a perturbation expansion. We use a procedure analogous to that of Rayleigh–Schrödinger perturbation theory in quantum mechanics. We replace $\tilde{E}(t)$ in Eq. (1.4.1) by $\lambda\tilde{E}(t)$, where λ is a parameter that ranges continuously between zero and one and that will be set equal to one at the end of the calculation. The expansion parameter λ thus characterizes the strength of the perturbation. Equation (1.4.1) then becomes

$$\ddot{\tilde{x}} + 2\gamma\dot{\tilde{x}} + \omega_0^2\tilde{x} + a\tilde{x}^2 = -\lambda e\tilde{E}(t)/m. \tag{1.4.5}$$

We now seek a solution to Eq. (1.4.5) in the form of a power series expansion in the strength λ of the perturbation, that is, a solution of the form

$$\tilde{x} = \lambda\tilde{x}^{(1)} + \lambda^2\tilde{x}^{(2)} + \lambda^3\tilde{x}^{(3)} + \cdots. \tag{1.4.6}$$

In order for Eq. (1.4.6) to be a solution to Eq. (1.4.5) for any value of the coupling strength λ, we require that the terms in Eq. (1.4.5) proportional to λ, λ^2, λ^3, etc., each satisfy the equation separately. We find that the terms proportional to λ, λ^2, and λ^3 lead respectively to the equations

$$\ddot{\tilde{x}}^{(1)} + 2\gamma\dot{\tilde{x}}^{(1)} + \omega_0^2\tilde{x}^{(1)} = -e\tilde{E}(t)/m, \tag{1.4.7a}$$

$$\ddot{\tilde{x}}^{(2)} + 2\gamma\dot{\tilde{x}}^{(2)} + \omega_0^2\tilde{x}^{(2)} + a[\tilde{x}^{(1)}]^2 = 0, \tag{1.4.7b}$$

$$\ddot{\tilde{x}}^{(3)} + 2\gamma\dot{\tilde{x}}^{(3)} + \omega_0^2\tilde{x}^{(3)} + 2a\tilde{x}^{(1)}\tilde{x}^{(2)} = 0. \tag{1.4.7c}$$

We see from Eq. (1.4.7a) that the lowest-order contribution $\tilde{x}^{(1)}$ is governed by the same equation as that of the conventional (i.e., linear Lorentz) model. Its steady-state solution is given by

$$\tilde{x}^{(1)}(t) = x^{(1)}(\omega_1)e^{-i\omega_1 t} + x^{(1)}(\omega_2)e^{-i\omega_2 t} + \text{c.c.}, \tag{1.4.8}$$

where the amplitudes $x^{(1)}(\omega_j)$ have the form

$$x^{(1)}(\omega_j) = -\frac{e}{m}\frac{E_j}{D(\omega_j)}, \tag{1.4.9}$$

where we have introduced the complex denominator function

$$D(\omega) = \omega_0^2 - \omega^2 - 2i\omega\gamma. \tag{1.4.10}$$

This expression for $\tilde{x}^{(1)}(t)$ is now squared and substituted into Eq. (1.4.7b), which is solved to obtain the lowest-order correction term $\tilde{x}^{(2)}$. The square of $\tilde{x}^{(1)}(t)$ contains the frequencies $\pm 2\omega_1$, $\pm 2\omega_2$, $\pm(\omega_1 + \omega_2)$, $\pm(\omega_1 - \omega_2)$, and 0. To determine the response at frequency $2\omega_1$, for instance, we must solve the equation

$$\ddot{\tilde{x}}^{(2)} + 2\gamma\dot{\tilde{x}}^{(2)} + \omega_0^2\tilde{x}^{(2)} = \frac{-a(eE_1/m)^2 e^{-2i\omega_1 t}}{D^2(\omega_1)}. \tag{1.4.11}$$

We seek a steady-state solution of the form

$$\tilde{x}^{(2)}(t) = x^{(2)}(2\omega_1)e^{-2i\omega_1 t}. \tag{1.4.12}$$

Substitution of Eq. (1.4.12) into Eq. (1.4.11) leads to the result

$$x^{(2)}(2\omega_1) = \frac{-a(e/m)^2 E_1^2}{D(2\omega_1)D^2(\omega_1)}, \tag{1.4.13}$$

where we have made use of the definition (1.4.10) of the function $D(\omega)$. Analogously, the amplitudes of the response at the other frequencies are found to be

$$x^{(2)}(2\omega_2) = \frac{-a(e/m)^2 E_2^2}{D(2\omega_2)D^2(\omega_2)}, \tag{1.4.14a}$$

$$x^{(2)}(\omega_1 + \omega_2) = \frac{-2a(e/m)^2 E_1 E_2}{D(\omega_1 + \omega_2)D(\omega_1)D(\omega_2)}, \tag{1.4.14b}$$

$$x^{(2)}(\omega_1 - \omega_2) = \frac{-2a(e/m)^2 E_1 E_2^*}{D(\omega_1 - \omega_2)D(\omega_1)D(-\omega_2)}, \tag{1.4.14c}$$

$$x^{(2)}(0) = \frac{-2a(e/m)^2 E_1 E_1^*}{D(0)D(\omega_1)D(-\omega_1)} + \frac{-2a(e/m)^2 E_2 E_2^*}{D(0)D(\omega_2)D(-\omega_2)}. \tag{1.4.14d}$$

We next express these results in terms of the linear ($\chi^{(1)}$) and nonlinear ($\chi^{(2)}$) susceptibilities. The linear susceptibility is defined through the relation

$$P^{(1)}(\omega_i) = \chi^{(1)}(\omega_i)E(\omega_i). \tag{1.4.15}$$

Since the linear contribution to the polarization is given by

$$P^{(1)}(\omega_i) = -Nex^{(1)}(\omega_i), \tag{1.4.16}$$

where N is the number density of atoms, we find using Eqs. (1.4.8) and (1.4.9) that the linear susceptibility is given by

$$\chi^{(1)}(\omega_i) = \frac{N(e^2/m)}{D(\omega_i)}. \tag{1.4.17}$$

The nonlinear susceptibilities are calculated in an analogous manner. The nonlinear susceptibility describing second-harmonic generation is defined by the relation

$$P^{(2)}(2\omega_1) = \chi^{(2)}(2\omega_1, \omega_1, \omega_1)E(\omega_1)^2, \tag{1.4.18}$$

where $P^{(2)}(2\omega_1)$ is the amplitude of the component of the nonlinear polarization oscillating at frequency $2\omega_1$ and is given by

$$P^{(2)}(2\omega_1) = -Nex^{(2)}(2\omega_1). \tag{1.4.19}$$

Comparison of these equations with Eq. (1.4.13) gives

$$\chi^{(2)}(2\omega_1, \omega_1, \omega_1) = \frac{N(e^3/m^2)a}{D(2\omega_1)D^2(\omega_1)}. \tag{1.4.20}$$

Through use of Eq. (1.4.17), this result can be written instead in terms of the product of linear susceptibilities as

$$\chi^{(2)}(2\omega_1, \omega_1, \omega_1) = \frac{ma}{N^2 e^3} \chi^{(1)}(2\omega_1)[\chi^{(1)}(\omega_1)]^2. \tag{1.4.21}$$

The nonlinear susceptibility for second-harmonic generation of the ω_2 field is obtained trivially from Eqs. (1.4.20) and (1.4.21) through the substitution $\omega_1 \to \omega_2$. The nonlinear susceptibility describing sum-frequency generation is obtained from the relations

$$P^{(2)}(\omega_1 + \omega_2) = 2\chi^{(2)}(\omega_1 + \omega_2, \omega_1, \omega_2)E(\omega_1)E(\omega_2) \tag{1.4.22}$$

and

$$P^{(2)}(\omega_1 + \omega_2) = -Nex^{(2)}(\omega_1 + \omega_2). \tag{1.4.23}$$

Note that in this case the relation defining the nonlinear susceptibility contains a factor of two because the two input fields are distinct, as discussed in relation to Eq. (1.3.20). By comparison of these equations with (1.4.14b), the nonlinear susceptibility is seen to be given by

$$\chi^{(2)}(\omega_1 + \omega_2, \omega_1, \omega_2) = \frac{N(e^3/m^2)a}{D(\omega_1 + \omega_2)D(\omega_1)D(\omega_2)}, \tag{1.4.24}$$

which can be expressed in terms of the product of linear susceptibilities as

$$\chi^{(2)}(\omega_1 + \omega_2, \omega_1, \omega_2) = \frac{ma}{N^2 e^3} \chi^{(1)}(\omega_1 + \omega_2)\chi^{(1)}(\omega_1)\chi^{(1)}(\omega_2). \tag{1.4.25}$$

It can be seen by comparison of Eqs. (1.4.20) and (1.4.24) that, as ω_2 approaches ω_1, $\chi^{(2)}(\omega_1 + \omega_2, \omega_1, \omega_2)$ approaches $\chi^{(2)}(2\omega_1, \omega_1, \omega_1)$.

The nonlinear susceptibilities describing the other second-order processes are obtained in an analogous manner. For difference-frequency generation we find that

$$\begin{aligned} \chi^{(2)}(\omega_1 - \omega_2, \omega_1, \omega_2) &= \frac{N(e^3/m^2)a}{D(\omega_1 - \omega_2)D(\omega_1)D(-\omega_2)} \\ &= \frac{ma}{N^2 e^3} \chi^{(1)}(\omega_1 - \omega_2)\chi^{(1)}(\omega_1)\chi^{(1)}(-\omega_2), \end{aligned} \tag{1.4.26}$$

and for optical rectification we find that

$$\begin{aligned} \chi^{(2)}(0, \omega_1, -\omega_1) &= \frac{N(e^3/m^2)a}{D(0)D(\omega_1)D(-\omega_1)} \\ &= \frac{ma}{N^2 e^3} \chi^{(1)}(0)\chi^{(1)}(\omega_1)\chi^{(1)}(-\omega_1). \end{aligned} \tag{1.4.27}$$

We have just seen that the lowest-order nonlinear contribution to the polarization of a noncentrosymmetric material is of second order in the applied field strength. The analysis described above can readily be extended to include higher-order nonlinearities. The solution to Eq. (1.4.7c), for example, leads to a third-order or $\chi^{(3)}$ susceptibility, and more generally terms proportional to λ^n in the expansion described by Eq. (1.4.6) lead to a $\chi^{(n)}$ susceptibility.

Miller's Rule

An empirical rule due to Miller (Miller, 1964; see also Garrett and Robinson, 1966) can be understood in terms of the calculation just presented. Miller noted that the quantity

$$\frac{\chi^{(2)}(\omega_1 + \omega_2, \omega_1, \omega_2)}{\chi^{(1)}(\omega_1 + \omega_2)\chi^{(1)}(\omega_1)\chi^{(1)}(\omega_2)} \tag{1.4.28}$$

is nearly constant for all noncentrosymmetric crystals. By comparison with Eq. (1.4.25), we see this quantity will be constant only if the quantity

$$\frac{ma}{N^2 e^3} \tag{1.4.29}$$

is nearly constant. In fact, the atomic number density N is nearly the same ($\sim 10^{22}$ cm^{-3}) for all condensed matter, and the parameters m and e are fundamental constants. We can estimate the size of the nonlinear coefficient a by noting that the linear and nonlinear contributions to the restoring force given by Eq. (1.4.2) would be expected to be comparable when the displacement $\tilde{x}$ of the electron from its equilibrium position is approximately equal to the size of the atom. This distance is of the order of the separation between atoms, that is, of the lattice constant d. Consequently, one would expect that, to order of magnitude, $m\omega_0^2 d = mad^2$, or that

$$a = \frac{\omega_0^2}{d}. \tag{1.4.30}$$

Since ω_0 and d are roughly the same for most solids, the quantity a would also be expected to be roughly the same for all noncentrosymmetric solids.

We can also use the estimate of the nonlinear coefficient a given by Eq. (1.4.30) to make an estimate of the size of the second-order susceptibility under nonresonant conditions. If we replace $D(\omega)$ by ω_0^2 in the denominator of Eq. (1.4.24), set N equal to $1/d^3$, and set a equal to ω_0^2/d, we find that $\chi^{(2)}$ is given approximately by

$$\chi^{(2)} = \frac{e^3}{m^2 \omega_0^4 d^4}. \tag{1.4.31}$$

Using the values $\omega_0 = 1 \times 10^{16}$ rad/s, $d = 3$ Å, $e = 4.8 \times 10^{-10}$ esu, and $m = 9.1 \times 10^{-28}$ g, we find that

$$\chi^{(2)} \simeq 3 \times 10^{-8} \text{ esu}, \tag{1.4.32}$$

which is in good agreement with the measured values presented in Table 1.5.3 in the next section.

Centrosymmetric Media

For the case of a centrosymmetric medium, we assume that the electronic restoring force is given not by Eq. (1.4.2) but rather by

$$\tilde{F}_{\text{restoring}} = -m\omega_0^2 \tilde{x} + mb\tilde{x}^3, \tag{1.4.33}$$

where b is a parameter that characterizes the strength of the nonlinearity. This restoring force corresponds to the potential energy function

$$U = -\int \tilde{F}_{\text{restoring}}\, d\tilde{x} = \tfrac{1}{2} m\omega_0^2 \tilde{x}^2 - \tfrac{1}{4} mb\tilde{x}^4. \tag{1.4.34}$$

This potential function is illustrated in the Fig. 1.4.2 (for the usual case in which b is positive) and is seen to be symmetric under the operation $\tilde{x} \to -\tilde{x}$, which it

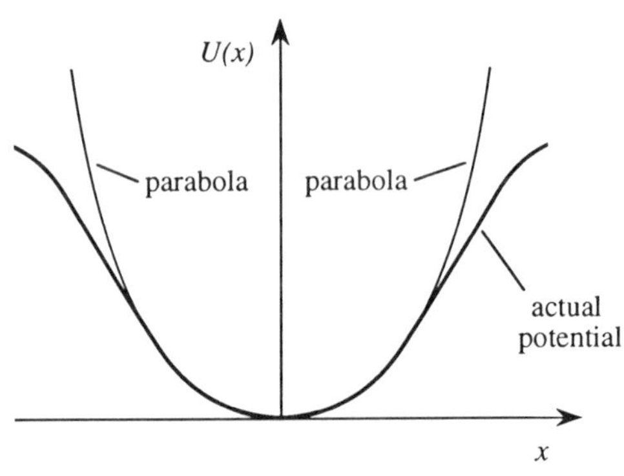

FIGURE 1.4.2 Potential energy function for a centrosymmetric medium.

must be for a medium that possesses a center of inversion symmetry. Note that $-mb\tilde{x}^4/4$ is simply the lowest-order correction term to the parabolic potential well described by the term $\frac{1}{2}m\omega_0^2\tilde{x}^2$. We assume that the electronic displacement $\tilde{x}$ never becomes so large that it is necessary to include higher-order terms in the potential.

We shall see below that the lowest-order nonlinear response resulting from the restoring force of Eq. (1.4.33) is a third-order contribution to the polarization, which can be described by a $\chi^{(3)}$ susceptibility. As in the case of noncentrosymmetric media, the tensor properties of this susceptibility cannot be specified unless the internal symmetries of the medium are completely known. One of the most important cases is that of a material which is isotropic (as well as being centrosymmetric). Examples of such materials are glasses and liquids. In such a case, we can take the restoring force to have the form

$$\tilde{\mathbf{F}}_{\text{restoring}} = -m\omega_0^2\tilde{\mathbf{r}} + mb(\tilde{\mathbf{r}}\cdot\tilde{\mathbf{r}})\tilde{\mathbf{r}}. \tag{1.4.35}$$

The second contribution to the restoring force must have the form shown because it is the only form that is third-order in the displacement $\tilde{\mathbf{r}}$ and is directed in the $\tilde{\mathbf{r}}$ direction, which is the only possible direction for an isotropic medium.

The equation of motion for the electron displacement from equilibrium is thus

$$\ddot{\tilde{\mathbf{r}}} + 2\gamma\dot{\tilde{\mathbf{r}}} + \omega_0^2\tilde{\mathbf{r}} - b(\tilde{\mathbf{r}}\cdot\tilde{\mathbf{r}})\tilde{\mathbf{r}} = -e\tilde{\mathbf{E}}(t)/m. \tag{1.4.36}$$

We assume that the applied field is given by

$$\tilde{\mathbf{E}}(t) = \mathbf{E}_1 e^{-i\omega_1 t} + \mathbf{E}_2 e^{-i\omega_2 t} + \mathbf{E}_3 e^{-i\omega_3 t} + \text{c.c.}; \tag{1.4.37}$$

we allow the field to have three distinct frequency components because this is the most general possibility for a third-order interaction. However, the algebra becomes very tedious if all three terms are written explicitly, and hence we express the applied field as

$$\tilde{\mathbf{E}}(t) = \sum_n \mathbf{E}(\omega_n)e^{-i\omega_n t}. \tag{1.4.38}$$

The method of solution is analogous to that used above for a noncentrosymmetric medium. We replace $\tilde{\mathbf{E}}(t)$ in Eq. (1.4.36) by $\lambda\tilde{\mathbf{E}}(t)$, where λ is a parameter that characterizes the strength of the perturbation and that is set equal to unity at the end of the calculation. We seek a solution to Eq. (1.4.36) having the form of a power series in the parameter λ:

$$\tilde{\mathbf{r}}(t) = \lambda\tilde{\mathbf{r}}^{(1)}(t) + \lambda^2\tilde{\mathbf{r}}^{(2)}(t) + \lambda^3\tilde{\mathbf{r}}^{(3)} + \cdots. \tag{1.4.39}$$

We insert Eq. (1.4.39) into Eq. (1.4.36) and require that the terms proportional to λ^n vanish separately for each value of n. We thereby find that

$$\ddot{\tilde{\mathbf{r}}}^{(1)} + 2\gamma\dot{\tilde{\mathbf{r}}}^{(1)} + \omega_0^2\tilde{\mathbf{r}}^{(1)} = -e\tilde{\mathbf{E}}(t)/m, \tag{1.4.40a}$$

$$\ddot{\tilde{\mathbf{r}}}^{(2)} + 2\gamma\dot{\tilde{\mathbf{r}}}^{(2)} + \omega_0^2\tilde{\mathbf{r}}^{(2)} = 0, \tag{1.4.40b}$$

$$\ddot{\tilde{\mathbf{r}}}^{(3)} + 2\gamma\dot{\tilde{\mathbf{r}}}^{(3)} + \omega_0^2\tilde{\mathbf{r}}^{(3)} - b(\tilde{\mathbf{r}}^{(1)}\cdot\tilde{\mathbf{r}}^{(1)})\tilde{\mathbf{r}}^{(1)} = 0 \tag{1.4.40c}$$

for $n = 1, 2,$ and 3 respectively. Equation (1.4.40a) is simply the vector version of Eq. (1.4.7a), encountered above. Its steady-state solution is

$$\tilde{\mathbf{r}}^{(1)}(t) = \sum_n \mathbf{r}^{(1)}(\omega_n)e^{-i\omega_n t}, \tag{1.4.41a}$$

where

$$\mathbf{r}^{(1)}(\omega_n) = \frac{-e\mathbf{E}(\omega_n)/m}{D(\omega_n)} \tag{1.4.41b}$$

with $D(\omega_n)$ given as before by $D(\omega_n) = \omega_0^2 - \omega_n^2 - 2i\omega_n\gamma$. Since the polarization at frequency ω_n is given by

$$\mathbf{P}^{(1)}(\omega_n) = -Ne\mathbf{r}^{(1)}(\omega_n), \tag{1.4.42}$$

we can describe the cartesian components of the polarization through the relation

$$P_i^{(1)}(\omega_n) = \sum_j \chi_{ij}^{(1)}(\omega_n)E_j(\omega_n), \tag{1.4.43a}$$

where the linear susceptibility is given by

$$\chi_{ij}^{(1)}(\omega_n) = \chi^{(1)}(\omega_n)\,\delta_{ij} \tag{1.4.43b}$$

with $\chi^{(1)}(\omega_n)$ given as before by

$$\chi^{(1)}(\omega_n) = \frac{Ne^2/m}{D(\omega_n)}, \tag{1.4.43c}$$

and where δ_{ij} is the Kronecker delta, which is defined such that $\delta_{ij} = 1$ for $i = j$ and $\delta_{ij} = 0$ for $i \neq j$.

The second-order response of the system is described by Eq. (1.4.40b). Since this equation is damped but not driven, its steady-state solution vanishes, that is,

$$\tilde{\mathbf{r}}^{(2)} = 0. \tag{1.4.44}$$

To calculate the third-order response, we substitute the expression for $\tilde{\mathbf{r}}^{(1)}(t)$ given by Eq. (1.4.41) into Eq. (1.4.40c), which becomes

$$\ddot{\tilde{\mathbf{r}}}^{(3)} + 2\gamma\dot{\tilde{\mathbf{r}}}^{(3)} + \omega_0^2\tilde{\mathbf{r}}^{(3)} = -\sum_{mnp}\frac{be^3[\mathbf{E}(\omega_m)\cdot\mathbf{E}(\omega_n)]\mathbf{E}(\omega_p)}{m^3D(\omega_m)D(\omega_n)D(\omega_p)} \times e^{-i(\omega_m+\omega_n+\omega_p)t}. \tag{1.4.45}$$

Due to the summation over m, n, and p, the right-hand side of this equation contains many different frequencies. We denote one of these frequencies by $\omega_q = \omega_m + \omega_n + \omega_p$. The solution to Eq. (1.4.45) can then be written in the form

$$\tilde{\mathbf{r}}^{(3)}(t) = \sum_q \mathbf{r}^{(3)}(\omega_q) e^{-i\omega_q t}. \tag{1.4.46}$$

We substitute Eq. (1.4.46) into Eq. (1.4.45) and find that $\mathbf{r}^{(3)}(\omega_q)$ is given by

$$(-\omega_q^2 - i\omega_q 2\gamma + \omega_0^2)\mathbf{r}^{(3)}(\omega_q) = -\sum_{(mnp)} \frac{be^3[\mathbf{E}(\omega_m) \cdot \mathbf{E}(\omega_n)]\mathbf{E}(\omega_p)}{m^3 D(\omega_m) D(\omega_n) D(\omega_p)}, \tag{1.4.47}$$

where the summation is to be carried out over frequencies ω_m, ω_n, and ω_p with the restriction that $\omega_m + \omega_n + \omega_p$ must equal ω_q. Since the coefficient of $\mathbf{r}^{(3)}(\omega_q)$ on the left-hand side is just $D(\omega_q)$, we obtain

$$\mathbf{r}^{(3)}(\omega_q) = -\sum_{(mnp)} \frac{be^3[\mathbf{E}(\omega_m) \cdot \mathbf{E}(\omega_n)]\mathbf{E}(\omega_p)}{m^3 D(\omega_q) D(\omega_m) D(\omega_n) D(\omega_p)}. \tag{1.4.48}$$

The amplitude of the polarization component oscillating at frequency ω_q then is given in terms of this amplitude by

$$\mathbf{P}^{(3)}(\omega_q) = -Ne\mathbf{r}^{(3)}(\omega_q). \tag{1.4.49}$$

We next recall the definition of the third-order nonlinear susceptibility (1.3.21),

$$P_i^{(3)}(\omega_q) = \sum_{jkl} \sum_{(mnp)} \chi_{ijkl}^{(3)}(\omega_q, \omega_m, \omega_n, \omega_p) E_j(\omega_m) E_k(\omega_n) E_l(\omega_p). \tag{1.4.50}$$

Since this equation contains a summation over the dummy variables m, n, and p, there is more than one possible choice for the expression for the nonlinear susceptibility. An obvious choice for this expression for the susceptibility, based on the way in which Eqs. (1.4.48) and (1.4.49) are written, is

$$\chi_{ijkl}^{(3)}(\omega_q, \omega_m, \omega_n, \omega_p) = \frac{Nbe^4 \, \delta_{jk} \, \delta_{il}}{m^3 D(\omega_q) D(\omega_m) D(\omega_n) D(\omega_p)}. \tag{1.4.51}$$

While Eq. (1.4.51) is a perfectly adequate expression for the nonlinear susceptibility, it does not explicitly show the full symmetry of the interaction in terms of the arbitrariness of which field we call $E_j(\omega_m)$, which we call $E_k(\omega_n)$, and which we call $E_l(\omega_p)$. It is conventional to define nonlinear susceptibilities in a manner that displays this symmetry, which is known as intrinsic permutation symmetry. Since there are six possible permutations of the orders in which $E_j(\omega_m)$, $E_k(\omega_n)$, and $E_l(\omega_p)$ may be taken, we define the third-order susceptibility to be one-sixth of the sum of the six expressions analogous to Eq. (1.4.51) with the input fields taken in all possible orders. When we carry out this prescription, we find that only three distinct contributions occur, and that the resulting form for the nonlinear susceptibility is given by

$$\chi_{ijkl}^{(3)}(\omega_q, \omega_m, \omega_n, \omega_p) = \frac{Nbe^4[\delta_{ij}\delta_{kl} + \delta_{ik}\delta_{jl} + \delta_{il}\delta_{jk}]}{3m^3 D(\omega_q)D(\omega_m)D(\omega_n)D(\omega_p)}. \tag{1.4.52}$$

This expression can be rewritten in terms of the linear susceptibilities at the four different frequencies ω_q, ω_m, ω_n, and ω_p by using Eq. (1.4.43c) to eliminate the resonance denominator factors $D(\omega)$. We thereby obtain

$$\chi_{ijkl}^{(3)}(\omega_q, \omega_m, \omega_n, \omega_p) = \frac{bm}{3N^3e^4}[\chi^{(1)}(\omega_q)\chi^{(1)}(\omega_m)\chi^{(1)}(\omega_n)\chi^{(1)}(\omega_p)] \times [\delta_{ij}\delta_{kl} + \delta_{ik}\delta_{jl} + \delta_{il}\delta_{jk}]. \tag{1.4.53}$$

We can estimate the value of the phenomenological constant b that appears in this result by means of an argument analogous to that used above (see Eq. (1.4.30)) to estimate the value of the constant a that appears in the expression for $\chi^{(2)}$. We assume that the linear and nonlinear contributions to the restoring force given by Eq. (1.4.33) will become comparable in magnitude when the displacement $\tilde{x}$ is comparable to the atomic dimension d, that is, when $m\omega_0^2 d = mbd^3$, which implies that

$$b = \frac{\omega_0^2}{d^2}. \tag{1.4.54}$$

Using this expression for b, we can now estimate the value of the nonlinear susceptibility. For the case of nonresonant excitation, $D(\omega)$ is approximately equal to ω_0^2, and hence from Eq. (1.4.52) we obtain

$$\chi^{(3)} \simeq \frac{Nbe^4}{m^3\omega_0^8} = \frac{e^4}{m^3\omega_0^6 d^5}. \tag{1.4.55}$$

Taking $d = 3$ Å and $\omega_0 = 7 \times 10^{15}$ rad/sec, we obtain

$$\chi^{(3)} \simeq 3 \times 10^{-14} \text{ esu}. \tag{1.4.56}$$

We shall see in Chapter 4 that this value is typical of the nonlinear susceptibility of many materials.

1.5. Properties of the Nonlinear Susceptibility

In this section we study some of the formal symmetry properties of the nonlinear susceptibility. Let us first see why it is important that we understand these symmetry properties. We consider the mutual interaction of three waves of frequencies ω_1, ω_2, and $\omega_3 = \omega_1 + \omega_2$. A complete description of the interaction of these waves requires that we know the nonlinear polarizations $\mathbf{P}(\omega_i)$ influencing each of them. Since these quantities are given in general (see also Eq. (1.3.13)) by the expression

$$P_i(\omega_n + \omega_m) = \sum_{jk} \sum_{(nm)} \chi_{ijk}^{(2)}(\omega_n + \omega_m, \omega_n, \omega_m) E_j(\omega_n) E_k(\omega_m), \tag{1.5.1}$$

we therefore need to determine the six tensors

$$\chi_{ijk}^{(2)}(\omega_1, \omega_3, -\omega_2), \qquad \chi_{ijk}^{(2)}(\omega_1, -\omega_2, \omega_3), \qquad \chi_{ijk}^{(2)}(\omega_2, \omega_3, -\omega_1),$$

$$\chi_{ijk}^{(2)}(\omega_2, -\omega_1, \omega_3), \qquad \chi_{ijk}^{(2)}(\omega_3, \omega_1, \omega_2), \quad \text{and} \quad \chi_{ijk}^{(2)}(\omega_3, \omega_2, \omega_1)$$

and six additional tensors in which each frequency is replaced by its negative. In these expressions, the indices i, j, and k can independently take on the values x, y, and z. Since each of these 12 tensors thus consists of 27 cartesian components, as many as 324 different (complex) numbers need to be specified in order to describe the interaction.

Fortunately, there are a number of restrictions resulting from symmetries that relate the various components of $\chi^{(2)}$, and hence far fewer than 324 numbers are usually needed in order to describe the nonlinear coupling. In this section, we shall study some of these formal properties of the nonlinear susceptibility. The discussion will deal primarily with the second-order $\chi^{(2)}$ susceptibility, but can readily be extended to $\chi^{(3)}$ and higher-order susceptibilities.

Reality of the Fields

Recall that the nonlinear polarization describing the sum-frequency response to input fields at frequencies ω_n and ω_m has been represented as

$$\tilde{P}_i(\mathbf{r}, t) = P_i(\omega_n + \omega_m) e^{-i(\omega_n + \omega_m)t} + P_i(-\omega_n - \omega_m) e^{i(\omega_n + \omega_m)t}. \tag{1.5.2}$$

Since $\tilde{P}_i(\mathbf{r}, t)$ is a physically measurable quantity, it must be purely real, and hence its positive- and negative-frequency components must be related by

$$P_i(-\omega_n - \omega_m) = P_i(\omega_n + \omega_m)^*. \tag{1.5.3}$$

The electric field must also be a real quantity, and its complex frequency components must obey the analogous conditions:

$$E_j(-\omega_n) = E_j(\omega_n)^*, \tag{1.5.4a}$$

$$E_k(-\omega_m) = E_k(\omega_m)^*. \tag{1.5.4b}$$

Since the fields and polarization are related to each other through the second-order susceptibility of Eq. (1.5.1), we conclude that the positive- and negative-frequency components of the susceptibility must be related according to

$$\chi_{ijk}^{(2)}(-\omega_n - \omega_m, -\omega_n, -\omega_m) = \chi_{ijk}^{(2)}(\omega_n + \omega_m, \omega_n, \omega_m)^*. \tag{1.5.5}$$

Intrinsic Permutation Symmetry

Earlier we introduced the concept of intrinsic permutation symmetry when we rewrote the expression (1.4.51) for the nonlinear susceptibility of a classical, anharmonic oscillator in the conventional form of Eq. (1.4.52). In the present section, we discuss the concept of intrinsic permutation symmetry from a more general point of view.

According to Eq. (1.5.1), one of the contributions to the nonlinear polarization $P_i(\omega_n + \omega_m)$ is the product $\chi_{ijk}^{(2)}(\omega_n + \omega_m, \omega_n, \omega_m)E_j(\omega_n)E_k(\omega_m)$. However, since j, k, n, and m are dummy indices, we could just as well have written this contribution with n interchanged with m and with j interchanged with k, that is, as $\chi_{ikj}^{(2)}(\omega_n + \omega_m, \omega_m, \omega_n)E_k(\omega_m)E_j(\omega_n)$. These two expressions are numerically equal if we require that the nonlinear susceptibility be unchanged by the simultaneous interchange of its last two frequency arguments and its last two cartesian indices:

$$\chi_{ijk}^{(2)}(\omega_n + \omega_m, \omega_n, \omega_m) = \chi_{ikj}^{(2)}(\omega_n + \omega_m, \omega_m, \omega_n). \tag{1.5.6}$$

This property is known as intrinsic permutation symmetry.

Note that this symmetry condition is introduced purely as a matter of convenience. For example, we could set one member of the pair of elements shown in Eq. (1.5.6) equal to zero and double the value of the other member. Then, when the double summation of Eq. (1.5.1) was carried out, the result for the physically meaningful quantity $P_i(\omega_n + \omega_m)$ would be left unchanged.

This symmetry condition can also be derived from a more general point of view using the concept of the nonlinear response function (Butcher, 1965; Flytzanis, 1975).

Symmetries for Lossless Media

Two additional symmetries of the nonlinear susceptibility tensor occur for the case of a lossless nonlinear medium.

The first of these conditions states that for a lossless medium all of the components of $\chi_{ijk}^{(2)}(\omega_n + \omega_m, \omega_n, \omega_m)$ are real. This result is obeyed for the classical anharmonic oscillator discussed in Section 1.4, as can be verified by evaluating the expression for $\chi^{(2)}$ in the limit in which all of the applied frequencies and their sums and differences are significantly different from the resonance frequency. The general proof that $\chi^{(2)}$ is real for a lossless medium is obtained by verifying that the quantum-mechanical expression for $\chi^{(2)}$ (which is derived in Chapter 3) is also purely real in this limit.

The second of these new symmetries is *full* permutation symmetry. This condition states that *all* of the frequency arguments of the nonlinear susceptibility can be freely interchanged, as long as the corresponding cartesian indices are interchanged simultaneously. In permuting the frequency arguments, it must be recalled that the first argument is always the sum of the latter two, and thus the signs of the frequencies must be inverted when the first frequency is interchanged with either of the latter two. Full permutation symmetry implies, for instance, that

$$\chi_{ijk}^{(2)}(\omega_3 = \omega_1 + \omega_2) = \chi_{jki}^{(2)}(-\omega_1 = \omega_2 - \omega_3). \tag{1.5.7}$$

However, according to Eq. (1.5.5) the right-hand side of this equation is equal to $\chi_{jki}^{(2)}(\omega_1 = -\omega_2 + \omega_3)^*$, which, due to the reality of $\chi^{(2)}$ for a lossless medium, is equal to $\chi_{jki}^{(2)}(\omega_1 = -\omega_2 + \omega_3)$. We hence conclude that

$$\chi_{ijk}^{(2)}(\omega_3 = \omega_1 + \omega_2) = \chi_{jki}^{(2)}(\omega_1 = -\omega_2 + \omega_3). \tag{1.5.8}$$

By an analogous procedure, one can show that

$$\chi_{ijk}^{(2)}(\omega_3 = \omega_1 + \omega_2) = \chi_{kij}^{(2)}(\omega_2 = \omega_3 - \omega_1). \tag{1.5.9}$$

A general proof of the validity of the condition of full permutation symmetry entails verifying that the quantum-mechanical expression for $\chi^{(2)}$ (which is derived in Chapter 3) obeys this condition when all of the optical frequencies are detuned many linewidths from the resonance frequencies of the optical medium. Full permutation symmetry can also be deduced from a consideration of the field energy density within a nonlinear medium, as shown below.

Field Energy Density for a Nonlinear Medium

The fact that the nonlinear susceptibility must possess full permutation symmetry for a lossless medium can be deduced from a consideration of the form of the electromagnetic field energy within a nonlinear medium. For the case of a linear medium, the energy density associated with the electric field

$$\tilde{E}_i(t) = \sum_n E_i(\omega_n) e^{-i\omega_n t} \tag{1.5.10}$$

is given according to Poynting's theorem as

$$U = \frac{1}{8\pi}\overline{\tilde{\mathbf{D}} \cdot \tilde{\mathbf{E}}} = \frac{1}{8\pi}\sum_i \overline{\tilde{D}_i \tilde{E}_i}, \tag{1.5.11}$$

where the overbar denotes a time average. Since the displacement vector is given by

$$\tilde{D}_i(t) = \sum_j \epsilon_{ij}\tilde{E}_j(t) = \sum_j \sum_n \epsilon_{ij}(\omega_n)E_j(\omega_n)e^{-i\omega_n t}, \tag{1.5.12}$$

where the dielectric tensor is given by

$$\epsilon_{ij}(\omega_n) = \delta_{ij} + 4\pi\chi_{ij}^{(1)}(\omega_n), \tag{1.5.13}$$

we can write the energy density as

$$U = \frac{1}{8\pi}\sum_i \sum_n E_i^*(\omega_n)E_i(\omega_n) + \frac{1}{2}\sum_{ij}\sum_n E_i^*(\omega_n)\chi_{ij}^{(1)}(\omega_n)E_j(\omega_n). \tag{1.5.14}$$

Here the first term represents the energy density associated with the electric field in vacuum and the second term represents the energy stored in the polarization of the medium.

For the case of a nonlinear medium, the expression for the electric field energy density (Kleinman, 1962; Armstrong *et al.*, 1962; Pershan, 1963) associated with the polarization of the medium takes the more general form

$$\begin{aligned} U = {} & \frac{1}{2}\sum_{ij}\sum_n \chi_{ij}^{(1)}(\omega_n)E_i^*(\omega_n)E_j(\omega_n) \\ & + \frac{1}{3}\sum_{ijk}\sum_{mn} \chi_{ijk}^{(2)\prime}(-\omega_n - \omega_m, \omega_m, \omega_n)E_i^*(\omega_m + \omega_n)E_j(\omega_m)E_k(\omega_n) \\ & + \frac{1}{4}\sum_{ijkl}\sum_{mno} \chi_{ijkl}^{(3)\prime}(-\omega_o - \omega_n - \omega_m, \omega_m, \omega_n, \omega_o) \\ & \times E_i^*(\omega_m + \omega_n + \omega_o)E_j(\omega_m)E_k(\omega_n)E_l(\omega_o) + \cdots . \end{aligned} \tag{1.5.15}$$

For the present, the quantities $\chi^{(2)\prime}$, $\chi^{(3)\prime}, \ldots$ are to be thought of simply as coefficients in the power series expansion of U in the amplitudes of the applied field; later these quantities will be related to the nonlinear susceptibilities. Since the order in which the fields are multiplied together in determining U is immaterial, the quantities $\chi^{(n)\prime}$ clearly possess full permutation symmetry, that is, their frequency arguments can be freely permuted as long as the corresponding indices are also permuted.

In order to relate the expression (1.5.15) for the energy density to the nonlinear polarization, and subsequently to the nonlinear susceptibility, we use the result that the polarization of a medium is given by (Pershan, 1963; Landau and Lifshitz, 1960) the expression

$$P_i(\omega_n) = \frac{\partial U}{\partial E_i^*(\omega_n)}. \tag{1.5.16}$$

Thus, by differentiation of Eq. (1.5.15), we obtain an expression for the linear polarization as

$$P_i^{(1)}(\omega_m) = \sum_j \chi_{ij}^{(1)}(\omega_m)E_j(\omega_m), \tag{1.5.17a}$$

and for the nonlinear polarizations as*

$$P_i^{(2)}(\omega_m + \omega_n) = \sum_{jk}\sum_{(mn)}{}' \chi_{ijk}^{(2)\prime}(-\omega_m - \omega_n, \omega_m, \omega_n)E_j(\omega_m)E_k(\omega_n) \tag{1.5.17b}$$

$$P_i^{(3)}(\omega_m + \omega_n + \omega_o) = \sum_{jkl}\sum_{(mno)}{}' \chi_{ijkl}^{(3)\prime}(-\omega_m - \omega_n - \omega_o, \omega_m, \omega_n, \omega_o) \times E_j(\omega_m)E_k(\omega_n)E_l(\omega_o). \tag{1.5.17c}$$

We note that these last two expressions are identical to Eqs. (1.3.13) and (1.3.21), which define the nonlinear susceptibilities (except for the unimportant fact that the quantities $\chi^{(n)}$ and $\chi^{(n)\prime}$ use opposite conventions regarding the sign of the first frequency argument). Since the quantities $\chi^{(n)\prime}$ possess full permutation symmetry, we conclude that the susceptibilities $\chi^{(n)}$ do also. Note that this demonstration is valid only for the case of a lossless medium, because only in this case is the internal energy a function of state.

Kleinman's Symmetry

Quite often nonlinear optical interactions involve optical waves whose frequencies ω_i are much smaller than the lowest resonance frequency of the material system. Under these conditions, the nonlinear susceptibility is essentially independent of frequency. For example, the expression (1.4.24) for the second-order susceptibility of an anharmonic oscillator predicts a value of the susceptibility that is essentially independent of the frequencies of the

* In performing the differentiation, the prefactors $\frac{1}{2}, \frac{1}{3}, \frac{1}{4}, \ldots$ of Eq. (1.5.15) disappear because 2, 3, 4,... equivalent terms appear as the result the summations over the frequency arguments.

applied waves whenever these frequencies are much smaller than the resonance frequency ω_0. Furthermore, under conditions of low-frequency excitation the system responds essentially instantaneously to the applied field, and we have seen in Section 1.2 that under such conditions the nonlinear polarization can be described in the time domain by the relation

$$\tilde{P}(t) = \chi^{(2)}\tilde{E}(t)^2, \tag{1.5.18}$$

where $\chi^{(2)}$ can be taken to be a constant.

Since the medium is necessarily lossless whenever the applied field frequencies ω_i are very much smaller than the resonance frequency ω_0, the condition of full permutation symmetry (1.5.7) must be valid under these circumstances. This condition states that the indices can be permuted as long as the frequencies are permuted simultaneously, and leads to the conclusion that

$$\begin{aligned}\chi_{ijk}^{(2)}(\omega_3 = \omega_1 + \omega_2) = \chi_{jki}^{(2)}(\omega_1 = -\omega_2 + \omega_3) &= \chi_{kij}^{(2)}(\omega_2 = \omega_3 - \omega_1)\\ = \chi_{ikj}^{(2)}(\omega_3 = \omega_2 + \omega_1) &= \chi_{jik}^{(2)}(\omega_1 = \omega_3 - \omega_2)\\ = \chi_{kji}^{(2)}(\omega_2 = -\omega_1 + \omega_3).&\end{aligned}$$

However, under the present conditions $\chi^{(2)}$ does not actually depend on the frequencies, and we can therefore permute the indices without permuting the frequencies, leading to the result

$$\begin{aligned}\chi_{ijk}^{(2)}(\omega_3 = \omega_1 + \omega_2) = \chi_{jki}^{(2)}(\omega_3 = \omega_1 + \omega_2) &= \chi_{kij}^{(2)}(\omega_3 = \omega_1 + \omega_2)\\ = \chi_{ikj}^{(2)}(\omega_3 = \omega_1 + \omega_2) &= \chi_{jik}^{(2)}(\omega_3 = \omega_1 + \omega_2)\\ = \chi_{kji}^{(2)}(\omega_3 = \omega_1 + \omega_2).&\end{aligned} \tag{1.5.19}$$

This result is known as the Kleinman symmetry condition.

Contracted Notation

We now introduce a notational device that is often used when the Kleinman symmetry condition is valid. We introduce the tensor

$$d_{ijk} = \tfrac{1}{2}\chi_{ijk}^{(2)} \tag{1.5.20}$$

and for simplicity suppress the frequency arguments. The nonlinear polarization can then be written as

$$P_i(\omega_n + \omega_m) = \sum_{jk}\sum_{(mn)} 2d_{ijk}E_j(\omega_n)E_k(\omega_m). \tag{1.5.21}$$

We now assume that d_{ijk} is symmetric in its last two indices. This assumption is valid whenever Kleinman's symmetry condition is valid and in addition is valid in general for second-harmonic generation, since in this case ω_n and ω_m are equal. We then simplify the notation by introducing a contracted matrix d_{il} according to the prescription

$$\begin{array}{lcccccc} jk: & 11 & 22 & 33 & 23,32 & 31,13 & 12,21 \\ l: & 1 & 2 & 3 & 4 & 5 & 6 \end{array} \tag{1.5.22}$$

The nonlinear susceptibility tensor can then be represented as the 3×6 matrix

$$d_{il} = \begin{bmatrix} d_{11} & d_{12} & d_{13} & d_{14} & d_{15} & d_{16} \\ d_{21} & d_{22} & d_{23} & d_{24} & d_{25} & d_{26} \\ d_{31} & d_{32} & d_{33} & d_{34} & d_{35} & d_{36} \end{bmatrix}. \tag{1.5.23}$$

If we now *explicitly* introduce the Kleinman symmetry condition, i.e., that the indices d_{ijk} can be freely permuted, we find that not all of the 18 elements of d_{il} are independent. For instance, we see that

$$d_{12} \equiv d_{122} = d_{212} \equiv d_{26} \tag{1.5.24a}$$

and that

$$d_{14} \equiv d_{123} = d_{213} \equiv d_{25}. \tag{1.5.24b}$$

By applying this type of argument systematically, we find that d_{il} has only 10 independent elements when the Kleinman symmetry condition is valid; the form of d_{il} under these conditions is

$$d_{il} = \begin{bmatrix} d_{11} & d_{12} & d_{13} & d_{14} & d_{15} & d_{16} \\ d_{16} & d_{22} & d_{23} & d_{24} & d_{14} & d_{12} \\ d_{15} & d_{24} & d_{33} & d_{23} & d_{13} & d_{14} \end{bmatrix}. \tag{1.5.25}$$

We can describe the nonlinear polarization leading to second-harmonic generation in terms of d_{il} by the matrix equation

$$\begin{bmatrix} P_x(2\omega) \\ P_y(2\omega) \\ P_z(2\omega) \end{bmatrix} = 2 \begin{bmatrix} d_{11} & d_{12} & d_{13} & d_{14} & d_{15} & d_{16} \\ d_{21} & d_{22} & d_{23} & d_{24} & d_{25} & d_{26} \\ d_{31} & d_{32} & d_{33} & d_{34} & d_{35} & d_{36} \end{bmatrix} \begin{bmatrix} E_x(\omega)^2 \\ E_y(\omega)^2 \\ E_z(\omega)^2 \\ 2E_y(\omega)E_z(\omega) \\ 2E_x(\omega)E_z(\omega) \\ 2E_x(\omega)E_y(\omega) \end{bmatrix}. \tag{1.5.26}$$

When the Kleinman symmetry condition is valid, we can describe the nonlinear polarization leading to sum-frequency generation (with $\omega_3 = \omega_1 + \omega_2$)

by the equation

$$\begin{bmatrix} P_x(\omega_3) \\ P_y(\omega_3) \\ P_z(\omega_3) \end{bmatrix} = 4 \begin{bmatrix} d_{11} & d_{12} & d_{13} & d_{14} & d_{15} & d_{16} \\ d_{21} & d_{22} & d_{23} & d_{24} & d_{25} & d_{26} \\ d_{31} & d_{32} & d_{33} & d_{34} & d_{35} & d_{36} \end{bmatrix} \times \begin{bmatrix} E_x(\omega_1)E_x(\omega_2) \\ E_y(\omega_1)E_y(\omega_2) \\ E_z(\omega_1)E_z(\omega_2) \\ E_y(\omega_1)E_z(\omega_2) + E_z(\omega_1)E_y(\omega_2) \\ E_x(\omega_1)E_z(\omega_2) + E_z(\omega_1)E_x(\omega_2) \\ E_x(\omega_1)E_y(\omega_2) + E_y(\omega_1)E_x(\omega_2) \end{bmatrix}. \tag{1.5.27}$$

As described above in relation to Eq. (1.3.19), the extra factor of 2 comes from the summation over n and m in Eq. (1.3.13).

Effective Value of d ($d_{\rm eff}$).

For a fixed geometry (i.e., for fixed propagation and polarization directions) it is possible to express the nonlinear polarization giving rise to sum-frequency generation by means of the scalar relationship

$$P(\omega_3) = 4d_{\rm eff}E(\omega_1)E(\omega_2), \tag{1.5.28}$$

and analogously for second-harmonic generation by

$$P(2\omega) = 2d_{\rm eff}E(\omega)^2. \tag{1.5.29}$$

In each case, $d_{\rm eff}$ is obtained by evaluation of the summation $\sum_{jk}$ in the general equation (1.3.13).

A general prescription for calculating $d_{\rm eff}$ for each of the crystal classes has been presented by Midwinter and Warner (1965); see also Table 3.1 of Zernike and Midwinter (1973). They show, for example, that for crystals of crystal class $3m$ the effective value of d is given by the expression

$$d_{\rm eff} = d_{31}\sin\theta - d_{22}\cos\theta\sin 3\phi \tag{1.5.30a}$$

under conditions (known as type I conditions) such that the two lower-frequency waves have the same polarization, and by

$$d_{\rm eff} = d_{22}\cos^2\theta\cos 3\theta \tag{1.5.30b}$$

under conditions (known as type II conditions) such that polarizations are orthogonal. In these equations, θ is the angle between the propagation vector and the crystalline z axis (the optic axis), and ϕ is the azimuthual angle between the propagation vector and the xz crystalline plane.

Spatial Symmetry of the Nonlinear Medium

The form of the nonlinear susceptibility tensor is constrained by the symmetry properties of the nonlinear optical medium. To see why this should be so, let us consider a crystal for which the x and y directions are equivalent but for which the z direction is different. By saying that the x and y directions are equivalent, we mean that if the crystal were rotated by 90 degrees about the z axis, the crystal structure would look identical after the rotation. The z axis is then said to be a fourfold axis of symmetry. For such a crystal, we would expect that the optical response would be the same for an applied optical field polarized in either the x or the y direction, and thus, for example, that the second-order susceptibility components $\chi^{(2)}_{zxx}$ and $\chi^{(2)}_{zyy}$ would be equal.

For any particular crystal, the form of the nonlinear optical susceptibility is determined by considering the consequences of all of the symmetry properties for that particular crystal. For this reason, it is necessary to determine what types of symmetry properties can occur in a crystalline medium. By means of the mathematical method known as group theory, crystallographers have found that all crystals can be classified as belonging to one of 32 possible crystal classes depending on what is called the point group symmetry of the crystal. The details of this classification scheme lie outside of the subject matter of the present text.* However, by way of examples, a crystal is said to belong to point group 4 if it possesses only a fourfold axis of symmetry, to point group 3 if it possesses only a threefold axis of symmetry, and to belong to point group $3m$ if it possesses a threefold axis of symmetry and in addition a plane of mirror symmetry perpendicular to this axis.

Inversion Symmetry

One of the symmetry properties that some but not all crystals possess is inversion symmetry. For a material system that is centrosymmetric (i.e., possesses a center of inversion) the $\chi^{(2)}$ nonlinear susceptibility must vanish identically. Since 11 of the 32 crystal classes possess inversion symmetry, this rule is very powerful, as it immediately eliminates all crystals belonging to these classes from consideration for second-order nonlinear optical interactions.

While the result that $\chi^{(2)}$ vanishes for a centrosymmetric medium is general in nature, we will demonstrate this fact only for the special case of second-harmonic generation in a medium that responds instantaneously to the

* The reader who is interested in the details should consult Buerger (1963) or any of the other books on group theory and crystal symmetry listed in the bibliography at the end of the present chapter.

applied optical field. We assume that the nonlinear polarization is given by

$$\tilde{P}(t) = \chi^{(2)} \tilde{E}^2(t), \tag{1.5.31}$$

where the applied field is given by

$$\tilde{E}(t) = \mathscr{E} \cos \omega t. \tag{1.5.32}$$

If we now change the sign of the applied electric field $\tilde{E}(t)$, the sign of the induced polarization $\tilde{P}(t)$ must also change, because we have assumed that the medium possesses inversion symmetry. Hence the relation (1.5.31) must be replaced by

$$-\tilde{P}(t) = \chi^{(2)} [-\tilde{E}(t)]^2, \tag{1.5.33}$$

which shows that

$$-\tilde{P}(t) = \chi^{(2)} \tilde{E}^2(t). \tag{1.5.34}$$

By comparison of this result with Eq. (1.5.31), we see that $\tilde{P}(t)$ must equal $-\tilde{P}(t)$, which can occur only if $\tilde{P}(t)$ vanishes identically. This result shows that

$$\chi^{(2)} = 0. \tag{1.5.35}$$

This result can be understood intuitively by considering the motion of an electron in a nonparabolic potential well. Due to the nonlinearity of the associated restoring force, the atomic response will show significant harmonic distortion. Part (a) of Fig. 1.5.1 shows the waveform of the incident monochromatic electromagnetic wave of frequency ω. For the case of a medium with linear response (part b), there is no distortion of the waveform associated with the polarization of the medium. Part (c) shows the induced polarization for the case of a nonlinear medium that possesses a center of symmetry and whose potential energy function has the form shown in Fig. 1.4.2. Although significant waveform distortion is evident, only odd harmonics of the fundamental frequency are present. For the case of a nonlinear, noncentrosymmetric medium having a potential energy function of the form shown in Fig. 1.4.1 (part d), both even and odd harmonics are present in the waveform associated with the atomic response. Note also the qualitative difference between the waveforms shown in parts (c) and (d). For the centrosymmetric medium (part c), the time-averaged response is zero, whereas for the noncentrosymmetric medium (part d) the time-averaged response is nonzero, because the medium responds differently to an electric field pointing, say, in the upward direction than to one pointing downward.*

* Parts (a) and (b) of Fig. 1.5.1 are plots of the function $\sin \omega t$, part (c) is a plot of the function $\sin \omega t - 0.25 \sin 3\omega t$, and part (d) is a plot of $-0.2 + \sin \omega t + 0.2 \cos 2\omega t$.

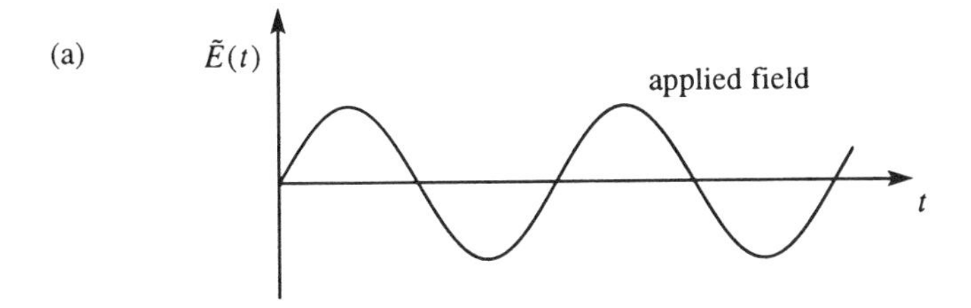

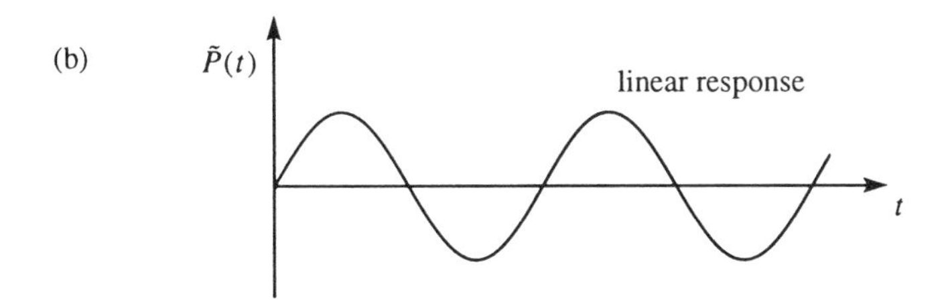

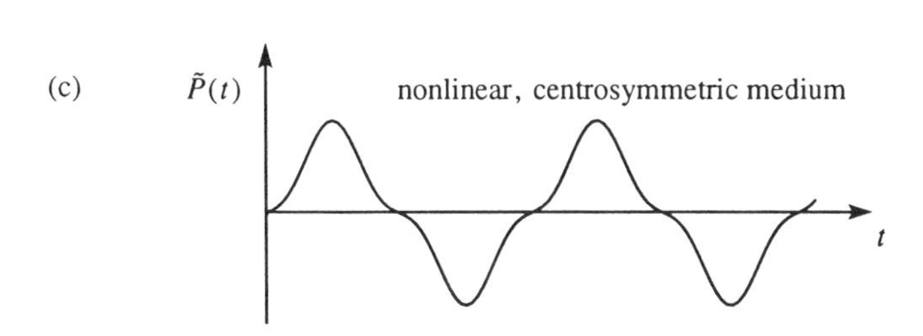

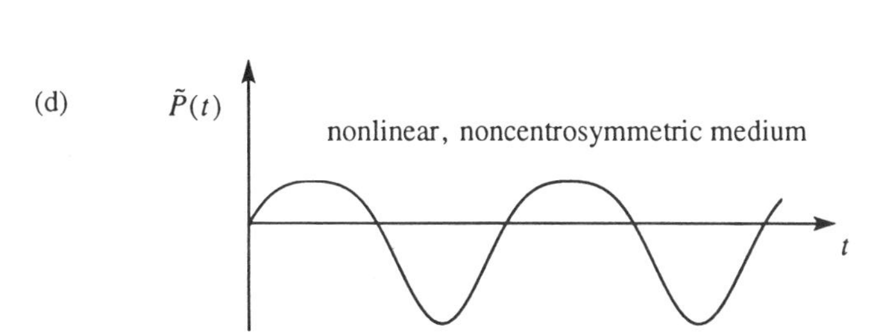

FIGURE 1.5.1 Waveforms associated with the atomic response.

TABLE 1.5.1 Form of the second-order susceptibility tensor for each of the crystal classes. Each element is denoted by its cartesian indices.

Crystal system	Crystal class	Nonvanishing tensor elements
Triclinic	1	All elements are independent and nonzero
	$\bar{1}$	Each element vanishes
Monoclinic	2	xyz, xzy, xxy, xyx, yxx, yyy, yzz, yzx, yxz, zyz, zzy, zxy, zyx (twofold axis parallel to $\hat{y}$)
	m	xxx, xyy, xzz, xzx, xxz, yyz, yzy, yxy, yyx, zxx, zyy, zzz, zzx, zxz (mirror plane perpendicular to $\hat{y}$)
	$2/m$	Each element vanishes
Orthorhombic	222	xyz, xzy, yzx, yxz, zxy, zyx
	$mm2$	xzx, xxz, yyz, yzy, zxx, zyy, zzz
	mmm	Each element vanishes
Tetragonal	4	$xyz = -yxz$, $xzy = -yzx$, $xzx = yzy$, $xxz = yyz$, $zxx = zyy$, zzz, $zxy = -zyx$
	$\bar{4}$	$xyz = yxz$, $xzy = yzx$, $xzx = -yzy$, $xxz = -yyz$, $zxx = -zyy$, $zxy = zyx$
	422	$xyz = -yxz$, $xzy = -yzx$, $zxy = -zyx$
	$4mm$	$xzx = yzy$, $xxz = yyz$, $zxx = zyy$, zzz
	$\bar{4}2m$	$xyz = yxz$, $xzy = yzx$, $zxy = zyx$
	$4/m$, $4/mmm$	Each element vanishes
Cubic	432	$xyz = -xzy = yzx = -yxz = zxy = -zyx$
	$\bar{4}3m$	$xyz = xzy = yzx = yxz = zxy = zyx$
	23	$xyz = yzx = zxy$, $xzy = yxz = zyx$
	$m3$, $m3m$	Each element vanishes
Trigonal	3	$xxx = -xyy = -yyz = -yxy$, $xyz = -yxz$, $xzy = -yzx$, $xzx = yzy$, $xxz = yyz$, $yyy = -yxx = -xxy = -xyx$, $zxx = zyy$, zzz, $zxy = -zyx$
	32	$xxx = -xyy = -yyx = -yxy$, $xyz = -yxz$, $xzy = -yzx$, $zxy = -zyx$
	$3m$	$xzx = yzy$, $xxz = yyz$, $zxx = zyy$, zzz, $yyy = -yxx = -xxy = -xyx$ (mirror plane perpendicular to $\hat{x}$)
	$\bar{3}$, $\bar{3}m$	Each element vanishes
Hexagonal	6	$xyz = -yxz$, $xzy = -yzx$, $xzx = yxy$, $xxz = yyz$, $zxx = zyy$, zzz, $zxy = -zyx$
	$\bar{6}$	$xxx = -xyy = -yxy = -yyx$, $yyy = -yxx = -xyx = -xxy$
	622	$xyz = -yxz$, $xzy = -yxz$, $zxy = -zyx$
	$6mm$	$xzx = yzy$, $xxz = yyz$, $zxx = zyy$, zzz
	$\bar{6}m2$	$yyy = -yxx = -xxy = -xyx$
	$6/m$, $6/mmm$	Each element vanishes

Additional Spatial Symmetries

Any additional symmetry properties of a nonlinear optical medium can impose additional restrictions on the form of the nonlinear susceptibility tensor. By explicit consideration of the symmetries of each of the 32 crystal classes, one can determine the allowed form of the susceptibility tensor for crystals of that class. The results of such a calculation, which was performed originally by Butcher (1965) and was later corrected by Shang and Hsu (1987), are presented in Table 1.5.1. Under those conditions (described following Eq. (1.5.21)) where the second-order susceptibility can be described using contracted notation, the results presented in Table 1.5.1 can usefully be displayed graphically. These results, as adapted from Zernike and Midwinter (1973), are presented in Fig. 1.5.2. Note that the influence of Kleinman symmetry is also described in the figure. As an example of how to use the table, for a crystal of class $3m$ the form of the d_{il} matrix is

$$d_{il} = \begin{bmatrix} 0 & 0 & 0 & 0 & d_{31} & -d_{22} \\ -d_{22} & d_{22} & 0 & d_{31} & 0 & 0 \\ d_{31} & d_{31} & d_{33} & 0 & 0 & 0 \end{bmatrix}$$

The spatial symmetry of the nonlinear optical medium also restricts the form of the third-order nonlinear optical susceptibility. The allowed form of the susceptibility has been calculated by Butcher (1965) and has been summarized by Hellwarth (1977); their results are presented in Table 1.5.2. Note that for the important special case of an isotropic optical material, the results presented in Table 1.5.2 agree with the result derived explicitly in the discussion of the nonlinear refractive index in Section 4.2.

The second-order nonlinear optical susceptibilities of a number of crystals are summarized in Table 1.5.3.

Number of Independent Elements of $\chi^{(2)}_{ijk}(\omega_3, \omega_2, \omega_1)$

We remarked above in relation to Eq. (1.5.1) that as many as 324 complex numbers must be specified in order to describe the general interaction of three optical waves. In practice, this number is often greatly reduced.

Due to the reality of the physical fields, only half of these numbers are independent (see Eq. (1.5.5)). Furthermore, the intrinsic permutation symmetry of $\chi^{(2)}$ (Eq. (1.5.6)) shows that there are only 81 independent parameters. For a lossless medium, all elements of $\chi^{(2)}$ are real and the condition of full permutation symmetry is valid, implying that only 27 of these numbers

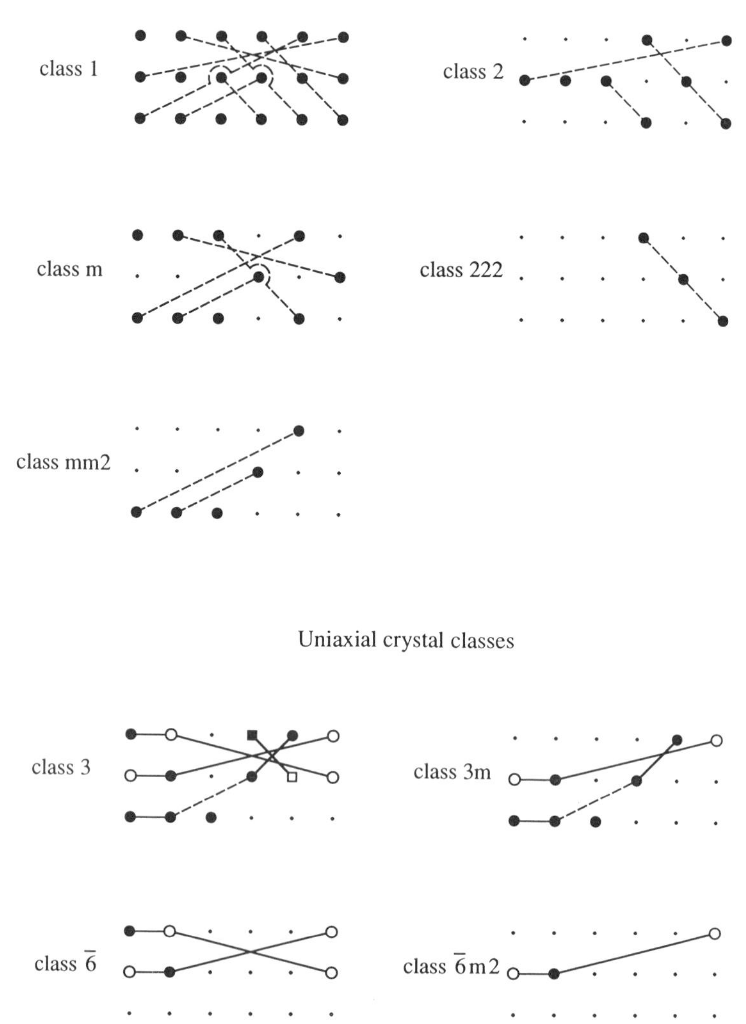

FIGURE 1.5.2 Form of the d_{il} matrix for the 21 crystal classes that lack inversion symmetry. Small dot: zero coefficient; large symbol: nonzero coefficient; square: coefficient that is zero when Kleinman's symmetry condition is valid; connected symbols: numerically equal coefficients, but the open-symbol coefficient is opposite in sign to the closed symbol to which it is joined. Dashed connections are valid only under Kleinman's symmetry conditions. (After Zernike and Midwinter, 1973.)

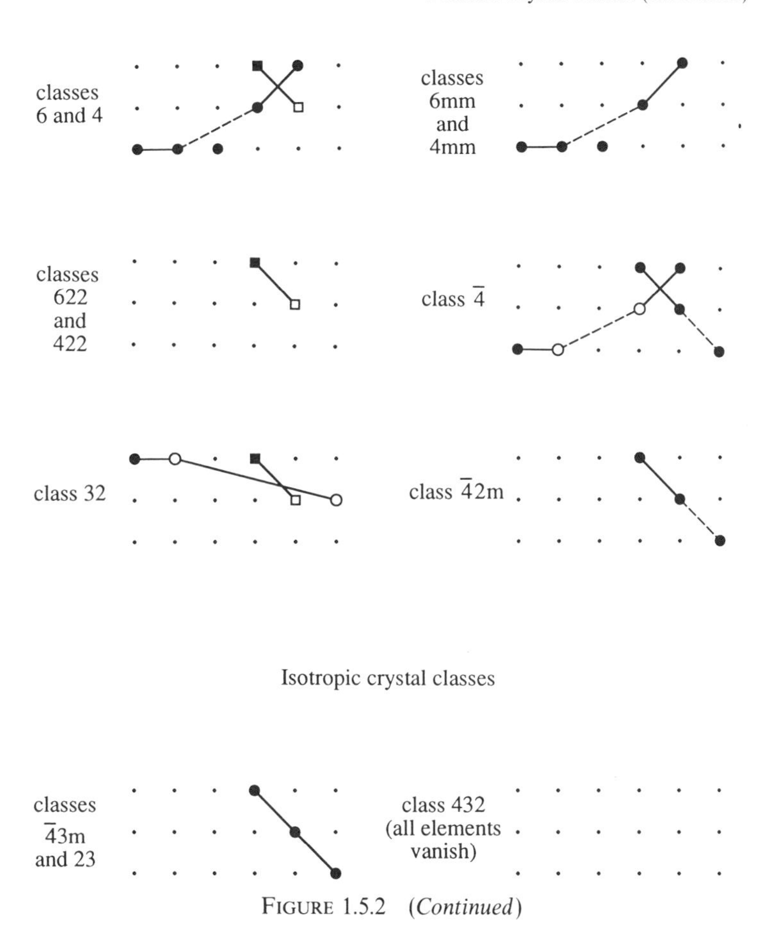

FIGURE 1.5.2 (*Continued*)

are independent. For second-harmonic generation, contracted notation can be used, and only 18 independent elements exist. When Kleinman's symmetry is valid, only 10 of these elements are independent. Furthermore, any crystalline symmetries of the nonlinear material can reduce this number still further.

Table 1.5.2 Form of the third-order susceptibility tensor $\chi^{(3)}$ for each of the crystal classes and for isotropic materials. Each element is denoted by its cartesian indices.

Isotropic

There are 21 nonzero elements, of which only 3 are independent. They are

$$yyzz = zzyy = zzxx = xxzz = xxyy = yyxx,$$

$$yzyz = zyzy = zxzx = xzxz = xyxy = yxyx,$$

$$yzzy = zyyz = zxxz = xzzx = xyyx = yxxy;$$

also

$$xxxx = yyyy = zzzz = xxyy + xyxy + xyyx.$$

Cubic

For the two classes 23 and $m3$, there are 21 nonzero elements, of which only 7 are independent. They are

$$xxxx = yyyy = zzzz,$$

$$yyzz = zzxx = xxyy,$$

$$zzyy = xxzz = yyxx,$$

$$yzyz = zxzx = xyxy,$$

$$zyzy = xzxz = yxyx,$$

$$yzzy = zxxz = xyyx,$$

$$zyyz = xzzx = yxxy.$$

For the three classes 432, $\bar{4}3m$, and $m3m$, there are 21 nonzero elements, of which only 4 are independent. They are

$$xxxx = yyyy = zzzz,$$

$$yyzz = zzyy = zzxx = xxzz = xxyy = yyxx,$$

$$yzyz = zyzy = zxzx = xzxz = xyxy = yxyx,$$

$$yzzy = zyyz = zxxz = xzzx = xyyx = yxxy.$$

Hexagonal

For the three classes 6, $\bar{6}$, and $6/m$, there are 41 nonzero elements, of which only 19 are independent. They are

$$zzzz, \qquad xxxx = yyyy = xxyy + xyyx + xyxy, \qquad \begin{cases} xxyy = yyxx, \\ xyyx = yxxy, \\ xyxy = yxyx, \end{cases}$$

TABLE 1.5.2 (*Continued*)

$$yyzz = xxzz, \quad xyzz = -yxzz,$$

$$zzyy = zzxx, \quad zzxy = -zzyx,$$

$$zyyz = zxxz, \quad zxyz = -zyxz,$$

$$yzzy = xzzx, \quad xzzy = -yzzx,$$

$$yzyz = xzxz, \quad xzyz = -yzxz,$$

$$zyzy = zxzx, \quad zxzy = -zyzx,$$

$$xxxy = -yyyx = yyxy + yxyy + xyyy, \quad \begin{cases} yyxy = -xxyx, \\ yxyy = -xyxx, \\ xyyy = -yxxx. \end{cases}$$

For the four classes 622, $6mm$, $6/mmm$, and $\bar{6}m2$, there are 21 nonzero elements, of which only 10 are independent. They are

$$zzzz,$$

$$xxxx = yyyy = xxyy + xyyx + xyxy, \quad \begin{cases} xxyy = yyxx, \\ xyyx = yxxy, \\ xyxy = yxyx, \end{cases}$$

$$yyzz = xxzz,$$

$$zzyy = zzxx,$$

$$zyyz = zxxz,$$

$$yzzy = xzzx,$$

$$yzyz = xzxz,$$

$$zyzy = zxzx.$$

Trigonal

For the two classes 3 and $\bar{3}$, there are 73 nonzero elements, of which only 27 are independent. They are

$$zzzz,$$

$$xxxx = yyyy = xxyy + xyyx + xyxy, \quad \begin{cases} xxyy = yyxx, \\ xyyx = yxxy, \\ xyxy = yxyx, \end{cases}$$

(*continues*)

TABLE 1.5.2 (*Continued*)

$$yyzz = xxzz, \qquad xyzz = -yxzz,$$
$$zzyy = zzxx, \qquad zzxy = -zzyx,$$
$$zyyz = zxxz, \qquad zxyz = -zyxz,$$
$$yzzy = xzzx, \qquad xzzy = -yzzx,$$
$$yzyz = xzxz, \qquad xzyz = -yzxz,$$
$$zyzy = zxzx, \qquad zxzy = -zyzx,$$
$$xxxy = -yyyx = yyxy + yxyy + xyyy, \quad \begin{cases} yyxy = -xxyx, \\ yxyy = -xyxx, \\ xyyy = -yxxx, \end{cases}$$
$$yyyz = -yxxz = -xyxz = -xxyz,$$
$$yyzy = -yxzx = -xyzx = -xxzy,$$
$$yzyy = -yzxx = -xzyx = -xzxy,$$
$$zyyy = -zyxx = -zxyx = -zxxy,$$
$$xxxz = -xyyz = -yxyz = -yyxz,$$
$$xxzx = -xyzy = -yxzy = -yyzx$$
$$xzxx = -yzxy = -yzyx = -xzyy,$$
$$zxxx = -zxyy = -zyxy = -zyyx.$$

For the three classes $3m$, $\bar{3}m$, and 32 there are 37 nonzero elements, of which only 14 are independent. They are

$$zzzz,$$
$$xxxx = yyyy = xxyy + xyyx + xyxy, \quad \begin{cases} xxyy = yyxx, \\ xyyx = yxxy, \\ xyxy = yxyx, \end{cases}$$
$$yyzz = xxzz, \qquad xxxz = -xyyz = -yxyz = -yyxz,$$
$$zzyy = zzxx, \qquad xxzx = -xyzy = -yxzy = -yyzx,$$
$$zyyz = zxxz, \qquad xzxx = -xzyy = -yzxy = -yzyx,$$
$$yzzy = xzzx, \qquad zxxx = -zxyy = -zyxy = -zyyx,$$
$$yzyz = xzxz,$$
$$zyzy = zxzx,$$

TABLE 1.5.2 (*Continued*)

Tetragonal

For the three classes 4, $\bar{4}$, and $4/m$, there are 41 nonzero elements, of which only 21 are independent. They are

$$xxxx = yyyy, \quad zzzz,$$

$$\begin{array}{llll} zzxx = zzyy, & xyzz = -yxzz, & xxyy = yyxx, & xxxy = -yyyx, \\ xxzz = yyzz, & zzxy = -zzyx, & xyxy = yxyx, & xxyx = -yyxy, \\ zxzx = zyzy, & xzyz = -yzxz, & xyyx = yxxy, & xyxx = -yxyy, \\ xzxz = yzyz, & zxzy = -zyzx, & & yxxx = -xyyy, \\ zxxz = zyyz, & zxyz = -zyxz, & & \\ xzzx = yzzy, & xzzy = -yzzx. & & \end{array}$$

For the four classes 422, $4mm$, $4/mmm$, and $\bar{4}2m$, there are 21 nonzero elements, of which only 11 are independent. They are

$$xxxx = yyyy, \quad zzzz,$$

$$\begin{array}{lll} yyzz = xxzz, & yzzy = xzzx & xxyy = yyxx, \\ zzyy = zzxx, & yzyz = xzxz & xyxy = yxyx, \\ zyyz = zxxz, & zyzy = zxzx & xyyx = yxxy. \end{array}$$

Monoclinic

For all three classes, 2, m, and $2/m$, there are 41 independent nonzero elements, consisting of

3 elements with suffixes all equal,
18 elements with suffixes equal in pairs,
12 elements with suffixes having two y's, one x, and one z,
4 elements with suffixes having three x's and one z,
4 elements with suffixes having three z's and one x.

Orthorhombic

For all three classes, 222, $mm2$, and mmm, there are 21 independent nonzero elements, consisting of

3 elements with suffixes all equal,
18 elements with suffixes equal in pairs.

Triclinic

For both classes, 1 and $\bar{1}$, there are 81 independent nonzero elements.

TABLE 1.5.3 Second-order nonlinear optical susceptibilities for several crystals

Material	Point group	d_{il} (10^{-9} esu)
Quartz	$32 = D_3$	$d_{11} = 0.96$ $d_{14} = 0.02$
$Ba_2NaNb_5O_{15}$	$mm2 = C_{2v}$	$d_{31} = -35$ $d_{32} = -35$ $d_{33} = -48$
$LiNbO_3$	$3m = C_{3v}$	$d_{22} = 7.4$ $d_{31} = 14$ $d_{33} = 98$
$BaTiO_3$	$4mm = C_{4v}$	$d_{15} = -41$ $d_{31} = -43$ $d_{33} = -16$
$NH_4H_2PO_4$ (ADP)	$\bar{4}2m = D_{2d}$	$d_{14} = 1.2$ $d_{36} = 1.2$
KH_2PO_4 (KDP)	$\bar{4}2m = D_{2d}$	$d_{14} = 1.2$ $d_{36} = 1.1$
$LiIO_3$	$6 = C_6$	$d_{35} = -13$ $d_{36} = -10$
CdSe	$6mm = C_{6v}$	$d_{15} = 74$ $d_{31} = 68$ $d_{33} = 130$
KD_2PO_4 (KD*P)	$\bar{4}2m = D_{2d}$	$d_{36} = 1.26$ $d_{14} = 1.26$
CdS	$6mm = C_{6v}$	$d_{33} = 86$ $d_{31} = 90$ $d_{36} = 100$
Ag_3AsS_3 (proustite)	$3m = C_{3v}$	$d_{22} = 68$ $d_{31} = 36$
$CdGeAs_2$	$\bar{4}2m = D_{2d}$	$d_{36} = 1090$
$AgGaSe_2$	$\bar{4}2m = D_{2d}$	$d_{36} = 81$
$AgSbS_3$ (pyrargyrite)	$3m = C_{3v}$	$d_{31} = 30$ $d_{22} = 32$
β-BaB_2O_4 (beta barium borate)		$d_{11} = 4.6$

Notes: Values are obtained from a variety of sources. One of the more complete tabulations is that of S. Singh in *Handbook of Lasers*, Chemical Rubber Company, Cleveland, Ohio 1971.

To convert to the MKS system using the convention that $P = dE^2$, multiply each entry by $4\pi\epsilon_0/(3 \times 10^4) = 3.71 \times 10^{-15}$ to obtain d in units of C/V^2.

To convert to the MKS system using the convention that $P = \epsilon_0 dE^2$, multiply each entry by $4\pi/(3 \times 10^4) = 4.189 \times 10^{-4}$ to obtain d in units of m/V.

In any system of units, $\chi^{(2)} = 2d$ by convention.

Problems

1. For proustite $\chi^{(2)}_{yyy}$ has the value 1.3×10^{-7} in gaussian units. What is its value in MKS units, assuming the convention that

$$\tilde{P}^{(2)}(t) = \epsilon_0 \chi^{(2)} \tilde{E}^2(t)?$$

[Ans.: 5.5×10^{-11} m/V.]

2. A laser beam of frequency ω carrying 1 W of power is focused to a spot size of 30-μm diameter in a crystal having a refractive index of $n = 2$ and a second-order susceptibility of $\chi^{(2)} = 1 \times 10^{-7}$ esu. Calculate numerically the amplitude $P(2\omega)$ of the component of the nonlinear polarization oscillating at frequency 2ω. Estimate numerically the amplitude of the dipole moment per atom $\mu(2\omega)$ oscillating at frequency 2ω. Compare this value with the atomic unit of dipole moment (ea_0, where a_0 is the Bohr radius) and with the linear response of the atom, that is, with the component $\mu(\omega)$ of the dipole moment oscillating at frequency ω. We shall see in the next chapter that, under the conditions stated above, nearly all of the incident power can be converted to the second harmonic for a 1-cm-long crystal.

[Ans.: $P(2\omega) = 1.5 \times 10^{-5}$ esu. Assuming that $N = 10^{22}$ atoms/cm^3, $\mu(2\omega) = 1.5 \times 10^{-27}$ esu $= 5.9 \times 10^{-10} ea_0$. By comparison, $P(\omega) = 2.9$ esu and $\mu(\omega) = 2.9 \times 10^{-22}$ esu $= 1.1 \times 10^{-4} ea_0$, which shows that $\mu(2\omega)/\mu(\omega) = 5.2 \times 10^{-6}$.]

3. Explain why it is unnecessary to include the term $\lambda^0 \tilde{x}^{(0)}$ in the power series of Eq. (1.4.6).

4. Verify Eqs. (1.5.30).

5. Through explicit consideration of the symmetry properties of each of the 32 point groups, verify the results presented in Tables 1.5.1 and 1.5.2 and in Fig. 1.5.2.

[Notes: This problem is lengthy and requires a more detailed knowledge of group theory and crystal symmetry than that presented in this text. For a list of recommended readings on these subjects, see the reference list to the present chapter. For a discussion of this problem, see also Butcher (1965).]

References

General

J. A. Armstrong, N. Bloembergen, J. Ducuing, and P. S. Pershan, *Phys. Rev.* **127**, 1918 (1962).

R. L. Byer and S. E. Harris, *Phys. Rev.* **168**, 1064 (1968).
C. Flytzanis, in *Quantum Electronics, a Treatise*, Vol. 1, Part A, edited by H. Rabin and C. L. Tang, Academic Press, New York, 1975.
P. A. Franken, A. E. Hill, C. W. Peters, and G. Weinreich, *Phys. Rev. Lett.* **7**, 118 (1961).
C. G. B. Garrett and F. N. H. Robinson, *IEEE J. Quantum Electron.* **2**, 328 (1966).
W. Kaiser and C. G. B. Garrett, *Phys. Rev. Lett.* **7**, 229 (1961).
D. A. Kleinman, *Phys. Rev.* **126**, 1977 (1962).
S. E. Harris, M. K. Oshman, and R. L. Byer, *Phys. Rev. Lett.* **18**, 732 (1967).
R. W. Hellwarth, *Prog. Quantum Electron.* **5**, 1 (1977).
L. D. Landau and E. M. Lifshitz, *Electrodynamics of Continuous Media*, Pergamon, New York, 1960, Section 10.
J. E. Midwinter and J. Warner, *Brit. J. Appl. Phys.* **16**, 1135 (1965).
R. C. Miller, *Appl. Phys. Lett.* **5**, 17 (1964).
A. Owyoung, *The Origins of the Nonlinear Refractive Indices of Liquids and Glasses*, Ph.D. dissertation, California Institute of Technology, 1971.
P. S. Pershan, *Phys. Rev.* **130**, 919 (1963).
C. C. Shang and H. Hsu, *IEEE J. Quantum Electron.* **23**, 177 (1987).
Y. R. Shen, *Phys. Rev.* **167**, 818 (1968).
S. Singh, in *Handbook of Lasers*, Chemical Rubber Co., Cleveland, Ohio, 1971.

Books on Nonlinear Optics

G. P. Agrawal, *Nonlinear Fiber Optics*, Academic Press, Boston, 1989.
S. A. Akhmanov and R. V. Khokhlov, *Problems of Nonlinear Optics*, Gordon and Breach, New York, 1972.
G. C. Baldwin, *An Introduction to Nonlinear Optics*, Plenum Press, New York, 1969.
N. Bloembergen, *Nonlinear Optics*, Benjamin, New York, 1964.
P. N. Butcher, *Nonlinear Optical Phenomena*, Ohio State University, 1965.
P. N. Butcher and D. Cotter, *The Elements of Nonlinear Optics*, Cambridge University Press, 1990.
N. B. Delone, *Fundamentals of Nonlinear Optics of Atomic Gases*, Wiley, New York, 1988.
D. C. Hannah, M. A. Yuratich, and D. Cotter, *Nonlinear Optics of Free Atoms and Molecules*, Springer-Verlag, Berlin, 1979.
F. A. Hopf and G. I. Stegeman, *Applied Classical Electrodynamics, Vol. I: Linear Optics*, Wiley, New York, 1985; *Vol. 2: Nonlinear Optics*, Wiley, New York, 1986.
D. N. Klyshko, *Photons and Nonlinear Optics*, Gordon and Breach, New York, 1988.
M. D. Levenson and S. Kano, *Introduction to Nonlinear Laser Spectroscopy*, Academic Press, Boston, 1988.
R. Loudon, *The Quantum Theory of Light*, Clarendon Press, Oxford, 1983.
J. F. Reintjes, *Nonlinear Optical Parametric Processes in Liquids and Gases*, Academic Press, Orlando, Fla., 1984.

M. Schubert and B. Wilhelmi, *Nonlinear Optics and Quantum Electronics*, Wiley, New York, 1986.

Y. R. Shen, *The Principles of Nonlinear Optics*, Wiley, New York, 1984.

A. Yariv, *Quantum Electronics*, Wiley, New York, 1975.

F. Zernike and J. E. Midwinter, *Applied Nonlinear Optics*, Wiley, New York, 1973.

Books on Group Theory and Crystal Symmetry

S. Bhagavantan, *Crystal Symmetry and Physical Properties*, Academic Press, London, 1966.

W. L. Bond, *Crystal Technology*, Wiley, New York, 1976.

M. J. Buerger, *Elementary Crystallography*, Wiley, New York, 1963.

J. F. Nye, *The Physical Properties of Crystals*, Clarendon Press, Oxford, 1985.

F. C. Phillips, *An Introduction to Crystallography*, Wiley, New York, 1976.

M. Tinkham, *Group Theory and Quantum Mechanics*, McGraw-Hill, New York, 1964.

Chapter 2

Wave-Equation Description of Nonlinear Optical Interactions

2.1. The Wave Equation for Nonlinear Optical Media

We have seen in the last chapter how nonlinearities in the response of a material system to an intense laser field can cause the polarization of the medium to develop new frequency components not present in the incident radiation field. These new frequency components of the polarization act as sources of new frequency components of the electromagnetic field. In the present chapter, we examine how Maxwell's equations describe the generation of these new components of the field, and more generally we see how the various frequency components of the field become coupled by the nonlinear interaction.

Before developing the mathematical theory of these effects, we shall give a simple physical picture of how these frequency components are generated. For definiteness, we consider the case of sum-frequency generation as shown in part (a) of Fig. 2.1.1, where the input fields are at frequencies ω_1 and ω_2. Due to nonlinearities in the atomic response, each atom develops an oscillating dipole moment which contains a component at frequency $\omega_1 + \omega_2$. An isolated atom would radiate at this frequency in the form of a dipole radiation pattern, as shown symbolically in part (b) of the figure. However, any material sample contains an enormous number N of atomic dipoles, each oscillating with a phase that is determined by the phases of the incident fields. If the relative phasing of these dipoles is correct, the field radiated by each dipole will add constructively in the forward direction, leading to radiation in

(a)

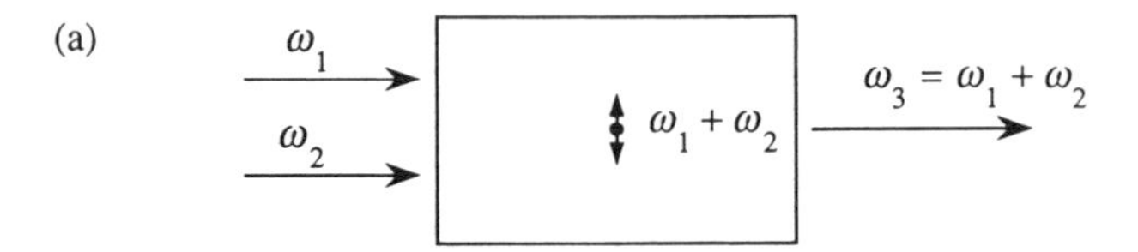

(b)

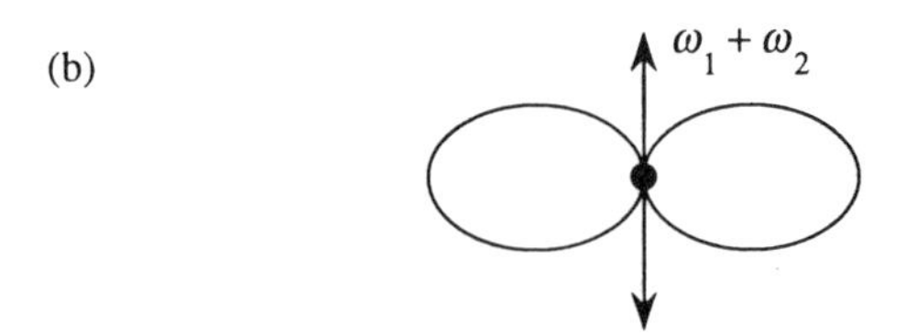

(c)

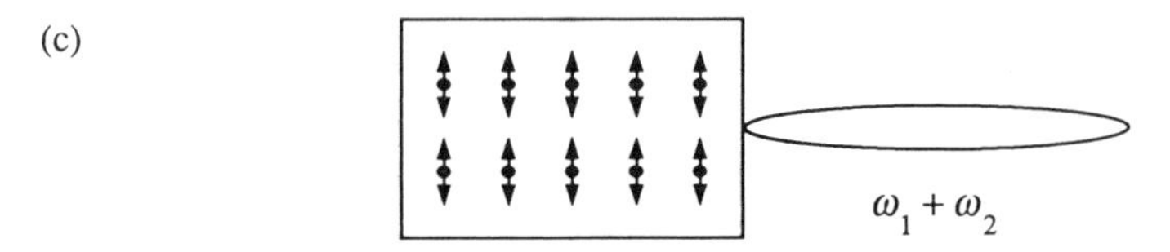

FIGURE 2.1.1 Sum-frequency generation.

the form of a well-defined beam, as illustrated in part (c) of the figure. The system will act as a phased array of dipoles when a certain condition, known as the phase-matching condition (see Eq. (2.2.15) in the next section), is satisfied. Under these conditions, the electric field strength of the radiation emitted in the forward direction will be N times larger than that due to any one atom, and consequently the intensity will be N^2 times as large.

Let us now consider the form of the wave equation for the propagation of light through a nonlinear optical medium. We begin with Maxwell's equations, which we write in gaussian units in the form*

$$\nabla \cdot \tilde{\mathbf{D}} = 4\pi\tilde{\rho}, \tag{2.1.1}$$

$$\nabla \cdot \tilde{\mathbf{B}} = 0, \tag{2.1.2}$$

* Throughout the text we use a tilde to denote a quantity that varies rapidly in time.

$$\nabla \times \tilde{\mathbf{E}} = -\frac{1}{c}\frac{\partial \tilde{\mathbf{B}}}{\partial t}, \tag{2.1.3}$$

$$\nabla \times \tilde{\mathbf{H}} = \frac{1}{c}\frac{\partial \tilde{\mathbf{D}}}{\partial t} + \frac{4\pi}{c}\tilde{\mathbf{J}}, \tag{2.1.4}$$

We are primarily interested in the solution of these equations in regions of space that contain no free charges, so that

$$\tilde{\rho} = 0, \tag{2.1.5}$$

and that contain no free currents, so that

$$\tilde{\mathbf{J}} = 0. \tag{2.1.6}$$

We assume that the material is nonmagnetic, so that

$$\tilde{\mathbf{B}} = \tilde{\mathbf{H}}. \tag{2.1.7}$$

However, we allow the material to be nonlinear in the sense that the fields $\tilde{\mathbf{D}}$ and $\tilde{\mathbf{E}}$ are related by

$$\tilde{\mathbf{D}} = \tilde{\mathbf{E}} + 4\pi\tilde{\mathbf{P}}, \tag{2.1.8}$$

where in general the polarization vector $\tilde{\mathbf{P}}$ depends nonlinearly upon the local value of the electric field strength $\tilde{\mathbf{E}}$.

We now proceed to derive the optical wave equation in the usual manner. We take the curl of the curl-$\tilde{\mathbf{E}}$ Maxwell equation (2.1.3), interchange the order of space and time derivatives on the right-hand side of the resulting equation, and use Eqs. (2.1.4), (2.1.6), and (2.1.7) to replace $\nabla \times \tilde{\mathbf{B}}$ by $(1/c)(\partial\tilde{\mathbf{D}}/\partial t)$, to obtain the equation

$$\nabla \times \nabla \times \tilde{\mathbf{E}} + \frac{1}{c^2}\frac{\partial^2}{\partial t^2}\tilde{\mathbf{D}} = 0. \tag{2.1.9a}$$

We now use Eq. (2.1.8) to eliminate $\tilde{\mathbf{D}}$ from this equation, and we thereby obtain the expression

$$\nabla \times \nabla \times \tilde{\mathbf{E}} + \frac{1}{c^2}\frac{\partial^2}{\partial t^2}\tilde{\mathbf{E}} = \frac{-4\pi}{c^2}\frac{\partial^2\tilde{\mathbf{P}}}{\partial t^2}. \tag{2.1.9b}$$

This is the most general form of the wave equation in nonlinear optics. Under certain conditions it can be simplified. For example, by using an identity from vector calculus, we can write the first term on the left-hand side of Eq. (2.1.9b) as

$$\nabla \times \nabla \times \tilde{\mathbf{E}} = \nabla(\nabla \cdot \tilde{\mathbf{E}}) - \nabla^2\tilde{\mathbf{E}}. \tag{2.1.10}$$

In the linear optics of isotropic source-free media, the first term on the right-hand side of this equation vanishes because the Maxwell equation $\nabla \cdot \tilde{\mathbf{D}} = 0$ implies that $\nabla \cdot \tilde{\mathbf{E}} = 0$. However, in nonlinear optics this term is generally nonvanishing even for isotropic materials, owing to the more general relation (2.1.8) between $\tilde{\mathbf{D}}$ and $\tilde{\mathbf{E}}$. Fortunately, in nonlinear optics the first term on the right-hand side of Eq. (2.1.10) can usually be dropped for cases of interest. For example, if $\tilde{\mathbf{E}}$ is of the form of a transverse, infinite plane wave, $\nabla \cdot \tilde{\mathbf{E}}$ vanishes identically. More generally, the first term can often be shown to be small, even when it does not vanish identically, especially when the slowly-varying-amplitude approximation (see Section 2.2) is valid.

It is often convenient to split $\tilde{\mathbf{P}}$ into its linear and nonlinear parts as

$$\tilde{\mathbf{P}} = \tilde{\mathbf{P}}^{(1)} + \tilde{\mathbf{P}}^{\mathrm{NL}}. \tag{2.1.11}$$

Here $\tilde{\mathbf{P}}^{(1)}$ is the part of $\tilde{\mathbf{P}}$ that depends linearly on the electric field strength $\tilde{\mathbf{E}}$. We can similarly decompose the displacement field $\tilde{\mathbf{D}}$ into its linear and nonlinear parts as

$$\tilde{\mathbf{D}} = \tilde{\mathbf{D}}^{(1)} + 4\pi\tilde{\mathbf{P}}^{\mathrm{NL}}, \tag{2.1.12a}$$

where the linear part is given by

$$\tilde{\mathbf{D}}^{(1)} = \tilde{\mathbf{E}} + 4\pi\tilde{\mathbf{P}}^{(1)}. \tag{2.1.12b}$$

In terms of this quantity, the wave equation (2.1.9) becomes

$$\nabla \times \nabla \times \tilde{\mathbf{E}} + \frac{1}{c^2}\frac{\partial^2 \tilde{\mathbf{D}}^{(1)}}{\partial t^2} = \frac{-4\pi}{c^2}\frac{\partial^2 \tilde{\mathbf{P}}^{\mathrm{NL}}}{\partial t^2}. \tag{2.1.13}$$

To see why this form of the wave equation is useful, let us first consider the case of a lossless, dispersionless medium. We can then express the relation between $\tilde{\mathbf{D}}^{(1)}$ and $\tilde{\mathbf{E}}$ in terms of a real, frequency-independent dielectric tensor $\boldsymbol{\epsilon}^{(1)}$ as

$$\tilde{\mathbf{D}}^{(1)} = \boldsymbol{\epsilon}^{(1)} \cdot \tilde{\mathbf{E}}. \tag{2.1.14a}$$

For the case of an isotropic material, this relation reduces simply to

$$\tilde{\mathbf{D}}^{(1)} = \epsilon^{(1)}\tilde{\mathbf{E}}, \tag{2.1.14b}$$

where $\epsilon^{(1)}$ is a scalar quantity. For this (simpler) case of an isotropic material, the wave equation (2.1.13) becomes

$$\nabla \times \nabla \times \tilde{\mathbf{E}} + \frac{\epsilon^{(1)}}{c^2}\frac{\partial^2 \tilde{\mathbf{E}}}{\partial t^2} = \frac{-4\pi}{c^2}\frac{\partial^2 \tilde{\mathbf{P}}^{\mathrm{NL}}}{\partial t^2}. \tag{2.1.15}$$

This equation has the form of a driven (i.e., inhomogeneous) wave equation; the nonlinear response of the medium acts as a source term which appears on the right-hand side of this equation. In the absence of this source term, Eq. (2.1.15) admits solutions of the form of free waves propagating with velocity c/n, where $n = [\epsilon^{(1)}]^{1/2}$ is the (linear) index of refraction.

For the case of a dispersive medium, we must consider each frequency component of the field separately. We represent the electric, linear displacement, and polarization fields as the sums of their various frequency components:

$$\tilde{\mathbf{E}}(\mathbf{r}, t) = \sum_n{}' \tilde{\mathbf{E}}_n(\mathbf{r}, t), \tag{2.1.16a}$$

$$\tilde{\mathbf{D}}^{(1)}(\mathbf{r}, t) = \sum_n{}' \tilde{\mathbf{D}}_n^{(1)}(\mathbf{r}, t), \tag{2.1.16b}$$

$$\tilde{\mathbf{P}}^{\mathrm{NL}}(\mathbf{r}, t) = \sum_n{}' \tilde{\mathbf{P}}_n^{\mathrm{NL}}(\mathbf{r}, t), \tag{2.1.16c}$$

where the summation is to be performed over positive field frequencies only, and we represent each frequency component in terms of its complex amplitude as

$$\tilde{\mathbf{E}}_n(\mathbf{r}, t) = \mathbf{E}_n(\mathbf{r}) e^{-i\omega_n t} + \mathrm{c.c.}, \tag{2.1.17a}$$

$$\tilde{\mathbf{D}}_n^{(1)}(\mathbf{r}, t) = \mathbf{D}_n^{(1)}(\mathbf{r}) e^{-i\omega_n t} + \mathrm{c.c.}, \tag{2.1.17b}$$

$$\tilde{\mathbf{P}}_n^{\mathrm{NL}}(\mathbf{r}, t) = \mathbf{P}_n^{\mathrm{NL}}(\mathbf{r}) e^{-i\omega_n t} + \mathrm{c.c.} \tag{2.1.17c}$$

If dissipation can be neglected, the relationship between $\tilde{\mathbf{D}}_n^{(1)}$ and $\tilde{\mathbf{E}}_n$ can be expressed in terms of a real, frequency-dependent dielectric tensor according to

$$\tilde{\mathbf{D}}_n^{(1)}(\mathbf{r}, t) = \boldsymbol{\epsilon}^{(1)}(\omega_n) \cdot \tilde{\mathbf{E}}_n(\mathbf{r}, t). \tag{2.1.18}$$

When Eqs. (2.1.16) through (2.1.18) are introduced into Eq. (2.1.13), we obtain a wave equation analogous to (2.1.15) that is valid for each frequency component of the field:

$$\nabla \times \nabla \times \tilde{\mathbf{E}}_n + \frac{\boldsymbol{\epsilon}^{(1)}(\omega_n)}{c^2} \cdot \frac{\partial^2 \tilde{\mathbf{E}}_n}{\partial t^2} = \frac{-4\pi}{c^2} \frac{\partial^2 \tilde{\mathbf{P}}_n^{\mathrm{NL}}}{\partial t^2}. \tag{2.1.19}$$

As mentioned above in connection with Eq. (2.1.10), the first term on the left-hand side of this equation can often be replaced by $-\nabla^2 \tilde{\mathbf{E}}_n$.

The general case of a dissipative medium is treated by allowing the dielectric tensor to be a complex quantity that relates the complex field amplitudes according to

$$\mathbf{D}_n^{(1)}(\mathbf{r}) = \boldsymbol{\epsilon}^{(1)}(\omega_n) \cdot \mathbf{E}_n(\mathbf{r}). \tag{2.1.20}$$

This expression, along with Eqs. (2.1.16) and (2.1.17), can be introduced into the wave equation (2.1.13), to obtain

$$\nabla \times \nabla \times \mathbf{E}_n(\mathbf{r}) - \frac{\omega_n^2}{c^2} \boldsymbol{\epsilon}^{(1)}(\omega_n) \cdot \mathbf{E}_n(\mathbf{r}) = \frac{4\pi\omega_n^2}{c^2} \mathbf{P}_n^{\mathrm{NL}}(\mathbf{r}). \tag{2.1.21}$$

When the first term on the left-hand side of this expression can be reduced simply to the negative of the Laplacian of $\tilde{\mathbf{E}}_n$ (see the discussion following Eq. (2.1.10)), this equation has the form of an inhomogeneous Helmholtz equation.

2.2. The Coupled-Wave Equations for Sum-Frequency Generation

We next study how the nonlinear optical wave equation that we derived in the previous section can be used to describe specific nonlinear optical interactions. In particular, we consider sum-frequency generation in a lossless nonlinear optical medium involving monochromatic, continuous-wave input beams. We assume the configuration shown in Fig. 2.2.1, where the applied waves fall onto the nonlinear medium at normal incidence. For simplicity, we ignore double refraction effects. The treatment given here can be generalized straightforwardly to include nonnormal incidence and double refraction.*

The wave equation in the form (2.1.19) must hold for each frequency component of the field and in particular for the sum-frequency component at frequency ω_3. In the absence of a nonlinear source term, the solution to this equation for a plane wave at frequency ω_3 propagating in the $+z$ direction is

$$\tilde{E}_3(z,t) = A_3 e^{i(k_3 z - \omega_3 t)} + \text{c.c.}, \tag{2.2.1}$$

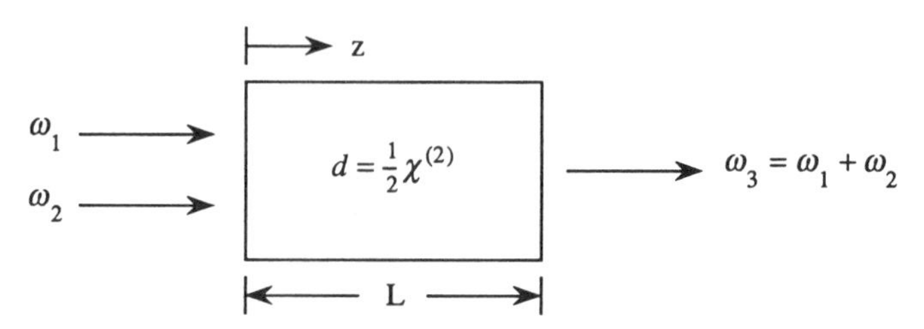

FIGURE 2.2.1 Sum-frequency generation.

* See, for example, Shen (1984), Chapter 6.

where

$$k_3 = \frac{n_3 \omega_3}{c}, \qquad n_3 = [\epsilon^{(1)}(\omega_3)]^{1/2}, \tag{2.2.2}$$

and where the amplitude of the wave A_3 is a *constant*.* We expect on physical grounds that, when the nonlinear source term is not too large, the solution to Eq. (2.1.19) will still be of the form of Eq. (2.2.1), except that A_3 will become a slowly varying function of z. We hence adopt Eq. (2.2.1) with A_3 taken to be a function of z as the form of the trial solution to the wave equation (2.1.19) in the presence of the nonlinear source term.

We represent the nonlinear source term appearing in Eq. (2.1.19) as

$$\tilde{P}_3(z,t) = P_3 e^{-i\omega_3 t} + \text{c.c.}, \tag{2.2.3}$$

where according to Eq. (1.5.28)

$$P_3 = 4\, d E_1 E_2. \tag{2.2.4}$$

For simplicity, we denote d_{eff} of Eq. (1.5.28) simply as d. If we represent the applied fields as

$$\tilde{E}_i(z,t) = E_i e^{-i\omega_i t} + \text{c.c.}, \qquad i = 1, 2, \tag{2.2.5}$$

with

$$E_i = A_i e^{ik_i z}, \qquad i = 1, 2, \tag{2.2.6}$$

the amplitude of the nonlinear polarization can be written as

$$P_3 = 4\, d A_1 A_2 e^{i(k_1 + k_2)z}. \tag{2.2.7}$$

We now substitute Eqs. (2.2.1), (2.2.3), and (2.2.7) into the wave equation (2.1.19) with $\nabla \times \nabla \times$ replaced by $-\nabla^2$ (see the discussion following Eq. (2.1.10)). Since the fields depend only on the spatial coordinate z, we can replace ∇^2 by $\partial^2/\partial z^2$. We then obtain

$$\left[\frac{\partial^2 A_3}{\partial z^2} + 2ik_3 \frac{\partial A_3}{\partial z} - k_3^2 A_3 + \frac{\epsilon^{(1)}(\omega_3)\omega_3^2 A_3}{c^2}\right] e^{i(k_3 z - \omega_3 t)} + \text{c.c.}$$
$$= \frac{-16\pi d \omega_3^2}{c^2} A_1 A_2 e^{i[(k_1 + k_2)z - \omega_3 t]} + \text{c.c.} \tag{2.2.8}$$

Since $k_3^2 = \epsilon^{(1)}(\omega_3)\omega_3^2/c^2$, the third and fourth terms on the left-hand side of this expression cancel. Note that we can drop the complex conjugate terms

* For convenience, we are working in the scalar field approximation; n_3 represents the refractive index appropriate to the state of polarization of the ω_3 wave.

from each side and still maintain the equality. We can then cancel the factor $\exp(-i\omega_3 t)$ on each side and write the resulting equation as

$$\frac{d^2A_3}{dz^2} + 2ik_3\frac{dA_3}{dz} = \frac{-16\pi\, d\omega_3^2}{c^2} A_1 A_2 e^{i(k_1+k_2-k_3)z}. \tag{2.2.9}$$

Note that we have replaced $\partial/\partial z$ by d/dz because the field amplitude A_3 is a function of z only. It is usually permissible to neglect the first term on the left-hand side of this equation on the grounds that it is very much smaller than the second. This approximation is known as the slowly-varying-amplitude approximation and is valid whenever

$$\left|\frac{d^2A_3}{dz^2}\right| \ll \left|k_3\frac{dA_3}{dz}\right|. \tag{2.2.10}$$

This condition requires that the fractional change in A_3 in a distance of the order of an optical wavelength must be much smaller than unity. When this approximation is made, Eq. (2.2.9) becomes

$$\frac{dA_3}{dz} = \frac{8\pi i\, d\omega_3^2}{k_3c^2} A_1 A_2 e^{i\Delta k z}, \tag{2.2.11}$$

where we have introduced the quantity

$$\Delta k = k_1 + k_2 - k_3, \tag{2.2.12}$$

which is called the wave vector (or momentum) mismatch. Equation (2.2.11) is known as a coupled-amplitude equation, because it shows how the amplitude of the ω_3 wave varies due to its coupling to the ω_1 and ω_2 waves. In general, the spatial variation of the ω_1 and ω_2 waves must also be taken into consideration, and we can derive analogous equations for the ω_1 and ω_2 fields by repeating the derivation given above for each of these frequencies. We hence find two additional coupled-amplitude equations given by

$$\frac{dA_1}{dz} = \frac{8\pi i\, d\omega_1^2}{k_1c^2} A_3 A_2^* e^{-i\Delta k z} \tag{2.2.13}$$

and

$$\frac{dA_2}{dz} = \frac{8\pi i\, d\omega_2^2}{k_2c^2} A_3 A_1^* e^{-i\Delta k z}. \tag{2.2.14}$$

Note that, in order to write these equations in the forms shown, we have had to make use of the assumption that the medium is lossless (1) so that no linear loss terms appear in these equations and (2) so that we could make use of the condition of full permutation symmetry (Eq. (1.5.8)) to deduce that the same coupling coefficient d appeared in each equation.

Phase-Matching Considerations

For simplicity, let us first assume that the amplitudes A_1 and A_2 of the input fields can be taken as constants on the right-hand side of Eq. (2.2.11). This assumption is valid whenever the conversion of the input fields into the sum-frequency field is not too large. We note that, for the special case

$$\Delta k = 0, \tag{2.2.15}$$

the amplitude A_3 of the sum-frequency wave increases linearly with z, and consequently that its intensity increases quadratically with z. The condition (2.2.15) is known as the condition of perfect phase matching. When this condition is fulfilled, the generated wave maintains a fixed phase relation with respect to the nonlinear polarization and is able to extract energy most efficiently from the incident waves. From a microscopic point of view, when the condition (2.2.15) is fulfilled the individual atomic dipoles that constitute the material system are properly phased so that the field emitted by each dipole adds coherently in the forward direction. The total power radiated by the ensemble of atomic dipoles thus scales as the square of the number of atoms that participate.

When the condition (2.2.15) is not satisfied, the intensity of the emitted radiation is smaller than for the case of $\Delta k = 0$. The amplitude of the sum-frequency (ω_3) field at the exit plane of the nonlinear medium is given in this case by integrating Eq. (2.2.11) from $z = 0$ to $z = L$, yielding

$$A_3(L) = \frac{8\pi i d\omega_3^2 A_1 A_2}{k_3 c^2} \int_0^L e^{i\Delta k z}\, dz = \frac{8\pi i d\omega_3^2 A_1 A_2}{k_3 c^2} \left(\frac{e^{i\Delta k L} - 1}{i\Delta k} \right). \tag{2.2.16}$$

The intensity of the ω_3 wave is given by the magnitude of the time-averaged Poynting vector, which for our definition of field amplitude is given by

$$I_i = \frac{n_i c}{2\pi} |A_i|^2, \qquad i = 1, 2, 3. \tag{2.2.17}$$

We thus obtain

$$I_3 = \frac{32\pi d^2 \omega_3^4 |A_1|^2 |A_2|^2 n_3}{k_3^2 c^3} \left| \frac{e^{i\Delta k L} - 1}{\Delta k} \right|^2. \tag{2.2.18}$$

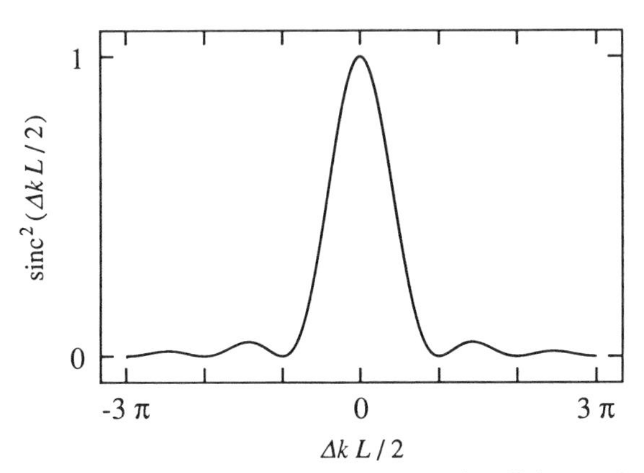

FIGURE 2.2.2 The effects of wave vector mismatch on the efficiency of sum-frequency generation.

The squared modulus that appears in this equation can be expressed as

$$\left|\frac{e^{i\Delta k L}-1}{\Delta k}\right|^2 = L^2\left(\frac{e^{i\Delta k L}-1}{\Delta k L}\right)\left(\frac{e^{-i\Delta k L}-1}{\Delta k L}\right) = 2L^2\frac{(1-\cos\Delta k\, L)}{(\Delta k\, L)^2}$$
$$= L^2\frac{\sin^2(\Delta k\, L/2)}{(\Delta k\, L/2)^2} \equiv L^2 \operatorname{sinc}^2(\Delta k\, L/2). \tag{2.2.19}$$

Finally, our expression for I_3 can be written in terms of the intensities of the incident fields by using Eq. (2.2.17) to express $|A_i|^2$ in terms of the intensities, yielding the result

$$I_3 = \frac{512\pi^5 d^2 I_1 I_2}{n_1 n_2 n_3 \lambda_3^2 c} L^2 \operatorname{sinc}^2(\Delta k\, L/2), \tag{2.2.20}$$

where $\lambda_3 = 2\pi c/\omega_3$ is the vacuum wavelength of the ω_3 wave. Note that the effect of wave vector mismatch is included entirely in the factor $\operatorname{sinc}^2(\Delta k\, L/2)$. This factor, which is known as the phase mismatch factor, is plotted in Fig. 2.2.2.

It should be noted that the efficiency of the three-wave mixing process decreases as $|\Delta k|L$ increases, with some oscillations occurring. The reason for this behavior is that if L is greater than approximately $1/\Delta k$, the output wave can get out of phase with its driving polarization, and power can flow from the ω_3 wave back into the ω_1 and ω_2 waves (see Eq. (2.2.11)). For this reason, one sometimes defines

$$L_c = 2/\Delta k \tag{2.2.21}$$

to be the *coherence length* of the interaction, so that the phase mismatch factor in Eq. (2.2.20) can be written as

$$\text{sinc}^2(L/L_c). \tag{2.2.22}$$

2.3. The Manley–Rowe Relations

Let us now consider, from a general point of view, the mutual interaction of three optical waves propagating through a lossless nonlinear optical medium, as illustrated in Fig. 2.3.1.

We have just derived the coupled-amplitude equations (Eqs. (2.2.11) through (2.2.14)) that describe the spatial variation of the amplitude of each wave. Let us now consider the spatial variation of the *intensity* associated with each of these waves. Since

$$I_i = \frac{n_i c}{2\pi} A_i A_i^*, \tag{2.3.1}$$

the variation of the intensity is described by

$$\frac{dI_i}{dz} = \frac{n_i c}{2\pi}\left(A_i^* \frac{dA_i}{dz} + A_i \frac{dA_i^*}{dz}\right). \tag{2.3.2}$$

Through use of this result and Eq. (2.2.13), we find that the spatial variation of the intensity of the wave at frequency ω_1 is given by

$$\begin{aligned}\frac{dI_1}{dz} &= \frac{n_1 c}{2\pi}\,\frac{8\pi d\omega_1^2}{k_1 c^2}(iA_1^* A_3 A_2^* e^{-i\Delta k z} + \text{c.c.})\\ &= 4\,d\omega_1(iA_3 A_1^* A_2^* e^{-i\Delta k z} + \text{c.c.})\end{aligned}$$

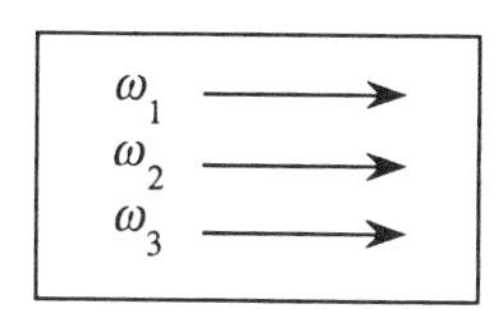

FIGURE 2.3.1 Optical waves of frequencies ω_1, ω_2, and $\omega_3 = \omega_1 + \omega_2$ interact in a lossless nonlinear optical medium.

or by

$$\frac{dI_1}{dz} = -8\,d\omega_1\,\mathrm{Im}(A_3 A_1^* A_2^* e^{-i\Delta k z}). \tag{2.3.3a}$$

We similarly find that the spatial variation of the intensities of the waves at frequencies ω_2 and ω_3 is given by

$$\frac{dI_2}{dz} = -8\,d\omega_2\,\mathrm{Im}(A_3 A_1^* A_2^* e^{-i\Delta k z}), \tag{2.3.3b}$$

$$\begin{aligned}\frac{dI_3}{dz} &= -8\,d\omega_3\,\mathrm{Im}(A_3^* A_1 A_2 e^{i\Delta k z}) \\ &= 8\,d\omega_3\,\mathrm{Im}(A_3 A_1^* A_2^* e^{-i\Delta k z}).\end{aligned} \tag{2.3.3c}$$

We see that the sign of dI_1/dz is the same as that of dI_2/dz but is opposite to that of dI_3/dz. We also see that the direction of energy flow depends on the relative phases of the three interacting fields.

The set of equations (2.3.3) shows that the total power flow is conserved, as expected for propagation through a lossless medium. To demonstrate this fact, we define the total intensity as

$$I = I_1 + I_2 + I_3. \tag{2.3.4}$$

We then find that the spatial variation of the total intensity is given by

$$\begin{aligned}\frac{dI}{dz} &= \frac{dI_1}{dz} + \frac{dI_2}{dz} + \frac{dI_3}{dz} \\ &= -8d(\omega_1 + \omega_2 - \omega_3)\,\mathrm{Im}(A_3 A_1^* A_2^* e^{-i\Delta k z}) = 0,\end{aligned} \tag{2.3.5}$$

where we have made use of Eqs. (2.3.3) and where the last equality follows from the fact that $\omega_3 = \omega_1 + \omega_2$.

The set of equations (2.3.3) also implies that

$$\frac{d}{dz}\left(\frac{I_1}{\omega_1}\right) = \frac{d}{dz}\left(\frac{I_2}{\omega_2}\right) = -\frac{d}{dz}\left(\frac{I_3}{\omega_3}\right), \tag{2.3.6}$$

as can be verified by inspection. These equalities are known as the Manley–Rowe relations (Manley and Rowe, 1959). Since the energy of a photon of frequency ω_i is $\hbar\omega_i$, the quantity I_i/ω_i that appears in these relations is proportional to the intensity of the wave measured in photons per unit area per unit time. The Manley–Rowe relations can alternatively be expressed as

$$\frac{d}{dz}\left(\frac{I_2}{\omega_2} + \frac{I_3}{\omega_3}\right) = 0, \quad \frac{d}{dz}\left(\frac{I_1}{\omega_1} + \frac{I_3}{\omega_3}\right) = 0, \quad \frac{d}{dz}\left(\frac{I_1}{\omega_1} - \frac{I_2}{\omega_2}\right) = 0. \tag{2.3.7}$$

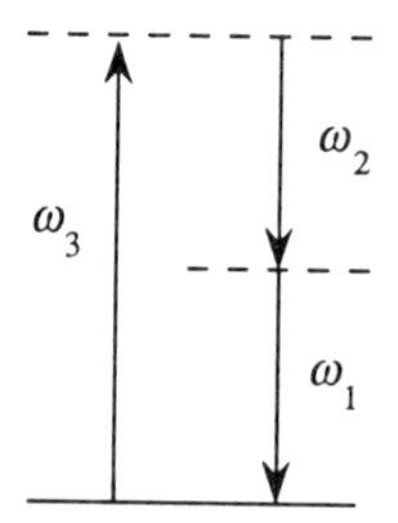

FIGURE 2.3.2 Photon description of the interaction of three optical waves.

These equations can be formally integrated to obtain the three conserved quantities (conserved in the sense that they are spatially invariant) M_1, M_2, and M_3, which are given by

$$\frac{I_2}{\omega_2} + \frac{I_3}{\omega_3} = M_1, \qquad \frac{I_1}{\omega_1} + \frac{I_3}{\omega_3} = M_2, \qquad \frac{I_1}{\omega_1} - \frac{I_2}{\omega_2} = M_3. \quad (2.3.8)$$

These relations tell us that the rate at which photons at frequency ω_1 are created is equal to the rate at which photons at frequency ω_2 are created and is equal to the rate at which photons at frequency ω_3 are destroyed. This result can be understood intuitively by means of the energy level description of a three-wave mixing process, which is shown in Figure 2.3.2. This diagram shows that, for a lossless medium, the creation of an ω_1 photon must be accompanied by the creation of an ω_2 photon and the annihilation of an ω_3 photon. It seems somewhat surprising that the Manley–Rowe relations should be consistent with this quantum-mechanical interpretation, when our derivation of these relations appears to be entirely classical. Note, however, that our derivation implicitly assumes that the nonlinear susceptibility possesses full permutation symmetry in that we have taken the coupling constant to have the same value in each of the coupled-amplitude equations (2.2.11), (2.2.13), and (2.2.14). We remarked earlier (following Eq. (1.5.9)) that in a sense the condition of full permutation symmetry is a consequence of the laws of quantum mechanics.

2.4. Sum-Frequency Generation

In Section 2.2, we treated the process of sum-frequency generation in the simple limit in which the two input fields are undepleted by the nonlinear interaction. In the present section, we treat this process more generally. We assume the configuration shown in Fig. 2.4.1.

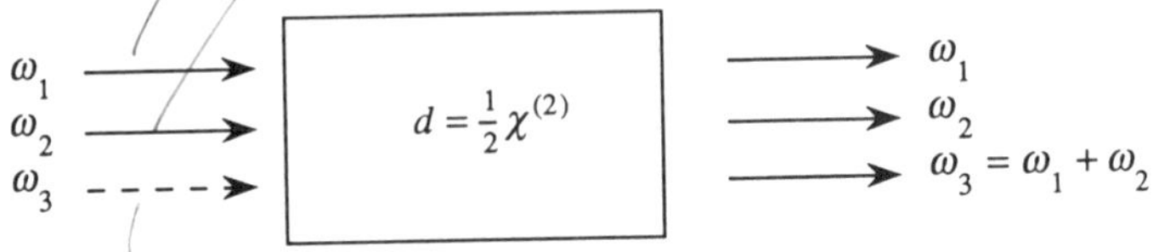

ω_1 → $d = \frac{1}{2}\chi^{(2)}$ → ω_1

ω_2 → → ω_2

ω_3 - - - → → $\omega_3 = \omega_1 + \omega_2$

FIGURE 2.4.1 Sum-frequency generation. Typically, no input field is applied at frequency ω_3.

The coupled-amplitude equations describing this interaction were derived above and appear as Eqs. (2.2.11) through (2.2.14). These equations can be solved exactly in terms of the Jacobi elliptic functions. We will not present the details of this solution, because the method is very similar to the one that we use in Section 2.6 to treat second-harmonic generation. Details can be found in Armstrong *et al.* (1962); see also Problem 2 at the end of this chapter.

Instead, we treat the somewhat simpler (but more illustrative) case in which one of the applied fields (taken to be at frequency ω_2) is strong, but the other field (at frequency ω_1) is weak. This case would apply to the conversion of a weak infrared signal of frequency ω_1 to a visible frequency ω_3 by mixing with an intense laser beam of frequency ω_2 (see, for example, Boyd and Townes, 1977). This process is known as upconversion. Since we can assume that the amplitude A_2 of the field at frequency ω_2 is unaffected by the interaction, we can take A_2 as a constant in the coupled-amplitude equations (Eqs. (2.2.11) through (2.2.14)), which then reduce to the simpler set

$$\frac{dA_1}{dz} = K_1 A_3 e^{-i\Delta k z}, \tag{2.4.1a}$$

$$\frac{dA_3}{dz} = K_3 A_1 e^{+i\Delta k z} \tag{2.4.1b}$$

where

$$K_1 = \frac{8\pi i \omega_1^2 d}{k_1 c^2} A_2^*, \qquad K_3 = \frac{8\pi i \omega_3^2 d}{k_3 c^2} A_2, \tag{2.4.1c}$$

and

$$\Delta k = k_1 + k_2 - k_3. \tag{2.4.2}$$

The solution to Eqs. (2.4.1) is particularly simple if we set $\Delta k = 0$, and we first treat this case. We take the derivative of Eq. (2.4.1a) to obtain

$$\frac{d^2 A_1}{dz^2} = K_1 \frac{dA_3}{dz}. \tag{2.4.3}$$

We now use Eq. (2.4.1b) to eliminate dA_3/dz from the right-hand side of this equation to obtain an equation involving only $A_1(z)$:

$$\frac{d^2A_1}{dz^2} = -\kappa^2 A_1, \tag{2.4.4}$$

where we have introduced the *positive* coupling coefficient κ^2 defined by

$$\kappa^2 \equiv -K_1K_3 = \frac{64\pi^2\omega_1^2\omega_3^2 d^2|A_2|^2}{k_1k_3c^4}. \tag{2.4.5}$$

The general solution to Eq. (2.4.4) is

$$A_1(z) = B\cos\kappa z + C\sin\kappa z. \tag{2.4.6a}$$

We now obtain the form of $A_3(z)$ through use of Eq. (2.4.1a), which shows that $A_3(z) = (dA_1/dz)/K_1$, or

$$A_3(z) = \frac{-B\kappa}{K_1}\sin\kappa z + \frac{C\kappa}{K_1}\cos\kappa z. \tag{2.4.6b}$$

We next find the solution that satisfies the appropriate boundary conditions. We assume that the ω_3 field is not present at the input, so that the boundary conditions become $A_3(0) = 0$ with $A_1(0)$ specified. We find from Eq. (2.4.6b) that the boundary condition $A_3(0) = 0$ implies that $C = 0$, and from Eq. (2.4.6a) that $B = A_1(0)$. The solution for the ω_1 field is thus given by

$$A_1(z) = A_1(0)\cos\kappa z \tag{2.4.7}$$

and for the ω_3 field by

$$A_3(z) = -A_1(0)\frac{\kappa}{K_1}\sin\kappa z. \tag{2.4.8}$$

To simplify the form of this expression, we express the ratio κ/K_1 as follows:

$$\frac{\kappa}{K_1} = \left(\frac{8\pi\omega_1\omega_3 d|A_2|}{(k_1k_3)^{1/2}c^2}\right)\left(\frac{k_1c^2}{8\pi i\omega_1^2 dA_2^*}\right) = -i\left(\frac{n_1\omega_3}{n_3\omega_1}\right)^{1/2}\frac{|A_2|}{A_2^*}.$$

The ratio $|A_2|/A_2^*$ can be represented as

$$\frac{|A_2|}{A_2^*} = \frac{A_2}{A_2}\frac{|A_2|}{A_2^*} = \frac{A_2|A_2|}{|A_2|^2} = \frac{A_2}{|A_2|} = e^{i\phi_2},$$

where ϕ_2 denotes the phase of A_2. We hence find that

$$A_3(z) = i\left(\frac{n_1\omega_3}{n_3\omega_1}\right)^{1/2} A_1(0)\sin\kappa z\, e^{i\phi_2}. \tag{2.4.9}$$

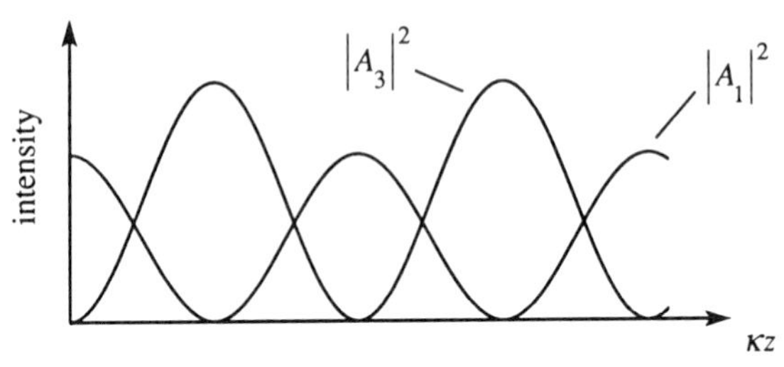

FIGURE 2.4.2 Variation of $|A_1|^2$ and $|A_3|^2$ for the case of perfect phase matching in the undepleted-pump approximation.

The nature of the solution given by Eqs. (2.4.7) and (2.4.9) is illustrated in Fig. 2.4.2.

Next we solve Eqs. (2.4.1) for the general case of arbitrary wave vector mismatch. We seek a solution to these equations of the form

$$A_1(z) = (Fe^{igz} + Ge^{-igz})e^{-i\,\Delta k\, z/2}, \tag{2.4.10}$$

$$A_3(z) = (Ce^{igz} + De^{-igz})e^{i\,\Delta k\, z/2}, \tag{2.4.11}$$

where g gives the rate of spatial variation of the fields and where C, D, F, and G are constants whose values depend on the boundary conditions. We take this form for the trial solution because we expect the ω_1 and ω_3 waves to display the same spatial variation, since they are coupled together. We separate out the factors $e^{\pm i\,\Delta k\, z/2}$ because doing so simplifies the final form of the solution. Equations (2.4.10) and (2.4.11) are now substituted into Eq. (2.4.1a), to obtain

$$\begin{aligned}(igFe^{igz} - igGe^{-igz})e^{-(1/2)i\,\Delta k\, z} - \tfrac{1}{2}i\,\Delta k(Fe^{igz} + Ge^{-igz})e^{-(1/2)i\,\Delta k\, z}\\ = (K_1Ce^{igz} + K_1De^{-igz})e^{-(1/2)i\,\Delta k\, z}.\end{aligned} \tag{2.4.12}$$

Since this equation must hold for all values of z, the terms that vary as e^{igz} and e^{-igz} must each maintain the equality separately; the coefficients of these terms thus must be related by

$$F(ig - \tfrac{1}{2}i\,\Delta k) = K_1C, \tag{2.4.13}$$

$$-G(ig + \tfrac{1}{2}i\,\Delta k) = K_1D. \tag{2.4.14}$$

In a similar fashion, we find by substituting the trial solution into Eq. (2.4.1b) that

$$\begin{aligned}(igCe^{igz} - igDe^{-igz})e^{(1/2)i\,\Delta k\, z} + \tfrac{1}{2}i\,\Delta k(Ce^{igz} + De^{-igz})e^{(1/2)i\,\Delta k\, z}\\ = (K_3Fe^{igz} + K_3Ge^{-igz})e^{(1/2)i\,\Delta k\, z},\end{aligned} \tag{2.4.15}$$

and in order for this equation to hold for all values of z, the coefficients must satisfy

$$C(ig + \tfrac{1}{2}i\,\Delta k) = K_3 F, \tag{2.4.16}$$

$$-D(ig - \tfrac{1}{2}i\,\Delta k) = K_3 G. \tag{2.4.17}$$

Equation (2.4.13) and (2.4.16) constitute simultaneous equations for F and C. We write these equations in matrix form as

$$\begin{bmatrix} i(g - \frac{1}{2}\Delta k) & -K_1 \\ -K_3 & i(g + \frac{1}{2}\Delta k) \end{bmatrix} \begin{bmatrix} F \\ C \end{bmatrix} = 0.$$

A solution to this set of equations exists only if the determinant of the matrix of coefficients vanishes, i.e., if

$$g^2 = -K_1 K_3 + \tfrac{1}{4}\Delta k^2. \tag{2.4.18}$$

As before (cf. Eq. (2.4.5)), we introduce the positive quantity $\kappa^2 = -K_1 K_3$, so that we can express the solution to Eq. (2.4.18) as

$$g = \sqrt{\kappa^2 + \tfrac{1}{4}\Delta k^2}. \tag{2.4.19}$$

In determining g we take only the positive square root in the above expression, since our trial solution (2.4.10) and (2.4.11) explicitly contains both the e^{+gz} and e^{-gz} spatial variation.

The general solution to our original set of equations (2.4.1) is given by Eqs. (2.4.10) and (2.4.11) with g given by Eq. (2.4.19). We evaluate the arbitrary constants C, D, F, and G appearing in the general solution by applying appropriate boundary conditions. We assume that the fields A_1 and A_3 are specified at the input plane $z = 0$ of the nonlinear medium, so that $A_1(0)$ and $A_3(0)$ are known. Then, by evaluating Eqs. (2.4.10) and (2.4.11) at $z = 0$, we find that

$$A_1(0) = F + G, \tag{2.4.20}$$

$$A_3(0) = C + D. \tag{2.4.21}$$

Equations (2.4.13) and (2.4.14) give two additional relations among the quantities C, D, F, and G. Consequently there are four independent linear equations relating the four quantities C, D, F, and G, and their simultaneous solution specifies these four quantities. The values of C, D, F, and G thereby obtained are introduced into the trial solution (2.4.10) and (2.4.11) to obtain the solution that meets the boundary conditions. This solution is given by

$$A_1(z) = \left[A_1(0) \cos gz + \left(\frac{K_1}{g} A_3(0) + \frac{i\,\Delta k}{2g} A_1(0) \right) \sin gz \right] \times e^{-(1/2)i\,\Delta k\, z}, \tag{2.4.22}$$

$$A_3(z) = \left[A_3(0) \cos gz + \left(\frac{-i\,\Delta k}{2g} A_3(0) + \frac{K_3}{g} A_1(0) \right) \sin gz \right] \times e^{(1/2)i\,\Delta k\, z}. \tag{2.4.23}$$

In order to interpret this result, let us consider the special case in which no sum-frequency field is incident on the medium, so that $A_3(0) = 0$. Equation (2.4.23) then reduces to

$$A_3(z) = \frac{K_3}{g} A_1(0) \sin gz\, e^{(1/2)i\,\Delta k\, z} \tag{2.4.24}$$

and the intensity of the generated wave is proportional to

$$|A_3(z)|^2 = |A_1(0)|^2 \frac{|K_3|^2}{g^2} \sin^2 gz. \tag{2.4.25}$$

where g is given as before by Eq. (2.4.19). We note that the characteristic scale length g^{-1} of the interaction becomes shorter as Δk increases. However, as Δk increases the maximum intensity of the generated wave decreases. Since, according to Eq. (2.4.25), the intensity of the generated wave is inversely proportional to g^2, we see that as Δk is increased the maximum intensity of the generated wave is decreased by the factor $\kappa^2/(\kappa^2 + \frac{1}{4}\Delta k^2)$. This sort of behavior is illustrated in Fig. 2.4.3, in which the predictions of Eq. (2.4.25) are displayed graphically.

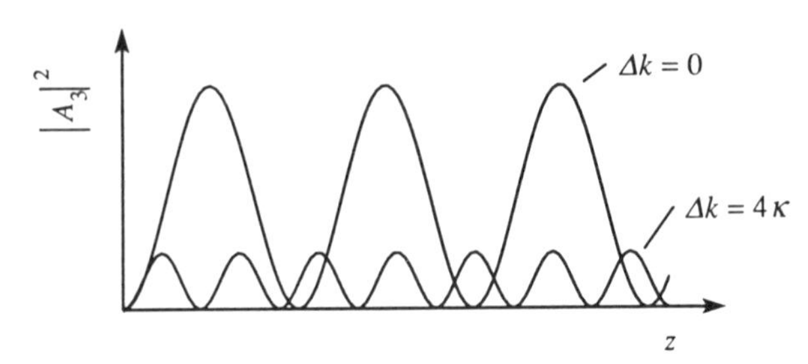

FIGURE 2.4.3 Spatial variation of the sum-frequency wave in the undepleted pump approximation.

2.5. Difference-Frequency Generation and Parametric Amplification

Let us now consider the situation shown in Fig. 2.5.1, in which optical waves at frequencies ω_3 and ω_1 interact in a lossless nonlinear optical medium to produce an output wave at the difference frequency $\omega_2 = \omega_3 - \omega_1$. For simplicity, we assume that the ω_3 wave is a strong wave (i.e., is undepleted by the nonlinear interaction, so that we can treat A_3 as being essentially constant), and for the present we assume that no field at frequency ω_2 is incident on the medium.

The coupled-amplitude equations describing this interaction are obtained by a method analogous to that used in Section 2.2 to obtain the equations describing sum-frequency generation, and have the form

$$\frac{dA_1}{dz} = \frac{8\pi i\omega_1^2 d}{k_1 c^2} A_3 A_2^* e^{i\Delta k z}, \tag{2.5.1a}$$

$$\frac{dA_2}{dz} = \frac{8\pi i\omega_2^2 d}{k_2 c^2} A_3 A_1^* e^{i\Delta k z}, \tag{2.5.1b}$$

where

$$\Delta k = k_3 - k_1 - k_2. \tag{2.5.2}$$

We first solve these equations for the case of perfect phase matching, that is, $\Delta k = 0$. We differentiate Eq. (2.5.1b) with respect to z and introduce the complex conjugate of Eq. (2.5.1a) to eliminate dA_1^*/dz from the right-hand side. We thereby obtain the equation

$$\frac{d^2A_2}{dz^2} = \frac{64\pi^2\omega_1^2\omega_2^2 d^2}{k_1 k_2 c^4} A_3 A_3^* A_2 \equiv \kappa^2 A_2, \tag{2.5.3}$$

where we have introduced the coupling constant

$$\kappa^2 = \frac{64\pi^2 d^2\omega_1^2\omega_2^2}{k_1 k_2 c^4} |A_3|^2. \tag{2.5.4}$$

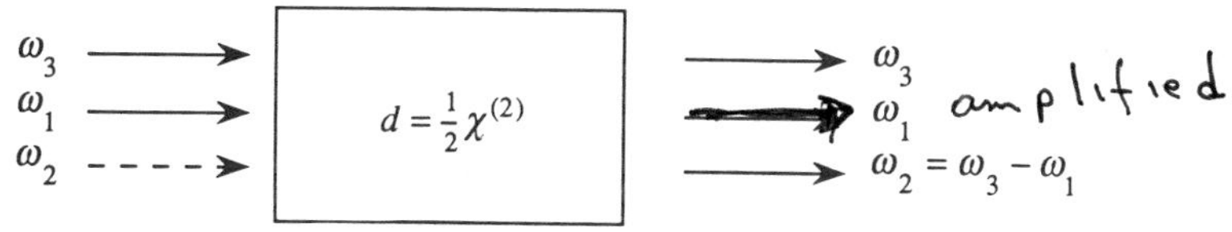

ω_3 →
ω_1 →
ω_2 - - - →

$d = \frac{1}{2}\chi^{(2)}$

→ ω_3
→ ω_1
→ $\omega_2 = \omega_3 - \omega_1$

FIGURE 2.5.1 Difference-frequency generation. Typically, no input field is applied at frequency ω_2.

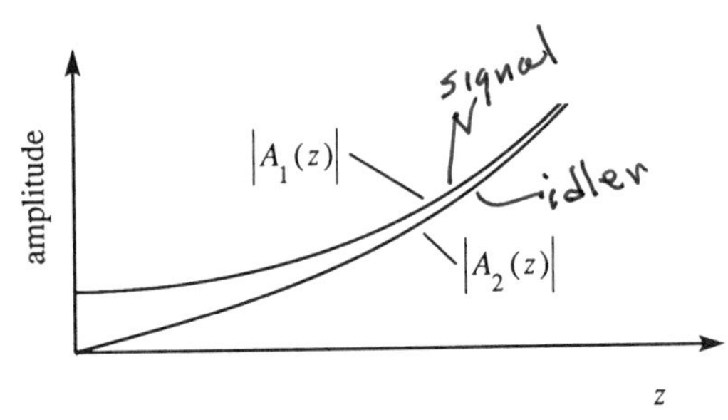

FIGURE 2.5.2 Spatial evolution of A_1 and A_2 for difference-frequency generation for the case $\Delta k = 0$ in the constant-pump approximation.

The general solution to this equation is

$$A_2(z) = C \sinh \kappa z + D \cosh \kappa z, \tag{2.5.5}$$

where C and D are integration constants whose value depends on the boundary conditions.

We now assume the boundary conditions

$$A_2(0) = 0, \qquad A_1(0) \text{ arbitrary}. \tag{2.5.6}$$

The solution to Eqs. (2.5.1) that meets these boundary conditions is readily found to be

$$A_1(z) = A_1(0) \cosh \kappa z, \tag{2.5.7}$$

$$A_2(z) = i \left(\frac{n_1 \omega_2}{n_2 \omega_1} \right)^{1/2} \frac{A_3}{|A_3|} A_1^*(0) \sinh \kappa z. \tag{2.5.8}$$

The nature of this solution is shown in Fig. 2.5.2. Note that both the ω_1 and the ω_2 fields experience monotonic growth and that each grows asymptotically (i.e., for $\kappa z \gg 1$) as $e^{\kappa z}$. We see from the form of the solution that the ω_1 field retains its initial phase and is simply amplified by the interaction, while the generated wave at frequency ω_2 has a phase that depends both on that of the pump wave and on that of the ω_1 wave. This behavior of monotonic growth of both waves is qualitatively dissimilar from that of sum-frequency generation, where oscillatory behavior occurs.

The reason for the different behavior in this case can be understood intuitively in terms of the energy level diagram shown in Fig. 2.5.3. We can think of diagram (a) as showing how the presence of a field at frequency ω_1 stimulates the downward transition that leads to the generation of the ω_2 field. Likewise, diagram (b) shows that the ω_2 field stimulates the generation of the

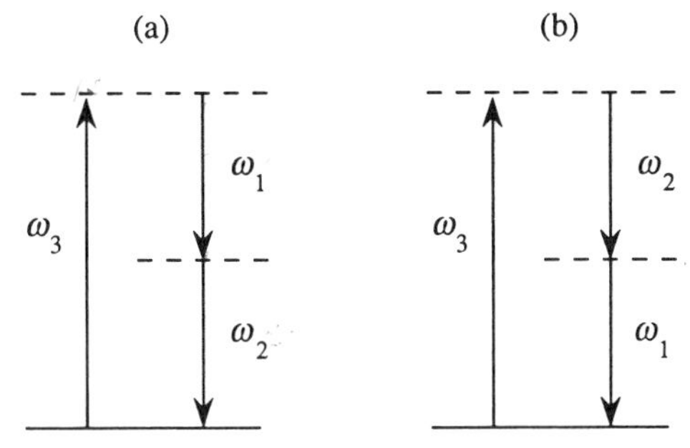

FIGURE 2.5.3 Difference-frequency generation.

ω_1 field. Hence the generation of the ω_1 field reinforces the generation of the ω_2 field, and vice versa, leading to the exponential growth of each wave.

Since the ω_1 field is amplified by the process of difference-frequency generation, which is a parametric process, this process is also known as parametric amplification. In this language, one says that the signal wave (the ω_1 wave) is amplified by the nonlinear mixing process, and an idler wave (at $\omega_2 = \omega_3 - \omega_1$) is generated by the process. If mirrors that are highly reflecting at frequencies ω_1 and/or ω_2 are placed on either side of the nonlinear medium to form an optical resonator, oscillation can occur due to the gain of the parametric amplification process. Such a device is known as a parametric oscillator.

The first cw optical parametric oscillator was built by Giordmaine and Miller in 1965. The theory of parametric amplification and parametric oscillators has been reviewed by Byer and Herbst (1977).

The solution to the coupled-amplitude equations (2.5.1) for the general case of arbitrary $\Delta k \neq 0$ makes a good exercise for the reader (see Problem 3 at the end of this chapter). The solution for the case of arbitrary boundary conditions (i.e., both $A_1(0)$ and $A_2(0)$ specified) is given by

$$A_1(z) = \left[A_1(0)\left(\cosh gz - \frac{i\Delta k}{2g} \sinh gz \right) + \frac{\kappa_1}{g} A_2^*(0) \sinh gz \right] e^{i\Delta k\, z/2}, \quad (2.5.9a)$$

$$A_2(z) = \left[A_2(0)\left(\cosh gz - \frac{i\Delta k}{2g} \sinh gz \right) + \frac{\kappa_2}{g} A_1^*(0) \sinh gz \right] e^{i\Delta k\, z/2}, \quad (2.5.9b)$$

where the coefficient g (which is not the same as that of Eq. (2.4.19)) is given by

$$g = \left[\kappa_1 \kappa_2^* - \left(\frac{\Delta k}{2} \right)^2 \right]^{1/2} \quad (2.5.9c)$$

with

$$\kappa_j = \frac{8\pi i \omega_j^2 \, d A_3}{k_j c^2}. \tag{2.5.9d}$$

2.6. Second-Harmonic Generation

In this section we present a mathematical description of the process of second-harmonic generation, shown symbolically in Fig. 2.6.1. We assume that the medium is lossless both at the fundamental frequency ω_1 and at the second-harmonic frequency $\omega_2 = 2\omega_1$, so that the nonlinear susceptibility obeys the condition of full permutation symmetry. Our discussion closely follows that of one of the first theoretical treatments of second-harmonic generation (Armstrong *et al.*, 1962).

We take the total electric field within the nonlinear medium to be given by

$$\tilde{E}(z,t) = \tilde{E}_1(z,t) + \tilde{E}_2(z,t), \tag{2.6.1}$$

where each component is expressed in terms of a complex amplitude $E_j(z)$ and slowly varying amplitude $A_j(z)$ according to

$$\tilde{E}_j(z,t) = E_j(z) e^{-i\omega_j t} + \text{c.c.}, \tag{2.6.2}$$

where

$$E_j(z) = A_j(z) e^{i k_j z}, \tag{2.6.3}$$

and where the wave number and refractive index are given by

$$k_j = n_j \omega_j / c, \qquad n_j = [\epsilon^{(1)}(\omega_j)]^{1/2}. \tag{2.6.4}$$

We assume that each frequency component of the electric field obeys the driven wave equation (see also Eq. 2.1.19)

$$\frac{\partial^2 \tilde{E}_j}{\partial z^2} - \frac{\epsilon^{(1)}(\omega_j)}{c^2} \frac{\partial^2 \tilde{E}_j}{\partial t^2} = \frac{4\pi}{c^2} \frac{\partial^2}{\partial t^2} \tilde{P}_j. \tag{2.6.5}$$

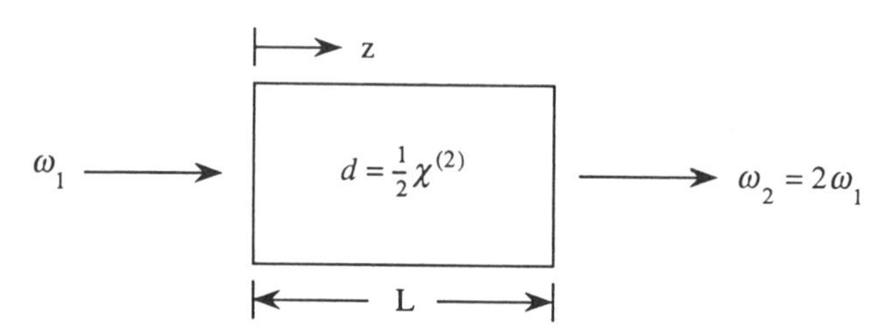

FIGURE 2.6.1 Second-harmonic generation.

The nonlinear polarization is represented as

$$\tilde{P}^{\mathrm{NL}}(z,t) = \tilde{P}_1(z,t) + \tilde{P}_2(z,t) \tag{2.6.6}$$

with

$$\tilde{P}_j(z,t) = P_j(z)e^{-i\omega_j t} + \mathrm{c.c.}, \qquad j = 1,2. \tag{2.6.7}$$

The expressions for P_j are given according to Eqs. (1.5.28) and (1.5.29) by

$$P_1(z) = 4\,dE_2E_1^* = 4\,dA_2A_1^*e^{i(k_2-k_1)z} \tag{2.6.8}$$

and

$$P_2(z) = 2\,dE_1^2 = 2\,dA_1^2e^{2ik_1z}. \tag{2.6.9}$$

Note that the degeneracy factors appearing in these two expressions are different. We obtain coupled-amplitude equations for the two frequency components by methods analogous to those used in Section 2.2 in deriving the coupled-amplitude equations for sum-frequency generation. We find that

$$\frac{dA_1}{dz} = \frac{8\pi i\omega_1^2 d}{k_1c^2}A_2A_1^*e^{-i\Delta k z} \tag{2.6.10}$$

and

$$\frac{dA_2}{dz} = \frac{4\pi i\omega_2^2 d}{k_2c^2}A_1^2e^{i\Delta k z}, \tag{2.6.11}$$

where

$$\Delta k = 2k_1 - k_2. \tag{2.6.12}$$

In the undepleted-pump approximation (i.e., A_1 constant), Eq. (2.6.11) can be integrated immediately to obtain an expression for the spatial dependence of the second-harmonic field amplitude. More generally, the pair of coupled equations must be solved simultaneously. To do so, it is convenient to work with the modulus and phase of each of the field amplitudes rather than with the complex quantities themselves. It is also convenient to express these amplitudes in dimensionless form. We hence write the complex, slowly varying field amplitudes as

$$A_1 = \left(\frac{2\pi I}{n_1c}\right)^{1/2}u_1e^{i\phi_1}, \tag{2.6.13}$$

$$A_2 = \left(\frac{2\pi I}{n_2c}\right)^{1/2}u_2e^{i\phi_2}, \tag{2.6.14}$$

Here we have introduced the total intensity of the two waves,

$$I = I_1 + I_2, \tag{2.6.15}$$

where the intensity of each wave is given by

$$I_j = \frac{n_j c}{2\pi} |A_j|^2. \tag{2.6.16}$$

As a consequence of the Manley–Rowe relations, the total intensity I is a constant. The new field amplitudes u_1 and u_2 are defined in such a manner that $u_1^2 + u_2^2$ is also a conserved (i.e., spatially invariant) quantity normalized such that

$$u_1(z)^2 + u_2(z)^2 = 1. \tag{2.6.17}$$

We next introduce a normalized distance parameter

$$\zeta = z/l, \tag{2.6.18}$$

where

$$l = \left(\frac{n_1^2 n_2 c^3}{2\pi I}\right)^{1/2} \frac{1}{8\pi\omega_1 d} \tag{2.6.19}$$

is the characteristic distance over which the fields exchange energy. We also introduce the relative phase of the interacting fields,

$$\theta = 2\phi_1 - \phi_2 + \Delta k\, z, \tag{2.6.20}$$

and a normalized momentum mismatch parameter

$$\Delta s = \Delta k\, l. \tag{2.6.21}$$

The quantities u_j, ϕ_j, ζ, and Δs defined in Eqs. (2.6.13) through (2.6.21) are now introduced into the coupled-amplitude equations (2.6.10) and (2.6.11), which reduce after straightforward (but lengthy) algebra to the set of coupled equations for the three real quantities u_1, u_2, and θ:

$$\frac{du_1}{d\zeta} = u_1 u_2 \sin\theta, \tag{2.6.22}$$

$$\frac{du_2}{d\zeta} = -u_1^2 \sin\theta, \tag{2.6.23}$$

$$\frac{d\theta}{d\zeta} = \Delta s + \frac{\cos\theta}{\sin\theta} \frac{d}{d\zeta}(\ln u_1^2 u_2). \tag{2.6.24}$$

This set of equations has been solved under general conditions by Armstrong *et al.* We will return later to a discussion of the general solution, but for now we assume the case of perfect phase matching so that Δk and hence Δs vanish. It is easy to verify by direct differentiation that, for $\Delta s = 0$, Eq. (2.6.24) can be rewritten as

$$\frac{d}{d\zeta}\ln(\cos\theta\, u_1^2 u_2) = 0. \tag{2.6.25}$$

Hence the quantity $\ln(\cos\theta\, u_1^2 u_2)$ is a constant, which we call $\ln\Gamma$, so that the solution to Eq. (2.6.25) can be expressed as

$$u_1^2 u_2 \cos\theta = \Gamma, \tag{2.6.26}$$

where the constant Γ is independent of the normalized propagation distance ζ. The value of Γ can be determined from the known values of u_1, u_2, and θ at the entrance face to the nonlinear medium, $\zeta = 0$.

We have now found two conserved quantities: $u_1^2 + u_2^2$ (according to Eq. 2.6.17) and $u_1^2 u_2 \cos\theta$ (according to Eq. 2.6.26). These conserved quantities can be used to decouple the set of equations (2.6.22)–(2.6.24). Equation (2.6.23), for instance, can be written using Eq. (2.6.17) and the identity $\sin^2\theta + \cos^2\theta = 1$ as

$$\frac{du_2}{d\zeta} = \pm(1 - u_2^2)(1 - \cos^2\theta)^{1/2}. \tag{2.6.27}$$

Equations (2.6.26) and (2.6.17) are next used to express $\cos^2\theta$ in terms of the conserved quantity Γ and the unknown function u_2; the resulting expression is substituted into Eq. (2.6.27), which becomes

$$\frac{du_2}{d\zeta} = \pm(1 - u_2^2)\left(1 - \frac{\Gamma^2}{u_1^4 u_2^2}\right)^{1/2} = \pm(1 - u_2^2)\left(1 - \frac{\Gamma^2}{(1 - u_2^2)^2 u_2^2}\right)^{1/2}. \tag{2.6.28}$$

This result is simplified algebraically to give

$$u_2\frac{du_2}{d\zeta} = \pm[(1 - u_2^2)^2 u_2^2 - \Gamma^2]^{1/2},$$

or

$$\frac{du_2^2}{d\zeta} = \pm 2[(1 - u_2^2)^2 u_2^2 - \Gamma^2]^{1/2}. \tag{2.6.29}$$

This equation is of a standard form, whose solution can be expressed in terms of the Jacobi elliptic functions. An example of the solution for one particular choice of initial conditions is illustrated in Fig. 2.6.2. Note that, in general, the fundamental and second-harmonic fields interchange energy periodically.

The solution of Eq. (2.6.29) becomes particularly simple for the special case in which the constant Γ is equal to zero. The condition $\Gamma = 0$ occurs whenever the amplitude of either of the two input fields is equal to zero or whenever the

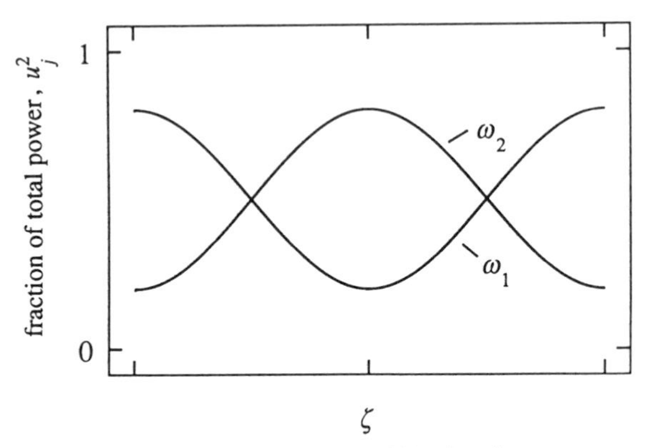

FIGURE 2.6.2 Typical solution to Eq. (2.6.29), after Armstrong *et al.* (1962).

fields are initially phased so that $\cos\theta = 0$. We note that since Γ is a conserved quantity, it is then equal to zero for all values of ζ, which in general requires (see Eq. (2.6.26)) that

$$\cos\theta = 0. \tag{2.6.30a}$$

For definiteness, we assume that

$$\sin\theta = -1 \tag{2.6.30b}$$

(rather than 1). We hence see that the relative phase of the interacting fields is spatially invariant for the case of $\Gamma = 0$. In addition, when $\Gamma = 0$ the coupled-amplitude equations take on the relatively simple forms

$$\frac{du_1}{d\zeta} = -u_1 u_2, \tag{2.6.31}$$

$$\frac{du_2}{d\zeta} = u_1^2. \tag{2.6.32}$$

This second equation can be transformed through use of Eq. (2.6.17) to obtain

$$\frac{du_2}{d\zeta} = 1 - u_2^2, \tag{2.6.33}$$

whose solution is

$$u_2 = \tanh(\zeta + \zeta_0), \tag{2.6.34}$$

where ζ_0 is a constant of integration.

We now assume that the initial conditions are

$$u_1(0) = 1, \qquad u_2(0) = 0. \tag{2.6.35}$$

These conditions imply that no second-harmonic light is incident on the nonlinear crystal, as is the case in most experiments. Then, since $\tanh 0 = 0$, we see that the integration constant ζ_0 is equal to 0 and hence that

$$u_2(\zeta) = \tanh \zeta. \tag{2.6.36}$$

The amplitude u_1 of the fundamental wave is found through use of Eq. (2.6.32) (or through use of Eq. (2.6.17)) to be given by

$$u_1 = \operatorname{sech} \zeta. \tag{2.6.37}$$

Recall that $\zeta = z/l$. For the case in which only the fundamental field is present at $z = 0$, the length parameter l of Eq. (2.6.19) is given by

$$l = \frac{(n_1 n_2)^{1/2} c}{8\pi\omega_1 d |A_1(0)|}. \tag{2.6.38}$$

The solution given by Eqs. (2.6.36) and (2.6.37) is shown graphically in Fig. 2.6.3. We see that all of the incident radiation is converted into the second harmonic in the limit $\zeta \to \infty$. In addition, we note that $\tanh(\zeta + \zeta_0)$ has the same asymptotic behavior for any finite value of ζ_0. Thus, whenever Γ is equal to zero, all of the radiation at the fundamental frequency will eventually be converted to the second harmonic, for any initial ratio of u_1 to u_2.

As mentioned above, Armstrong *et al.* have also solved the coupled-amplitude equations describing second-harmonic generation for arbitrary Δk. They find that the solution can be expressed in terms of elliptic integrals. We

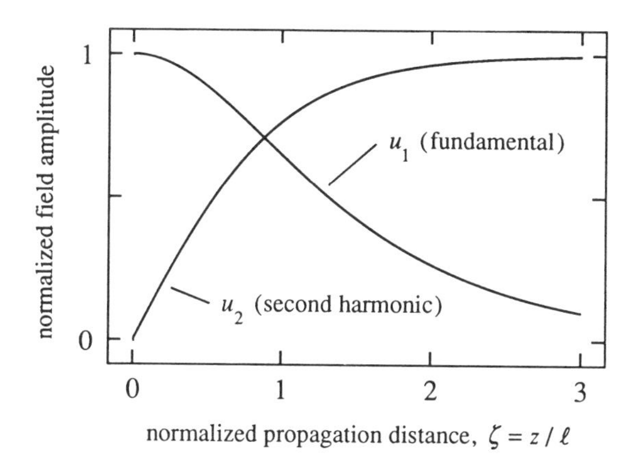

FIGURE 2.6.3 Spatial variations of the fundamental and second-harmonic field amplitudes for the case of perfect phase matching and the boundary condition $u_2(0) = 0$.

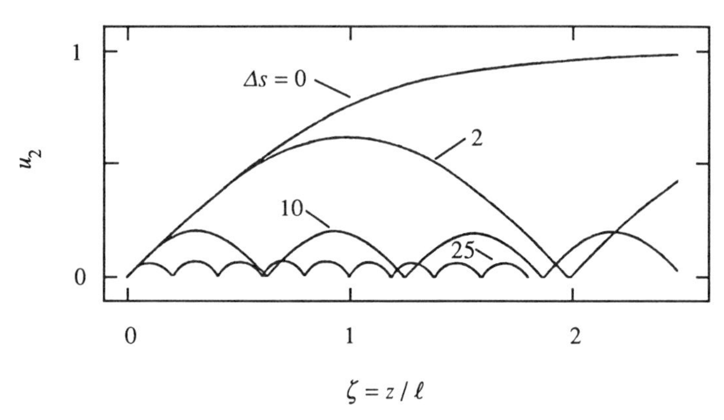

FIGURE 2.6.4 Effect of wave vector mismatch on the efficiency of second-harmonic generation.

shall not reproduce their derivation here; instead we summarize their results graphically for the case in which no radiation is incident at the second-harmonic frequency. We see from Fig. 2.6.4 that the effect of a nonzero propagation vector mismatch is to lower the conversion efficiency.

As an illustration of how to apply the formulas derived in this section, we estimate the conversion efficiency for second-harmonic generation attainable using typical cw lasers. We first estimate the numerical value of the parameter ζ given by Eqs. (2.6.18) and (2.6.38) at the plane $z = L$, where L is the length of the nonlinear crystal. We assume that the incident laser beam carries power P and is focused to a spot size w_0 at the center of the crystal. The field strength A_1 can then be estimated by the expression

$$I_1 = \frac{n_1 c}{2\pi} A_1^2 = \frac{P}{\pi w_0^2}. \tag{2.6.39}$$

We assume that the beam is optimally focused in the sense that the focal spot size w_0 is chosen so that the depth b of the focal region is equal to the length L of the crystal, that is,*

$$b \equiv \frac{2\pi w_0^2 n_1}{\lambda_1} = L, \tag{2.6.40}$$

where λ_1 denotes the wavelength of the incident wave in vacuum. From Eqs. (2.6.39) and (2.6.40), the laser field amplitude under conditions of opti-

* See also the discussion of nonlinear interactions involving focused gaussian beams presented in Section 2.8.

mum focusing is seen to be given by

$$A_1 = \left(\frac{4\pi P}{c\lambda_1 L}\right)^{1/2}, \tag{2.6.41}$$

and hence the parameter $\zeta = L/l$ is given through use of Eq. (2.6.38) by

$$\zeta = \left(\frac{1024\pi^5 d^2 LP}{n_1 n_2 c\lambda_1^3}\right)^{1/2}. \tag{2.6.42}$$

Typical values of the parameters appearing in this equation are $d = 1 \times 10^{-8}$ esu, $L = 1$ cm, $P = 1$ W $= 1 \times 10^7$ erg/sec, $\lambda = 0.5 \times 10^{-4}$ cm, and $n = 2$, which lead to the value $\zeta = 0.14$. The efficiency η for conversion of power from the ω_1 wave to the ω_2 wave can be defined by

$$\eta = \frac{u_2^2(L)}{u_1^2(0)}, \tag{2.6.43}$$

and from Eq. (2.6.36), we see that for the values given above, η is of the order of 2%.

2.7. Phase-Matching Considerations

We saw in Section 2.2 that for sum-frequency generation involving undepleted input beams, the intensity of the generated field at frequency $\omega_3 = \omega_1 + \omega_2$ varied with the wave vector mismatch

$$\Delta k = k_1 + k_2 - k_3 \tag{2.7.1}$$

according to

$$I_3 = I_3(\text{max}) \frac{\sin^2(\Delta k\, L/2)}{(\Delta k\, L/2)^2}. \tag{2.7.2}$$

This expression predicts a dramatic decrease in the efficiency of the sum-frequency generation when the condition of perfect phase matching, $\Delta k = 0$, is not satisfied.

Behavior of the sort predicted by Eq. (2.7.2) was first observed experimentally by P. D. Maker *et al.* (1962) and is illustrated in Fig. 2.7.1. Their experiment involved focusing the output of a pulsed ruby laser into a single crystal of quartz and measuring how the intensity of the second-harmonic signal varied as the crystal was rotated, thus varying the effective path length L

(a)

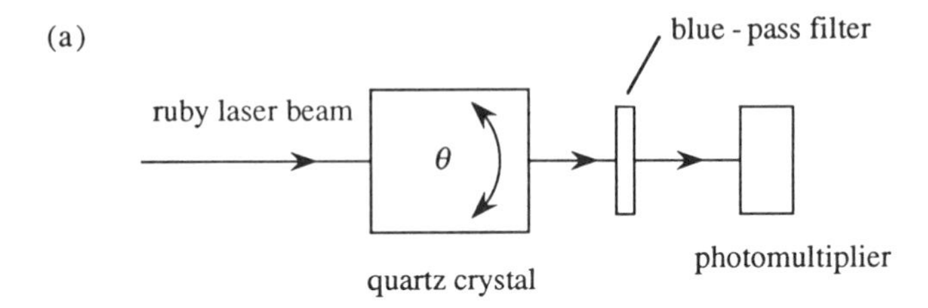

(b)

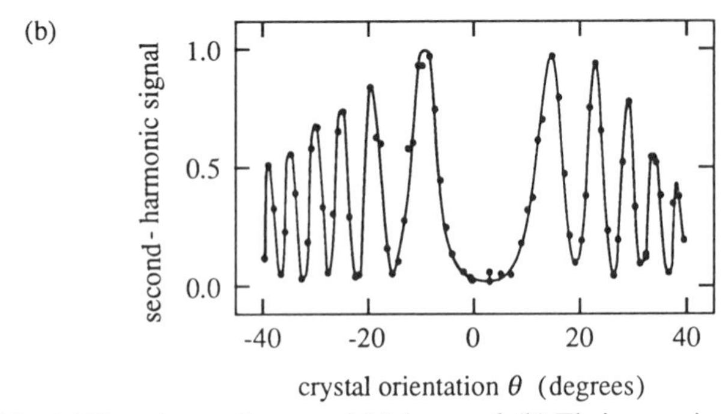

FIGURE 2.7.1 (a) Experimental setup of Maker *et al.* (b) Their experimental results.

through the crystal. The wave vector mismatch Δk was nonzero and approimately the same for all orientations used in their experiment.

For nonlinear mixing processes sufficiently efficient to lead to depletion of the input beams, the functional dependence of the efficiency of the process on the phase mismatch is no longer given by Eq. (2.7.2). However, even in this case the efficient generation of the output field requires that the condition $\Delta k = 0$ be maintained.

The phase-matching condition $\Delta k = 0$ is often difficult to achieve because the refractive index of materials that are lossless in the range ω_1 to ω_3 (we assume that $\omega_1 \leq \omega_2 < \omega_3$) shows an effect known as normal dispersion: the refractive index is an increasing function of frequency. As a result, the condition for perfect phase matching with collinear beams,

$$n_1\omega_1 + n_2\omega_2 = n_3\omega_3, \tag{2.7.3}$$

where

$$\omega_1 + \omega_2 = \omega_3, \tag{2.7.4}$$

cannot be achieved. For the case of second-harmonic generation, with $\omega_1 = \omega_2$, $\omega_3 = 2\omega_1$, these conditions require that

$$n(\omega_1) = n(2\omega_1), \tag{2.7.5}$$

which is clearly not possible when $n(\omega)$ increases monotonically with ω. For the case of sum-frequency generation, the argument is slightly more complicated, but the conclusion is the same. To show that phase matching is not possible, we first rewrite Eq. (2.7.3) as

$$n_3 = \frac{n_1\omega_1 + n_2\omega_2}{\omega_3}. \tag{2.7.6}$$

This result is now used to express the refractive index difference $n_3 - n_2$ as

$$n_3 - n_2 = \frac{n_1\omega_1 + n_2\omega_2 - n_2\omega_3}{\omega_3} = \frac{n_1\omega_1 - n_2(\omega_3 - \omega_2)}{\omega_3} = \frac{n_1\omega_1 - n_2\omega_1}{\omega_3},$$

or finally as

$$n_3 - n_2 = (n_1 - n_2)\frac{\omega_1}{\omega_3}. \tag{2.7.7}$$

For normal dispersion, n_3 must be greater than n_2, and hence the left-hand side of this equation must be positive. However, n_2 must also be greater than n_1, showing that the right-hand side must be negative, which demonstrates that Eq. (2.7.7) cannot possess a solution.

In principle, it is possible to achieve the phase-matching condition by making use of anomalous dispersion, that is, the decrease in refractive index with increasing frequency that occurs near an absorption feature. However, the most common procedure for achieving phase matching is to make use of the birefringence displayed by many crystals. Birefringence is the dependence of the refractive index on the direction of polarization of the optical radiation. Not all crystals display birefringence; in particular, crystals belonging to the cubic crystal system are optically isotropic (i.e., show no birefringence) and thus are not phase-matchable.

The linear optical properties of the various crystal systems are summarized in Table 2.7.1.

In order to achieve phase matching through the use of birefringent crystals, the highest-frequency wave $\omega_3 = \omega_1 + \omega_2$ is polarized in the direction that gives it the lower of the two possible refractive indices. For the case of a negative uniaxial crystal, as in the example shown in Figure 2.7.2, this choice corresponds to the extraordinary polarization. There are two choices for the

TABLE 2.7.1 Linear optical classification of the various crystal systems

System	Linear optical classification
Triclinic, monoclinic, orthorhombic	Biaxial
Trigonal, tetragonal, hexagonal	Uniaxial
Cubic	Isotropic

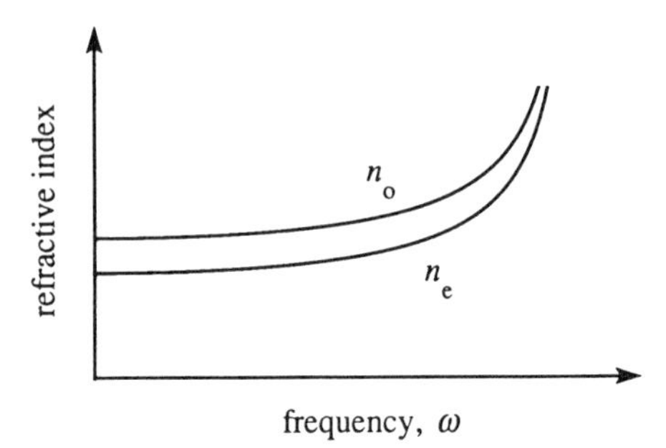

FIGURE 2.7.2 Dispersion of the refractive indices of a negative uniaxial crystal. For the case (not shown) of a positive uniaxial crystal the extraordinary index n_e is greater than the ordinary index n_o.

polarizations of the lower-frequency waves. Midwinter and Warner (1965) define type I phase matching to be the case where the lower-frequency waves have the same polarization, and type II to be the case where the polarizations are orthogonal. The possibilities are summarized in Table 2.7.2. No assumptions regarding the relative magnitudes of ω_1 and ω_2 are implied by this classification scheme. However, for type II phase matching it is easier to achieve the phase-matching condition (i.e., less birefringence is required) if

TABLE 2.7.2 Phase-matching methods for uniaxial crystals

	Positive uniaxial ($n_e > n_o$)	Negative uniaxial ($n_e < n_o$)
Type I	$n_3^o\omega_3 = n_1^e\omega_1 + n_2^e\omega_2$	$n_3^e\omega_3 = n_1^o\omega_1 + n_2^o\omega_2$
Type II	$n_3^o\omega_3 = n_1^o\omega_1 + n_2^e\omega_2$	$n_3^e\omega_3 = n_1^e\omega_1 + n_2^o\omega_2$

$\omega_2 > \omega_1$ for the choice of ω_1 and ω_2 used in writing the table. Also, independent of the relative magnitudes of ω_1 and ω_2, type I phase matching is easier to achieve than type II.

Careful control of the refractive indices at each of the three optical frequencies is required in order to establish the phase-matching condition ($\Delta k = 0$). In practice, one of two methods is used to achieve phase matching.

Angle Tuning

This method involves precise angular orientation of the crystal with respect to the propagation direction of the incident light. It is most simply described for the case of a uniaxial crystal, and the following discussion will be restricted to this case. Uniaxial crystals are characterized by a particular direction known as the optic axis (or c axis or z axis). Light polarized perpendicular to the plane containing the propagation vector **k** and the optic axis is said to have the ordinary polarization. Such light experiences a refractive index n_o called the ordinary refractive index. Light polarized in the plane containing **k** and the optic axis is said to have the extraordinary polarization and experiences a refractive index $n_e(\theta)$ that depends on the angle θ between the optic axis and **k** according to the relation*

$$\frac{1}{n_e(\theta)^2} = \frac{\sin^2\theta}{\bar{n}_e^2} + \frac{\cos^2\theta}{n_o^2}. \tag{2.7.8}$$

Here $\bar{n}_e$ is the principal value of the extraordinary refractive index. Note that $n_e(\theta)$ is equal to the principal value $\bar{n}_e$ for $\theta = 90$ degrees and is equal to n_o for $\theta = 0$. Phase matching is achieved by adjusting the angle θ to achieve the value of $n_e(\theta)$ for which the condition $\Delta k = 0$ is satisfied.

As an illustration of angle phase matching, we consider the case of type I second-harmonic generation in a negative uniaxial crystal, as shown in Figure 2.7.3. Since n_e is less than n_o for a negative uniaxial crystal, we choose

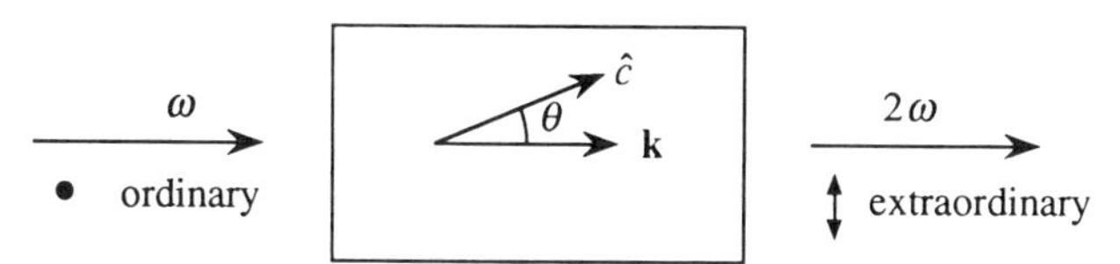

FIGURE 2.7.3 Geometry for angle-tuned phase matching of second-harmonic generation.

* For a derivation of this relation, see, for example, Born and Wolf (1975), Section 14.3; Klein (1970), Eq. (11.160a); or Zernike and Midwinter (1973), Eq. (1.26).

the fundamental to be an ordinary wave and the second harmonic to be an extraordinary wave, in order that the birefringence of the material can compensate for the dispersion. The phase-matching condition (2.7.5) then becomes

$$n_e(2\omega, \theta) = n_o(\omega), \tag{2.7.9}$$

or

$$\frac{\sin^2\theta}{\bar{n}_e(2\omega)^2} + \frac{\cos^2\theta}{n_o(2\omega)^2} = \frac{1}{n_o(\omega)^2}. \tag{2.7.10}$$

In order to simplify this equation, we replace $\cos^2\theta$ by 1-$\sin^2\theta$ and solve for $\sin^2\theta$ to obtain

$$\sin^2\theta = \frac{\dfrac{1}{n_o(\omega)^2} - \dfrac{1}{n_o(2\omega)^2}}{\dfrac{1}{\bar{n}_e(2\omega)^2} - \dfrac{1}{n_o(2\omega)^2}}. \tag{2.7.11}$$

This equation shows how the crystal should be oriented in order to achieve the phase matching condition.

Temperature Tuning

There is one serious drawback to the use of angle tuning. Whenever the angle θ between the propagation direction and the optic axis has a value other than 0 or 90 degrees, the Poynting vector **S** and propagation vector **k** are not parallel for extraordinary rays. As a result, ordinary and extraordinary rays with parallel propagation vectors quickly diverge from one another as they propagate through the crystal. This walkoff effect limits the spatial overlap of the two waves and decreases the efficiency of any nonlinear mixing process involving such waves.

For some crystals, notably lithium niobate, the amount of birefringence is strongly temperature-dependent. As a result, it is possible to phase-match the mixing process by holding θ fixed at 90 degrees and varying the temperature of the crystal. The temperature dependence of the refractive indices of lithium niobate has been given by Hobden and Warner (1966).

2.8. Nonlinear Optical Interactions with Focused Gaussian Beams

In the past several sections we have treated nonlinear optical interactions in the approximation in which all of the interacting waves are taken to be infinite plane waves. However, in practice, the incident radiation is usually focused into the nonlinear optical medium in order to increase its intensity and hence to increase the efficiency of the nonlinear optical process. This section explores the nature of nonlinear optical interactions that are excited by focused laser beams.

Paraxial Wave Equation

We begin with the wave equation in the form of Eq. (2.1.19) and assume that it is valid to replace $\nabla \times \nabla \times \tilde{\mathbf{E}}_n$ by $-\nabla^2 \tilde{\mathbf{E}}_n$, so that the wave equation for the part of the wave at frequency ω_n is given by

$$\nabla^2 \tilde{\mathbf{E}}_n - \frac{1}{(c/n)^2} \frac{\partial^2 \tilde{\mathbf{E}}_n}{\partial t^2} = \frac{4\pi}{c^2} \frac{\partial^2 \tilde{\mathbf{P}}_n}{\partial t^2}. \tag{2.8.1}$$

We next represent the electric field $\tilde{\mathbf{E}}_n$ and polarization $\tilde{\mathbf{P}}_n$ as

$$\tilde{\mathbf{E}}_n(\mathbf{r}, t) = \mathbf{A}_n(\mathbf{r}) e^{i(k_n z - \omega_n t)} + \text{c.c.}, \tag{2.8.2a}$$

$$\tilde{\mathbf{P}}_n(\mathbf{r}, t) = \mathbf{p}_n(\mathbf{r}) e^{i(k_n' z - \omega_n t)} + \text{c.c.} \tag{2.8.2b}$$

Here we have allowed $\tilde{\mathbf{E}}_n$ and $\tilde{\mathbf{P}}_n$ to represent nonplane waves by allowing the complex amplitudes $\mathbf{A}_n$ and $\mathbf{p}_n$ to be spatially varying quantities. In addition, we have allowed the possibility of a wave vector mismatch by allowing the wave vector of $\tilde{\mathbf{P}}_n$ to be different from that of $\tilde{\mathbf{E}}_n$. We next substitute Eqs. (2.8.2) into (2.8.1). Since we have specified the z direction as the dominant direction of propagation of the wave $\tilde{\mathbf{E}}_n$, it is useful to express the laplacian operator as $\nabla^2 = \partial^2/\partial z^2 + \nabla_T^2$, where the transverse laplacian is given by $\nabla_T^2 = \partial^2/\partial x^2 + \partial^2/\partial y^2$ in rectangular coordinates and is given by $\nabla_T^2 = (1/r)(\partial/\partial r)(r\,\partial/\partial r) + (1/r)^2\,\partial^2/\partial\phi^2$ in cylindrical coordinates. As in the derivation of Eq. (2.2.11), we now make the slowly-varying-amplitude approximation, that is, we assume that the variation of $\mathbf{A}_n$ with z occurs only over distances much larger than an optical wavelength. We hence find that Eq. (2.8.1) becomes

$$2ik_n \frac{\partial \mathbf{A}_n}{\partial z} + \nabla_T^2 \mathbf{A}_n = -\frac{4\pi\omega_n^2}{c^2} \mathbf{p}_n e^{i\,\Delta k\, z}, \tag{2.8.3}$$

where $\Delta k = k_n' - k_n$. This result is known as the paraxial wave equation, because the approximation of neglecting the contribution $\partial^2 A/\partial z^2$ on the left-hand side is justifiable insofar as the wave $\mathbf{E}_n$ is propagating primarily along the z axis.

Gaussian Beams

Let us first study the nature of the solution to Eq. (2.8.3) for the case of the free propagation of an optical wave, that is, for the case in which the source term containing p_n vanishes. The paraxial wave equation is solved in such a case by a beam having a transverse intensity distribution that is everywhere a gaussian and that can be represented as (Kogelnik and Li, 1966)

$$A(r,z) = \mathscr{A}\frac{w_0}{w(z)} e^{-r^2/w(z)^2} e^{ikr^2/2R(z)} e^{i\Phi(z)}, \tag{2.8.4a}$$

where

$$w(z) = w_0[1 + (\lambda z/\pi w_0^2)^2]^{1/2} \tag{2.8.4b}$$

represents the $1/e$ radius of the field distribution, where

$$R(z) = z[1 + (\pi w_0^2/\lambda z)^2] \tag{2.8.4c}$$

represents the radius of curvature of the optical wavefront, and where

$$\Phi(z) = -\arctan(\lambda z/\pi w_0^2) \tag{2.8.4d}$$

represents the spatial variation of the phase of the wave (measured with respect to that of an infinite plane wave). In these formulas, w_0 represents the beam waist radius (that is, the value of w at the plane $z = 0$), and $\lambda = 2\pi c/n\omega$ represents the wavelength of the radiation in the medium. This angular divergence of the beam in the far field is given by $\theta_{\text{ff}} = \lambda/\pi w_0$. The nature of this solution is illustrated in Fig. 2.8.1. For theoretical work it is often convenient to represent the gaussian beam in the more compact (but less intuitive) form (see Problem 6 at the end of the chapter)

$$A(r,z) = \frac{\mathscr{A}}{1 + i\zeta} e^{-r^2/w_0^2(1+i\zeta)}. \tag{2.8.5a}$$

Here

$$\zeta = 2z/b \tag{2.8.5b}$$

is a dimensionless longitudinal coordinate defined in terms of the confocal parameter

$$b \equiv 2\pi w_0^2/\lambda = kw_0^2, \tag{2.8.5c}$$

which, as illustrated in part (c) of Fig. 2.8.1, is a measure of the longitudinal extent of the focal region of the gaussian beam. The total power $\mathscr{P}$ carried by a gaussian laser beam can be calculated by integrating over the transverse intensity distribution of the beam. Since $\mathscr{P} = \int I\, 2\pi r\, dr$, where the intensity is given by $I = (nc/2\pi)|A|^2$, we find that

$$\mathscr{P} = \tfrac{1}{4} ncw_0^2 |\mathscr{A}|^2. \tag{2.8.6}$$

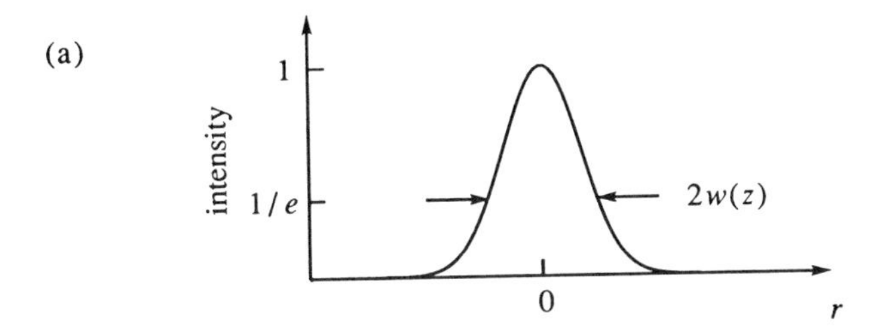

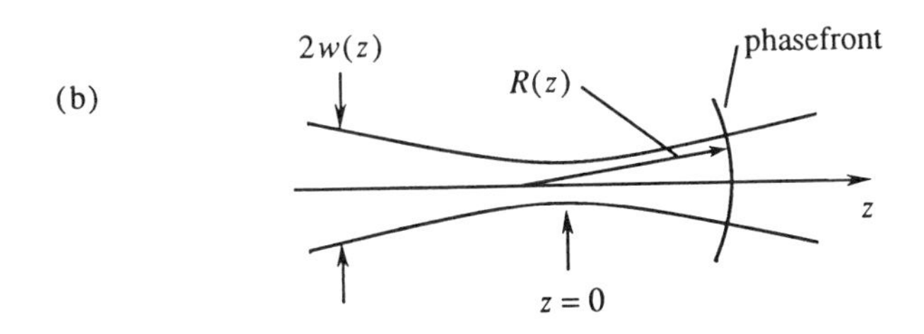

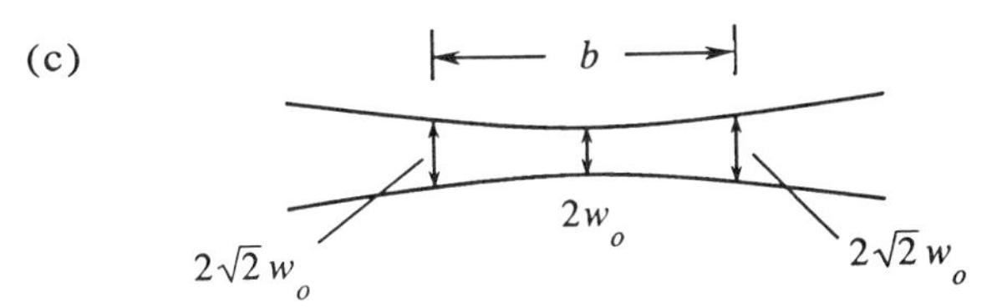

FIGURE 2.8.1 (a) Intensity distribution of a gaussian laser beam. (b) Variation of the beam radius w and wavefront radius of curvature R with position z. (c) Relation between the beam waist radius w_0 and the confocal parameter b.

Harmonic Generation Using Focused Gaussian Beams

Let us now treat harmonic generation excited by a gaussian fundamental beam. For generality, we consider the generation of the qth harmonic. According to Eq. (2.8.3), the amplitude A_q of the $q\omega$ frequency component of the optical field must obey the equation

$$2ik_q \frac{\partial A_q}{\partial z} + \nabla_T^2 A_q = -\frac{4\pi\omega_q^2}{c^2} \chi^{(q)} A_1^q e^{i\Delta k z} \tag{2.8.7}$$

where $\Delta k = qk_1 - k_q$ and where we have set the complex amplitude p_q of the nonlinear polarization equal to $p_q = \chi^{(q)} A_1^q$. Here $\chi^{(q)}$ is the nonlinear susceptibility describing qth-harmonic generation, i.e., $\chi^{(q)} = \chi^{(q)}(q\omega = \omega + \omega + \cdots + \omega)$, and A_1 is the complex amplitude of the fundamental wave, which according to Eq. (2.8.5a) can be represented as

$$A_1(r, z) = \frac{\mathscr{A}_1}{1 + i\zeta} e^{-r^2/w_0^2(1 + i\zeta)}. \tag{2.8.8}$$

We work in the constant-pump approximation. We solve Eq. (2.8.7) by adopting the trial solution

$$A_q(r, z) = \frac{\mathscr{A}_q(z)}{1 + i\zeta} e^{-qr^2/w_0^2(1 + i\zeta)}, \tag{2.8.9}$$

where $\mathscr{A}_q(z)$ is a function of z. One might guess this form for the trial solution because its radial dependence is identical to that of the source term in Eq. (2.8.7). Note also that (ignoring the spatial variation of $\mathscr{A}_q(z)$) the trial solution corresponds to a beam with the same confocal parameter as the fundamental beam (2.8.8); this behavior makes sense in that the harmonic wave is generated coherently over a region whose longitudinal extent is equal to that of the fundamental wave. If the trial solution (2.8.9) is substituted into Eq. (2.8.7), we find that it satisfies this equation so long as $\mathscr{A}_q(z)$ obeys the (ordinary) differential equation

$$\frac{d\mathscr{A}_q}{dz} = \frac{i2\pi q\omega}{nc} \chi^{(q)} \mathscr{A}_1^q \frac{e^{i\Delta k z}}{(1 + i\zeta)^{q-1}}. \tag{2.8.10}$$

This equation can be integrated directly to obtain

$$\mathscr{A}_q(z) = \frac{i2\pi q\omega}{nc} \chi^{(q)} \mathscr{A}_1^q J_q(\Delta k, z_0, z), \tag{2.8.11a}$$

where

$$J_q(\Delta k, z_0, z) = \int_{z_0}^{z} \frac{e^{i\Delta k z'}\, dz'}{(1 + 2iz'/b)^{q-1}}, \tag{2.8.11b}$$

and where z_0 represents the value of z at the entrance to the nonlinear medium. We see that the harmonic radiation is generated with a confocal parameter equal to that of the incident laser beam. Hence the beam waist radius of the qth harmonic radiation is $q^{1/2}$ times smaller than that of the incident beam, and the far-field diffraction angle $\theta_{\text{ff}} = \lambda/\pi w_0$ is $q^{1/2}$ times smaller than that of the incident laser beam. We have solved Eq. (2.8.7) by guessing the correct form (Eq. 2.8.9) for the trial solution; a constructive solution to Eq. (2.8.7) has been presented by Kleinman *et al.* (1966) for second-harmonic generation and by Ward and New (1969) for the general case of qth-harmonic generation.

The integral appearing in Eq. (2.8.11b) can be evaluated analytically for certain special cases. One such case is the plane-wave limit, where $b \gg |z_0|, |z|$. In this limit the integral reduces to

$$J_q(\Delta k, z_0, z) = \int_{z_0}^{z} e^{i\Delta k z'}\, dz' = \frac{e^{i\Delta k z} - e^{i\Delta k z_0}}{i\,\Delta k}, \tag{2.8.12a}$$

which implies that

$$|J_q(\Delta k, z_0, z)|^2 = L^2 \operatorname{sinc}^2\left(\frac{\Delta k\, L}{2}\right), \tag{2.812b}$$

where $L = z - z_0$ is the length of the interaction region.

The opposite limiting case is that in which the fundamental wave is focused tightly within the interior of the nonlinear medium; this condition implies that $z_0 = -|z_0|$, $z = |z|$, and $b \ll |z_0|, |z|$. In this limit the integral in Eq. (2.8.11b) can be approximated as

$$J_q(\Delta k, z_0, z) = \int_{-\infty}^{\infty} \frac{e^{i\Delta k z'}\, dz'}{(1 + 2iz'/b)^{q-1}}. \tag{2.8.13a}$$

This integral can be evaluated by means of a straightforward contour integration. One finds that

$$J_q(\Delta k, z_0, z) = \begin{cases} 0, & \Delta k \le 0, \\ \dfrac{b}{2}\dfrac{2\pi}{(q-2)!}\left(\dfrac{b\,\Delta k}{2}\right)^{q-2} e^{-b\,\Delta k/2}, & \Delta k > 0. \end{cases} \tag{2.8.13b}$$

This functional form is illustrated for the case of third-harmonic generation ($q = 3$) in Fig. 2.8.2. We find the somewhat surprising result that the efficiency of third-harmonic generation in the tight-focusing limit vanishes identically for the case of perfect phase matching ($\Delta k = 0$) and is maximized through the use of a positive wave vector mismatch. This behavior can be understood in terms of the phase shift of π radians that any beam of light experiences in passing through its focus. This effect is known as the phase anomaly and was first studied systematically by Gouy (1890). For the case of nonlinear optics, this effect has important consequences over and above the phase shift imparted to the transmitted light beam, because in general the nonlinear polarization $p = \chi^{(q)} A_1^q$ will experience a phase shift that is q times larger than that experienced by the incident wave of amplitude A_1. Consequently the nonlinear polarization will be unable to couple efficiently to the generated

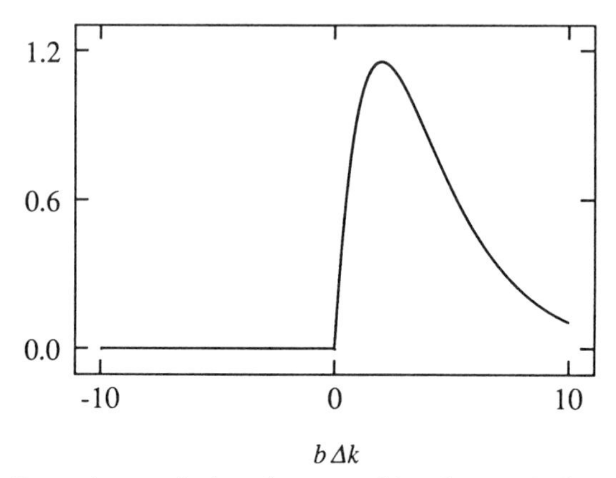

FIGURE 2.8.2 Dependence of the phase-matching factor J_3 for third-harmonic generation, normalized by the confocal parameter b, on the normalized wave vector mismatch $b\,\Delta k$, in the tight focusing limit.

wave of amplitude A_q unless a wave vector mismatch Δk is introduced to compensate for the phase shift due to the passage of the incident wave through its focus. The reason why Δk should be positive in order for this compensation to occur can be understood intuitively in terms of the argument presented in Fig. 2.8.3.

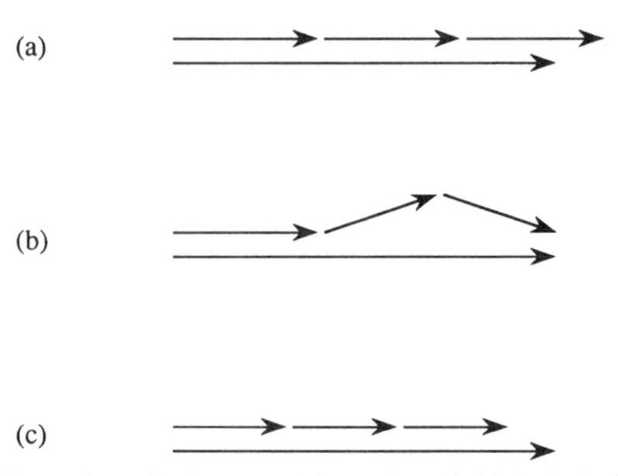

FIGURE 2.8.3 Illustration of why a positive value of Δk is desirable for harmonic generation with focused laser beams. (a) Wave vector diagram for third-harmonic generation with Δk positive. Even though this process is phase-mismatched, the fundamental beam will contain an angular spread of wave vectors and the phase-matched process illustrated in (b) can occur with high efficiency. (c) Conversely, for Δk negative, efficient harmonic generation cannot occur.

Boyd and Kleinman (1968) have considered how to adjust the focus of the incident laser beam in order to optimize the efficiency of second-harmonic generation. They find that the highest efficiency is obtained when beam walkoff effects (mentioned in Section 2.7) are negligible, when the incident laser beam is focused so that the beam waist is located at the center of the crystal and the ratio L/b is equal to 2.84, and when the wave vector mismatch is set equal to $\Delta k = 3.2/L$. In this case, the power generated at the second-harmonic frequency is equal to

$$\mathscr{P}_{2\omega} = 1.068\left[\frac{128\pi^2\omega_1^3 d^2 L}{c^4 n_1 n_2}\right]\mathscr{P}_{\omega}^2. \tag{2.8.14}$$

In addition, Boyd and Kleinman show heuristically that other parametric processes, such as sum- and difference-frequency generation, are optimized by choosing the same confocal parameter for both input waves and applying the same criteria used to optimize second-harmonic generation.

Problems

1. One means of detecting infrared radiation is first to convert the infrared radiation to the visible by the process of sum-frequency generation. Assume that infrared radiation of frequency ω_1 is mixed with an intense laser beam of frequency ω_2 to form the upconverted signal at frequency $\omega_3 = \omega_1 + \omega_2$. Derive a formula that shows how the quantum efficiency for converting infrared photons to visible photons depends on the length L and nonlinear coefficient d of the mixing crystal, and on the phase mismatch Δk. Estimate numerically the value of the quantum efficiency for upconversion of 10-μm infrared radiation using a 1-cm-long proustite crystal, 1 W of laser power at a wavelength of 0.65 μm, and the case of perfect phase matching and optimum focusing.

   [Ans: $\eta_Q = 2\%$.]

2. Solve the coupled-wave equations describing sum-frequency generation (Eqs. (2.2.11) through (2.2.14)) for the case of perfect phase matching ($\Delta k = 0$) but without making the approximation of Section 2.4 that the amplitude of the ω_2 wave can be taken to be constant.

   [Hint: This problem is very challenging. For help, see Armstrong *et al.* (1962).]

3. Solve the coupled-amplitude equations describing difference-frequency generation in the constant-pump limit, and thereby verify Eqs. (2.5.9) of the

text. Assume that $\omega_1 + \omega_2 = \omega_3$, where the amplitude A_3 of the ω_3 pump wave is constant, that the medium is lossless at each of the optical frequencies, that the momentum mismatch Δk is arbitrary, and that in general there can be an input signal at each of the frequencies ω_1 and ω_2. Interpret your results by sketching representative cases of the solution and by taking special limiting cases such as that of perfect phase matching and of only two input fields.

4. Verify that Eq. (2.6.29) possesses solutions of the sort shown in Fig. 2.6.2.

5. Solve the coupled amplitude equations for the case of second-harmonic generation with the initial conditions $A_2 = 0$ but A_1 arbitrary at $z = 0$. Assume that Δk is arbitrary. Sketch how $|A(2\omega)|^2$ varies with z for several values of Δk, and hence verify the results shown in Fig. 2.6.4.

6. Verify that Eq. (2.8.4) and (2.8.5) are equivalent descriptions of a gaussian laser beam, and verify that they satisfy the paraxial wave equation (2.8.3).

7. Verify the statement made in the text that the trial solution given by Eq. (2.8.9) satisfies the paraxial wave equation in the form of Eq. (2.8.7) if the amplitude $\mathscr{A}_q(z)$ satisfies the ordinary differential equation (2.8.10).

8. Evaluate the integral appearing in Eq. (2.8.13a) and thereby verify Eq. (2.8.13b).

References

Sections 2.1 through 2.7

J. A. Armstrong, N. Bloembergen, J. Ducuing, and P. S. Pershan, *Phys. Rev.* **127**, 1918 (1962).

M. Born and E. Wolf, *Principles of Optics*, Pergamon Press, Oxford, 1975.

R. W. Boyd and C. H. Townes, *Appl. Phys. Lett.* **31**, 440 (1977).

R. L. Byer and R. L. Herbst, in *Tunable Infrared Generation*, edited by Y. R. Shen, Springer-Verlag, Berlin, 1977.

J. A. Giordmaine and R. C. Miller, *Phys. Rev. Lett.* **14**, 973 (1965); *Appl. Phys. Lett.* **9**, 298 (1966).

M. V. Hobden and J. Warner, *Phys. Lett.* **22**, 243 (1966).

M. V. Klein, *Optics*, Wiley, New York, 1970.

P. D. Maker, R. W. Terhune, M. Nisenoff, and C. M. Savage, *Phys. Rev. Lett.* **8**, 21 (1962).

J. M. Manley and H. E. Rowe, *Proc. IRE* **47**, 2115 (1959).

J. E. Midwinter and J. Warner, *Brit. J. Appl. Phys.* **16**, 1135 (1965).

Y. R. Shen, *The Principles of Nonlinear Optics*, Wiley, New York, 1984.
F. Zernike and J. E. Midwinter, *Applied Nonlinear Optics*, Wiley, New York, 1973.

Nonlinear Optical Interactions with Focused Gaussian Beams

G. C. Bjorklund, *IEEE J. Quantum Electron.* **QE-11**, 287 (1975).
G. D. Boyd and D. A. Kleinman, *J. Appl. Phys.* **39**, 3597 (1968).
Gouy, *C. R. Acad. Sci. Paris* **110**, 1251 (1890).
D. A. Kleinman, A. Ashkin, and G. D. Boyd, *Phys. Rev.* **145**, 338 (1966).
H. Kogelnik and T. Li, *Appl. Opt.* **5**, 1550 (1966).
R. B. Miles and S. E. Harris, *IEEE J. Quantum Electron.* **QE-9**, 470 (1973).
J. F. Ward and G. H. C. New, *Phys. Rev.* **185**, 57 (1969).

Chapter 3

Quantum-Mechanical Theory of the Nonlinear Optical Susceptibility

3.1. Introduction

In this chapter, we use the laws of quantum mechanics to calculate explicit expressions for the nonlinear optical susceptibility. The motivation for obtaining these expressions is threefold: (1) these expressions display the functional form of the nonlinear optical susceptibility and hence show how the susceptibility depends on material parameters such as dipole transition moments and atomic energy levels, (2) these expressions show the internal symmetries of the susceptibility, and (3) these expressions can be used to obtain numerical values of the nonlinear susceptibilities. These numerical predictions are particularly reliable for the case of atomic vapors, because the atomic parameters (such as atomic energy levels and dipole transition moments) that appear in the quantum-mechanical expressions are usually known with high accuracy. In addition, since the energy levels of free atoms are very sharp (as opposed to the case of most solids, where allowed energies have the form of broad bands), it is possible to obtain very large values of the nonlinear susceptibility through the techniques of resonance enhancement. The idea behind resonance enhancement of the nonlinear optical susceptibility is shown schematically in Fig. 3.1.1 for the case of third-harmonic generation. In part (a) of this figure, we show the third-harmonic generation process in terms of the virtual levels introduced in Chapter 1. In part (b) we also show real atomic levels, indicated by solid horizontal lines. If one of the real atomic levels is nearly coincident with one of the virtual levels of the indicated process, the coupling between the

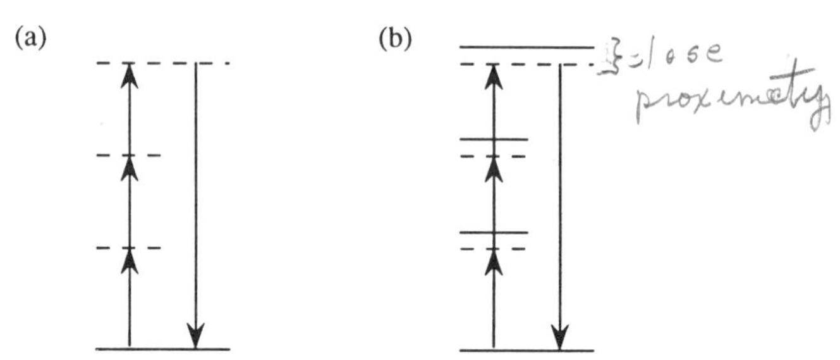

FIGURE 3.1.1 Third-harmonic generation described in terms of virtual levels (a) and with real atomic levels indicated (b).

radiation and the atom is particularly strong and the nonlinear optical susceptibility becomes large.

Three possible strategies for enhancing the efficiency of third-harmonic generation through the technique of resonance enhancement are illustrated in Fig. 3.1.2. In part (a), the one-photon transition is nearly resonant, in part (b) the two-photon transition is nearly resonant, and in part (c) the three-photon transition is nearly resonant. The formulas derived later in this chapter demonstrate that all three procedures are equally effective at increasing the value of the third-order nonlinear susceptibility. However, the method shown in part (b) is usually the preferred way in which to generate the third-harmonic field with high efficiency, for the following reason: For the case of a one-photon resonance (part a), the incident field experiences linear absorption and can be rapidly attenuated as it propagates through the medium. Similarly, for the case of the three-photon resonance (part c), the generated field experiences linear absorption. However, for the case of a two-photon resonance (part b), attenuation occurs only as the result of two-photon absorption processes, which occur with much lower efficiency than one-photon processes.

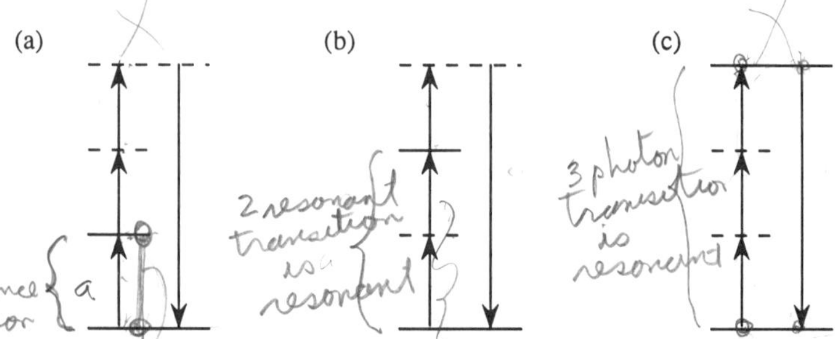

(a) (b) (c)

FIGURE 3.1.2 Three strategies for enhancing the process of third-harmonic generation.

Schrödinger Equation Calculation of the Nonlinear Optical Susceptibility

section, we present a derivation of the nonlinear optical susceptibility ›n quantum-mechanical perturbation theory of the atomic wave func- ıe expressions that we derive using this formalism can be used to make e predictions of the *nonresonant* response of atomic and molecular sys- elaxation processes, which are important for the case of near-resonant on, cannot be adequately described by this formalism. Relaxation pro- re discussed later in this chapter in connection with the density matrix ıtion of the theory of the nonlinear optical susceptibility. Even though the density matrix formalism provides results that are more generally valid, the calculation of the nonlinear susceptibilities is much more complicated when performed using this method. For this reason, we first present a calculation of the nonlinear susceptibility based on the properties of the atomic wave function, since this method is somewhat simpler and for this reason gives a clearer picture of the underlying physics of the nonlinear interaction.

We assume that all of the properties of the atomic system can be described in terms of the atomic wave function $\psi(\mathbf{r}, t)$, which is the solution to the time-dependent Schrödinger equation

$$i\hbar \frac{\partial \psi}{\partial t} = \hat{H}\psi. \tag{3.2.1}$$

Here $\hat{H}$ is the Hamiltonian operator

$$\hat{H} = \hat{H}_0 + \hat{V}(t), \tag{3.2.2}$$

which is written as the sum of the Hamiltonian $\hat{H}_0$ for a free atom and an interaction Hamiltonian, $\hat{V}(t)$, which describes the interaction of the atom with the electromagnetic field. We usually take the interaction Hamiltonian to be of the form

$$\hat{V}(t) = -\hat{\boldsymbol{\mu}} \cdot \tilde{\mathbf{E}}(t), \tag{3.2.3}$$

where $\hat{\boldsymbol{\mu}} = -e\hat{\mathbf{r}}(t)$ is the electric dipole moment operator, $-e$ is the charge of the electron, and where we assume that $\tilde{\mathbf{E}}(t)$ can be represented as a discrete sum of (positive and negative) frequency components as

$$\tilde{\mathbf{E}}(t) = \sum_p \mathbf{E}(\omega_p) e^{-i\omega_p t}. \tag{3.2.4}$$

Energy Eigenstates

For the case in which no external field is applied to the atom, the Hamiltonian $\hat{H}$ is simply equal to $\hat{H}_0$, and Schrödinger's equation (3.2.1) possesses solutions in the form of stationary states. These solutions correspond to energy eigenstates, that is, to states of well-defined energy, and have the form

$$\psi_n(\mathbf{r}, t) = u_n(\mathbf{r})e^{-i\omega_n t}. \tag{3.2.5a}$$

By substituting this form into the Schrödinger equation (3.2.1), we find that the spatially varying part of the wave function $u_n(\mathbf{r})$ must satisfy the eigenvalue equation (known as the time-independent Schrödinger equation)

$$\hat{H}_0 u_n(\mathbf{r}) = E_n u_n(\mathbf{r}), \tag{3.2.5b}$$

where $E_n = \hbar\omega_n$. For future convenience, we assume that these solutions are chosen in such a manner that they constitute a complete, orthonormal set satisfying the condition

$$\int u_m^* u_n \, d^3r = \delta_{mn}. \tag{3.2.6}$$

Perturbation Solution to Schrödinger's Equation

For the general case in which the atom is exposed to an electromagnetic field, Schrödinger's equation (3.2.1) usually cannot be solved exactly. In such cases, it is often adequate to solve Schrödinger's equation through the use of perturbation theory. In order to solve Eq. (3.2.1) systematically in terms of a perturbation expansion, we replace the Hamiltonian (3.2.2) by

$$\hat{H} = \hat{H}_0 + \lambda\hat{V}(t), \tag{3.2.7}$$

where λ is a continuously varying parameter ranging from zero to unity that characterizes the strength of the interaction; the value $\lambda = 1$ corresponds to the actual physical situation. We now seek a solution to Schrödinger's equation in the form of a power series in λ:

$$\psi(\mathbf{r}, t) = \psi^{(0)}(\mathbf{r}, t) + \lambda\psi^{(1)}(\mathbf{r}, t) + \lambda^2\psi^{(2)}(\mathbf{r}, t) + \cdots. \tag{3.2.8}$$

By requiring that the solution be of this form for any value of λ, we assure that $\psi^{(N)}$ will be that part of the solution which is of order N in the interaction energy V. We now introduce Eq. (3.2.8) into Eq. (3.2.1) and require that all terms that are proportional to λ^N satisfy the equality separately. We thereby obtain the set of equations

$$i\hbar\frac{\partial\psi^{(0)}}{\partial t} = \hat{H}_0\psi^{(0)}, \tag{3.2.9a}$$

$$i\hbar\frac{\partial\psi^{(N)}}{\partial t} = \hat{H}_0\psi^{(N)} + \hat{V}\psi^{(N-1)}, \qquad N = 1, 2, 3, \ldots. \tag{3.2.9b}$$

Eq. (3.2.9a) is simply Schrödinger's equation for the atom in the absence of its interaction with the applied field; we assume for definiteness that initially the atom is in state g (typically the ground state) so that the solution to this equation is

$$\psi^{(0)}(\mathbf{r}, t) = u_g(\mathbf{r})e^{-iE_g t/\hbar}. \tag{3.2.10}$$

The remaining equations in the perturbation expansion (Eqs. 3.2.9b) are solved by making use of the fact that the energy eigenfunctions for the free atom constitute a complete set of basis functions, in terms of which any function can be expanded. In particular, we represent the Nth-order contribution to the wave function $\psi^{(N)}(\mathbf{r}, t)$ as the sum

$$\psi^{(N)}(\mathbf{r}, t) = \sum_l a_l^{(N)}(t)u_l(\mathbf{r})e^{-i\omega_l t}. \tag{3.2.11}$$

Here $a_l^{(N)}(t)$ gives the probability amplitude that, to Nth order in the perturbation, the atom is in energy eigenstate l at time t. If Eq. (3.2.11) is substituted into Eq. (3.2.9b), we find that the probability amplitudes obey the system of equations

$$i\hbar\sum_l \dot{a}_l^{(N)}u_l(\mathbf{r})e^{-i\omega_l t} = \sum_l a_l^{(N-1)}\hat{V}u_l(\mathbf{r})e^{-i\omega_l t}, \tag{3.2.12}$$

where the dot denotes a total time derivative. This equation relates all of the probability amplitudes of order N to all of the amplitudes of order $N - 1$. To simplify this equation, we multiply each side from the left by u_m^* and we integrate the resulting equation over all space. Then, through use of the orthonormality condition (3.2.6), we obtain the equation

$$\dot{a}_m^{(N)} = (i\hbar)^{-1}\sum_l a_l^{(N-1)}V_{ml}e^{i\omega_{ml}t}, \tag{3.2.13}$$

where $\omega_{ml} \equiv \omega_m - \omega_l$ and where we have introduced the matrix elements of the perturbing Hamiltonian, which are defined by

$$V_{ml} \equiv \langle u_m|\hat{V}|u_l\rangle = \int u_m^*\hat{V}u_l\, d^3r. \tag{3.2.14}$$

The form of Eq. (3.2.13) demonstrates the usefulness of the perturbation technique; once the probability amplitudes of order $N - 1$ are determined, the amplitudes of the next higher order (N) can be obtained by straightforward time integration. In particular, we find that

$$a_m^{(N)}(t) = (i\hbar)^{-1} \sum_l \int_{-\infty}^{t} dt' \, V_{ml}(t') a_l^{(N-1)}(t') e^{i\omega_{ml}t'} \tag{3.2.15}$$

We shall eventually be interested in determining the linear, second-order, and third-order optical susceptibilities. To do so, we shall require explicit expressions for the probability amplitudes up to third order in the perturbation expansion. We now determine the form of these amplitudes.

To determine the first-order amplitudes $a_m^{(1)}(t)$, we set $a_l^{(0)}$ in Eq. (3.2.15) equal to δ_{lg} (corresponding to an atom known to be in state g in zeroth order) and, through use of Eqs. (3.2.3) and (3.2.4), replace $V_{ml}(t')$ by $-\sum_p \boldsymbol{\mu}_{ml} \cdot \mathbf{E}(\omega_p) \exp(-i\omega_p t')$, where $\boldsymbol{\mu}_{ml} = \int u_m^* \hat{\boldsymbol{\mu}} u_l \, d^3r$ is known as the dipole transition moment. We next evaluate the integral appearing in Eq. (3.2.15) and assume that the contribution from the lower limit of integration vanishes; we thereby find that

$$a_m^{(1)}(t) = \frac{1}{\hbar} \sum_p \frac{\boldsymbol{\mu}_{mg} \cdot \mathbf{E}(\omega_p)}{\omega_{mg} - \omega_p} e^{i(\omega_{mg} - \omega_p)t}. \tag{3.2.16}$$

We next determine the second-order correction to the probability amplitude by using Eq. (3.2.15) once again, but with N set equal to 2. We introduce Eq. (3.2.16) for $a_m^{(1)}$ into the right-hand side of this equation and perform the integration to find that

$$a_n^{(2)}(t) = \frac{1}{\hbar^2} \sum_{pq} \sum_m \frac{[\boldsymbol{\mu}_{nm} \cdot \mathbf{E}(\omega_q)][\boldsymbol{\mu}_{mg} \cdot \mathbf{E}(\omega_p)]}{(\omega_{ng} - \omega_p - \omega_q)(\omega_{mg} - \omega_p)} e^{i(\omega_{ng} - \omega_p - \omega_q)t}. \tag{3.2.17}$$

Analogously, through an additional use of Eq. (3.2.15), we find that the third-order correction to the probability amplitude is given by

$$a_\nu^{(3)}(t) = \frac{1}{\hbar^3} \sum_{pqr} \sum_{mn} \frac{[\boldsymbol{\mu}_{\nu n} \cdot \mathbf{E}(\omega_r)][\boldsymbol{\mu}_{nm} \cdot \mathbf{E}(\omega_q)][\boldsymbol{\mu}_{mg} \cdot \mathbf{E}(\omega_p)]}{(\omega_{\nu g} - \omega_p - \omega_q - \omega_r)(\omega_{ng} - \omega_p - \omega_q)(\omega_{mg} - \omega_p)} \times e^{i(\omega_{\nu g} - \omega_p - \omega_q - \omega_r)t} \tag{3.2.18}$$

Linear Susceptibility

Let us now use the results just obtained to determine the linear optical properties of a material system. The expectation value of the electric dipole moment is given by

$$\langle \tilde{\mathbf{p}} \rangle = \langle \psi | \hat{\boldsymbol{\mu}} | \psi \rangle, \tag{3.2.19}$$

where ψ is given by the perturbation expansion (3.2.8) with λ set equal to one. We find that the lowest-order contribution to $\langle \tilde{\mathbf{p}} \rangle$ (i.e., the contribution linear in the applied field amplitude) is given by

$$\langle \tilde{\mathbf{p}}^{(1)} \rangle = \langle \psi^{(0)} | \hat{\mu} | \psi^{(1)} \rangle + \langle \psi^{(1)} | \hat{\mu} | \psi^{(0)} \rangle, \tag{3.2.20}$$

where $\psi^{(0)}$ is given by Eq. (3.2.10) and $\psi^{(1)}$ is given by Eqs. (3.2.11) and (3.2.16). By substituting these forms into (3.2.20), we find that

$$\langle \tilde{\mathbf{p}}^{(1)} \rangle = \frac{1}{\hbar} \sum_p \sum_m \left(\frac{\boldsymbol{\mu}_{gm} [\boldsymbol{\mu}_{mg} \cdot \mathbf{E}(\omega_p)]}{\omega_{mg} - \omega_p} e^{-i\omega_p t} + \frac{[\boldsymbol{\mu}_{mg} \cdot \mathbf{E}(\omega_p)]^* \boldsymbol{\mu}_{mg}}{\omega_{mg}^* - \omega_p} e^{i\omega_p t} \right). \tag{3.2.21}$$

In obtaining the expression (3.2.21), we have had to calculate the complex conjugate of the amplitude $a_m^{(1)}$ given by Eq. (3.2.16). In writing Eq. (3.2.21) in the form shown, we have formally allowed the possibility that the transition frequency ω_{mg} is a complex quantity. We have done this because a crude way of incorporating damping phenomena into the theory is to take ω_{mg} to be the complex quantity $\omega_{mg} = \omega_{mg}^0 - i\Gamma_m/2$, where ω_{mg}^0 is the (real) transition frequency and Γ_m is the population decay rate of the upper level m. This procedure is not totally acceptable, because it cannot describe the cascade of population among the excited states nor can it describe dephasing processes that are not accompanied by the transfer of population. However, for the remainder of the present section, we will allow the transition frequency to be a complex quantity in order to provide an indication of how damping effects could be incorporated into the present theory.

Equation (3.1.21) is written as a summation over all positive and negative field frequencies ω_p. The result is easier to interpret if we formally replace ω_p by $-\omega_p$ in the second term, in which case the expression becomes

$$\langle \tilde{\mathbf{p}}^{(1)} \rangle = \frac{1}{\hbar} \sum_p \sum_m \left(\frac{\boldsymbol{\mu}_{gm} [\boldsymbol{\mu}_{mg} \cdot \mathbf{E}(\omega_p)]}{\omega_{mg} - \omega_p} + \frac{[\boldsymbol{\mu}_{gm} \cdot \mathbf{E}(\omega_p)] \boldsymbol{\mu}_{mg}}{\omega_{mg}^* + \omega_p} \right) e^{-i\omega_p t}. \tag{3.2.22}$$

We next use this result to calculate the form of the linear susceptibility. We take the linear polarization to be $\tilde{\mathbf{P}}^{(1)} = N \langle \tilde{\mathbf{p}}^{(1)} \rangle$, where N is the number density of atoms. We next express the polarization in terms of its complex amplitude as $\tilde{\mathbf{P}}^{(1)} = \sum_p \mathbf{P}^{(1)}(\omega_p) \exp(-i\omega_p t)$. Finally, we introduce the linear susceptibility defined through the relation $P_i^{(1)}(\omega_p) = \sum_j \chi_{ij}^{(1)} E_j(\omega_p)$. We thereby find that

$$\chi_{ij}^{(1)}(\omega_p) = \frac{N}{\hbar} \sum_m \left(\frac{\mu_{gm}^i \mu_{mg}^j}{\omega_{mg} - \omega_p} + \frac{\mu_{gm}^j \mu_{mg}^i}{\omega_{mg}^* + \omega_p} \right). \tag{3.2.23}$$

The first and second terms in Eq. (3.2.23) can be interpreted as the resonant and antiresonant contributions to the susceptibility, as illustrated in Fig. 3.2.1. In this figure we have indicated where level m would have to be located in order for each of the terms to become resonant. Note that if g denotes the ground state, it is impossible for the second term to become resonant, which is why it is called the antiresonant contribution.

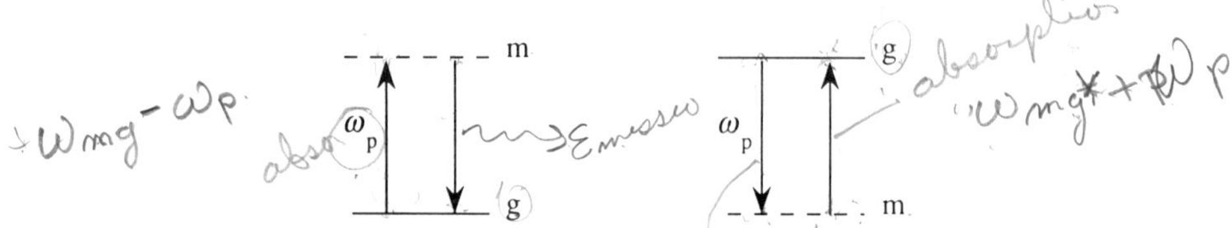

FIGURE 3.2.1 The resonant and antiresonant contributions to the linear susceptibility of Eq. (3.2.23).

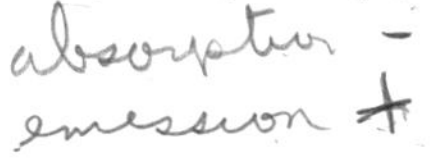

Second-Order Susceptibility

The expression for the second-order susceptibility is derived in a manner analogous to that used for the linear susceptibility. The second-order contribution (i.e., the contribution second order in $\hat{V}$) to the induced dipole moment per atom is given by

$$\langle \tilde{\mathbf{p}}^{(2)} \rangle = \langle \psi^{(0)} | \hat{\boldsymbol{\mu}} | \psi^{(2)} \rangle + \langle \psi^{(1)} | \hat{\boldsymbol{\mu}} | \psi^{(1)} \rangle + \langle \psi^{(2)} | \hat{\boldsymbol{\mu}} | \psi^{(0)} \rangle, \tag{3.2.24}$$

where $\psi^{(0)}$ is given by Eq. (3.2.10), and $\psi^{(1)}$ and $\psi^{(2)}$ are given by Eqs. (3.2.11), (3.2.16), and (3.2.17). We find that $\langle \tilde{\mathbf{p}}^{(2)} \rangle$ is given explicitly by

$$\begin{aligned}\langle \tilde{\mathbf{p}}^{(2)} \rangle = \frac{1}{\hbar^2} \sum_{pq} \sum_{mn} \Bigg(& \frac{\boldsymbol{\mu}_{gn} [\boldsymbol{\mu}_{nm} \cdot \mathbf{E}(\omega_q)] [\boldsymbol{\mu}_{mg} \cdot \mathbf{E}(\omega_p)]}{(\omega_{ng} - \omega_p - \omega_q)(\omega_{mg} - \omega_p)} e^{-i(\omega_p + \omega_q)t} \\ & + \frac{[\boldsymbol{\mu}_{ng} \cdot \mathbf{E}(\omega_q)]^* \boldsymbol{\mu}_{nm} [\boldsymbol{\mu}_{mg} \cdot \mathbf{E}(\omega_p)]}{(\omega_{ng}^* - \omega_q)(\omega_{mg} - \omega_p)} e^{-i(\omega_p - \omega_q)t} \\ & + \frac{[\boldsymbol{\mu}_{ng} \cdot \mathbf{E}(\omega_q)]^* [\boldsymbol{\mu}_{mn} \cdot \mathbf{E}(\omega_p)]^* \boldsymbol{\mu}_{mg}}{(\omega_{ng}^* - \omega_q)(\omega_{mg}^* - \omega_p - \omega_q)} e^{i(\omega_p + \omega_q)t} \Bigg).\end{aligned} \tag{3.2.25}$$

As in the case of the linear susceptibility, this equation can be rendered more transparent by replacing ω_q by $-\omega_q$ in the second term and by replacing ω_q by $-\omega_q$ and ω_p by $-\omega_p$ in the third term; these substitutions are permissible because the expression is to be summed over frequencies ω_p and ω_q. We thereby obtain the result

$$\begin{aligned}\langle \tilde{\mathbf{p}}^{(2)} \rangle = \frac{1}{\hbar^2} \sum_{pq} \sum_{mn} \Bigg(& \frac{\boldsymbol{\mu}_{gn} [\boldsymbol{\mu}_{nm} \cdot \mathbf{E}(\omega_q)] [\boldsymbol{\mu}_{mg} \cdot \mathbf{E}(\omega_p)]}{(\omega_{ng} - \omega_p - \omega_q)(\omega_{mg} - \omega_p)} \\ & + \frac{[\boldsymbol{\mu}_{gn} \cdot \mathbf{E}(\omega_q)] \boldsymbol{\mu}_{nm} [\boldsymbol{\mu}_{mg} \cdot \mathbf{E}(\omega_p)]}{(\omega_{ng}^* + \omega_q)(\omega_{mg} - \omega_p)} \\ & + \frac{[\boldsymbol{\mu}_{gn} \cdot \mathbf{E}(\omega_q)] [\boldsymbol{\mu}_{nm} \cdot \mathbf{E}(\omega_p)] \boldsymbol{\mu}_{mg}}{(\omega_{ng}^* + \omega_q)(\omega_{mg}^* + \omega_p + \omega_q)} \Bigg) e^{-i(\omega_p + \omega_q)t}.\end{aligned} \tag{3.2.26}$$

We next take the second-order polarization to be $\tilde{\mathbf{P}}^{(2)} = N\langle\tilde{\mathbf{p}}^{(2)}\rangle$ and represent it in terms of its frequency components as $\tilde{\mathbf{P}}^{(2)} = \sum_r \mathbf{P}^{(2)}(\omega_r)\exp(-i\omega_r t)$. We also introduce the standard definition of the second-order susceptibility (see also Eq. (1.3.13)):

$$P_i^{(2)} = \sum_{jk}\sum_{(pq)} \chi_{ijk}^{(2)}(\omega_p + \omega_q, \omega_q, \omega_p)E_j(\omega_q)E_k(\omega_p),$$

and find that the second-order susceptibility is given by

$$\chi_{ijk}^{(2)}(\omega_p + \omega_q, \omega_q, \omega_p) = \frac{N}{\hbar^2}\mathscr{P}_I \sum_{mn}\left(\frac{\mu_{gn}^i\mu_{nm}^j\mu_{mg}^k}{(\omega_{ng} - \omega_p - \omega_q)(\omega_{mg} - \omega_p)} + \frac{\mu_{gn}^j\mu_{nm}^i\mu_{mg}^k}{(\omega_{ng}^* + \omega_q)(\omega_{mg} - \omega_p)} + \frac{\mu_{gn}^j\mu_{nm}^k\mu_{mg}^i}{(\omega_{ng}^* + \omega_q)(\omega_{mg}^* + \omega_p + \omega_q)}\right). \tag{3.2.27}$$

In this expression, the symbol $\mathscr{P}_I$ denotes the intrinsic permutation operator. This operator tells us to average the expression that follows it over both permutations of the frequencies ω_p and ω_q of the applied fields. The cartesian indices j and k are to be permuted simultaneously. We introduce the intrinsic permutation operator into Eq. (3.2.27) to ensure that the resulting expression obeys the condition of intrinsic permutation symmetry, as described in the discussion of Eqs. (1.4.52) and (1.5.6). The nature of the expression (3.2.27) for the second-order susceptibility can be understood in terms of the energy level diagrams shown in Fig. 3.2.2, which show where the levels m and n would have to be located in order for each term in the expression to become resonant.

The quantum-mechanical expression for the second-order susceptibility actually comprises six terms; through use of the intrinsic permutation operator $\mathscr{P}_I$, we have been able to express the susceptibility in the form (3.2.27), in

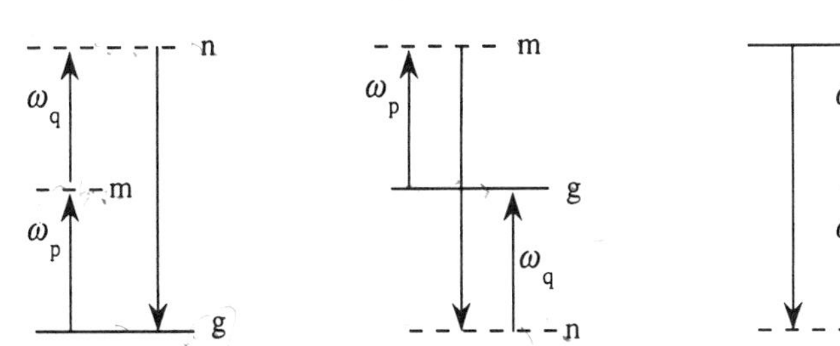

FIGURE 3.2.2 Resonant structure of the three terms of the second-order susceptibility of Eq. (3.2.27).

which only three terms are displayed explicitly. For the case of highly nonresonant excitation, such that the resonance frequencies ω_{mg} and ω_{ng} can be taken to be real quantities, the expression for $\chi^{(2)}$ can be simplified still further. In particular, under such circumstances Eq. (3.2.27) can be expressed as

$$\chi_{ijk}^{(2)}(\omega_\sigma, \omega_q, \omega_p) = \frac{N}{\hbar^2} \mathscr{P}_{\mathrm{F}} \sum_{mn} \frac{\mu_{gn}^i \mu_{nm}^j \mu_{mg}^k}{(\omega_{ng} - \omega_\sigma)(\omega_{mg} - \omega_p)} \tag{3.2.28}$$

where $\omega_\sigma = \omega_p + \omega_q$. Here we have introduced the full permutation operator, $\mathscr{P}_{\mathrm{F}}$, defined such that the expression that follows it is to be summed over all permutations of the frequencies ω_p, ω_q, and $-\omega_\sigma$, that is, over all input and output frequencies. The cartesian indices are to be permuted along with the frequencies. The final result is then to be divided by the number of permutations of the input frequencies. The equivalence of Eqs. (3.2.27) and (3.2.28) can be verified by explicitly expanding the right-hand side of each equation into all six terms. The six permuations denoted by the operator $\mathscr{P}_{\mathrm{F}}$ are

$$(-\omega_\sigma, \omega_q, \omega_p) \to (-\omega_\sigma, \omega_p, \omega_q), (\omega_q, -\omega_\sigma, \omega_p), (\omega_q, \omega_p, -\omega_\sigma),$$
$$(\omega_p, -\omega_\sigma, \omega_q), (\omega_p, \omega_q, -\omega_\sigma).$$

Since we can express the nonlinear susceptibility in the form of Eq. (3.2.28), we have proven the statement made in Section 1.5 that the nonlinear susceptibility of a lossless medium possesses full permutation symmetry.

Third-Order Susceptibility

We now calculate the third-order susceptibility. The dipole moment per atom, correct to third order in perturbabion theory, is given by

$$\begin{aligned}\langle \tilde{\mathbf{p}}^{(3)} \rangle = \langle \psi^{(0)} | \hat{\boldsymbol{\mu}} | \psi^{(3)} \rangle + \langle \psi^{(1)} | \hat{\boldsymbol{\mu}} | \psi^{(2)} \rangle + \langle \psi^{(2)} | \hat{\boldsymbol{\mu}} | \psi^{(1)} \rangle \\ + \langle \psi^{(3)} | \hat{\boldsymbol{\mu}} | \psi^{(0)} \rangle.\end{aligned} \tag{3.2.29}$$

Formulas for $\psi^{(0)}$, $\psi^{(1)}$, $\psi^{(2)}$, $\psi^{(3)}$, are given by Eqs. (3.2.10), (3.2.11), (3.2.16), (3.2.17), and (3.2.18). We find that

$$\begin{aligned}\langle \tilde{\mathbf{p}}^{(3)} \rangle = \frac{1}{\hbar^3} \sum_{pqr} \sum_{mnv} \\ \times \bigg(\frac{\boldsymbol{\mu}_{gv} [\boldsymbol{\mu}_{vn} \cdot \mathbf{E}(\omega_r)][\boldsymbol{\mu}_{nm} \cdot \mathbf{E}(\omega_q)][\boldsymbol{\mu}_{mg} \cdot \mathbf{E}(\omega_p)]}{(\omega_{vg} - \omega_r - \omega_q - \omega_p)(\omega_{ng} - \omega_q - \omega_p)(\omega_{mg} - \omega_p)} \\ \times e^{-i(\omega_p + \omega_q + \omega_r)t}\end{aligned}$$

$$+\frac{[\boldsymbol{\mu}_{vg}\cdot\mathbf{E}(\omega_r)]^*\boldsymbol{\mu}_{vn}[\boldsymbol{\mu}_{nm}\cdot\mathbf{E}(\omega_q)][\boldsymbol{\mu}_{mg}\cdot\mathbf{E}(\omega_p)]}{(\omega_{vg}^*-\omega_r)(\omega_{ng}-\omega_q-\omega_p)(\omega_{mg}-\omega_p)}$$

$$\times e^{-i(\omega_p+\omega_q-\omega_r)t} \tag{3.2.30}$$

$$+\frac{[\boldsymbol{\mu}_{vg}\cdot\mathbf{E}(\omega_r)]^*[\boldsymbol{\mu}_{nv}\cdot\mathbf{E}(\omega_q)]^*\boldsymbol{\mu}_{nm}[\boldsymbol{\mu}_{mg}\cdot\mathbf{E}(\omega_p)]}{(\omega_{vg}^*-\omega_r)(\omega_{ng}^*-\omega_r-\omega_q)(\omega_{mg}-\omega_p)}$$

$$\times e^{-i(\omega_p-\omega_q-\omega_r)t}$$

$$+\frac{[\boldsymbol{\mu}_{vg}\cdot\mathbf{E}(\omega_r)]^*[\boldsymbol{\mu}_{nv}\cdot\mathbf{E}(\omega_q)]^*[\boldsymbol{\mu}_{mn}\cdot\mathbf{E}(\omega_p)]^*\boldsymbol{\mu}_{mg}}{(\omega_{vg}^*-\omega_r)(\omega_{ng}^*-\omega_r-\omega_q)(\omega_{mg}^*-\omega_r-\omega_q-\omega_p)}$$

$$\left.\times e^{+i(\omega_p+\omega_q+\omega_r)t}\right).$$

Since the expression is summed over all positive and negative values of ω_p, ω_q, and ω_r, we can replace these quantities by their negatives in those expressions where the complex conjugate of a field amplitude appears. We thereby obtain the expression

$$\langle\tilde{\mathbf{p}}^{(3)}\rangle=\frac{1}{\hbar^3}\sum_{pqr}\sum_{mnv}$$

$$\times\left(\frac{\boldsymbol{\mu}_{gv}[\boldsymbol{\mu}_{vn}\cdot\mathbf{E}(\omega_r)][\boldsymbol{\mu}_{nm}\cdot\mathbf{E}(\omega_q)][\boldsymbol{\mu}_{mg}\cdot\mathbf{E}(\omega_p)]}{(\omega_{vg}-\omega_r-\omega_q-\omega_p)(\omega_{ng}-\omega_q-\omega_p)(\omega_{mg}-\omega_p)}\right.$$

$$+\frac{[\boldsymbol{\mu}_{gv}\cdot\mathbf{E}(\omega_r)]\boldsymbol{\mu}_{vn}[\boldsymbol{\mu}_{ml}\cdot\mathbf{E}(\omega_q)][\boldsymbol{\mu}_{mq}\cdot\mathbf{E}(\omega_p)]}{(\omega_{vg}^*+\omega_r)(\omega_{ng}-\omega_q-\omega_p)(\omega_{mg}-\omega_p)} \tag{3.2.31}$$

$$+\frac{[\boldsymbol{\mu}_{gv}\cdot\mathbf{E}(\omega_r)][\boldsymbol{\mu}_{vn}\cdot\mathbf{E}(\omega_q)]\boldsymbol{\mu}_{nm}[\boldsymbol{\mu}_{mg}\cdot\mathbf{E}(\omega_p)]}{(\omega_{vg}^*+\omega_r)(\omega_{ng}^*+\omega_r+\omega_q)(\omega_{mg}-\omega_p)}$$

$$\left.+\frac{[\boldsymbol{\mu}_{gv}\cdot\mathbf{E}(\omega_r)][\boldsymbol{\mu}_{vn}\cdot\mathbf{E}(\omega_q)][\boldsymbol{\mu}_{nm}\cdot\mathbf{E}(\omega_p)]\boldsymbol{\mu}_{mg}}{(\omega_{vg}^*+\omega_r)(\omega_{ng}^*+\omega_r+\omega_q)(\omega_{mg}^*+\omega_r+\omega_q+\omega_p)}\right)$$

$$\times e^{-i(\omega_p+\omega_q+\omega_r)t}.$$

We now use this result to calculate the third-order susceptibility: We let $\tilde{\mathbf{P}}^{(3)}=N\langle\tilde{\mathbf{p}}^{(3)}\rangle=\sum_s\mathbf{P}^{(3)}(\omega_s)\exp(-i\omega_s t)$ and introduce the definition (1.3.21) of the third-order susceptibility:

$$P_k(\omega_p+\omega_q+\omega_r)=\sum_{hij}\sum_{(pqr)}\chi_{kjih}^{(3)}(\omega_\sigma,\omega_r,\omega_q,\omega_p)E_j(\omega_r)E_i(\omega_q)E_h(\omega_p).$$

We thereby obtain the result

$$
\begin{aligned}
\chi^{(3)}_{kjih}(\omega_\sigma, \omega_r, \omega_q, \omega_p) = \frac{N}{\hbar^3} \mathscr{P}_{\mathrm{I}}
& \times \sum_{mnv} \left(\frac{\mu^k_{gv} \mu^j_{vn} \mu^i_{nm} \mu^h_{mg}}{(\omega_{vg} - \omega_r - \omega_q - \omega_p)(\omega_{ng} - \omega_q - \omega_p)(\omega_{mg} - \omega_p)} \right. \\
& + \frac{\mu^j_{gv} \mu^k_{vn} \mu^i_{nm} \mu^h_{mg}}{(\omega^*_{vg} + \omega_r)(\omega_{ng} - \omega_q - \omega_p)(\omega_{mg} - \omega_p)} \\
& + \frac{\mu^j_{gv} \mu^i_{vn} \mu^k_{nm} \mu^h_{mg}}{(\omega^*_{vg} + \omega_r)(\omega^*_{ng} + \omega_r + \omega_q)(\omega_{mg} - \omega_p)} \\
& \left. + \frac{\mu^j_{gv} \mu^i_{vn} \mu^h_{nm} \mu^k_{mg}}{(\omega^*_{vg} + \omega_r)(\omega^*_{ng} + \omega_r + \omega_q)(\omega^*_{mg} + \omega_r + \omega_q + \omega_p)} \right)
\end{aligned}
\tag{3.2.32}
$$

Here we have again made use of the intrinsic permutation operator $\mathscr{P}_{\mathrm{I}}$ defined following Eq. (3.2.27). The complete expression for the third-order susceptibility actually contains twenty-four terms, of which only four are displayed explicitly in Eq. (3.2.32); the others can be obtained through permutations of the frequencies (and cartesian indices) of the applied fields. The locations of the resonances in the displayed terms of this expression are illustrated in Fig. 3.2.3.

As in the case of the second-order susceptibility, the expression for $\chi^{(3)}$ can be written very compactly for the case of highly nonresonant excitation such that the imaginary parts of the resonant frequencies (recall that $\omega_{lg} = \omega^0_{lg} - i\Gamma_l/2$) can be ignored. In this case, the expression for $\chi^{(3)}$ can be

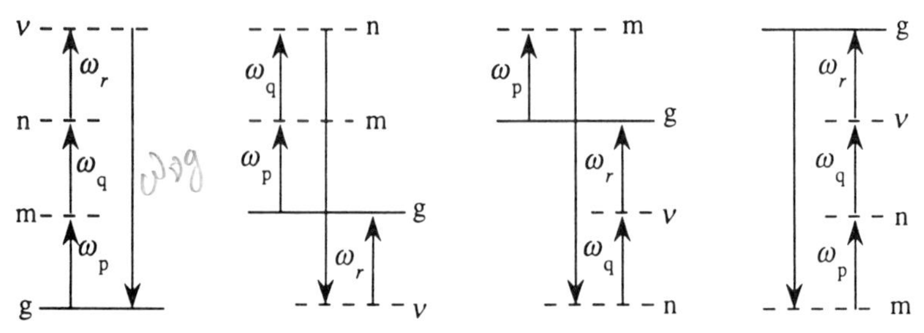

FIGURE 3.2.3 Locations of the resonances of each term in the expression (3.2.32) for the third-order susceptibility.

written as

$$\chi^{(3)}_{kjih}(\omega_\sigma, \omega_r, \omega_q, \omega_p) = \frac{N}{\hbar^3}\mathscr{P}_F \sum_{mnv} \frac{\mu^k_{gv}\mu^j_{vn}\mu^i_{nm}\mu^h_{mg}}{(\omega_{vg} - \omega_\sigma)(\omega_{ng} - \omega_q - \omega_p)(\omega_{mg} - \omega_p)}, \tag{3.2.33}$$

where $\omega_\sigma = \omega_p + \omega_q + \omega_r$ and where we have made use of the full permutation operator $\mathscr{P}_F$ defined following Eq. (3.2.28).

Third-Harmonic Generation in Alkali Metal Vapors

As an example of the use of Eq. (3.2.32), we next calculate the nonlinear optical susceptibility describing third-harmonic generation in a vapor of sodium atoms. Except for minor changes in notation, our treatment is similar to the original treatment by Miles and Harris (1973). We assume that the incident radiation is linearly polarized in the z direction. Consequently, the nonlinear polarization will have only a z component, and we can suppress the tensor nature of the nonlinear interaction. If we represent the applied field as

$$\tilde{E}(\mathbf{r}, t) = E_1(\mathbf{r})e^{-i\omega t} + \text{c.c.}, \tag{3.2.34}$$

we find that the nonlinear polarization can be represented as

$$\tilde{P}(\mathbf{r}, t) = P_3(\mathbf{r})e^{-i3\omega t} + \text{c.c.}, \tag{3.2.35}$$

where

$$P_3(\mathbf{r}) = \chi^{(3)}(3\omega)E_1^3 \tag{3.2.36}$$

and where the nonlinear susceptibility describing third-harmonic generation is given, ignoring damping effects, by

$$\begin{aligned}\chi^{(3)}(3\omega) = \frac{N}{\hbar^3}\sum_{mnv} \mu_{gv}\mu_{vn}\mu_{nm}\mu_{mg} &\times \left[\frac{1}{(\omega_{vg} - 3\omega)(\omega_{ng} - 2\omega)(\omega_{mg} - \omega)}\right. \\ &+ \frac{1}{(\omega_{vg} + \omega)(\omega_{ng} - 2\omega)(\omega_{mg} - \omega)} \\ &\times \frac{1}{(\omega_{vg} + \omega)(\omega_{ng} + 2\omega)(\omega_{mg} - \omega)} \\ &+ \left.\frac{1}{(\omega_{vg} + \omega)(\omega_{ng} + 2\omega)(\omega_{mg} + 3\omega)}\right].\end{aligned} \tag{3.2.37}$$

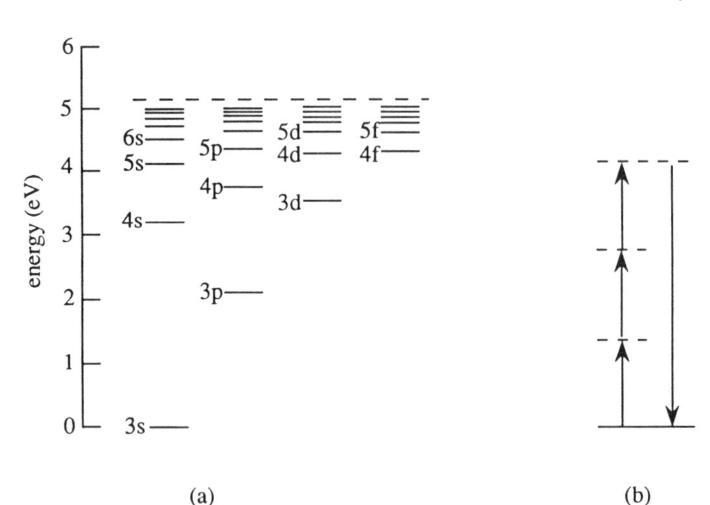

FIGURE 3.2.4 (a) Energy level diagram of the sodium atom. (b) The third-harmonic generation process.

Equation (3.2.37) can be readily evaluated through use of the known energy level structure and dipole transition moments of the sodium atom. Figure 3.2.4 shows an energy level diagram of the low-lying states of the sodium atom and a photon energy level diagram describing the process of third-harmonic generation. We see that only the first contribution to Eq. (3.2.37) can become fully resonant. This term becomes fully resonant when ω is nearly equal to ω_{mg}, 2ω is nearly equal to ω_{ng}, and 3ω is nearly equal to ω_{vg}. In performing the summation over excited levels m, n, and v, the only levels that contribute are those that obey the selection rules for

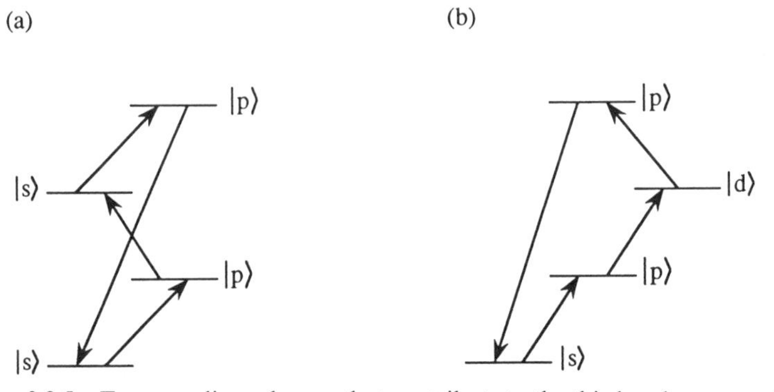

FIGURE 3.2.5 Two coupling schemes that contribute to the third-order susceptibility.

electric dipole transitions. In particular, since the ground state is an s state, the matrix element μ_{mg} will be nonzero only if m denotes a p state. Similarly, since m denotes a p state, the matrix element μ_{nm} will be nonzero only if n denotes an s state or a d state. In either case, v must denote a p state, since only in this case can both μ_{vn} and μ_{gv} be nonzero. The two types of coupling schemes that contribute to $\chi^{(3)}$ are shown in Fig. 3.2.5.

Through use of tabulated values of the matrix elements for the sodium atom, Miles and Harris have calculated numerically the value of $\chi^{(3)}$ as a function of the frequency ω of the incident laser field. The results of this calculation are shown in Fig. 3.2.6. A number of strong resonances in the nonlinear susceptibility are evident. Each such resonance is labeled by the quantum number of the level and the type of resonance that leads to the resonance enhancement. The peak labeled 3p (3ω), for example, is due to a three-photon resonance with the 3p level of sodium.

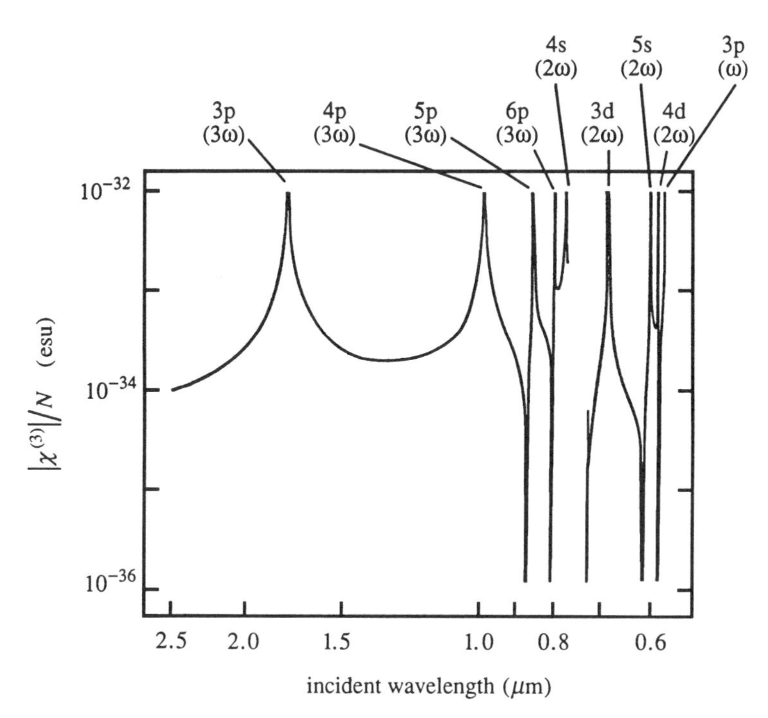

FIGURE 3.2.6 The nonlinear susceptibility describing third-harmonic generation in atomic sodium vapor plotted versus the wavelength of the fundamental radiation (after Miles and Harris, 1973).

3.3. Density Matrix Formalism of Quantum Mechanics

In the present section through Section 3.7, we calculate the nonlinear optical susceptibility through use of the density matrix formulation of quantum mechanics. We use this formalism because it is capable of treating effects, such as collisional broadening of the atomic resonances, that cannot be treated by the simpler theoretical formalism based on the atomic wave function. It is important that the formalism be capable of treating collisional effects for the following reason: We saw in the last section that the nonlinear response will be particularly large when the incident laser frequency or one of its harmonics (or, more generally, when the sum or difference of the applied field frequencies) is tuned close to one of the resonance frequencies of the atomic system. For those cases in which the detuning from resonance is comparable to or less than the width of the resonance, it is necessary that the theory include a treatment of the physical processes that produce the line broadening.

Let us begin by reviewing how the density matrix formalism follows from the basic laws of quantum mechanics.* If a quantum-mechanical system (such as an atom) is known to be in a particular quantum-mechanical state which we designate s, we can describe all of the physical properties of the system in terms of the wave function $\psi_s(\mathbf{r}, t)$ appropriate to this state. This wave function obeys the Schrödinger equation

$$i\hbar \frac{\partial \psi_s(\mathbf{r}, t)}{\partial t} = \hat{H}\psi_s(\mathbf{r}, t), \tag{3.3.1}$$

where $\hat{H}$ denotes the Hamiltonian operator of the system. We assume that $\hat{H}$ can be represented as

$$\hat{H} = \hat{H}_0 + \hat{V}(t), \tag{3.3.2}$$

where $\hat{H}_0$ is the Hamiltonian for a free atom and $\hat{V}(t)$ represents the interaction energy. In order to determine how the wave function evolves in time, it is often helpful to make explicit use of the fact that the energy eigenstates of the free-atom Hamiltonian $\hat{H}_0$ form a complete set of basis functions. We can hence represent the wave function of state s as

$$\psi_s(\mathbf{r}, t) = \sum_n C_n^s(t) u_n(\mathbf{r}), \tag{3.3.3}$$

where, as in Section 3.2, the functions $u_n(\mathbf{r})$ are the energy eigensolutions to the

* The reader who is already familiar with the density matrix formalism can skip directly to Section 3.4.

time-independent Schrödinger equation

$$\hat{H}_0 u_n(\mathbf{r}) = E_n u_n(\mathbf{r}) \tag{3.3.4}$$

and are assumed to be orthonormal in that they obey the relation

$$\int u_m^*(\mathbf{r}) u_n(\mathbf{r})\, d^3r = \delta_{mn}. \tag{3.3.5}$$

The expansion coefficient $C_n^s(t)$ gives the probability amplitude that the atom, which is known to be in state s, is in energy eigenstate n at time t. The time evolution of $\psi_s(\mathbf{r}, t)$ can be specified in terms of the time evolution of each of the expansion coefficients $C_n^s(t)$. To determine how these coefficients evolve in time, we introduce the expansion (3.3.3) into Schrödinger's equation (3.3.1) to obtain

$$i\hbar \sum_n \frac{dC_n^s(t)}{dt} u_n(\mathbf{r}) = \sum_n C_n^s(t) \hat{H} u_n(\mathbf{r}). \tag{3.3.6}$$

Each side of this equation involves a summation over all of the energy eigenstates of the system. In order to simplify this equation, we multiply each side from the left by $u_m^*(\mathbf{r})$ and integrate over all space. The summation on the left-hand side of the resulting equation reduces to a single term through use of the orthogonality condition of Eq. (3.3.5). The right-hand side is simplified by introducing the matrix elements of the Hamiltonian operator $\hat{H}$, defined through

$$H_{mn} = \int u_m^*(\mathbf{r}) \hat{H} u_n(\mathbf{r})\, d^3r. \tag{3.3.7}$$

We thereby obtain the result

$$i\hbar \frac{d}{dt} C_m^s(t) = \sum_n H_{mn} C_n^s(t). \tag{3.3.8}$$

This equation is entirely equivalent to the Schrödinger equation (3.3.1), but is written in terms of the probability amplitudes $C_n^s(t)$.

The expectation value of any observable quantity can be calculated in terms of the wave function of the system. A basic postulate of quantum mechanics states that any observable quantity A is associated with a Hermitian operator $\hat{A}$. The expectation value of A is then obtained according to the prescription

$$\langle A \rangle = \int \psi_s^* \hat{A} \psi_s\, d^3r. \tag{3.3.9}$$

Here the angular brackets denote a quantum-mechanical average. This relationship is conveniently written in Dirac notation as

$$\langle A \rangle = \langle \psi_s | \hat{A} | \psi_s \rangle = \langle s | \hat{A} | s \rangle, \tag{3.3.10}$$

where we will alternatively use $|\psi_s\rangle$ or $|s\rangle$ to note the state s. The expectation value $\langle A \rangle$ can also be expressed in terms of the probability amplitudes $C_n^s(t)$ by introducing Eq. (3.3.3) into Eq. (3.3.9) to obtain

$$\langle A \rangle = \sum_{mn} C_m^{s*} C_n^s A_{mn}, \tag{3.3.11}$$

where we have introduced the matrix elements A_{mn} of the operator $\hat{A}$, defined through

$$A_{mn} = \langle u_m | \hat{A} | u_n \rangle = \int u_m^* \hat{A} u_n \, d^3r. \tag{3.3.12}$$

As long as the initial state of the system and the Hamiltonian operator $\hat{H}$ for the system are known, the formalism described above by Eqs. (3.2.1)–(3.3.12) is capable of providing a complete description of the time evolution of the system and of all of its observable properties. However, there are circumstances under which the state of the system is not known in a precise manner. An example is a collection of atoms in an atomic vapor, where the atoms can interact with one another by means of collisions. Each time a collision occurs, the wave function of each interacting atom is modified. If the collisions are sufficiently weak, the modification may involve only an overall change in the phase of the wave function. However, since it is computationally infeasible to keep track of the phase of each atom within the atomic vapor, from a practical point of view the state of each atom is not known.

Under such circumstances, where the precise state of the system is unknown, the density matrix formalism can be used to describe the system in a statistical sense. Let us denote by $p(s)$ the probability that the system is in the state s. The quantity $p(s)$ is to be understood as a classical rather than a quantum-mechanical probability. Hence $p(s)$ simply reflects our lack of knowledge of the actual quantum-mechanical state of the system; it is not a consequence of any sort of quantum-mechanical uncertainty relation. In terms of $p(s)$, we define the elements of the density matrix of the system by

$$\rho_{nm} = \sum_s p(s) C_m^{s*} C_n^s. \tag{3.3.13}$$

This relation can also be written symbolically as

$$\rho_{nm} = \overline{C_m^* C_n}, \tag{3.3.14}$$

where the overbar denotes an ensemble average, that is, an average over all of the possible states of the system. In either form, the indices n and m are understood to run over all of the energy eigenstates of the system.

The elements of the density matrix have the following physical interpretation: The diagonal elements ρ_{nn} give the probability that the system is in energy eigenstate n. The off-diagonal elements have a somewhat more abstract interpretation: ρ_{nm} gives the "coherence" between levels n and m, in the sense that ρ_{nm} will be nonzero only if the system is in a coherent superposition of energy eigenstates n and m. We show below that the off-diagonal elements of the density matrix are, in certain circumstances, proportional to the induced electric dipole moment of the atom.

The density matrix is useful because it can be used to calculate the expectation value of any observable quantity. Since the expectation value of an observable quantity A for a system known to be in the quantum state s is given according to Eq. (3.3.11) by $\langle A \rangle = \sum_{mn} C_m^{s*} C_n^s A_{mn}$, the expectation value for the case in which the exact state of the system is not known is obtained by averaging Eq. (3.3.11) over all possible states of the system, to yield

$$\overline{\langle A \rangle} = \sum_s p(s) \sum_{nm} C_m^{s*} C_n^s A_{mn}. \tag{3.3.15}$$

The notation used on the left-hand side of this equation means that we are calculating the ensemble average of the quantum-mechanical expectation value of the observable quantity A.* Through use of Eq. (3.3.13), this quantity can alternatively be expressed as

$$\overline{\langle A \rangle} = \sum_{nm} \rho_{nm} A_{mn}. \tag{3.3.16}$$

The double summation in the equation can be simplified as follows:

$$\sum_{nm} \rho_{nm} A_{mn} = \sum_n \left(\sum_m \rho_{nm} A_{mn} \right) = \sum_n (\hat{\rho}\hat{A})_{nn} \equiv \operatorname{tr}(\hat{\rho}\hat{A}),$$

where we have introduced the trace operation, which is defined for any operator $\hat{M}$ by $\operatorname{tr} \hat{M} = \sum_n M_{nn}$. The expectation value of A is hence given by

$$\overline{\langle A \rangle} = \operatorname{tr}(\hat{\rho}\hat{A}). \tag{3.3.17}$$

The notation used in these equations is that $\hat{\rho}$ denotes the density operator, whose n, m matrix component is denoted ρ_{nm}; $\hat{\rho}\hat{A}$ denotes the product of $\hat{\rho}$

* In later sections of this chapter, we shall follow conventional notation and omit the overbar from expressions such as $\overline{\langle A \rangle}$, allowing the angular brackets to denote both a quantum and a classical average.

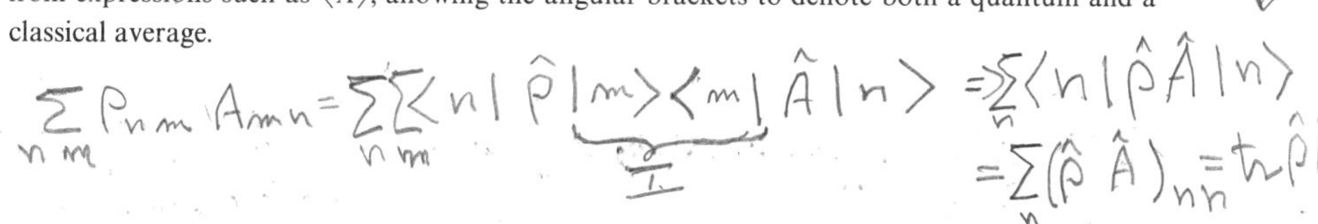

with the operator $\hat{A}$; and $(\hat{\rho}\hat{A})_{nn}$ denotes the n, n component of the matrix representation of this product.

We have just seen that the expectation value of any observable quantity can be determined straightforwardly in terms of the density matrix. In order to determine how any expectation value evolves in time, it is necessary only to determine how the density matrix itself evolves in time. By direct time differentiation of Eq. (3.3.13), we obtain

$$\dot{\rho}_{nm} = \sum_s \frac{dp(s)}{dt} C_m^{s*} C_n^s + \sum_s p(s)\left(C_m^{s*} \frac{dC_n^s}{dt} + \frac{dC_m^{s*}}{dt} C_n^s \right). \tag{3.3.18}$$

For the present, let us assume that $p(s)$ does not vary in time, so that the first term in this expression vanishes. We can then evaluate the second term straightforwardly by using Schrödinger's equation for the time evolution of the probability amplitudes (Eq. (3.3.8)). From this equation we obtain the expressions

$$C_m^{s*} \frac{dC_n^s}{dt} = \frac{-i}{\hbar} C_m^{s*} \sum_\nu H_{n\nu} C_\nu^s,$$

$$C_n^s \frac{dC_m^{s*}}{dt} = \frac{i}{\hbar} C_n^s \sum_\nu H_{m\nu}^* C_\nu^{s*} = \frac{i}{\hbar} C_n^s \sum_\nu H_{\nu m} C_\nu^{s*}.$$

These results are now substituted into Eq. (3.3.18) (with the first term on the right-hand side omitted) to obtain

$$\dot{\rho}_{nm} = \sum_s p(s) \frac{i}{\hbar} \sum_\nu (C_n^s C_\nu^{s*} H_{\nu m} - C_m^{s*} C_\nu^s H_{n\nu}). \tag{3.3.19}$$

The right-hand side of this equation can be written more compactly by introducing the form (3.3.13) for the density matrix to obtain

$$\dot{\rho}_{nm} = \frac{i}{\hbar} \sum_\nu (\rho_{n\nu} H_{\nu m} - H_{n\nu} \rho_{\nu m}). \tag{3.3.20}$$

Finally, the summation over ν can be performed formally to write this result as

$$\dot{\rho}_{nm} = \frac{i}{\hbar} (\hat{\rho}\hat{H} - \hat{H}\hat{\rho})_{nm} = \frac{-i}{\hbar} [\hat{H}, \hat{\rho}]_{nm}. \tag{3.3.21}$$

We have written the last form in terms of the commutator, defined for any two operators $\hat{A}$ and $\hat{B}$ by $[\hat{A}, \hat{B}] = \hat{A}\hat{B} - \hat{B}\hat{A}$.

Equation (3.3.21) describes how the density matrix evolves in time as the result of interactions that are included in the Hamiltonian $\hat{H}$. However, as mentioned above, there are certain interactions (such as those resulting from collisions between atoms) that cannot conveniently be included in a Hamiltonian description. Such interactions can lead to a change in the state of the system, and hence to a nonvanishing value of $dp(s)/dt$. We include such effects in the formalism by adding phenomenological damping terms to the equation of motion (3.3.21). There is more than one way to model such decay processes. For the most part, we shall model such processes by taking the density matrix equations to have the form

$$\dot{\rho}_{nm} = \frac{-i}{\hbar}[\hat{H}, \hat{\rho}]_{nm} - \gamma_{nm}(\rho_{nm} - \rho_{nm}^{\text{eq}}). \tag{3.3.22}$$

Here the second term on the right-hand side is a phenomenological damping term, which indicates that ρ_{nm} relaxes to its equilibrium value ρ_{nm}^{eq} at rate γ_{nm}. Since γ_{nm} is a decay rate, we assume that $\gamma_{nm} = \gamma_{mn}$. In addition, we make the physical assumption that

$$\rho_{nm}^{\text{eq}} = 0 \qquad \text{for} \quad n \neq m. \tag{3.3.23}$$

We are hence assuming that in thermal equilibrium the excited states of the system may contain population (i.e., ρ_{nn}^{eq} can be nonzero) but that thermal excitation, which is expected to be an incoherent process, cannot produce any coherent superpositions of atomic states ($\rho_{nm}^{\text{eq}} = 0$ for $n \neq m$).

An alternative method of describing decay phenomena is to assume that the off-diagonal elements of the density matrix are damped in the manner described above, but to describe the damping of the diagonal elements by allowing population to decay from higher-lying levels to lower-lying levels. In such a case, the density matrix equations of motion are given by

$$\dot{\rho}_{nm} = -i\hbar^{-1}[\hat{H}, \hat{\rho}]_{nm} - \gamma_{nm}\rho_{nm}, \qquad n \neq m, \tag{3.3.24a}$$

$$\dot{\rho}_{nn} = -i\hbar^{-1}[\hat{H}, \hat{\rho}]_{nn} + \sum_{E_m > E_n} \Gamma_{nm}\rho_{mm} - \sum_{E_m < E_n} \Gamma_{mn}\rho_{nn}. \tag{3.3.24b}$$

Here, Γ_{nm} gives the rate per atom at which population decays from level m to level n, and, as above, γ_{nm} gives the damping rate of the ρ_{nm} coherence.

The damping rates γ_{nm} for the off-diagonal elements of the density matrix are not entirely independent of the damping rates of the diagonal elements. In fact, under quite general conditions the off-diagonal elements can be represented as

$$\gamma_{nm} = \tfrac{1}{2}(\Gamma_n + \Gamma_m) + \gamma_{nm}^{\text{col}}. \tag{3.3.25}$$

Here, Γ_n and Γ_m denote the total decay rates of population out of levels n and m, respectively. In the notation of Eq. (3.3.24b), for example, Γ_n is given by the expression

$$\Gamma_n = \sum_{E_n' < E_n} \Gamma_{n'n}. \tag{3.3.26}$$

The quantity γ_{nm}^{col} in Eq. (3.3.25) is the dipole dephasing rate due to processes (such as elastic collisions) that are not associated with the transfer of population; γ_{nm}^{col} is sometimes called the proper dephasing rate. To see why Eq. (3.3.25) depends upon the population decay rates in the manner indicated, we note that if level n has lifetime $\tau_n = 1/\Gamma_n$, the probability to be in level n must decay as

$$|C_n(t)|^2 = |C_n(0)|^2 e^{-\Gamma_n t}, \tag{3.3.27}$$

and hence the probability amplitude must vary in time as

$$C_n(t) = C_n(0) e^{-i\omega_n t} e^{-\Gamma_n t/2}. \tag{3.3.28}$$

Likewise, the probability amplitude of being in level m must vary as

$$C_m(t) = C_m(0) e^{-i\omega_m t} e^{-\Gamma_m t/2}. \tag{3.3.29}$$

Thus the coherence between the two levels must vary as

$$C_n^*(t) C_m(t) = C_n^*(0) C_m(0) e^{-i\omega_{mn} t} e^{-(\Gamma_n + \Gamma_m)t/2}. \tag{3.3.30}$$

But since the ensemble average of $C_n^* C_m$ is just ρ_{mn}, whose damping rate is denoted γ_{mn}, it follows that

$$\gamma_{mn} = \tfrac{1}{2}(\Gamma_n + \Gamma_m). \tag{3.3.31}$$

Example: Two-Level Atom

As an example of the use of the density matrix formalism, we apply it to the simple case illustrated in Fig. 3.3.1, in which only the two atomic states a and b interact appreciably with the incident optical field. The wave function describing state s of such an atom is given by

$$\psi_s(\mathbf{r}, t) = C_a^s(t) u_a(\mathbf{r}) + C_b^s(t) u_b(\mathbf{r}), \tag{3.3.32}$$

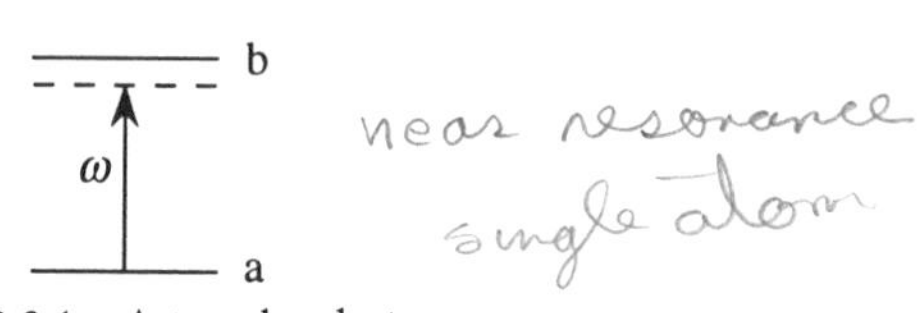

FIGURE 3.3.1 A two-level atom.

and hence the density matrix describing the atom is the two-by-two matrix given explicitly by

$$\begin{bmatrix} \rho_{aa} & \rho_{ab} \\ \rho_{ba} & \rho_{bb} \end{bmatrix} = \begin{bmatrix} \overline{C_a C_a^*} & \overline{C_a C_b^*} \\ \overline{C_b C_a^*} & \overline{C_b C_b^*} \end{bmatrix}. \tag{3.3.33}$$

The matrix representation of the dipole moment operator is

$$\hat{\mu} \Rightarrow \begin{bmatrix} 0 & \mu_{ab} \\ \mu_{ba} & 0 \end{bmatrix}, \tag{3.3.34}$$

where $\mu_{ij} = \mu_{ji}^* = \langle i | -e\hat{z} | j \rangle$, $-e$ is the electron charge, and $\hat{z}$ is the position operator for the electron. We have set the diagonal elements of the dipole moment operator equal to zero on the basis of the implicit assumption that states a and b have definite parity, in which case $\langle a|\hat{r}|a\rangle$ and $\langle b|\hat{r}|b\rangle$ vanish identically due to symmetry considerations. The expectation value of the dipole moment is given according to Eq. (3.3.17) by $\langle \hat{\mu} \rangle = \mathrm{tr}(\hat{\rho}\hat{\mu})$. Explicitly, $\hat{\rho}\hat{\mu}$ is represented as

$$\hat{\rho}\hat{\mu} \Rightarrow \begin{bmatrix} \rho_{aa} & \rho_{ab} \\ \rho_{ba} & \rho_{bb} \end{bmatrix} \begin{bmatrix} 0 & \mu_{ab} \\ \mu_{ba} & 0 \end{bmatrix} = \begin{bmatrix} \rho_{ab}\mu_{ba} & \rho_{aa}\mu_{ab} \\ \rho_{bb}\mu_{ba} & \rho_{ba}\mu_{ab} \end{bmatrix}, \tag{3.3.35}$$

and hence the expectation value of the induced dipole moment is given by

$$\langle \mu \rangle = \mathrm{tr}(\hat{\rho}\hat{\mu}) = \rho_{ab}\mu_{ba} + \rho_{ba}\mu_{ab}. \tag{3.3.36}$$

As stated above in connection with Eq. (3.3.14), the expectation value of the dipole moment is seen to depend upon the off-diagonal elements of the density matrix.

The density matrix treatment of the two-level atom is developed more fully in Chapter 5.

3.4. Perturbation Solution of the Density Matrix Equation of Motion

In the last section, we saw that the density matrix equation of motion with the phenomenological inclusion of damping is given by

$$\dot{\rho}_{nm} = \frac{-i}{\hbar}[\hat{H}, \hat{\rho}]_{nm} - \gamma_{nm}(\rho_{nm} - \rho_{nm}^{\mathrm{eq}}). \tag{3.4.1}$$

In general, this equation cannot be solved exactly for physical systems of interest, and for this reason we develop a perturbative technique for solving it.

This technique presupposes that, as in Eq. (3.3.2) in the preceding section, the Hamiltonian can be split into two parts as

$$\hat{H} = \hat{H}_0 + \hat{V}(t), \tag{3.4.2}$$

where $\hat{H}_0$ represents the Hamiltonian of the free atom and $\hat{V}(t)$ represents the energy of interaction of the atom with the externally applied radiation field. This interaction is assumed to be weak in the sense that the expectation value and matrix elements of $\hat{V}$ are much smaller than the expectation value of $\hat{H}_0$. We usually assume that this interaction energy is given adequately by the electric dipole approximation as

$$\hat{V} = -\hat{\boldsymbol{\mu}} \cdot \tilde{\mathbf{E}}(t), \tag{3.4.3}$$

where $\hat{\boldsymbol{\mu}} = -e\hat{\mathbf{r}}$ denotes the electric dipole moment operator of the atom. However, for generality and for compactness of notation, we will introduce Eq. (3.4.3) only when necessary.

When Eq. (3.4.2) is introduced into Eq. (3.4.1), the commutator $[\hat{H}, \hat{\rho}]$ splits into two terms. We examine first the commutator of $\hat{H}_0$ with $\hat{\rho}$. We assume that the states n represent the energy eigenfunctions u_n of the unperturbed Hamiltonian $\hat{H}_0$ and hence satisfy the equation $\hat{H}_0 u_n = E_n u_n$. (See also Eq. (3.3.4).) As a consequence, the matrix representation of $\hat{H}_0$ is diagonal, that is,

$$H_{0,nm} = E_n \delta_{nm}. \tag{3.4.4}$$

The commutator can thus be expanded as

$$\begin{aligned} [\hat{H}_0, \hat{\rho}]_{nm} &= (\hat{H}_0\hat{\rho} - \hat{\rho}\hat{H}_0)_{nm} = \sum_\nu (H_{0,n\nu}\rho_{\nu m} - \rho_{n\nu}H_{0,\nu m}) \\ &= \sum_\nu (E_n \delta_{n\nu}\rho_{\nu m} - \rho_{n\nu}\delta_{\nu m}E_m) \\ &= E_n\rho_{nm} - E_m\rho_{nm} = (E_n - E_m)\rho_{nm}. \end{aligned} \tag{3.4.5}$$

For future convenience, we define the transition frequency (in angular frequency units) as

$$\omega_{nm} = \frac{E_n - E_m}{\hbar}. \tag{3.4.6}$$

Through use of Eqs. (3.4.2), (3.4.5), and (3.4.6), the density matrix equation of motion (3.4.1) becomes

$$\dot{\rho}_{nm} = -i\omega_{nm}\rho_{nm} - \frac{i}{\hbar}[\hat{V}, \hat{\rho}]_{nm} - \gamma_{nm}(\rho_{nm} - \rho_{nm}^{\text{eq}}). \tag{3.4.7}$$

We can also expand the commutator of $\hat{V}$ with $\hat{\rho}$ to obtain the density matrix equation of motion in the form*

$$\dot{\rho}_{nm} = -i\omega_{nm}\rho_{nm} - \frac{i}{\hbar}\sum_{\nu}(V_{n\nu}\rho_{\nu m} - \rho_{n\nu}V_{\nu m}) - \gamma_{nm}(\rho_{nm} - \rho_{nm}^{\text{eq}}). \tag{3.4.8}$$

For most problems of physical interest, Eq. (3.4.8) cannot be solved analytically. We therefore seek a solution in the form of a perturbation expansion. In order to carry out this procedure, we replace V_{ij} in (3.4.8) by λV_{ij}, where λ is a parameter ranging between zero and one that characterizes the strength of the perturbation. The value $\lambda = 1$ is taken to represent the actual physical situation. We now seek a solution to Eq. (3.4.8) in the form of a power series in λ, that is,

$$\rho_{nm} = \rho_{nm}^{(0)} + \lambda\rho_{nm}^{(1)} + \lambda^2\rho_{nm}^{(2)} + \cdots. \tag{3.4.9}$$

We require that Eq. (3.4.9) be a solution of Eq. (3.4.8) for any value of the parameter λ. In order for this to occur, the coefficients of each power of λ must satisfy Eq. (3.4.8) separately. We thereby obtain the set of equations

$$\dot{\rho}_{nm}^{(0)} = -i\omega_{nm}\rho_{nm}^{(0)} - \gamma_{nm}(\rho_{nm}^{(0)} - \rho_{nm}^{\text{eq}}), \tag{3.4.10a}$$

$$\dot{\rho}_{nm}^{(1)} = -(i\omega_{nm} + \gamma_{nm})\rho_{nm}^{(1)} - \frac{i}{\hbar}[\hat{V}, \hat{\rho}^{(0)}]_{nm}, \tag{3.4.10b}$$

$$\dot{\rho}_{nm}^{(2)} = -(i\omega_{nm} + \gamma_{nm})\rho_{nm}^{(2)} - \frac{i}{\hbar}[\hat{V}, \hat{\rho}^{(1)}]_{nm}, \tag{3.4.10c}$$

etc. This system of equations can now be integrated directly, since, if the set of equations is solved in the order shown, each equation contains only linear homogeneous terms and inhomogeneous terms that are already known.

Equation (3.4.10a) describes the time evolution of the system in the absence

* In this section, we are describing the time evolution of the system in the Schrödinger picture. It is sometimes convenient to describe the time evolution instead in the interaction picture. To find the analogous equation of motion in the interaction picture, we define new quantities σ_{nm} and σ_{nm}^{eq} through

$$\rho_{nm} = \sigma_{nm}e^{-i\omega_{nm}t} \quad \rho_{nm}^{\text{eq}} = \sigma_{nm}^{\text{eq}}e^{-i\omega_{nm}t}.$$

In terms of these new quantities, Eq. (3.4.8) becomes

$$\dot{\sigma}_{nm} = -\frac{i}{\hbar}\sum_{\nu}[V_{n\nu}\sigma_{\nu m}e^{i\omega_{n\nu}t} - \sigma_{n\nu}e^{i\omega_{\nu m}t}V_{\nu m}] - \gamma_{nm}(\sigma_{nm} - \sigma_{nm}^{\text{eq}}).$$

of any external field. We take the steady-state solution to this equation to be

$$\rho_{nm}^{(0)} = \rho_{nm}^{\mathrm{eq}} \tag{3.4.11a}$$

where (for reasons given above; see Eq. (3.3.23))

$$\rho_{nm}^{\mathrm{eq}} = 0 \qquad \text{for} \quad n \neq m. \tag{3.4.11b}$$

Now that $\rho_{nm}^{(0)}$ is known, Eq. (3.4.10b) can be integrated. In order to do so, we make a change of variables by representing $\rho_{nm}^{(1)}$ as

$$\rho_{nm}^{(1)}(t) = S_{nm}^{(1)}(t)e^{-(i\omega_{nm}+\gamma_{nm})t}. \tag{3.4.12}$$

The derivative $\dot{\rho}_{nm}^{(1)}$ can be represented in terms of $S_{nm}^{(1)}$ as

$$\dot{\rho}_{nm}^{(1)} = -(i\omega_{nm}+\gamma_{nm})S_{nm}^{(1)}e^{-(i\omega_{nm}+\gamma_{nm})t} + \dot{S}_{nm}^{(1)}e^{-(i\omega_{nm}+\gamma_{nm})t}. \tag{3.4.13}$$

These forms are substituted into Eq. (3.4.10b), which then becomes

$$\dot{S}_{nm}^{(1)} = \frac{-i}{\hbar}[\hat{V}, \hat{\rho}^{(0)}]_{nm} e^{(i\omega_{nm}+\gamma_{nm})t}, \tag{3.4.14}$$

and which can be integrated to give

$$S_{nm}^{(1)} = \int_{-\infty}^{t} \frac{-i}{\hbar}[\hat{V}(t'), \rho^{(0)}]_{nm} e^{(i\omega_{nm}+\gamma_{nm})t'}\, dt'. \tag{3.4.15}$$

This expression is now substituted back into Eq. (3.4.12) to obtain

$$\rho_{nm}^{(1)}(t) = \int_{-\infty}^{t} \frac{-i}{\hbar}[\hat{V}(t'), \hat{\rho}^{(0)}]_{nm} e^{(i\omega_{nm}+\gamma_{nm})(t'-t)}\, dt'. \tag{3.4.16}$$

In similar fashion, all of the higher-order corrections to the density matrix can be obtained. These expressions are formally identical to Eq. (3.4.16). The expression for $\rho_{nm}^{(q)}$, for example, is obtained by replacing $\hat{\rho}^{(0)}$ with $\hat{\rho}^{(q-1)}$ on the right-hand side of Eq. (3.4.16).

3.5. Density Matrix Calculation of the Linear Susceptibility

As a first application of the perturbation solution to the density matrix equations of motion, we now calculate the linear susceptibility of an atomic system. The relevant starting equation for this calculation is Eq. (3.4.16), which we write in the form

$$\rho_{nm}^{(1)}(t) = e^{-(i\omega_{nm}+\gamma_{nm})t} \int_{-\infty}^{t} dt' \frac{-i}{\hbar}[\hat{V}(t'), \hat{\rho}^{(0)}]_{nm} e^{(i\omega_{nm}+\gamma_{nm})t'}. \tag{3.5.1}$$

As before, the interaction Hamiltonian is given by Eq. (3.4.3) as

$$\hat{V}(t') = -\hat{\boldsymbol{\mu}} \cdot \tilde{\mathbf{E}}(t'), \tag{3.5.2}$$

and we assume that the unperturbed density matrix is given by (see also Eq. (3.4.11))

$$\rho_{nm}^{(0)} = 0 \qquad \text{for} \quad n \neq m. \tag{3.5.3}$$

We represent the applied field as

$$\tilde{\mathbf{E}}(t) = \sum_p \mathbf{E}(\omega_p) e^{-i\omega_p t}. \tag{3.5.4}$$

The first step is to obtain an explicit expression for the commutator appearing in Eq. (3.5.1):

$$\begin{aligned} [\hat{V}(t), \hat{\rho}^{(0)}]_{nm} &= \sum_\nu [V(t)_{n\nu} \rho_{\nu m}^{(0)} - \rho_{n\nu}^{(0)} V(t)_{\nu m}] \\ &= -\sum_\nu [\boldsymbol{\mu}_{n\nu} \rho_{\nu m}^{(0)} - \rho_{n\nu}^{(0)} \boldsymbol{\mu}_{\nu m}] \cdot \tilde{\mathbf{E}}(t) \\ &= -(\rho_{mm}^{(0)} - \rho_{nn}^{(0)}) \boldsymbol{\mu}_{nm} \cdot \tilde{\mathbf{E}}(t). \end{aligned} \tag{3.5.5}$$

Here the second form is obtained by introducing $\hat{V}(t)$ explicitly from (3.5.2), and the third form is obtained by performing the summation over all ν and utilizing the condition (3.5.3). This expression for the commutator is introduced into Eq. (3.5.1) to obtain

$$\rho_{nm}^{(1)}(t) = \frac{i}{\hbar}(\rho_{mm}^{(0)} - \rho_{nn}^{(0)}) \boldsymbol{\mu}_{nm} \cdot e^{-(i\omega_{nm} + \gamma_{nm})t} \int_{-\infty}^{t} \tilde{\mathbf{E}}(t') e^{(i\omega_{nm} + \gamma_{nm})t'} \, dt'. \tag{3.5.6}$$

We next introduce Eq. (3.5.4) for $\tilde{\mathbf{E}}(t)$ to obtain

$$\begin{aligned} \rho_{nm}^{(1)}(t) = {} & \frac{i}{\hbar}(\rho_{mm}^{(0)} - \rho_{nn}^{(0)}) \boldsymbol{\mu}_{nm} \cdot \sum_p \mathbf{E}(\omega_p) \\ & \times e^{-(i\omega_{nm} + \gamma_{nm})t} \int_{-\infty}^{t} e^{[i(\omega_{nm} - \omega_p) + \gamma_{nm}]t'} \, dt'. \end{aligned} \tag{3.5.7}$$

The second line of this expression can be evaluated explicitly as

$$e^{-(i\omega_{nm} + \gamma_{nm})t} \left(\frac{e^{[i(\omega_{nm} - \omega_p) + \gamma_{nm}]t'}}{i(\omega_{nm} - \omega_p) + \gamma_{nm}} \right) \Bigg|_{-\infty}^{t} = \frac{e^{-i\omega_p t}}{i(\omega_{nm} - \omega_p) + \gamma_{nm}}, \tag{3.5.8}$$

and $\rho_{nm}^{(1)}$ is hence seen to be given by

$$\rho_{nm}^{(1)}(t) = \hbar^{-1}(\rho_{mm}^{(0)} - \rho_{nn}^{(0)}) \sum_p \frac{\boldsymbol{\mu}_{nm} \cdot \mathbf{E}(\omega_p) e^{-i\omega_p t}}{(\omega_{nm} - \omega_p) - i\gamma_{nm}}. \tag{3.5.9}$$

We next use this result to calculate the expectation value of the induced dipole moment:*

$$\begin{aligned}\langle\tilde{\boldsymbol{\mu}}(t)\rangle &= \operatorname{tr}(\hat{\rho}^{(1)}\hat{\boldsymbol{\mu}}) = \sum_{nm}\rho_{nm}^{(1)}\boldsymbol{\mu}_{mn} \\ &= \sum_{nm}\hbar^{-1}(\rho_{mm}^{(0)}-\rho_{nn}^{(0)})\sum_{p}\frac{\boldsymbol{\mu}_{mn}[\boldsymbol{\mu}_{nm}\cdot\mathbf{E}(\omega_p)]e^{-i\omega_p t}}{(\omega_{nm}-\omega_p)-i\gamma_{nm}}.\end{aligned} \tag{3.5.10}$$

We decompose $\langle\tilde{\boldsymbol{\mu}}(t)\rangle$ into its frequency components according to

$$\langle\tilde{\boldsymbol{\mu}}(t)\rangle = \sum_{p}\langle\boldsymbol{\mu}(\omega_p)\rangle e^{-i\omega_p t}, \tag{3.5.11}$$

and define the linear susceptibility $\boldsymbol{\chi}^{(1)}(\omega)$ by the equation

$$\mathbf{P}(\omega_p) = N\langle\boldsymbol{\mu}(\omega_p)\rangle = \boldsymbol{\chi}^{(1)}(\omega_p)\cdot\mathbf{E}(\omega_p), \tag{3.5.12}$$

where N denotes the atomic number density. By comparing this equation with Eq. (3.5.10), we find that the linear susceptibility is given by

$$\boldsymbol{\chi}^{(1)}(\omega_p) = \frac{N}{\hbar}\sum_{nm}(\rho_{mm}^{(0)}-\rho_{nn}^{(0)})\frac{\boldsymbol{\mu}_{mn}\boldsymbol{\mu}_{nm}}{(\omega_{nm}-\omega_p)-i\gamma_{nm}}. \tag{3.5.13}$$

The result given by Eqs. (3.5.12) and (3.5.13) can be written in cartesian component form as

$$P_i(\omega_p) = N\langle\mu_i(\omega_p)\rangle = \sum_{j}\chi_{ij}^{(1)}(\omega_p)E_j(\omega_p) \tag{3.5.14}$$

with

$$\chi_{ij}^{(1)}(\omega_p) = \frac{N}{\hbar}\sum_{nm}(\rho_{mm}^{(0)}-\rho_{nn}^{(0)})\frac{\mu_{mn}^{i}\mu_{nm}^{j}}{(\omega_{nm}-\omega_p)-i\gamma_{nm}}. \tag{3.5.15}$$

We see that the linear susceptibility is proportional to the population difference $\rho_{mm}^{(0)}-\rho_{nn}^{(0)}$; hence if levels m and n contain equal populations, the $m\to n$ transition does not contribute to the linear susceptibility.

Equation (3.5.15) is an extremely compact way of representing the linear susceptibility. At times it is more intuitive to express the susceptibility in an expanded form. We first rewrite Eq. (3.5.15) as

$$\chi_{ij}^{(1)}(\omega_p) = \frac{N}{\hbar}\sum_{nm}\rho_{mm}^{(0)}\frac{\mu_{mn}^{i}\mu_{nm}^{j}}{(\omega_{nm}-\omega_p)-i\gamma_{nm}} - \frac{N}{\hbar}\sum_{nm}\rho_{nn}^{(0)}\frac{\mu_{mn}^{i}\mu_{nm}^{j}}{(\omega_{nm}-\omega_p)-i\gamma_{nm}}. \tag{3.5.16}$$

* Here and throughout the remainder of this chapter we are omitting the bar over quantities such as $\langle\boldsymbol{\mu}\rangle$ for simplicity of notation. Hence the angular brackets are meant to imply both a quantum and an ensemble average.

We next interchange the dummy indices n and m in the second summation, so that the two summations can be recombined as

$$\chi_{ij}^{(1)}(\omega_p) = \frac{N}{\hbar} \sum_{nm} \rho_{mm}^{(0)} \left[\frac{\mu_{mn}^i \mu_{nm}^j}{(\omega_{nm} - \omega_p) - i\gamma_{nm}} - \frac{\mu_{nm}^i \mu_{mn}^j}{(\omega_{mn} - \omega_p) - i\gamma_{mn}} \right]. \tag{3.5.17}$$

We now use the fact that $\omega_{mn} = -\omega_{nm}$ and $\gamma_{nm} = \gamma_{mn}$ to write this result as

$$\chi_{ij}^{(1)}(\omega_p) = \frac{N}{\hbar} \sum_{nm} \rho_{mm}^{(0)} \left[\frac{\mu_{mn}^i \mu_{nm}^j}{(\omega_{nm} - \omega_p) - i\gamma_{nm}} + \frac{\mu_{nm}^i \mu_{mn}^j}{(\omega_{nm} + \omega_p) + i\gamma_{nm}} \right]. \tag{3.5.18}$$

In order to interpret this result, let us first make the simplifying assumption that all of the population is in one level (typically the ground state), which we denote as level a. Mathematically, this assumption can be stated as

$$\rho_{aa}^{(0)} = 1, \qquad \rho_{mm}^{(0)} = 0 \quad \text{for } m \neq a. \tag{3.5.19}$$

We now perform the summation over m in Eq. (3.5.18) to obtain

$$\chi_{ij}^{(1)}(\omega_p) = \frac{N}{\hbar} \sum_{n} \left[\frac{\mu_{an}^i \mu_{na}^j}{(\omega_{na} - \omega_p) - i\gamma_{na}} + \frac{\mu_{na}^i \mu_{an}^j}{(\omega_{na} + \omega_p) + i\gamma_{na}} \right]. \tag{3.5.20}$$

We see that for positive frequencies (i.e., for $\omega_p > 0$), only the first term can become resonant. The second term is known as the antiresonant or counter-rotating term. We can often drop the second term, especially when ω_p is close to one of the resonance frequencies of the atom. Let us assume that ω_p is nearly resonant with the transition frequency ω_{na}. Then to good approximation the linear susceptibility is given by

$$\chi_{ij}^{(1)}(\omega_p) = \frac{N}{\hbar} \frac{\mu_{an}^i \mu_{na}^j}{(\omega_{na} - \omega_p) - i\gamma_{na}} = \frac{N}{\hbar} \mu_{an}^i \mu_{na}^j \frac{(\omega_{na} - \omega_p) + i\gamma_{na}}{(\omega_{na} - \omega_p)^2 + \gamma_{na}^2}. \tag{3.5.21}$$

The real and imaginary parts of this expression are shown in Fig. 3.5.1. We see that the imaginary part of χ_{ij} has the form of a Lorentzian line shape with a linewidth (full width at half maximum) equal to $2\gamma_{na}$.

Linear Dispersion Theory

Since linear dispersion theory plays a key role in our understanding of optical phenomena, the remainder of this section is devoted to relating the results derived above for the linear susceptibility to several other important concepts. Let us first specialize our results to the case of an isotropic material. As a consequence of symmetry considerations, **P** must be parallel to **E** in such a medium, and we can therefore express the linear susceptibility as the scalar

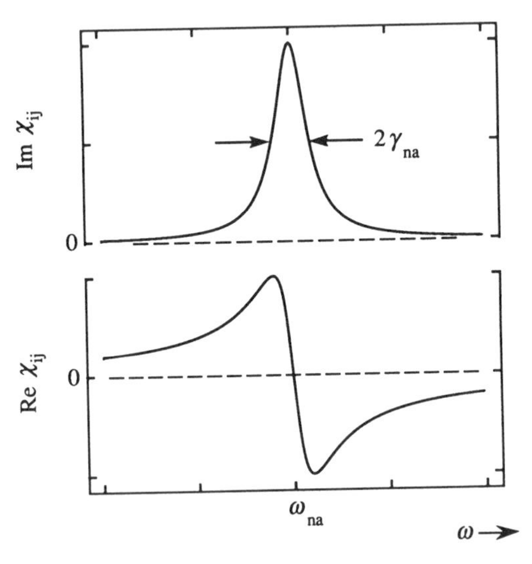

FIGURE 3.5.1 Resonance nature of the linear susceptibility.

quantity $\chi^{(1)}(\omega)$ defined through $\mathbf{P}(\omega) = \chi^{(1)}(\omega)\mathbf{E}(\omega)$, which is given by

$$\chi^{(1)}(\omega) = N\hbar^{-1}\sum_n \tfrac{1}{3}|\boldsymbol{\mu}_{na}|^2\left[\frac{1}{(\omega_{na}-\omega)-i\gamma_{na}} + \frac{1}{(\omega_{na}+\omega)+i\gamma_{na}}\right]. \tag{3.5.22}$$

For simplicity we are assuming the case of a nondegenerate ground state (e.g., $J = 0$). We have included the factor of $\frac{1}{3}$ in the numerator of this expression for the following reason: The summation over n includes all of the magnetic sublevels of the atomic excited states. However, on average only one-third of the $a \to n$ transitions will have their dipole transition moments parallel to the polarization vector of the incident field, and hence only one-third of these transitions contribute effectively to the susceptibility.

It is useful to introduce the *oscillator strength* of the $a \to n$ transition. This quantity is defined by

$$f_{na} = \frac{2m\omega_{na}|\boldsymbol{\mu}_{na}|^2}{3\hbar e^2}. \tag{3.5.23}$$

Standard books on quantum mechanics (see, for example, Bethe and Salpeter, 1977) show that this quantity obeys the *oscillator strength sum rule*, that is

$$\sum_n f_{na} = 1. \tag{3.5.24}$$

If a is the atomic ground state, the frequency ω_{na} is necessarily positive, and the sum rule hence shows that the oscillator strength is a positive quantity bounded by unity, that is, $0 \leq f_{na} \leq 1$. The expression (3.5.22) for the linear susceptibility can be written in terms of the oscillator strength as

$$\chi^{(1)}(\omega) = \sum_n \frac{N f_{na} e^2}{2m\omega_{na}} \left[\frac{1}{(\omega_{na} - \omega) - i\gamma_{na}} + \frac{1}{(\omega_{na} + \omega) + i\gamma_{na}} \right]$$
$$\simeq \sum_n f_{na} \left[\frac{Ne^2/m}{\omega_{na}^2 - \omega^2 - 2i\omega\gamma_{na}} \right]. \tag{3.5.25}$$

In the latter form, the expression in square brackets is formally identical to the expression for the linear susceptibility predicted by the classical Lorentz model of the atom (cf. Eq. (1.4.17)). We see that the quantum-mechanical prediction differs from that of the Lorentz model only in that in the quantum-mechanical theory there can be more than one resonance frequency ω_{na}. The strength of each such transition is given by the value of the oscillator strength.

Let us next see how to calculate the refractive index and absorption coefficient. The refractive index $n(\omega)$ is related to the linear dielectric constant $\epsilon(\omega)$ and linear susceptibility $\chi^{(1)}(\omega)$ through

$$n(\omega) = \sqrt{\epsilon(\omega)} = \sqrt{1 + 4\pi\chi^{(1)}(\omega)} \simeq 1 + 2\pi\chi^{(1)}(\omega). \tag{3.5.26}$$

In obtaining the last expression, we have assumed that the medium is sufficiently dilute (i.e., N sufficiently small) that $4\pi\chi^{(1)}(\omega) \ll 1$. For the remainder of this section, we will assume that this assumption is valid, both so that we can use Eq. (3.5.26) as written and also so that we can ignore local-field corrections (cf. Section 3.8). The significance of the refractive index $n(\omega)$ is that the propagation of a plane wave through the material system is described by

$$\tilde{E}(z, t) = E_0 e^{i(kz - \omega t)} + \text{c.c.}, \tag{3.5.27}$$

where the propagation constant k is given by

$$k = n(\omega)\omega/c. \tag{3.5.28}$$

Hence the intensity $I = (nc/4\pi)\langle \tilde{E}(z,t)^2 \rangle$ of this wave varies with position in the medium according to

$$I(z) = I_0 e^{-\alpha z}, \tag{3.5.29}$$

where the *absorption coefficient* α is given by

$$\alpha = 2n''\omega/c, \tag{3.5.30}$$

and where we have defined the real and imaginary parts of $n(\omega)$ as $n(\omega) = n' + in''$. Alternatively, through use of Eq. (3.5.26), we can represent the absorption coefficient in terms of the susceptibility as

$$\alpha = 4\pi\chi^{(1)''}\omega/c, \tag{3.5.31a}$$

where $\chi^{(1)}(\omega) = \chi^{(1)'} + i\chi^{(1)''}$. Through use of Eq. (3.5.25), we find that the absorption coefficient of the material system is given by

$$\alpha = \sum_n \frac{2\pi f_{na} N e^2}{m c \gamma_{na}} \left[\frac{\gamma_{na}^2}{(\omega_{na} - \omega)^2 + \gamma_{na}^2} \right]. \tag{3.5.31b}$$

It is often useful to describe the response of a material system to an applied field in terms of microscopic rather than macroscopic quantities. We define the atomic polarizability $\alpha^{(1)}(\omega)$ as the coefficient relating the induced dipole moment $\langle \boldsymbol{\mu}(\omega) \rangle$ and the applied field $\mathbf{E}(\omega)$:*

$$\langle \boldsymbol{\mu}(\omega) \rangle = \alpha^{(1)}(\omega)\mathbf{E}(\omega). \tag{3.5.32}$$

The susceptibility and polarizability are related (when local-field corrections can be ignored) through

$$\chi^{(1)}(\omega) = N\alpha^{(1)}(\omega), \tag{3.5.33}$$

and we hence find from Eq. (3.5.22) that the polarizability is given by

$$\alpha^{(1)}(\omega) = \hbar^{-1} \sum_n \tfrac{1}{3} |\boldsymbol{\mu}_{na}|^2 \left[\frac{1}{(\omega_{na} - \omega) - i\gamma_{na}} + \frac{1}{(\omega_{na} + \omega) + i\gamma_{na}} \right]. \tag{3.5.34}$$

Another microscopic quantity that is often encountered is the absorption cross section σ, which is defined through the relation

$$\sigma = \alpha/N. \tag{3.5.35}$$

The cross section can hence be interpreted as the effective area of an atom for removing radiation from an incident beam of light. By comparison with Eqs. (3.5.31) and (3.5.33), we see that the absorption cross section is related to the atomic polarizability $\alpha^{(1)} = \alpha^{(1)'} + i\alpha^{(1)''}$ through

$$\sigma = 4\pi\alpha^{(1)''}\omega/c. \tag{3.5.36}$$

Equation (3.5.34) shows how the polarizability can be calculated in terms of the transition frequencies ω_{na}, the dipole transition moments $\boldsymbol{\mu}_{na}$, and the dipole dephasing rates γ_{na}. The transition frequencies and dipole moments are inherent properties of any atomic system, and can be obtained either by

* Note that α denotes the absorption coefficient and $\alpha^{(1)}$ denotes the polarizability.

solving Schrödinger's equation for the atom or through laboratory measurement. The dipole dephasing rate, however, depends not only on the inherent atomic properties but also upon the local environment. We saw in Eq. (3.3.25) that the dipole dephasing rate γ_{nm} can be represented as

$$\gamma_{nm} = \tfrac{1}{2}(\Gamma_n + \Gamma_m) + \gamma_{nm}^{\text{col}}. \tag{3.5.37}$$

Next we calculate the maximum values that the polarizability and absorption cross section can attain. We consider the case of resonant excitation ($\omega = \omega_{na}$) of some excited level n. We find, through use of Eq. (3.5.34) and dropping the nonresonant contribution, that the polarizability is purely imaginary and is given by

$$\alpha_{\text{res}}^{(1)} = \frac{i|\mu_{n'a}|^2}{\hbar\gamma_{n'a}}. \tag{3.5.38}$$

We have let n' designate the state associated with level n that is excited by the incident light. Note that the factor of $\frac{1}{3}$ no longer appears in Eq. (3.5.38), because we have performed the summation over all states (i.e., magnetic sublevels) of level n. The polarizability will take on its maximum possible value if $\gamma_{n'a}$ is as small as possible, which according to Eq. (3.5.37) occurs when $\gamma_{n'a}^{\text{col}} = 0$. If a is the atomic ground state, as we have been assuming, its decay rate Γ_a must vanish, and hence the minimum possible value of $\gamma_{n'a}$ is $\frac{1}{2}\Gamma_{n'}$.

The population decay rate out of state n' is usually dominated by spontaneous emission. If state n' can decay only to the ground state, this decay rate is equal to the Einstein A coefficient and is given by

$$\Gamma_{n'} = \frac{4\omega_{na}^3|\boldsymbol{\mu}_{n'a}|^2}{3\hbar c^3}. \tag{3.5.39}$$

If $\gamma_{n'a} = \frac{1}{2}\Gamma_{n'}$ is inserted into Eq. (3.5.38), we find that the maximum possible value that the polarizability can possess is

$$\alpha_{\text{max}}^{(1)} = i\tfrac{3}{2}\left(\frac{\lambda}{2\pi}\right)^3. \tag{3.5.40}$$

We find the value of the absorption cross section associated with this value of the polarizability through use of Eq. (3.5.36):

$$\sigma_{\text{max}} = \frac{3\lambda^2}{2\pi}. \tag{3.5.41}$$

These results show that under resonant excitation an atomic system possesses an effective linear dimension approximately equal to an optical wavelength.

3.6. Density Matrix Calculation of the Second-Order Susceptibility

In this section we calculate the second-order (i.e., $\chi^{(2)}$) susceptibility of an atomic system. We present the calculation in considerable detail, for the following two reasons: (1) the second-order susceptibility is intrinsically important for many applications; (2) the calculation of the third-order susceptibility proceeds along lines that are analogous to those followed by the present derivation. However, the expression for the third-order susceptibility $\chi^{(3)}$ is so complicated (it contains 48 terms) that it is infeasible to treat $\chi^{(3)}$ in as much detail as we can treat $\chi^{(2)}$.

From the perturbation expansion (3.4.16), the general result for the second-order correction to $\hat{\rho}$ is given by

$$\rho_{nm}^{(2)} = e^{-(i\omega_{nm}+\gamma_{nm})t} \int_{-\infty}^{t} \frac{-i}{\hbar} [\hat{V}, \hat{\rho}^{(1)}]_{nm} e^{(i\omega_{nm}+\gamma_{nm})t'} \, dt', \tag{3.6.1}$$

where the commutator can be expressed (by analogy with Eq. 3.5.5) as

$$[\hat{V}, \hat{\rho}^{(1)}]_{nm} = -\sum_{\nu} (\boldsymbol{\mu}_{n\nu}\rho_{\nu m}^{(1)} - \rho_{n\nu}^{(1)}\boldsymbol{\mu}_{\nu m}) \cdot \tilde{\mathbf{E}}(t). \tag{3.6.2}$$

In order to evaluate this commutator, the first-order solution given by Eq. (3.5.9) is written with changes in the dummy indices as

$$\rho_{\nu m}^{(1)} = \hbar^{-1}(\rho_{mm}^{(0)} - \rho_{\nu\nu}^{(0)}) \sum_{p} \frac{\boldsymbol{\mu}_{\nu m} \cdot \mathbf{E}(\omega_p)}{(\omega_{\nu m} - \omega_p) - i\gamma_{\nu m}} e^{-i\omega_p t} \tag{3.6.3}$$

and as

$$\rho_{n\nu}^{(1)} = \hbar^{-1}(\rho_{\nu\nu}^{(0)} - \rho_{nn}^{(0)}) \sum_{p} \frac{\boldsymbol{\mu}_{n\nu} \cdot \mathbf{E}(\omega_p)}{(\omega_{n\nu} - \omega_p) - i\gamma_{n\nu}} e^{-i\omega_p t}. \tag{3.6.4}$$

The applied optical field $\tilde{\mathbf{E}}(t)$ is expressed as

$$\tilde{\mathbf{E}}(t) = \sum_{q} \mathbf{E}(\omega_q) e^{-i\omega_q t}. \tag{3.6.5}$$

The commutator of Eq. (3.6.2) thus becomes

$$\begin{aligned}
[\hat{V}, \hat{\rho}^{(1)}]_{nm} = & -\hbar^{-1} \sum_{\nu} (\rho_{mm}^{(0)} - \rho_{\nu\nu}^{(0)}) \\
& \times \sum_{pq} \frac{[\boldsymbol{\mu}_{n\nu} \cdot \mathbf{E}(\omega_q)][\boldsymbol{\mu}_{\nu m} \cdot \mathbf{E}(\omega_p)]}{(\omega_{\nu m} - \omega_p) - i\gamma_{\nu m}} e^{-i(\omega_p+\omega_q)t} \\
& + \hbar^{-1} \sum_{\nu} (\rho_{\nu\nu}^{(0)} - \rho_{nn}^{(0)}) \\
& \times \sum_{pq} \frac{[\boldsymbol{\mu}_{n\nu} \cdot \mathbf{E}(\omega_p)][\boldsymbol{\mu}_{\nu m} \cdot \mathbf{E}(\omega_q)]}{(\omega_{n\nu} - \omega_p) - i\gamma_{n\nu}} e^{-i(\omega_p+\omega_q)t}.
\end{aligned} \tag{3.6.6}$$

This expression is now inserted into Eq. (3.6.1), and the integration is performed to obtain

$$\rho_{nm}^{(2)} = \sum_{\nu} \sum_{pq} \times \left\{ \frac{\rho_{mm}^{(0)} - \rho_{\nu\nu}^{(0)}}{\hbar^2} \frac{[\boldsymbol{\mu}_{n\nu} \cdot \mathbf{E}(\omega_q)][\boldsymbol{\mu}_{\nu m} \cdot \mathbf{E}(\omega_p)]}{[(\omega_{nm} - \omega_p - \omega_q) - i\gamma_{nm}][(\omega_{\nu m} - \omega_p) - i\gamma_{\nu m}]} - \frac{\rho_{\nu\nu}^{(0)} - \rho_{nn}^{(0)}}{\hbar^2} \frac{[\boldsymbol{\mu}_{n\nu} \cdot \mathbf{E}(\omega_p)][\boldsymbol{\mu}_{\nu m} \cdot \mathbf{E}(\omega_q)]}{[(\omega_{nm} - \omega_p - \omega_q) - i\gamma_{nm}][(\omega_{n\nu} - \omega_p) - i\gamma_{n\nu}]} \right\} \times e^{-i(\omega_p + \omega_q)t} \equiv \sum_{\nu} \sum_{pq} K_{nm\nu} e^{-i(\omega_p + \omega_q)t}. \tag{3.6.7}$$

We have given the complicated expression in curly braces the label $K_{nm\nu}$ because this expression will appear in subsequent equations.

We next calculate the expectation value of the atomic dipole moment, which (according to Eq. (3.3.16)) is given by

$$\langle \tilde{\boldsymbol{\mu}} \rangle = \sum_{nm} \rho_{nm} \boldsymbol{\mu}_{mn}. \tag{3.6.8}$$

We are interested in the various frequency components of $\langle \tilde{\boldsymbol{\mu}} \rangle$, whose complex amplitudes $\langle \boldsymbol{\mu}(\omega_r) \rangle$ are defined through

$$\langle \tilde{\boldsymbol{\mu}} \rangle = \sum_{r} \langle \boldsymbol{\mu}(\omega_r) \rangle e^{-i\omega_r t}. \tag{3.6.9}$$

Then, in particular, the complex amplitude of the component of the atomic dipole moment oscillating at frequency $\omega_p + \omega_q$ is given by

$$\langle \boldsymbol{\mu}(\omega_p + \omega_q) \rangle = \sum_{nm\nu} \sum_{(pq)} K_{nm\nu} \boldsymbol{\mu}_{mn}, \tag{3.6.10}$$

and consequently the complex amplitude of the component of the nonlinear polarization oscillating at frequency $\omega_p + \omega_q$ is given by

$$\mathbf{P}^{(2)}(\omega_p + \omega_q) = N \langle \boldsymbol{\mu}(\omega_p + \omega_q) \rangle = N \sum_{nm\nu} \sum_{(pq)} K_{nm\nu} \boldsymbol{\mu}_{mn}. \tag{3.6.11}$$

We define the nonlinear susceptibility through the equation

$$P_i^{(2)}(\omega_p + \omega_q) = \sum_{jk} \sum_{(pq)} \chi_{ijk}^{(2)}(\omega_p + \omega_q, \omega_q, \omega_p) E_j(\omega_q) E_k(\omega_p), \tag{3.6.12}$$

using the same notation as that used earlier (see also Eq. (1.3.13)). By comparison of Eqs. (3.6.7), (3.6.11), and (3.6.12), we obtain a tentative expression for the susceptibility tensor given by

$$\chi_{ijk}^{(2)\prime}(\omega_p+\omega_q,\omega_q,\omega_p)=\frac{N}{\hbar^2}$$

$$\times\sum_{mn\nu}\left\{(\rho_{mm}^{(0)}-\rho_{\nu\nu}^{(0)})\frac{\mu_{mn}^{i}\mu_{n\nu}^{j}\mu_{\nu m}^{k}}{[(\omega_{nm}-\omega_p-\omega_q)-i\gamma_{nm}][(\omega_{\nu m}-\omega_p)-i\gamma_{\nu m}]}\right. \quad \text{(a)}$$

$$\left.-(\rho_{\nu\nu}^{(0)}-\rho_{nn}^{(0)})\frac{\mu_{mn}^{i}\mu_{\nu m}^{j}\mu_{n\nu}^{k}}{[(\omega_{nm}-\omega_p-\omega_q)-i\gamma_{nm}][(\omega_{n\nu}-\omega_p)-i\gamma_{n\nu}]}\right\}. \quad \text{(b)}$$

(3.6.13)

We have labeled the two terms that appear in this expression (a) and (b) so that we can keep track of how these terms contribute to our final expression for the second-order susceptibility.

Equation (3.6.13) can be used in conjunction with Eq. (3.6.12) to make proper predictions of the nonlinear polarization, which is a physically meaningful quantity. However, Eq. (3.6.13) does not possess intrinsic permutation symmetry (cf. Section 1.5), which we require the susceptibility to possess. We therefore define the nonlinear susceptibility to be one-half the sum of the right-hand side of Eq. (3.6.13) with an analogous expression obtained by simultaneously interchanging ω_p with ω_q and j with k. We thereby obtain the result

$$\chi_{ijk}^{(2)}(\omega_p+\omega_q,\omega_q,\omega_p)=\frac{N}{2\hbar^2}$$

$$\times\sum_{mn\nu}\left\{(\rho_{mm}^{(0)}-\rho_{\nu\nu}^{(0)})\left[\frac{\mu_{mn}^{i}\mu_{n\nu}^{j}\mu_{\nu m}^{k}}{[(\omega_{nm}-\omega_p-\omega_q)-i\gamma_{nm}][(\omega_{\nu m}-\omega_p)-i\gamma_{\nu m}]}\right.\right. \quad (\text{a}_1)$$

$$\left.+\frac{\mu_{mn}^{i}\mu_{n\nu}^{k}\mu_{\nu m}^{j}}{[(\omega_{nm}-\omega_p-\omega_q)-i\gamma_{nm}][(\omega_{\nu m}-\omega_q)-i\gamma_{\nu m}]}\right] \quad (\text{a}_2)$$

$$-(\rho_{\nu\nu}^{(0)}-\rho_{nn}^{(0)})\left[\frac{\mu_{mn}^{i}\mu_{\nu m}^{j}\mu_{n\nu}^{k}}{[(\omega_{nm}-\omega_p-\omega_q)-i\gamma_{nm}][(\omega_{n\nu}-\omega_p)-i\gamma_{n\nu}]}\right. \quad (\text{b}_1)$$

$$\left.\left.+\frac{\mu_{mn}^{i}\mu_{\nu m}^{k}\mu_{n\nu}^{j}}{[(\omega_{nm}-\omega_p-\omega_q)-i\gamma_{nm}][(\omega_{n\nu}-\omega_q)-i\gamma_{n\nu}]}\right]\right\} \quad (\text{b}_2).$$

(3.6.14)

This expression displays intrinsic permutation symmetry and gives the nonlinear susceptibility in a reasonably compact fashion. It is clear from its form that certain contributions to the susceptibility vanish when two of the levels associated with the contribution contain equal populations. We shall examine the nature of this cancellation in greater detail below (see Eq. (3.6.17)). Note that the population differences that appear in this expression are always associated with the two levels separated by a one-photon resonance, as we can see by inspection of the detuning factors that appear in the denominator.

The expression for the second-order nonlinear susceptibility can be rewritten in several different forms, all of which are equivalent, but which provide different insights into the resonant nature of the nonlinear coupling. Since the indices m, n, and v are summed over, they constitute dummy indices. We can therefore replace the indices v, n, and m in the last two terms of Eq. (3.6.14) by m, v, and n, respectively, so that the population difference term is the same as that of the first two terms. We thereby recast the second-order susceptibility into the form

$$\chi_{ijk}^{(2)}(\omega_p+\omega_q,\omega_q,\omega_p)=\frac{N}{2\hbar^2}\sum_{mnv}(\rho_{mm}^{(0)}-\rho_{vv}^{(0)})$$
$$\times\left\{\frac{\mu_{mn}^i\mu_{nv}^j\mu_{vm}^k}{[(\omega_{nm}-\omega_p-\omega_q)-i\gamma_{nm}][(\omega_{vm}-\omega_p)-i\gamma_{vm}]}\quad (\mathrm{a}_1)\right.$$
$$+\frac{\mu_{mn}^i\mu_{nv}^k\mu_{vm}^j}{[(\omega_{nm}-\omega_p-\omega_q)-i\gamma_{nm}][(\omega_{vm}-\omega_q)-i\gamma_{vm}]}\quad (\mathrm{a}_2) \qquad (3.6.15)$$
$$-\frac{\mu_{nv}^i\mu_{mn}^j\mu_{vm}^k}{[(\omega_{vn}-\omega_p-\omega_q)-i\gamma_{vn}][(\omega_{vm}-\omega_p)-i\gamma_{vm}]}\quad (\mathrm{b}_1)$$
$$\left.-\frac{\mu_{nv}^i\mu_{mn}^k\mu_{vm}^j}{[(\omega_{vn}-\omega_p-\omega_q)-i\gamma_{vn}][(\omega_{vm}-\omega_q)-i\gamma_{vm}]}\right\}\quad (\mathrm{b}_2).$$

We can make this result more transparent by making another change in dummy indices: we replace indices m, v, and n by l, m, and n, respectively. In addition, we replace ω_{lm}, ω_{ln}, and ω_{mn} by $-\omega_{ml}$, $-\omega_{nl}$, and $-\omega_{nm}$, respectively, whenever one of them appears. Also, we reorder the product of matrix elements in the numerator so that the subscripts n, m, and l are "chained" in the sense shown, and thereby obtain the result

$$\chi_{ijk}^{(2)}(\omega_p+\omega_q,\omega_q,\omega_p)=\frac{N}{2\hbar^2}\sum_{lmn}(\rho_{ll}^{(0)}-\rho_{mm}^{(0)})$$
$$\times\left\{\frac{\mu_{ln}^i\mu_{nm}^j\mu_{ml}^k}{[(\omega_{nl}-\omega_p-\omega_q)-i\gamma_{nl}][(\omega_{ml}-\omega_p)-i\gamma_{ml}]}\quad (\mathrm{a}_1)\right.$$
$$+\frac{\mu_{ln}^i\mu_{nm}^k\mu_{ml}^j}{[(\omega_{nl}-\omega_p-\omega_q)-i\gamma_{nl}][(\omega_{ml}-\omega_q)-i\gamma_{ml}]}\quad (\mathrm{a}_2) \qquad (3.6.16)$$
$$+\frac{\mu_{ln}^j\mu_{nm}^i\mu_{ml}^k}{[(\omega_{nm}+\omega_p+\omega_q)+i\gamma_{nm}][(\omega_{ml}-\omega_p)-i\gamma_{ml}]}\quad (\mathrm{b}_1)$$
$$\left.+\frac{\mu_{ln}^k\mu_{nm}^i\mu_{ml}^j}{[(\omega_{nm}+\omega_p+\omega_q)+i\gamma_{nm}][(\omega_{ml}-\omega_q)-i\gamma_{ml}]}\right\}\quad (\mathrm{b}_2).$$

One way of interpreting this result is to consider where levels l, m, and n would have to be located in order for each of the terms to become resonant. The positions of these energies are illustrated in Fig. 3.6.1. For definiteness, we have drawn the figure with ω_p and ω_q positive. In each case the magnitude of the contribution to the nonlinear susceptibility is proportional to the population difference between levels l and m.

In order to illustrate how to apply Eq. (3.6.16) and in order to examine the nature of the cancellation that can occur when more than one of the atomic levels contains population, we consider the simple three-level atomic system illustrated in Fig. 3.6.2. We assume that only levels a, b, and c interact appreciably with the optical fields, and that the applied field at frequency ω_1 is nearly resonant with the $a \rightarrow b$ transition, the applied field at frequency ω_2 is nearly resonant with the $b \rightarrow c$ transition, and the generated field frequency $\omega_3 = \omega_1 + \omega_2$ is nearly resonant with the $a \rightarrow c$ transition. If we now perform the summation over the dummy indices l, m, and n in Eq. (3.6.16) and retain only those terms in which both factors in the denominator are resonant, we find that the nonlinear susceptibility is given by

$$\chi_{ijk}^{(2)}(\omega_3, \omega_2, \omega_1) = \frac{N}{2\hbar^2}\left\{(\rho_{aa}^{(0)} - \rho_{bb}^{(0)})\left[\frac{\mu_{ac}^i \mu_{cb}^j \mu_{ba}^k}{[(\omega_{ca} - \omega_3) - i\gamma_{ca}][(\omega_{ba} - \omega_1) - i\gamma_{ba}]}\right]\right.$$
$$\left. + (\rho_{cc}^{(0)} - \rho_{bb}^{(0)})\left[\frac{\mu_{ac}^i \mu_{cb}^j \mu_{ba}^k}{[(\omega_{ca} - \omega_3) - i\gamma_{ca}][(\omega_{cb} - \omega_2) - i\gamma_{cb}]}\right]\right\}. \tag{3.6.17}$$

Here the first term comes from the first term in Eq. (3.6.16), and the second term comes from the last (fourth) term in Eq. (3.6.16). Note that the first term vanishes if $\rho_{aa}^{(0)} = \rho_{bb}^{(0)}$ and that the second term vanishes if $\rho_{bb}^{(0)} = \rho_{cc}^{(0)}$. If all three populations are equal, the resonant contribution vanishes identically.

For some purposes it is useful to express the general result (3.6.16) for the

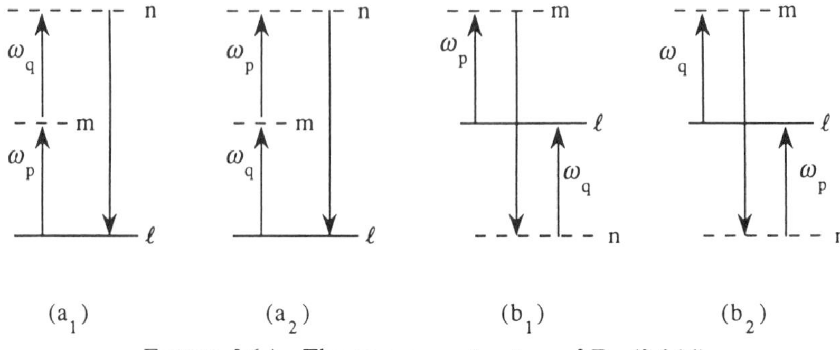

FIGURE 3.6.1 The resonance structure of Eq. (3.6.16).

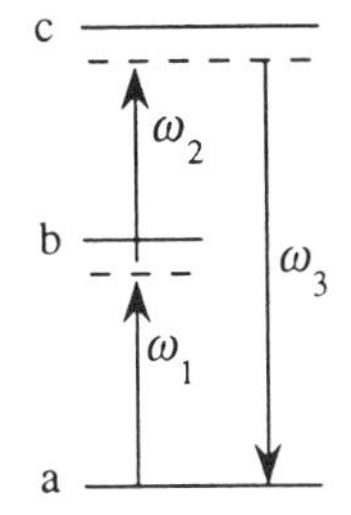

FIGURE 3.6.2 Three-level atomic system.

second-order susceptibility in terms of a summation over populations rather than a summation over population differences. In order to cast the susceptibility in such a form, we change the dummy indices l, m, and n to n, l, and m in the summation containing $\rho_{mm}^{(0)}$, but leave them unchanged in the summation containing $\rho_{ll}^{(0)}$. We thereby obtain the result

$$\chi_{ijk}^{(2)}(\omega_p+\omega_q,\omega_q,\omega_p)=\frac{N}{2\hbar^2}\sum_{lmn}\rho_{ll}^{(0)}$$

$$\times\left\{\frac{\mu_{ln}^i\mu_{nm}^j\mu_{ml}^k}{[(\omega_{nl}-\omega_p-\omega_q)-i\gamma_{nl}][(\omega_{ml}-\omega_p)-i\gamma_{ml}]}\right. \quad (\mathrm{a}_1)$$

$$+\frac{\mu_{ln}^i\mu_{nm}^k\mu_{ml}^j}{[(\omega_{nl}-\omega_p-\omega_q)-i\gamma_{nl}][(\omega_{ml}-\omega_q)-i\gamma_{ml}]} \quad (\mathrm{a}_2)$$

$$+\frac{\mu_{ln}^k\mu_{nm}^i\mu_{ml}^j}{[(\omega_{mn}-\omega_p-\omega_q)-i\gamma_{mn}][(\omega_{nl}+\omega_p)+i\gamma_{nl}]} \quad (\mathrm{a}_1')$$

$$+\frac{\mu_{ln}^j\mu_{nm}^i\mu_{ml}^k}{[(\omega_{mn}-\omega_p-\omega_q)-i\gamma_{mn}][(\omega_{nl}+\omega_q)+i\gamma_{nl}]} \quad (\mathrm{a}_2')$$

$$+\frac{\mu_{ln}^j\mu_{nm}^i\mu_{ml}^k}{[(\omega_{nm}+\omega_p+\omega_q)+i\gamma_{nm}][(\omega_{ml}-\omega_p)-i\gamma_{ml}]} \quad (\mathrm{b}_1)$$

$$+\frac{\mu_{ln}^k\mu_{nm}^i\mu_{ml}^j}{[(\omega_{nm}+\omega_p+\omega_q)+i\gamma_{nm}][(\omega_{ml}-\omega_q)-i\gamma_{ml}]} \quad (\mathrm{b}_2)$$

$$+\frac{\mu_{ln}^k\mu_{nm}^j\mu_{ml}^i}{[(\omega_{ml}+\omega_p+\omega_q)+i\gamma_{ml}][(\omega_{nl}+\omega_p)+i\gamma_{nl}]} \quad (\mathrm{b}_1')$$

$$\left.+\frac{\mu_{ln}^j\mu_{nm}^k\mu_{ml}^i}{[(\omega_{ml}+\omega_p+\omega_q)+i\gamma_{ml}][\omega_{nl}+\omega_q)+i\gamma_{nl}]}\right\} \quad (\mathrm{b}_2').$$

(3.6.18)

As above, we can interpret this result by considering the conditions under which each term of the equation can become resonant. Figure 3.6.3 shows where the energy levels l, m, and n would have to be located in order for each

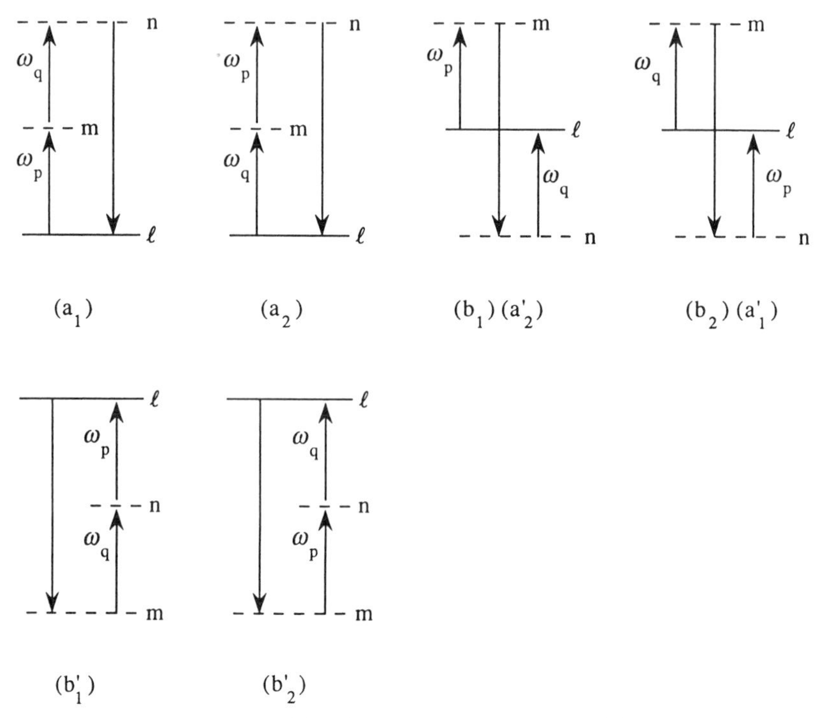

FIGURE 3.6.3 The resonances of Eq. (3.6.18).

term to become resonant, under the assumption that ω_p and ω_q are both positive. Note that the unprimed diagrams are the same as those of Fig. 3.6.1 (which represents Eq. (3.6.16), but that diagrams (b'_1) and (b'_2) represent new resonances not present in Fig. 3.6.1.

Another way of making sense of the general eight-term expression for $\chi^{(2)}$ (Eq. (3.6.18)) is to keep track of how the density matrix is modified in each order of perturbation theory. Through examination of Eqs. (3.6.1) through (3.6.7), we find that the terms of type (a), (a′), (b), (b′) occur as the result of the following perturbation expansion:

$$\begin{aligned} &(\mathrm{a}): \quad \rho_{mm}^{(0)} \rightarrow \rho_{\nu m}^{(1)} \rightarrow \rho_{nm}^{(2)}, \qquad (\mathrm{a}'): \quad \rho_{\nu\nu}^{(0)} \rightarrow \rho_{\nu m}^{(1)} \rightarrow \rho_{nm}^{(2)}, \\ &(\mathrm{b}): \quad \rho_{\nu\nu}^{(0)} \rightarrow \rho_{n\nu}^{(1)} \rightarrow \rho_{nm}^{(2)}, \qquad (\mathrm{b}'): \quad \rho_{nn}^{(0)} \rightarrow \rho_{n\nu}^{(1)} \rightarrow \rho_{nm}^{(2)}. \end{aligned} \tag{3.6.19}$$

However, in writing Eq. (3.6.18) in the displayed form, we have changed the dummy indices appearing in it. In terms of these new indices, the perturbation expansion is

$$\begin{aligned} &(\mathrm{a}): \quad \rho_{ll}^{(0)} \rightarrow \rho_{ml}^{(1)} \rightarrow \rho_{nl}^{(2)}, \qquad (\mathrm{a}'): \quad \rho_{ll}^{(0)} \rightarrow \rho_{ln}^{(1)} \rightarrow \rho_{mn}^{(2)}, \\ &(\mathrm{b}): \quad \rho_{ll}^{(0)} \rightarrow \rho_{ml}^{(1)} \rightarrow \rho_{mn}^{(2)}, \qquad (\mathrm{b}'): \quad \rho_{ll}^{(0)} \rightarrow \rho_{ln}^{(1)} \rightarrow \rho_{lm}^{(2)}. \end{aligned} \tag{3.6.20}$$

Note that the various terms differ in whether it is the left or right index that is changed by each elementary interaction and by the order in which such a modification occurs.

A convenient way of keeping track of the order in which the elementary interactions occur is by means of double-sided Feynman diagrams. These diagrams represent the way in which the *density operator* is modified by the interaction of the atom with the laser field. We represent the density operator as

$$\hat{\rho} = \overline{|\psi\rangle\langle\psi|}, \tag{3.6.21}$$

where $|\psi\rangle$ represents the ket vector for some state of the system, $\langle\psi|$ (the bra vector) represents the Hermitian adjoint of $\langle\psi|$, and the overbar represents an ensemble average. The elements of the density matrix are related to the density operator $\hat{\rho}$ through the equation

$$\rho_{nm} = \langle n|\hat{\rho}|m\rangle. \tag{3.6.22}$$

Figure 3.6.4 gives a pictorial description of the modification of the density matrix as indicated by the expressions (3.6.20). The left-hand side of each diagram indicates the time evolution of $|\psi\rangle$, and the right-hand side indicates the time evolution $\langle\psi|$, with time increasing vertically upward. Each

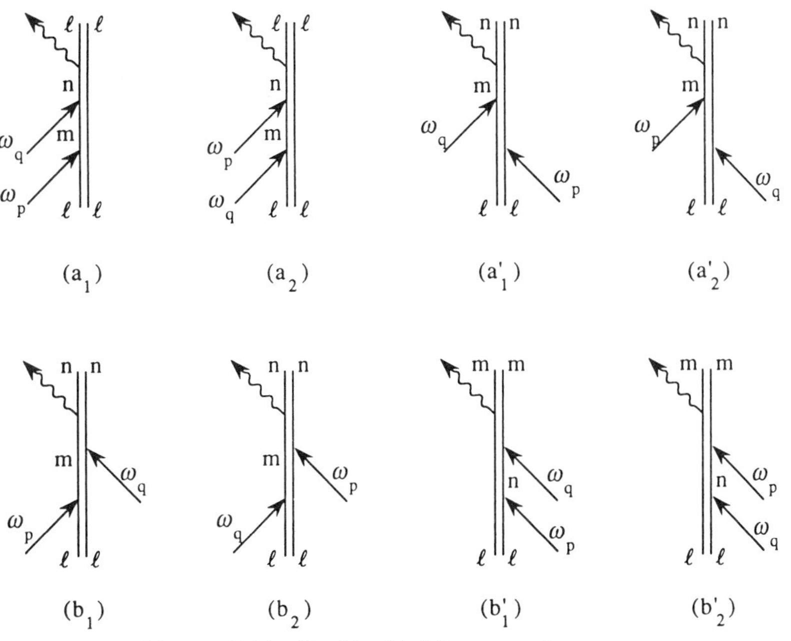

FIGURE 3.6.4 Double-sided Feynman diagrams.

interaction with the applied field is indicated by a solid arrow labeled by the field frequency. The trace operation, which corresponds to calculating the output field, is indicated by the wavy arrow.* It should be noted that there are several different conventions concerning the rules for drawing double-sided Feynman diagrams (Yee and Gustafson, 1978; Prior, 1984; Boyd and Mukamel, 1984).

$\chi^{(2)}$ *in the Limit of Nonresonant Excitation*

When all of the frequencies ω_p, ω_q, and $\omega_p + \omega_q$ differ significantly from any resonance frequency of the atomic system, the imaginary contributions to the denominators in Eq. (3.6.18) can be ignored. In this case, the expression for $\chi^{(2)}$ can be simplified. In particular, terms (a_2') and (b_1) can be combined into a single term, and similarly for terms (a_1') and (b_2). We note that the numerators of terms (a_2') and (b_1) are identical, and that their denominators can be combined as follows:

$$
\begin{aligned}
&\frac{1}{(\omega_{mn} - \omega_p - \omega_q)(\omega_{nl} + \omega_q)} + \frac{1}{(-\omega_{mn} + \omega_p + \omega_q)(\omega_{ml} - \omega_p)} \\
&\quad = \frac{1}{(\omega_{mn} - \omega_p - \omega_q)} \left[\frac{1}{\omega_{nl} + \omega_q} - \frac{1}{\omega_{ml} - \omega_p} \right] \\
&\quad = \frac{1}{\omega_{mn} - \omega_p - \omega_q} \left[\frac{\omega_{ml} - \omega_p - \omega_{nl} - \omega_q}{(\omega_{nl} + \omega_q)(\omega_{ml} - \omega_p)} \right] \\
&\quad = \frac{1}{\omega_{mn} - \omega_p - \omega_q} \left[\frac{\omega_{mn} - \omega_p - \omega_q}{(\omega_{nl} + \omega_q)(\omega_{ml} - \omega_p)} \right] \\
&\quad = \frac{1}{(\omega_{nl} + \omega_q)(\omega_{ml} - \omega_p)}.
\end{aligned} \tag{3.6.23}
$$

The same procedure can be performed on terms (a_1') and (b_2); the only difference between this case and the one treated in Eq. (3.6.23) is that ω_p and ω_q have switched roles. The frequency dependence is thus

$$
\frac{1}{(\omega_{nl} + \omega_p)(\omega_{ml} - \omega_q)}. \tag{3.6.24}
$$

* In drawing Fig. 3.6.4, we have implicitly assumed that all of the applied field frequencies are positive, which corresponds to the absorption of an incident photon. The interaction with a negative field frequency which corresponds to the emission of a photon, is sometimes indicated by a solid arrow pointing diagonally upward and away from (rather than towards) the central double line.

The expression for $\chi^{(2)}$ in the off-resonance case thus becomes

$$\chi^{(2)}_{ijk}(\omega_p + \omega_q, \omega_q, \omega_p) = \frac{N}{2\hbar^2} \sum_{lmn} \rho^{(0)}_{ll}$$

$$\times \left\{ \frac{\mu^i_{ln}\mu^j_{nm}\mu^k_{ml}}{(\omega_{nl} - \omega_p - \omega_q)(\omega_{ml} - \omega_p)} \quad (\mathrm{a}_1) \right.$$

$$+ \frac{\mu^i_{ln}\mu^k_{nm}\mu^j_{ml}}{(\omega_{nl} - \omega_p - \omega_q)(\omega_{ml} - \omega_q)} \quad (\mathrm{a}_2)$$

$$+ \frac{\mu^j_{ln}\mu^i_{nm}\mu^k_{ml}}{(\omega_{nl} + \omega_q)(\omega_{ml} - \omega_p)} \quad (\mathrm{b}_1), (\mathrm{a}'_2)$$

$$+ \frac{\mu^k_{ln}\mu^i_{nm}\mu^j_{ml}}{(\omega_{nl} + \omega_p)(\omega_{ml} - \omega_q)} \quad (\mathrm{b}_2), (\mathrm{a}'_1)$$

$$+ \frac{\mu^k_{ln}\mu^j_{nm}\mu^i_{ml}}{(\omega_{ml} + \omega_p + \omega_q)(\omega_{nl} + \omega_p)} \quad (\mathrm{b}'_1)$$

$$\left. + \frac{\mu^j_{ln}\mu^k_{nm}\mu^i_{ml}}{(\omega_{ml} + \omega_p + \omega_q)(\omega_{nl} + \omega_q)} \right\}. \quad (\mathrm{b}'_2) \qquad (3.6.25)$$

Note that only six terms appear in this expression for the off-resonance susceptibility, whereas eight terms appear in the general expression of Eq. (3.6.18). One can verify by explicit calculation that Eq. (3.6.25) satisfies the condition of full permutation symmetry (see also Eq. (1.5.7)). In addition, one can see by inspection that Eq. (3.6.25) is identical to the result obtained above (Eq. (3.2.27)) based on perturbation theory of the atomic wave function.

There are several diagramatic methods that can be used to interpret this expression. One of the simplest is to plot the photon energies on an atomic energy level diagram. This method displays the conditions under which each contribution can become resonant. The results of such an analysis give exactly the same diagrams displayed above in Fig. 3.6.3. Equation (3.6.25) can also be understood in terms of a diagrammatic approach introduced by Ward (1965).

3.7. Density Matrix Calculation of the Third-Order Susceptibility

The third-order correction to the density matrix is given by the perturbation expansion of Eq. (3.4.16) as

$$\rho^{(3)}_{nm} = e^{-(i\omega_{nm} + \gamma_{nm})t} \int_{-\infty}^{t} \frac{-i}{\hbar} [\hat{V}, \hat{\rho}^{(2)}]_{nm} e^{(i\omega_{nm} + \gamma_{nm})t'} \, dt', \qquad (3.7.1)$$

where the commutator is given by

$$[\hat{V}, \hat{\rho}^{(2)}]_{nm} = -\sum_{\nu} (\boldsymbol{\mu}_{n\nu}\rho_{\nu m}^{(2)} - \rho_{n\nu}^{(2)}\boldsymbol{\mu}_{\nu m}) \cdot \tilde{\mathbf{E}}(t). \tag{3.7.2}$$

Expressions for $\rho_{\nu m}^{(2)}$ and $\rho_{n\nu}^{(2)}$ are available from Eq. (3.6.7). Since these expressions are very complicated, we use the abbreviated notation introduced there:

$$\rho_{\nu m}^{(2)} = \sum_{l} \sum_{pq} K_{\nu ml} e^{-i(\omega_p + \omega_q)t}, \tag{3.7.3}$$

where

$$\begin{aligned} K_{\nu ml} = {} & \frac{\rho_{mm}^{(0)} - \rho_{ll}^{(0)}}{\hbar^2} \frac{[\boldsymbol{\mu}_{\nu l} \cdot \mathbf{E}(\omega_q)][\boldsymbol{\mu}_{lm} \cdot \mathbf{E}(\omega_p)]}{[(\omega_{\nu m} - \omega_p - \omega_q) - i\gamma_{\nu m}][(\omega_{lm} - \omega_p) - i\gamma_{lm}]} \\ & - \frac{\rho_{ll}^{(0)} - \rho_{\nu\nu}^{(0)}}{\hbar^2} \frac{[\boldsymbol{\mu}_{\nu l} \cdot \mathbf{E}(\omega_p)][\boldsymbol{\mu}_{lm} \cdot \mathbf{E}(\omega_q)]}{[(\omega_{\nu m} - \omega_p - \omega_q) - i\gamma_{\nu m}][(\omega_{\nu l} - \omega_p) - i\gamma_{\nu l}]}, \end{aligned} \tag{3.7.4}$$

and

$$\rho_{n\nu}^{(2)} = \sum_{l} \sum_{pq} K_{n\nu l} e^{-i(\omega_p + \omega_q)t}, \tag{3.7.5}$$

where

$$\begin{aligned} K_{n\nu l} = {} & \frac{\rho_{\nu\nu}^{(0)} - \rho_{ll}^{(0)}}{\hbar^2} \frac{[\boldsymbol{\mu}_{nl} \cdot \mathbf{E}(\omega_q)][\boldsymbol{\mu}_{l\nu} \cdot \mathbf{E}(\omega_p)]}{[(\omega_{n\nu} - \omega_p - \omega_q) - i\gamma_{n\nu}][(\omega_{l\nu} - \omega_p) - i\gamma_{l\nu}]} \\ & - \frac{\rho_{ll}^{(0)} - \rho_{nn}^{(0)}}{\hbar^2} \frac{[\boldsymbol{\mu}_{nl} \cdot \mathbf{E}(\omega_p)][\boldsymbol{\mu}_{l\nu} \cdot \mathbf{E}(\omega_q)]}{[(\omega_{n\nu} - \omega_p - \omega_q) - i\gamma_{n\nu}][(\omega_{nl} - \omega_p) - i\gamma_{nl}]}. \end{aligned} \tag{3.7.6}$$

We also designate the electric field by

$$\tilde{\mathbf{E}}(t) = \sum_{r} \mathbf{E}(\omega_r) e^{-i\omega_r t}. \tag{3.7.7}$$

The commutator thus becomes

$$\begin{aligned} [\hat{V}, \hat{\rho}^{(2)}]_{nm} = {} & -\sum_{\nu l} \sum_{pqr} [\boldsymbol{\mu}_{n\nu} \cdot \mathbf{E}(\omega_r)] K_{\nu ml} e^{-i(\omega_p + \omega_q + \omega_r)t} \\ & + \sum_{\nu l} \sum_{pqr} [\boldsymbol{\mu}_{\nu m} \cdot \mathbf{E}(\omega_r)] K_{n\nu l} e^{-i(\omega_p + \omega_q + \omega_r)t}. \end{aligned} \tag{3.7.8}$$

The integration of (3.7.1) with the commutator given by (3.7.8) can now be performed. We obtain

$$\begin{aligned} \rho_{nm}^{(3)} = \frac{1}{\hbar} \sum_{\nu l} \sum_{pqr} \Bigg\{ & \frac{[\boldsymbol{\mu}_{n\nu} \cdot \mathbf{E}(\omega_r)] K_{\nu ml}}{(\omega_{nm} - \omega_p - \omega_q - \omega_r) - i\gamma_{nm}} \\ & - \frac{[\boldsymbol{\mu}_{\nu m} \cdot \mathbf{E}(\omega_r)] K_{n\nu l}}{(\omega_{nm} - \omega_p - \omega_q - \omega_r) - i\gamma_{nm}} \Bigg\} e^{-i(\omega_p + \omega_q + \omega_r)t} \end{aligned} \tag{3.7.9}$$

The nonlinear polarization oscillating at frequency $\omega_p + \omega_q + \omega_r$ is given by

$$\mathbf{P}(\omega_p + \omega_q + \omega_r) = N\langle \boldsymbol{\mu}(\omega_p + \omega_q + \omega_r)\rangle, \tag{3.7.10}$$

where

$$\langle \tilde{\boldsymbol{\mu}} \rangle = \sum_{nm} \rho_{nm} \boldsymbol{\mu}_{mn} \equiv \sum_{s} \langle \boldsymbol{\mu}(\omega_s)\rangle e^{-i\omega_s t}. \tag{3.7.11}$$

We express the nonlinear polarization in terms of the third-order susceptibility defined by (see also Eq. (1.3.21))

$$\begin{aligned} P_k(\omega_p + \omega_q + \omega_r) = \sum_{hij} \sum_{(pqr)} & \chi^{(3)}_{kjih}(\omega_p + \omega_q + \omega_r, \omega_r, \omega_q, \omega_p) \\ & \times E_j(\omega_r) E_i(\omega_q) E_h(\omega_p). \end{aligned} \tag{3.7.12}$$

By combining Eqs. (3.7.9) through (3.7.12), we find that the third-order susceptibility is given by

$$\begin{aligned}
&\chi^{(3)}_{kjih}(\omega_p+\omega_q+\omega_r, \omega_r, \omega_q, \omega_p) = \frac{N}{\hbar^3} \mathscr{P}_I \\
&\times \sum_{nmvl} \Bigg\{ (\rho^{(0)}_{mm} - \rho^{(0)}_{ll}) \\
&\times \frac{\mu^k_{mn}\mu^j_{nv}\mu^i_{vl}\mu^h_{lm}}{[(\omega_{nm}-\omega_p-\omega_q-\omega_r)-i\gamma_{nm}][(\omega_{vm}-\omega_p-\omega_q)-i\gamma_{vm}][(\omega_{lm}-\omega_p)-i\gamma_{lm}]} && \text{(a)} \\
&-(\rho^{(0)}_{ll} - \rho^{(0)}_{vv}) \\
&\times \frac{\mu^k_{mn}\mu^j_{nv}\mu^i_{lm}\mu^h_{vl}}{[(\omega_{nm}-\omega_p-\omega_q-\omega_r)-i\gamma_{nm}][(\omega_{vm}-\omega_p-\omega_q)-i\gamma_{vm}][(\omega_{vl}-\omega_p)-i\gamma_{vl}]} && \text{(b)} \\
&-(\rho^{(0)}_{vv} - \rho^{(0)}_{ll}) \\
&\times \frac{\mu^k_{mn}\mu^j_{vm}\mu^i_{nl}\mu^h_{lv}}{[(\omega_{nm}-\omega_p-\omega_q-\omega_r)-i\gamma_{nm}][(\omega_{nv}-\omega_p-\omega_q)-i\gamma_{nv}][(\omega_{lv}-\omega_p)-i\gamma_{lv}]} && \text{(c)} \\
&+(\rho^{(0)}_{ll} - \rho^{(0)}_{nn}) \\
&\times \frac{\mu^k_{mn}\mu^j_{vm}\mu^i_{lv}\mu^h_{nl}}{[(\omega_{nm}-\omega_p-\omega_q-\omega_r)-i\gamma_{nm}][(\omega_{nv}-\omega_p-\omega_q)-i\gamma_{nv}][(\omega_{nl}-\omega_p)-i\gamma_{nl}]} \Bigg\}. && \text{(d)}
\end{aligned} \tag{3.7.13}$$

Here we have again made use of the intrinsic permutation operator $\mathscr{P}_I$, whose meaning is that everything to the right of it is to be averaged over all possible permutations of the input frequencies ω_p, ω_q, and ω_r, with the cartesian indices h, i, j permuted simultaneously. Next, we rewrite this equation as eight separate terms by changing the dummy indices so that l is always the index of $\rho_{ii}^{(0)}$. We also require that only positive resonance frequencies appear if the energies are ordered so that $E_v > E_n > E_m > E_l$, and we arrange the matrix elements so that they appear in "natural" order, $l \to m \to n \to v$ (reading right to left). We obtain

$$\chi_{kjih}^{(3)}(\omega_p+\omega_q+\omega_r, \omega_r, \omega_q, \omega_p) = \frac{N}{\hbar^3}\mathscr{P}_1 \sum_{vnml} \rho_{ll}^{(0)}$$

$$\times \left\{ \frac{\mu_{lv}^{k}\mu_{vn}^{j}\mu_{nm}^{i}\mu_{ml}^{h}}{[(\omega_{vl}-\omega_p-\omega_q-\omega_r)-i\gamma_{vl}][(\omega_{nl}-\omega_p-\omega_q)-i\gamma_{nl}][(\omega_{ml}-\omega_p)-i\gamma_{ml}]} \right. \quad (\mathrm{a}_1)$$

$$+\frac{\mu_{lv}^{h}\mu_{vn}^{k}\mu_{nm}^{j}\mu_{ml}^{i}}{[(\omega_{nv}-\omega_p-\omega_q-\omega_r)-i\gamma_{nv}][(\omega_{mv}-\omega_p-\omega_q)-i\gamma_{mv}][(\omega_{vl}+\omega_p)+i\gamma_{vl}]} \quad (\mathrm{a}_2)$$

$$+\frac{\mu_{lv}^{i}\mu_{vn}^{k}\mu_{nm}^{j}\mu_{ml}^{h}}{[(\omega_{nv}-\omega_p-\omega_q-\omega_r)-i\gamma_{nv}][(\omega_{vm}+\omega_p+\omega_q)+i\gamma_{vm}][(\omega_{ml}-\omega_p)-i\gamma_{ml}]} \quad (\mathrm{b}_1)$$

$$+\frac{\mu_{lv}^{h}\mu_{vn}^{i}\mu_{nm}^{k}\mu_{ml}^{j}}{[(\omega_{mv}-\omega_p-\omega_q-\omega_r)-i\gamma_{mn}][(\omega_{nl}+\omega_p+\omega_q)+i\gamma_{nl}][(\omega_{vl}+\omega_p)+i\gamma_{vl}]} \quad (\mathrm{b}_2)$$

$$+\frac{\mu_{lv}^{j}\mu_{vn}^{k}\mu_{nm}^{i}\mu_{ml}^{h}}{[(\omega_{vn}+\omega_p+\omega_q+\omega_r)+i\gamma_{vn}][(\omega_{nl}-\omega_p-\omega_q)-i\gamma_{nl}][(\omega_{ml}-\omega_p)-i\gamma_{ml}]} \quad (\mathrm{c}_1)$$

$$+\frac{\mu_{lv}^{h}\mu_{vn}^{j}\mu_{nm}^{k}\mu_{ml}^{i}}{[(\omega_{nm}+\omega_p+\omega_q+\omega_r)+i\gamma_{nm}][(\omega_{mv}-\omega_p-\omega_q)-i\gamma_{mv}][(\omega_{vl}+\omega_p)+i\gamma_{vl}]} \quad (\mathrm{c}_2)$$

$$+\frac{\mu_{lv}^{i}\mu_{vn}^{j}\mu_{nm}^{k}\mu_{ml}^{h}}{[(\omega_{nm}+\omega_p+\omega_q+\omega_r)+i\gamma_{nm}][\omega_{vm}+\omega_p+\omega_q)+i\gamma_{vm}][(\omega_{ml}-\omega_p)-i\gamma_{ml}]} \quad (\mathrm{d}_1)$$

$$\left. +\frac{\mu_{lv}^{h}\mu_{vn}^{i}\mu_{nm}^{j}\mu_{ml}^{k}}{[(\omega_{ml}+\omega_p+\omega_q+\omega_r)+i\gamma_{ml}][(\omega_{nl}+\omega_p+\omega_q)+i\gamma_{nl}][(\omega_{vl}+\omega_p)+i\gamma_{vl}]} \right\}. \quad (\mathrm{d}_2)$$

$$(3.7.14)$$

For the general case in which ω_p, ω_q, and ω_r are distinct, six permutations of the field frequencies occur, and hence the expression for $\chi^{(3)}$ consists of 48 different terms once the permutation operator $\mathscr{P}_1$ is expanded. The resonance structure of this expression can be understood in terms of the energy level diagrams shown in Fig. 3.7.1. Furthermore, the nature of the perturbation expansion leading to Eq. (3.7.14) can be understood in terms of the double-sided Feynman diagrams shown in Fig. 3.7.2.

We saw in Section 3.2 that the general expression for the third-order susceptibility calculated using perturbation theory applied to the atomic wave function contained 24 terms. Equation (3.2.32) shows four of these terms

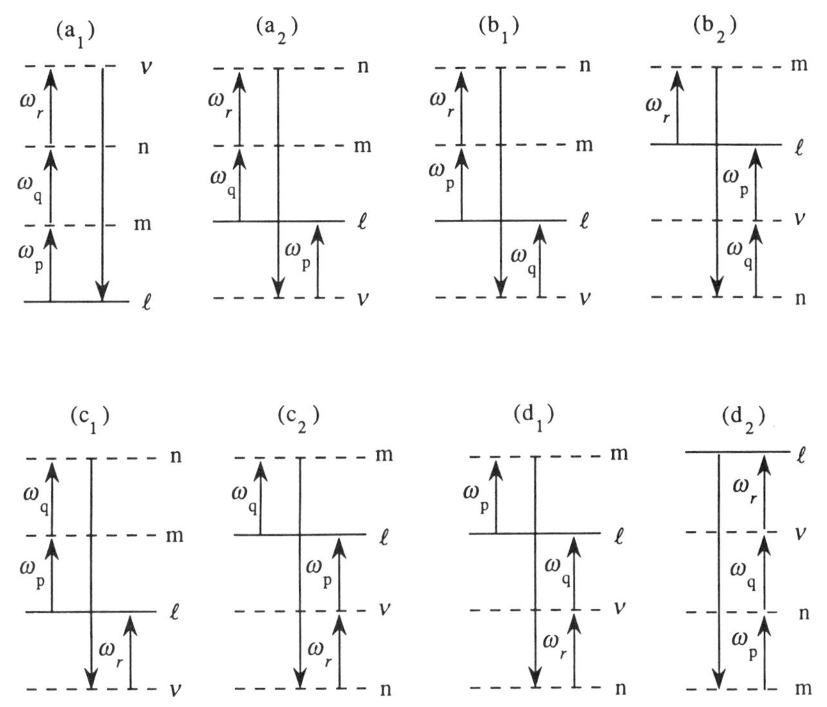

FIGURE 3.7.1 The resonance structure of the expression (3.7.14) for the third-order nonlinear susceptibility.

explicitly; the other terms are obtained from the six permutations of the frequencies of the applied field. It can be shown that Eq. (3.7.14) reduces to Eq. (3.2.32) in the limit of nonresonant excitation, where the imaginary contributions ($i\gamma_{\alpha\beta}$) appearing in Eq. (3.7.14) can be ignored. One can demonstrate this fact by means of a calculation similar to that used to derive Eq. (3.6.24), which applies to the case of the second-order susceptibility (see Problem 5 at the end of this chapter).

In fact, even in the general case in which the imaginary contributions $i\gamma_{\alpha\beta}$ appearing in Eq. (3.7.14) are retained, it is possible to rewrite the 48-term expression (3.7.14) in the form of the 24-term expression (3.2.32), by allowing the coefficient of each of the 24 terms to be weakly frequency-dependent. These frequency-dependent coefficients usually display resonances at frequencies other than those that appear in Fig. 3.7.1, and these new resonances occur only if the line broading mechanism is collisional (rather than radiative). The nature of these *collision-induced resonances* has been discussed by Bloembergen *et al.* (1978), Prior (1984), and Rothberg (1987).

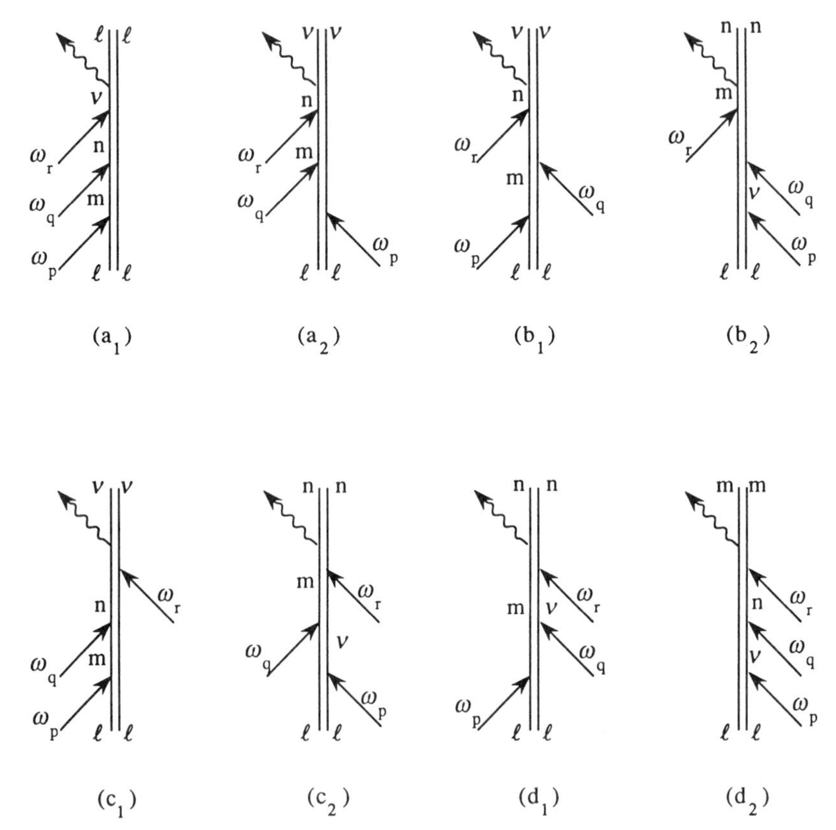

FIGURE 3.7.2 Double-sided Feynman diagrams associated with the various terms in Eq. (3.7.14).

3.8. Local-Field Corrections to the Nonlinear Optical Susceptibility

The treatment of the nonlinear optical susceptibility presented thus far has made the implicit assumption that the electric field acting on each atom or molecule is the macroscopic electric field that appears in Maxwell's equations. In general, one has to distinguish between the macroscopic electric field and the effective electric field that each atom experiences, which is also known as the Lorentz local field. The distinction between these two fields is important except for the case of a medium that is so dilute that its linear dielectric constant is nearly equal to unity.

Local-Field Effects in Linear Optics

Let us first review the theory of local-field effects in linear optics. The electric field $\tilde{\mathbf{E}}$ that appears in Maxwell's equations in the form of Eqs. (2.1.1) through (2.1.8) is known as the macroscopic or Maxwell field. This field is obtained by performing a spatial average of the actual (or microscopic) electric field over a region of space whose linear dimensions are of the order of at least several atomic diameters. It is useful to perform such an average to smooth out the wild variations in electric field that occur in the immediate vicinity of the atomic nuclei and electrons. The macroscopic electric field thus has contributions due to sources external to the material system and due to the dipole moments of all of the dipoles that constitute the system.

Let us now see how to calculate the dipole moment induced in a representative molecule contained within the material system. We assume for simplicity that the medium is lossless, so that we can conveniently represent the fields as time-varying quantities rather than having to introduce complex field amplitudes. We let $\tilde{\mathbf{E}}$ represent the macroscopic field and $\tilde{\mathbf{P}}$ the polarization within the bulk of the material. Furthermore, we represent the dipole moment induced in a typical molecule as

$$\tilde{\mathbf{p}} = \alpha \tilde{\mathbf{E}}_{\text{loc}}, \tag{3.8.1}$$

where α is the usual linear polarizability and where $\tilde{\mathbf{E}}_{\text{loc}}$ is the local field, that is, the effective electric field that acts on the molecule. The local field is the field due to all external sources and to all molecules within the sample *except* the one under consideration.

We calculate this field through use of a procedure described by Lorentz (1952). We imagine drawing a small sphere centered on the molecule under consideration, as shown in Fig. 3.8.1. This sphere is assumed to be sufficiently large that it contains many atoms. The electric field produced at the center of the sphere by molecules contained within the sphere (not including the molecule at the center) will tend to cancel, and for the case of a liquid, gas, or cubic crystal, this cancellation can be shown to be exact. We can then imagine

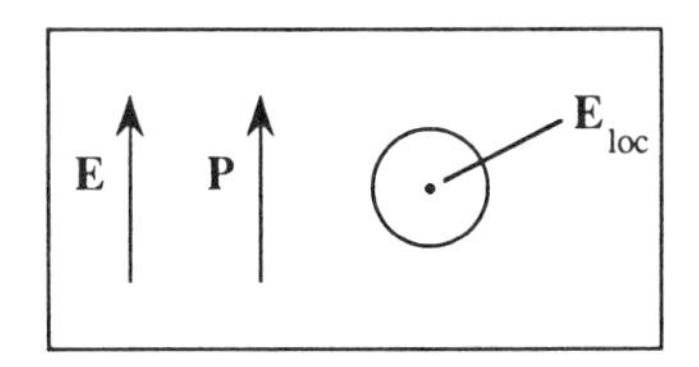

FIGURE 3.8.1 Calculation of the Lorentz local field.

removing these molecules from the sphere, leaving only the molecule under consideration, which is then located at the center of an evacuated sphere within an otherwise uniformly polarized medium. It is then a simple problem in electrostatics to calculate the value of the field at the center of the sphere. The field, which we identify as the Lorentz local field, is given by (see also Born and Wolf (1975), Section 2.3, or Jackson (1975), Section 4.5)

$$\tilde{\mathbf{E}}_{\text{loc}} = \tilde{\mathbf{E}} + \tfrac{4}{3}\pi\tilde{\mathbf{P}}. \tag{3.8.2}$$

By definition, the polarization of the material is given by

$$\tilde{\mathbf{P}} = N\tilde{\mathbf{p}}, \tag{3.8.3}$$

where N is the number density of atoms and $\tilde{\mathbf{p}}$ is the dipole moment per atom, which under the present circumstances is given by Eq. (3.8.1). By combining Eqs. (3.8.1) through (3.8.3), we find that the polarization and macroscopic field are related by

$$\tilde{\mathbf{P}} = N\alpha(\tilde{\mathbf{E}} + \tfrac{4}{3}\pi\tilde{\mathbf{P}}). \tag{3.8.4}$$

It is useful to express this result in terms of the linear susceptibility $\chi^{(1)}$, defined by

$$\tilde{\mathbf{P}} = \chi^{(1)}\tilde{\mathbf{E}}. \tag{3.8.5}$$

If we substitute this expression for $\tilde{\mathbf{P}}$ into Eq. (3.8.4) and solve the resulting equation for $\chi^{(1)}$, we find that

$$\chi^{(1)} = \frac{N\alpha}{1 - \frac{4}{3}\pi N\alpha}. \tag{3.8.6}$$

For the usual case in which the polarizability α is positive, we see that the susceptibility is larger than the value $N\alpha$ predicted by theories that ignore local-field corrections. We also see that the susceptibility increases with N more rapidly than linearly.

Alternatively, we can express the result given by Eq. (3.8.6) in terms of the linear dielectric constant

$$\epsilon^{(1)} = 1 + 4\pi\chi^{(1)}. \tag{3.8.7}$$

If the left-hand side of Eq. (3.8.6) is replaced by $\chi^{(1)} = (\epsilon^{(1)} - 1)/4\pi$ and the resulting equation is rearranged so that its right-hand side is linear in α, we find that the dielectric constant is given by the expression

$$\frac{\epsilon^{(1)} - 1}{\epsilon^{(1)} + 2} = \tfrac{4}{3}\pi N\alpha. \tag{3.8.8a}$$

This equation (often with $\epsilon^{(1)}$ replaced by n^2) is known as the Lorentz–Lorenz law. Note that, through rearrangement, Eq. (3.8.8a) can be written as

$$\frac{\epsilon^{(1)}+2}{3}=\frac{1}{1-\frac{4}{3}\pi N\alpha}. \tag{3.8.8b}$$

Equation (3.8.6) can thus be expressed as

$$\chi^{(1)}=\frac{\epsilon^{(1)}+2}{3}N\alpha. \tag{3.8.8c}$$

This result shows that $\chi^{(1)}$ is larger than $N\alpha$ by the factor $(\epsilon^{(1)}+2)/3$. The factor $(\epsilon^{(1)}+2)/3$ can thus be interpreted as the local-field correction factor for the linear susceptibility.

Local-Field Corrections in Nonlinear Optics

In the nonlinear-optical case, the Lorentz local field is still given by Eq. (3.8.2), but the polarization now has both linear and nonlinear contributions:

$$\tilde{\mathbf{P}}=\tilde{\mathbf{P}}^{\mathrm{L}}+\tilde{\mathbf{P}}^{\mathrm{NL}}. \tag{3.8.9}$$

We represent the linear contribution as

$$\tilde{\mathbf{P}}^{\mathrm{L}}=N\alpha\tilde{\mathbf{E}}_{\mathrm{loc}}. \tag{3.8.10}$$

Note that this contribution is linear in the sense that it is linear in the strength of the local field. In general it is not linear in the strength of the macroscopic field. We next introduce Eqs. (3.8.2) and (3.8.9) into this equation to obtain

$$\tilde{\mathbf{P}}^{L}=N\alpha(\tilde{\mathbf{E}}+\tfrac{4}{3}\pi\tilde{\mathbf{P}}^{\mathrm{L}}+\tfrac{4}{3}\pi\tilde{\mathbf{P}}^{\mathrm{NL}}). \tag{3.8.11}$$

We now solve this equation for $\tilde{\mathbf{P}}^{\mathrm{L}}$ and use Eqs. (3.8.6) and (3.8.7) to express the factor $N\alpha$ that appears in the resulting expression in terms of the linear dielectric constant. We thereby obtain

$$\tilde{\mathbf{P}}^{\mathrm{L}}=\frac{\epsilon^{(1)}-1}{4\pi}(\tilde{\mathbf{E}}+\tfrac{4}{3}\pi\tilde{\mathbf{P}}^{\mathrm{NL}}). \tag{3.8.12}$$

Next we consider the displacement vector

$$\tilde{\mathbf{D}}=\tilde{\mathbf{E}}+4\pi\tilde{\mathbf{P}}=\tilde{\mathbf{E}}+4\pi\tilde{\mathbf{P}}^{\mathrm{L}}+4\pi\tilde{\mathbf{P}}^{\mathrm{NL}}. \tag{3.8.13}$$

If the expression (3.8.12) for the linear polarization is substituted into this expression, we obtain

$$\tilde{\mathbf{D}}=\epsilon^{(1)}\tilde{\mathbf{E}}+4\pi\left(\frac{\epsilon^{(1)}+2}{3}\right)\tilde{\mathbf{P}}^{\mathrm{NL}}. \tag{3.8.14}$$

We see that the second term is not simply $4\pi\tilde{\mathbf{P}}^{\mathrm{NL}}$, as might have been expected, but that the nonlinear polarization appears multiplied by the factor $(\epsilon^{(1)} + 2)/3$. We recall that in the derivation of the polarization-driven wave equation for nonlinear optics, a nonlinear source term appears when the second time derivative of $\tilde{\mathbf{D}}$ is calculated (see, for example, Eq. (2.1.9a)). As a consequence of Eq. (3.8.14), we see that the nonlinear source term is actually the nonlinear polarization $\tilde{\mathbf{P}}^{\mathrm{NL}}$ multiplied by the factor $(\epsilon^{(1)} + 2)/3$. To emphasize this point, Bloembergen (1965) introduces the *nonlinear source polarization* defined by

$$\tilde{\mathbf{P}}^{\mathrm{NLS}} = \frac{\epsilon^{(1)} + 2}{3}\tilde{\mathbf{P}}^{\mathrm{NL}}, \tag{3.8.15}$$

so that Eq. (3.8.14) can be expressed as

$$\tilde{\mathbf{D}} = \epsilon^{(1)}\tilde{\mathbf{E}} + 4\pi\tilde{\mathbf{P}}^{\mathrm{NLS}}. \tag{3.8.16}$$

When the derivation of the wave equation is carried out as in Section 2.1 using this expression for $\tilde{\mathbf{D}}$, we obtain the result

$$\nabla \times \nabla \times \tilde{\mathbf{E}} + \frac{\epsilon^{(1)}}{c^2}\frac{\partial^2 \tilde{\mathbf{E}}}{\partial t^2} = -\frac{4\pi}{c^2}\frac{\partial^2 \tilde{\mathbf{P}}^{\mathrm{NLS}}}{\partial t^2}. \tag{3.8.17}$$

This result shows how local-field corrections are incorporated into the wave equation.

The distinction between the local and macroscopic fields also arises in that the field that induces a dipole moment in each atom is the local field, whereas by definition the nonlinear susceptibility relates the nonlinear source polarization to the macroscopic field. To good approximation, we can relate the local and macroscopic fields by replacing $\tilde{\mathbf{P}}$ by $\tilde{\mathbf{P}}^{\mathrm{L}}$ in Eq. (3.8.2) to obtain

$$\tilde{\mathbf{E}}_{\mathrm{loc}} = \tilde{\mathbf{E}} + \tfrac{4}{3}\pi\chi^{(1)}\tilde{\mathbf{E}} = \left(1 + \frac{4\pi}{3}\frac{\epsilon^{(1)} - 1}{4\pi}\right)\tilde{\mathbf{E}},$$

or

$$\tilde{\mathbf{E}}_{\mathrm{loc}} = \frac{\epsilon^{(1)} + 2}{3}\tilde{\mathbf{E}}. \tag{3.8.18}$$

We now apply the results of Eqs. (3.8.17) and (3.8.18) to the case of second-order nonlinear interactions. We define the nonlinear susceptibility by means of the equation (see also Eq. (1.3.13))

$$P_i^{\mathrm{NLS}}(\omega_m + \omega_n) = \sum_{jk}\sum_{(mn)} \chi_{ijk}^{(2)}(\omega_m + \omega_n, \omega_m, \omega_n)E_j(\omega_m)E_k(\omega_n), \tag{3.8.19}$$

where

$$P_i^{\mathrm{NLS}}(\omega_m + \omega_n) = \frac{\epsilon^{(1)}(\omega_m + \omega_n) + 2}{3} P_i^{\mathrm{NL}}(\omega_m + \omega_n) \tag{3.8.20}$$

and where the quantities $E_j(\omega_m)$ represent macroscopic fields. The nonlinear polarization (i.e., the second-order contribution to the dipole moment per unit volume) can be represented as

$$P_i^{\mathrm{NL}}(\omega_m + \omega_n) = N \sum_{jk} \sum_{(mn)} \beta_{ijk}(\omega_m + \omega_n, \omega_m, \omega_n) E_j^{\mathrm{loc}}(\omega_m) E_k^{\mathrm{loc}}(\omega_n), \tag{3.8.21}$$

where the proportionality constant β_{ijk} is known as the second-order hyperpolarizability. The local fields appearing in this expression are related to the macroscopic fields according to Eq. (3.8.18), which we now rewrite as

$$E_j^{\mathrm{loc}}(\omega_m) = \frac{\epsilon^{(1)}(\omega_m) + 2}{3} E_j(\omega_m). \tag{3.8.22}$$

By combining Eqs. (3.8.19) through (3.8.22), we find that the nonlinear susceptibility can be represented as

$$\begin{aligned} \chi_{ijk}^{(2)}(\omega_m + \omega_n, \omega_m, \omega_n) = {} & \mathscr{L}^{(2)}(\omega_m + \omega_n, \omega_m, \omega_n) \\ & \times N \beta_{ijk}(\omega_m + \omega_n, \omega_m, \omega_n), \end{aligned} \tag{3.8.23}$$

where

$$\begin{aligned} & \mathscr{L}^{(2)}(\omega_m + \omega_n, \omega_m, \omega_n) \\ & \quad = \left[\frac{\epsilon^{(1)}(\omega_m + \omega_n) + 2}{3}\right]\left[\frac{\epsilon^{(1)}(\omega_m) + 2}{3}\right]\left[\frac{\epsilon^{(1)}(\omega_n) + 2}{3}\right] \end{aligned} \tag{3.8.24}$$

gives the local-field correction factor for the second-order susceptibility. For example, Eq. (3.6.18) for $\chi^{(2)}$ should be multiplied by this factor to obtain the correct expression including local-field effects.

This result is readily generalized to higher-order nonlinear interaction. For example, the expression for $\chi^{(3)}$ obtained ignoring local-field corrections should be multiplied by the factor

$$\begin{aligned} & \mathscr{L}^{(3)}(\omega_l + \omega_m + \omega_n, \omega_l, \omega_m, \omega_n) \\ & \quad = \left[\frac{\epsilon^{(1)}(\omega_l + \omega_m + \omega_n) + 2}{3}\right]\left[\frac{\epsilon^{(1)}(\omega_l) + 2}{3}\right] \\ & \qquad \times \left[\frac{\epsilon^{(1)}(\omega_m) + 2}{3}\right]\left[\frac{\epsilon^{(1)}(\omega_n) + 2}{3}\right]. \end{aligned} \tag{3.8.25}$$

Our derivation of the form of the local-field correction factor has essentially followed the procedure of Bloembergen (1965). The nature of local-field corrections in nonlinear optics can be understood from a very different point of view introduced by Mizrahi and Sipe (1986). This method has the desirable feature that, unlike the procedure described above, it does not require that we maintain the somewhat arbitrary distinction between the nonlinear polarization and the nonlinear source polarization. For simplicity, we describe this procedure only for the case of third-harmonic generation in the scalar field approximation. We assume that the total polarization (including both linear and nonlinear contributions) at the third-harmonic frequency is given by

$$P(3\omega) = N\alpha(3\omega)E_{\mathrm{loc}}(3\omega) + N\gamma(3\omega,\omega,\omega,\omega)E_{\mathrm{loc}}(\omega)^3, \tag{3.8.26}$$

where $\alpha(3\omega)$ is the linear polarizability for radiation at frequency 3ω and where $\gamma(3\omega,\omega,\omega,\omega)$ is the hyperpolarizability leading to third-harmonic generation. We next use Eqs. (3.8.2) and (3.8.18) to rewrite Eq. (3.8.26) as

$$\begin{aligned} P(3\omega) &= N\alpha(3\omega)[E(3\omega) + \tfrac{4}{3}\pi P(3\omega)] \\ &\quad + N\gamma(3\omega,\omega,\omega,\omega)\left[\frac{\epsilon^{(1)}(\omega)+2}{3}\right]^3 E(\omega)^3. \end{aligned} \tag{3.8.27}$$

This equation is now solved algebraically for $P(3\omega)$ to obtain

$$P(3\omega) = \frac{N\alpha(3\omega)E(3\omega)}{1-\frac{4}{3}\pi N\alpha(3\omega)} + \frac{N\gamma(3\omega,\omega,\omega,\omega)}{1-\frac{4}{3}\pi N\alpha(3\omega)}\left[\frac{\epsilon^{(1)}(\omega)+2}{3}\right]^3 E(\omega)^3. \tag{3.8.28}$$

We can identify the first and second terms of this expression as the linear and third-order polarizations, which we represent as

$$P(3\omega) = \chi^{(1)}(3\omega)E(3\omega) + \chi^{(3)}(3\omega,\omega,\omega,\omega)E(\omega)^3, \tag{3.8.29}$$

where (in agreement with the usual Lorentz–Lorenz Law) the linear susceptibility is given by

$$\chi^{(1)}(3\omega) = \frac{N\alpha(3\omega)}{1-\frac{4}{3}\pi N\alpha(3\omega)}, \tag{3.8.30}$$

and where the third-order susceptibility is given by

$$\chi^{(3)}(3\omega,\omega,\omega,\omega) = \left[\frac{\epsilon^{(1)}(\omega)+2}{3}\right]^3 \frac{\epsilon^{(1)}(3\omega)+2}{3} N\gamma(3\omega,\omega,\omega,\omega). \tag{3.8.31}$$

We have made use of Eq. (3.8.8b) in writing Eq. (3.8.31) in the form shown. Note that the result (3.8.31) agrees with the previous result described by Eq. (3.8.25).

Problems

1. Show that Eq. (3.6.25) possesses full permutation symmetry.

2. Derive, using the density matrix formalism, an expression for the resonant contribution to the third-order susceptibility $\chi^{(3)}$ describing third-harmonic generation as illustrated below. Assume that, in thermal equilibrium, all of the population resides in the ground state. Note that since all input frequencies are equal and since only the resonant contribution is required, the answer will consist of one term and not 48 terms, which occur for the most general case of $\chi^{(3)}$. Work this problem by starting with the perturbation expansion (3.4.16) derived in the text and specializing the ensuing derivation to the interaction shown in the figure.

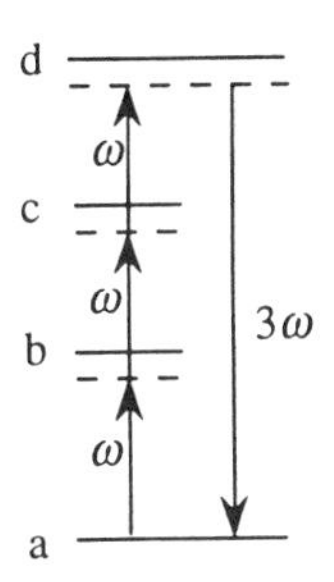

$$\left[\text{Ans: } \chi^{(3)}_{kjih}(3\omega,\omega,\omega,\omega)=\frac{N}{\hbar^3}\frac{\mu^k_{ad}\mu^j_{dc}\mu^i_{cb}\mu^h_{ba}}{[(\omega_{da}-3\omega)-i\gamma_{da}][(\omega_{ca}-2\omega)-i\gamma_{ca}][(\omega_{ba}-\omega)-i\gamma_{ba}]}\right].$$

3. Consider the mutual interaction of four optical fields as illustrated in the figure below. Assume that all of the fields have the same linear polarization and that in thermal equilibrium all of the population is contained in level a. Assume that the waves are tuned sufficiently closely to the indicated resonances that only these contributions to the nonlinear interaction need be taken into account

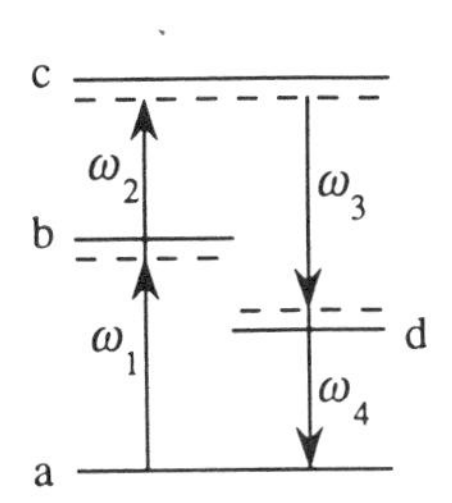

You may work this problem either by specializing the general result of

Eq. (3.7.14) to the interaction shown in the figure or by repeating the derivation given in the text and specializing at each step to this interaction.
(a) Calculate the four nonlinear susceptibilities

$$\chi^{(3)}(\omega_4 = \omega_1 + \omega_2 - \omega_3), \qquad \chi^{(3)}(\omega_3 = \omega_1 + \omega_2 - \omega_4),$$

$$\chi^{(3)}(\omega_1 = \omega_3 + \omega_4 - \omega_2), \qquad \chi^{(3)}(\omega_2 = \omega_3 + \omega_4 - \omega_1),$$

that describe the four-wave mixing process, and determine the conditions under which these quantities are equal.
(b) In addition, calculate the nonlinear susceptibilities

$$\chi^{(3)}(\omega_1 = \omega_1 + \omega_2 - \omega_2), \qquad \chi^{(3)}(\omega_2 = \omega_2 + \omega_1 - \omega_1)$$

that describe two-photon absorption of the ω_1 and ω_2 fields, and determine the conditions under which they are equal.

4. Repeat the calcuation of the resonant contributions to $\chi^{(3)}$ for the cases studied in Problems 2 and 3 for the more general situation in which each of the levels can contain population in thermal equilibrium. Interpret your results.

[Note: The solution to this problem is very lengthy.]

5. Verify the statement made in the text that Eq. (3.7.14) reduces to Eq. (3.2.32) in the limit in which damping effects are negligible. Show also that, even when damping is not negligible, the general 48-term expression for $\chi^{(3)}$ can be cast into an expression containing 24 terms, 12 of which contain "pressure-induced" resonances.

References

Quantum Mechanics

H. A. Bethe and E. E. Salpeter, *Quantum Mechanics of One- and Two-Electron Atoms*, Plenum, New York, 1977.

E. Merzbacher, *Quantum Mechanics*, Wiley, New York, 1970.

M. Sargent III, M. O. Scully, and W. E. Lamb, Jr., *Laser Physics*, Addison-Wesley, Reading, Mass., 1974.

Quantum-Mechanical Theories of the Nonlinear Optical Susceptibility

J. A. Armstrong, N. Bloembergen, J. Ducuing, and P. S. Pershan, *Phys. Rev.* **127**, 1918 (1962).

N. Bloembergen, *Nonlinear Optics*, Benjamin, New York, 1965.
N. Bloembergen and Y. R. Shen, *Phys. Rev.* **133**, A37 (1964).
N. Bloembergen, H. Lotem, and R. T. Lynch, Jr., *Indian J. Pure Appl. Phys.* **16**, 151 (1978).
R. W. Boyd and S. Mukamel, *Phys. Rev. A* **29**, 1973 (1984).
P. N. Butcher, *Nonlinear Optical Phenomena*, Ohio State University, 1965.
J. Ducuing, "Nonlinear Optical Processes," in *Quantum Optics*, edited by R. J. Glauber, Academic Press, New York, 1969.
C. Flytzanis, in *Quantum Electronics, a Treatise*, Vol. 1, Part A, edited by H. Rabin and C. L. Tang, Academic Press, New York, 1975.
D. C. Hanna, M. A. Yuratich, and D. Cotter, *Nonlinear Optics of Free Atoms and Molecules*, Springer-Verlag, Berlin, 1979.
D. Marcuse, *Principles of Quantum Electronics*, Academic Press, New York, 1980.
R. B. Miles and S. E. Harris, *IEEE J. Quantum Electron.* **9**, 470 (1973).
B. J. Orr and J. F. Ward, *Mol. Phys.* **20**, 513 (1971).
Y. Prior, *IEEE J. Quantum Electron.* **20**, 37 (1984).
L. Rothberg, "Dephasing-Induced Coherent Phenomena," in *Progress in Optics XXIV*, edited by E. Wolf, Elsevier, 1987.
Y. R. Shen, *Principles of Nonlinear Optics*, Wiley, New York, 1984.
J. F. Ward, *Rev. Mod. Phys.* **37**, 1 (1965).
T. K. Yee and T. K. Gustafson, *Phys. Rev. A* **18**, 1597 (1978).

Local-Field Corrections

M. Born and E. Wolf, *Principles of Optics*, Pergamon Press, Oxford, 1975.
J. D. Jackson, *Classical Electrodynamics*, Wiley, New York, 1975.
H. A. Lorentz, *The Theory of Electrons*, Dover, New York, 1952.
V. Mizrahi and J. E. Sipe, *Phys. Rev. B* **34**, 3700 (1986).

Chapter 4

The Intensity-Dependent Refractive Index

The refractive index of many optical materials depends upon the intensity of the light propagating through the material. In this chapter, we examine some of the mathematical descriptions of the nonlinear refractive index and examine some of the physical processes that give rise to this effect. In the following chapter, we study the intensity-dependent refractive index resulting from the resonant response of an atomic system, and in Chapter 6 we study some physical processes that result from the nonlinear refractive index.

4.1. Descriptions of the Intensity-Dependent Refractive Index

The refractive index of many materials can be described by the relation

$$n = n_0 + \bar{n}_2 \langle \tilde{E}^2 \rangle, \tag{4.1.1}$$

where n_0 represents the usual, weak-field refractive index and $\bar{n}_2$ is a new optical constant (sometimes called the second-order index of refraction) that gives the rate at which the refractive index increases with increasing optical intensity.* The angular brackets surrounding the quantity $\tilde{E}^2$ represent a time

* We place a bar over the symbol n_2 to prevent confusion with a different definition of n_2, introduced in Eq. (4.1.15) below. In accordance with conventional usage, the bar will be omitted in cases where little chance of confusion is likely.

average. For example, if the optical field is of the form

$$\tilde{E}(t) = E(\omega)e^{-i\omega t} + \text{c.c.}, \tag{4.1.2}$$

so that

$$\langle \tilde{E}(t)^2 \rangle = 2E(\omega)E(\omega)^* = 2|E(\omega)|^2, \tag{4.1.3}$$

we find that

$$n = n_0 + 2\bar{n}_2|E(\omega)|^2. \tag{4.1.4}$$

The change in refractive index described by Eq. (4.1.1) or (4.1.4) is sometimes called the optical Kerr effect, by analogy with the traditional Kerr electrooptic effect, in which the refractive index of a material changes by an amount that is proportional to the square of the strength of an applied static field.

Of course, the interaction of a beam of light with a nonlinear optical medium can also be described in terms of the nonlinear polarization. The part of the nonlinear polarization that influences the propagation of a beam of frequency ω is

$$P^{\mathrm{NL}}(\omega) = 3\chi^{(3)}(\omega = \omega + \omega - \omega)|E(\omega)|^2 E(\omega). \tag{4.1.5}$$

For simplicity we are assuming here that the light is linearly polarized and are suppressing the tensor indices of $\chi^{(3)}$; the tensor nature of $\chi^{(3)}$ is addressed explicitly in the next section. The total polarization of the material system is then described by

$$P^{\mathrm{TOT}}(\omega) = \chi^{(1)}E(\omega) + 3\chi^{(3)}|E(\omega)|^2 E(\omega) \equiv \chi_{\mathrm{eff}} E(\omega), \tag{4.1.6}$$

where we have introduced the effective susceptibility

$$\chi_{\mathrm{eff}} = \chi^{(1)} + 3\chi^{(3)}|E(\omega)|^2. \tag{4.1.7}$$

In order to relate the nonlinear susceptibility $\chi^{(3)}$ to the nonlinear refractive index n_2, we note that it is generally true that

$$n^2 = 1 + 4\pi\chi_{\mathrm{eff}}, \tag{4.1.8}$$

and by introducing Eq. (4.1.4) on the left-hand side and Eq. (4.1.7) on the right-hand side of this equation we find that

$$[n_0 + 2\bar{n}_2|E(\omega)|^2]^2 = 1 + 4\pi\chi^{(1)} + 12\pi\chi^{(3)}|E(\omega)|^2. \tag{4.1.9}$$

Correct to terms of order $|E(\omega)|^2$, this expression when expanded becomes $n_0^2 + 4n_0\bar{n}_2|E(\omega)|^2 = (1 + 4\pi\chi^{(1)}) + [12\pi\chi^{(3)}|E(\omega)|^2]$, which shows that the linear and nonlinear refractive indices are related to the linear and nonlinear susceptibilities by

$$n_0 = (1 + 4\pi\chi^{(1)})^{1/2} \tag{4.1.10}$$

and

$$\bar{n}_2 = \frac{3\pi\chi^{(3)}}{n_0}. \tag{4.1.11}$$

The discussion given above has implicitly assumed that the refractive index is measured using a single laser beam, as shown in Fig. 4.1.1(a). Another way of measuring the intensity-dependent refractive index is to use two separate beams, as illustrated in Fig. 4.1.1(b). Here the presence of the strong beam of amplitude $E(\omega)$ leads to a modification of the refractive index experienced by a weak probe wave of amplitude $E(\omega')$. The nonlinear polarization affecting the probe wave is given by

$$P^{\mathrm{NL}}(\omega') = 6\chi^{(3)}(\omega' = \omega' + \omega - \omega)|E(\omega)|^2 E(\omega'). \tag{4.1.12}$$

Note that the degeneracy factor (6) for this case is twice as large as that for the single-beam case of Eq. (4.1.5). In fact, for the two-beam case the degeneracy factor is equal to 6 even if ω' is equal to ω, because the probe beam is physically distinguishable from the strong pump beam owing to its different direction of propagation. The probe wave hence experiences a refractive index given by

$$n = n_0 + 2\bar{n}_2^{(\mathrm{weak})}|E(\omega)|^2, \tag{4.1.13}$$

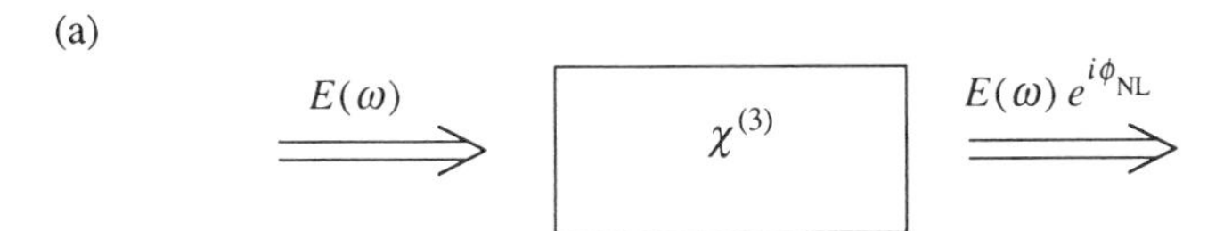

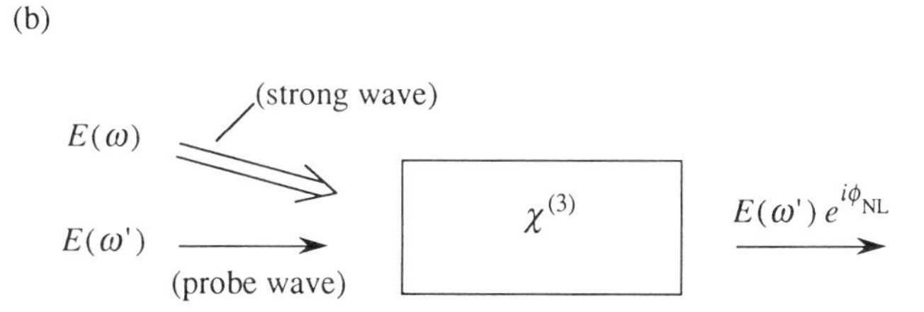

FIGURE 4.1.1 Two ways of measuring the nonlinear refractive index. In each case ϕ_{NL} represents the nonlinear contribution to the phase shift experienced in propagating through the optical material.

where

$$\bar{n}_2^{(\text{weak})} = \frac{6\pi\chi^{(3)}}{n_0}. \tag{4.1.14}$$

Note that $\bar{n}_2^{(\text{weak})}$ is twice as large as $\bar{n}_2$ of Eq. (4.1.11). Hence a strong wave affects the refractive index of a weak wave of the same frequency twice as much as it affects its own refractive index. This effect (for the case in which n_2 is positive) is known as weak-wave retardation (Chiao *et al.*, 1966).

An alternative way of defining the intensity-dependent refractive index* is by means of the equation

$$n = n_0 + n_2 I, \tag{4.1.15}$$

where I denotes the time-averaged intensity of the optical field, given by

$$I = \frac{n_0 c}{2\pi}|E(\omega)|^2. \tag{4.1.16}$$

Since the total refractive index n must be the same using either description of the nonlinear contribution, we see by comparing Eqs. (4.1.4) and (4.1.15) that

$$2\bar{n}_2|E(\omega)|^2 = n_2 I, \tag{4.1.17}$$

and hence that $\bar{n}_2$ and n_2 are related by

$$n_2 = \frac{4\pi}{n_0 c}\bar{n}_2, \tag{4.1.18}$$

where we have made use of Eq. (4.1.16). If Eq. (4.1.11) is introduced into this expression, we find that n_2 is related to $\chi^{(3)}$ by

$$n_2 = \frac{12\pi^2}{n_0^2 c}\chi^{(3)}. \tag{4.1.19}$$

It is often convenient to measure I in units of W/cm^2, in which case n_2 is measured in units of cm^2/W. We then find that numerically

$$n_2\left(\frac{\text{cm}^2}{\text{W}}\right) = \frac{12\pi^2}{n_0^2 c}10^7\chi^{(3)}(\text{esu}) = \frac{0.0395}{n_0^2}\chi^{(3)}(\text{esu}). \tag{4.1.20}$$

Some of the physical processes that can produce a nonlinear change in the refractive index are listed in Table 4.1.1, along with typical values of n_2, of $\chi^{(3)}$, and of the characteristic time scale for the nonlinear response to develop. Electronic polarization and molecular orientation are discussed in the present

* For definiteness, we are treating the single-beam case of Fig. 4.1.1(a). The extension to the two-beam case is straightforward.

TABLE 4.1.1 Typical values of the nonlinear refractive index*

Mechanism	n_2 (cm^2/W)	$\chi^{(3)}_{1111}$ (esu)	Response time (sec)
Electronic polarization	10^{-16}	10^{-14}	10^{-15}
Molecular orientation	10^{-14}	10^{-12}	10^{-12}
Electrostriction	10^{-14}	10^{-12}	10^{-9}
Saturated atomic absorption	10^{-10}	10^{-8}	10^{-8}
Thermal effects	10^{-6}	10^{-4}	10^{-3}
Photorefractive effect†	(large)	(large)	(intensity-dependent)

* For linearly polarized light, n_2 and $\chi^{(3)}$ are accurately related by Eq. (4.1.20).

† The photorefractive effect often leads to a very strong nonlinear response. This response usually cannot be described in terms of a $\chi^{(3)}$ (or an n_2) nonlinear susceptibility, because the nonlinear polarization does not depend on the applied field strength in the same manner as the other mechanisms listed.

chapter, saturated absorption is discussed in Chapter 6, electrostriction and thermal effects are discussed in Chapter 8, and the photorefractive effect is described in Chapter 10.

In Table 4.1.2 the experimentally measured values of nonlinear susceptibility are presented for several materials. Some of the methods that are used to measure the nonlinear susceptibility have been reviewed by Hellwarth (1977). As an example of the use of Table 4.1.2, note that for carbon disulfide

TABLE 4.1.2 Third-order nonlinear susceptibilities of various materials*

Material	$\chi^{(3)}_{1111}$ (esu)	Response time
Air (20°C)	1.2×10^{-17}	
Carbon disulfide	1.9×10^{-12}	2 ps
GaAs (bulk, excitonic, room temperature)	6.5×10^{-4}	20 ns
GaAs/GaAlAs (MQW)	0.04	20 ns
Indium antimonide (77 K, 5.4 μm)	0.3	400 ns
Semiconductor-doped glass (containing CdSe)	10^{-8}	30 ps
Optical glasses	$(1\text{–}100) \times 10^{-14}$	Very fast
Polydiacetylene:		
Nonresonant	2.5×10^{-10}	Very fast
At peak of exciton	7.5×10^{-6}	2 ps

* The value of n_2 defined by $n = n_0 + n_2 I$ in units of cm^2/W can be obtained by multiplying the value of $\chi^{(3)}_{1111}$ by $0.0395/n_0^2$. The value of $\bar{n}_2$ defined by $n = n_0 + 2\bar{n}_2|E|^2$ in units of cm^3/erg can be obtained by multiplying the value of $\chi^{(3)}_{1111}$ by $3\pi/n_0$.

the value of n_2 is approximately $3 \times 10^{-14}\ \mathrm{cm^2/W}$. Thus, a laser beam of intensity $I = 1\ \mathrm{MW/cm^2}$ can produce a refractive index change of 3×10^{-8}. Even though this change is rather small, refractive index changes of this order of magnitude can lead to dramatic nonlinear optical effects (some of which are described in Chapter 6) for the case of phase-matched nonlinear optical interactions.

4.2. Tensor Nature of the Third-Order Susceptibility

The third-order susceptibility $\chi^{(3)}_{ijkl}$ is a fourth-rank tensor, and thus is described in terms of 81 separate elements. For crystalline solids with low symmetry, all 81 of these elements are independent and can be nonzero (Butcher, 1965). However, for materials possessing a higher degree of spatial symmetry, the number of independent elements is very much reduced; as we show below, there are only three independent elements for an isotropic material.

Let us see how to determine the tensor nature of the third-order susceptibility for the case of an isotropic material such as a glass, a liquid, or a vapor. We begin by considering the general case in which the applied frequencies are arbitrary, and represent the susceptibility as $\chi_{ijkl} \equiv \chi^{(3)}_{ijkl}(\omega_4 = \omega_1 + \omega_2 + \omega_3)$. Since each of the coordinate axes must be equivalent in an isotropic material, it is clear that the susceptibility possesses the following symmetry properties:

$$\chi_{1111} = \chi_{2222} = \chi_{3333}, \tag{4.2.1a}$$

$$\chi_{1122} = \chi_{1133} = \chi_{2211} = \chi_{2233} = \chi_{3311} = \chi_{3322}, \tag{4.2.1b}$$

$$\chi_{1212} = \chi_{1313} = \chi_{2323} = \chi_{2121} = \chi_{3131} = \chi_{3232}, \tag{4.2.1c}$$

$$\chi_{1221} = \chi_{1331} = \chi_{2112} = \chi_{2332} = \chi_{3113} = \chi_{3223}. \tag{4.2.1d}$$

One can also see that the 21 elements listed above are the only nonzero elements of $\chi^{(3)}$, because these are the only elements that possess the property that any cartesian index (1, 2, or 3) that appears at least once appears an even number of times. An index cannot appear an odd number of times, because, for example, χ_{1222} would give the response in the $\hat{x}_1$ direction due to a field applied in the $\hat{x}_2$ direction. This response must vanish in an isotropic material, because there is no reason why the response should be in the $+\hat{x}_1$ direction rather than in the $-\hat{x}_1$ direction.

The four types of nonzero elements appearing in Eq. (4.2.1) are not independent of one another, and in fact are related by the equation

$$\chi_{1111} = \chi_{1122} + \chi_{1212} + \chi_{1221}. \tag{4.2.2}$$

One can obtain this result by requiring that the predicted value of the nonlinear polarization be the same when calculated in two different coordinate systems that are rotated with respect to each other by an arbitrary amount. A rotation of 45 degrees about the $\hat{x}_3$ axis is a convenient choice for deriving this result. The results given by Eqs. (4.2.1) and (4.2.2) can be used to express the nonlinear susceptibility in the compact form

$$\chi_{ijkl} = \chi_{1122}\,\delta_{ij}\,\delta_{kl} + \chi_{1212}\,\delta_{ik}\,\delta_{jl} + \chi_{1221}\,\delta_{il}\,\delta_{jk}. \tag{4.2.3}$$

This form shows that the third-order susceptibility has three independent elements for the general case in which the field frequencies are arbitrary.

Let us first specialize this result to the case of third-harmonic generation, where the frequency dependence of the susceptibility is taken as $\chi_{ijkl}(3\omega = \omega + \omega + \omega)$. As a consequence of the intrinsic permutation symmetry of the nonlinear susceptibility, the elements of the susceptibility tensor are related by $\chi_{1122} = \chi_{1212} = \chi_{1221}$ and hence Eq. (4.2.3) becomes

$$\begin{aligned}\chi_{ijkl}(3\omega = \omega + \omega + \omega) = \chi_{1122}&(3\omega = \omega + \omega + \omega)\\ &\times (\delta_{ij}\,\delta_{kl} + \delta_{ik}\,\delta_{jl} + \delta_{il}\,\delta_{jk}).\end{aligned} \tag{4.2.4}$$

Hence there is only one independent element of the susceptibility tensor describing third-harmonic generation.

We next apply the result given in Eq. (4.2.3) to the nonlinear refractive index, that is, we consider the choice of frequencies given by $\chi_{ijkl}(\omega = \omega + \omega - \omega)$. For this choice of frequencies, the condition of intrinsic permutation symmetry requires that χ_{1122} be equal to χ_{1212}, and hence χ_{ijkl} can be represented by

$$\begin{aligned}\chi_{ijkl}(\omega = \omega + \omega - \omega) = \chi_{1122}&(\omega = \omega + \omega - \omega)(\delta_{ij}\,\delta_{kl} + \delta_{ik}\,\delta_{jl})\\ &+ \chi_{1221}(\omega = \omega + \omega - \omega)(\delta_{il}\,\delta_{jk}).\end{aligned} \tag{4.2.5}$$

The nonlinear polarization leading to the nonlinear refractive index is given in terms of the nonlinear susceptibility by

$$P_i(\omega) = 3\sum_{jkl} \chi_{ijkl}(\omega = \omega + \omega - \omega)E_j(\omega)E_k(\omega)E_l(-\omega). \tag{4.2.6}$$

If we introduce Eq. (4.2.5) into this equation, we find that

$$P_i = 6\chi_{1122}E_i(\mathbf{E}\cdot\mathbf{E}^*) + 3\chi_{1221}E_i^*(\mathbf{E}\cdot\mathbf{E}). \tag{4.2.7}$$

This equation can be written entirely in vector form as

$$\mathbf{P} = 6\chi_{1122}(\mathbf{E}\cdot\mathbf{E}^*)\mathbf{E} + 3\chi_{1221}(\mathbf{E}\cdot\mathbf{E})\mathbf{E}^*. \tag{4.2.8}$$

Following the notation of Maker and Terhune (1965), we introduce the coefficients

$$A = 6\chi_{1122} \, (=3\chi_{1122} + 3\chi_{1212}) \tag{4.2.9a}$$

and

$$B = 6\chi_{1221}, \tag{4.2.9b}$$

in terms of which the nonlinear polarization of Eq. (4.2.8) can be written as

$$\mathbf{P} = A(\mathbf{E} \cdot \mathbf{E}^*)\mathbf{E} + \tfrac{1}{2}B(\mathbf{E} \cdot \mathbf{E})\mathbf{E}^*. \tag{4.2.10}$$

We see that the nonlinear polarization consists of two contributions. These contributions have very different physical characters, since the first contribution has the vector nature of **E** whereas the second contribution has the vector nature of **E***. The first contribution thus produces a nonlinear polarization with the same handedness as **E**, whereas the second contribution produces a nonlinear polarization with the opposite handedness. The consequences of this behavior on the propagation of a beam of light through a nonlinear optical medium are described below.

The origin of the different physical characters of the two contributions to **P** can be understood in terms of the energy level diagrams shown in Fig. 4.2.1. Here part (a) illustrates one-photon-resonant contributions to the nonlinear coupling. We show below in Eq. (4.3.14) that processes of this sort contribute only to the coefficient A. Part (b) of the figure illustrates two-photon-resonant processes, which in general contribute to both the coefficients A and B (see Eqs. (4.3.13) and (4.3.14)). However, under certain circumstances, such as those described in connection with Fig. 6.1.9, two-photon-resonant processes contribute only to the coefficient B.

For some purposes, it is useful to describe the nonlinear polarization not by Eq. (4.2.10) but rather in terms of an effective linear susceptibility defined by means of the relationship

$$P_i = \sum_j \chi_{ij} E_j. \tag{4.2.11}$$

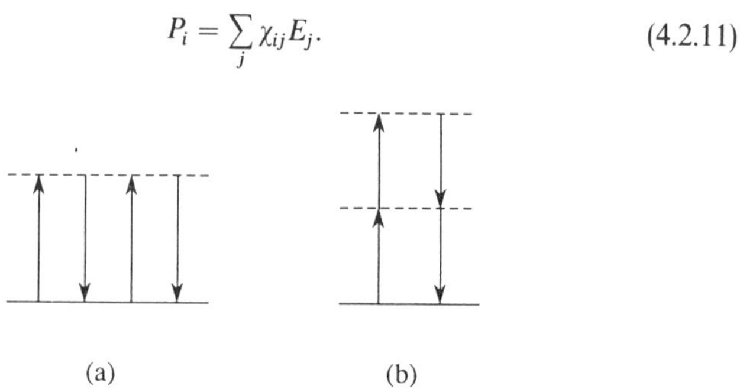

FIGURE 4.2.1 Diagrams (a) and (b) represent the resonant contributions to the nonlinear coefficients A and B, respectively.

Then, as can be verified by direct substitution, Eqs. (4.2.10) and (4.2.11) lead to identical predictions for the nonlinear polarization if the effective linear susceptibility is given by

$$\chi_{ij} = A'(\mathbf{E} \cdot \mathbf{E}^*)\,\delta_{ij} + \tfrac{1}{2}B'(E_i E_j^* + E_i^* E_j), \tag{4.2.12a}$$

where

$$A' = A - \tfrac{1}{2}B = 6\chi_{1122} - 3\chi_{1221} \tag{4.2.12b}$$

and

$$B' = B = 6\chi_{1221}. \tag{4.2.12c}$$

The results given in Eq. (4.2.10) or in Eqs. (4.2.12) show that the nonlinear susceptibility tensor describing the nonlinear refractive index of an isotropic material possesses only two independent elements. The relative magnitude of these two coefficients depends upon the nature of the physical process that produces the optical nonlinearity. For some of the physical mechanisms leading to a nonlinear refractive index, these ratios are given by

$$B/A = 6, \quad B'/A' = -3 \qquad \text{for molecular orientation,} \tag{4.2.13a}$$

$$B/A = 1, \quad B'/A' = 2 \qquad \text{for nonresonant electronic response,} \tag{4.2.13b}$$

$$B/A = 0, \quad B'/A' = 0 \qquad \text{for electrostriction.} \tag{4.2.13c}$$

These conclusions will be justified in the discussion that follows; see especially Eq. (4.4.37) for the case of molecular orientation, Eq. (4.3.14) for nonresonant electronic response of bound electrons, and Eq. (8.2.15) for electrostriction. Note also that A is equal to B by definition whenever the Kleinman symmetry condition is valid.

The trace of the effective susceptibility is given by

$$\operatorname{tr}\chi_{ij} \equiv \sum_i \chi_{ii} = (3A' + B')\mathbf{E} \cdot \mathbf{E}^*. \tag{4.2.14}$$

Hence, $\operatorname{tr}\chi_{ij}$ vanishes for the molecular orientation mechanism; the reason for this behavior is discussed below in connection with Eq. (4.4.56). For the resonant response of an atomic transition, the ratio of B to A depends upon the angular momentum quantum numbers of the two atomic levels. Formulas for A and B for such a case have been presented by Saikan and Kiguchi (1982).

Propagation through Isotropic Nonlinear Media

Let us next consider the propagation of a beam of light through a material whose nonlinear optical properties are described by Eq. (4.2.10). As we show below, only linearly or circularly polarized light is transmitted through such a medium with its state of polarization unchanged. When elliptically polarized light propagates through such a medium, the orientation of the polarization ellipse rotates as a function of position due to the nonlinear interaction.

Let us consider a beam of arbitrary polarization propagating in the positive z direction. The electric field vector of such a beam can always be decomposed into a linear combination of left- and right-hand circular components as

$$\mathbf{E} = E_+\hat{\boldsymbol{\sigma}}_+ + E_-\hat{\boldsymbol{\sigma}}_-, \tag{4.2.15}$$

where the circular-polarization unit vectors are illustrated in Fig. 4.2.2 and are defined by

$$\hat{\boldsymbol{\sigma}}_\pm = \frac{\hat{\mathbf{x}} \pm i\hat{\mathbf{y}}}{\sqrt{2}}. \tag{4.2.16}$$

By convention, $\hat{\boldsymbol{\sigma}}_+$ corresponds to left-hand circular and $\hat{\boldsymbol{\sigma}}_-$ to right-hand circular polarization (for a beam propagating in the positive z direction).

We now introduce the decomposition (4.2.15) into Eq. (4.2.10). We find, using the identities

$$\hat{\boldsymbol{\sigma}}_\pm^* = \hat{\boldsymbol{\sigma}}_\mp, \qquad \hat{\boldsymbol{\sigma}}_\pm \cdot \hat{\boldsymbol{\sigma}}_\pm = 0, \qquad \hat{\boldsymbol{\sigma}}_\pm \cdot \hat{\boldsymbol{\sigma}}_\mp = 1,$$

that the products $\mathbf{E}^* \cdot \mathbf{E}$ and $\mathbf{E} \cdot \mathbf{E}$ become

$$\mathbf{E}^* \cdot \mathbf{E} = (E_+^*\hat{\boldsymbol{\sigma}}_+^* + E_-^*\hat{\boldsymbol{\sigma}}_-^*) \cdot (E_+\hat{\boldsymbol{\sigma}}_+ + E_-\hat{\boldsymbol{\sigma}}_-) = E_+^*E_+ + E_-^*E_- = |E_+|^2 + |E_-|^2,$$

and

$$\mathbf{E} \cdot \mathbf{E} = (E_+\hat{\boldsymbol{\sigma}}_+ + E_-\hat{\boldsymbol{\sigma}}_-) \cdot (E_+\hat{\boldsymbol{\sigma}}_+ + E_-\hat{\boldsymbol{\sigma}}_-) = E_+E_- + E_-E_+ = 2E_+E_-,$$

so that Eq. (4.2.10) can be written as

$$\mathbf{P}^{\mathrm{NL}} = A(|E_+|^2 + |E_-|^2)\mathbf{E} + B(E_+E_-)\mathbf{E}^*. \tag{4.2.17}$$

If we now represent $\mathbf{P}^{\mathrm{NL}}$ in terms of its circular components as

$$\mathbf{P}^{\mathrm{NL}} = P_+\hat{\boldsymbol{\sigma}}_+ + P_-\hat{\boldsymbol{\sigma}}_-, \tag{4.2.18}$$

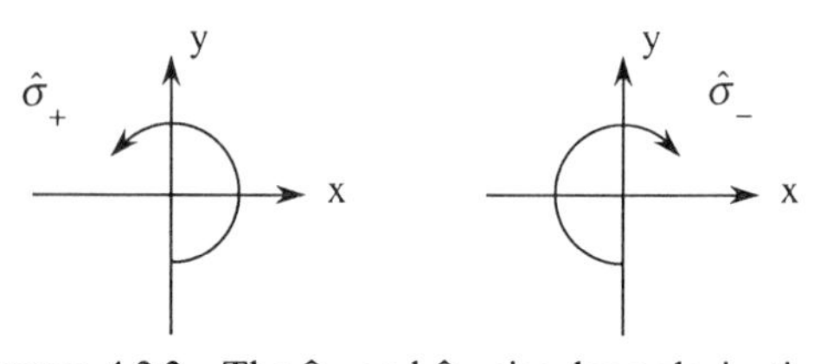

FIGURE 4.2.2 The $\hat{\boldsymbol{\sigma}}_+$ and $\hat{\boldsymbol{\sigma}}_-$ circular polarizations.

we find that the coefficient P_+ is given by

$$\begin{aligned} P_+ &= A(|E_+|^2 + |E_-|^2)E_+ + B(E_+E_-)E_-^* \\ &= A(|E_+|^2 + |E_-|^2)E_+ + B|E_-|^2E_+ \\ &= A|E_+|^2E_+ + (A + B)|E_-|^2E_+ \end{aligned} \tag{4.2.19a}$$

and similarly that

$$P_- = A|E_-|^2E_- + (A + B)|E_+|^2E_-. \tag{4.2.19b}$$

These results can be summarized as

$$P_\pm \equiv \chi_\pm^{\text{NL}} E_\pm, \tag{4.2.20a}$$

where we have introduced the effective nonlinear susceptibilities

$$\chi_\pm^{\text{NL}} = A|E_\pm|^2 + (A + B)|E_\mp|^2. \tag{4.2.20b}$$

The expressions (4.2.15) and (4.2.18) for the field and nonlinear polarization are now introduced into the wave equation,

$$\nabla^2 \mathbf{E}(z, t) = \frac{\epsilon^{(1)}}{c^2} \frac{\partial^2 \mathbf{E}(z, t)}{\partial t^2} + \frac{4\pi}{c^2} \frac{\partial^2}{\partial t^2} \mathbf{P}^{\text{NL}}, \tag{4.2.21}$$

which we decompose into its $\hat{\boldsymbol{\sigma}}_+$ and $\hat{\boldsymbol{\sigma}}_-$ components. Since, according to Eq. (4.2.20a), $P_\pm$ is proportional to $E_\pm$, the two terms on the right-hand side of the resulting equation can be combined into a single term, so that the wave equation for each circular component becomes

$$\nabla^2 E_\pm(z, t) = \frac{\epsilon_\pm^{\text{eff}}}{c^2} \frac{\partial^2 E_\pm(z, t)}{\partial t^2}, \tag{4.2.22a}$$

where

$$\epsilon_\pm^{\text{eff}} = \epsilon^{(1)} + 4\pi\chi_\pm^{\text{NL}}. \tag{4.2.22b}$$

This equation possesses solutions of the form of plane waves propagating with the phase velocity $c/n^\pm$, where $n_\pm = [\epsilon_\pm^{\text{eff}}]^{1/2}$. Letting $n_0^2 = \epsilon^{(1)}$, we find that

$$\begin{aligned} n_\pm^2 &= n_0^2 + 4\pi\chi_\pm^{\text{NL}} = n_0^2 + 4\pi[A|E_\pm|^2 + (A + B)|E_\mp|^2] \\ &= n_0^2\left(1 + \frac{4\pi}{n_0^2}[A|E_\pm|^2 + (A + B)|E_\mp|^2]\right), \end{aligned}$$

and hence that

$$n_\pm \simeq n_0 + \frac{2\pi}{n_0}[A|E_\pm|^2 + (A + B)|E_\mp|^2]. \tag{4.2.23}$$

We see that the left- and right-circular components of the beam propagate with different phase velocities. The difference in their refractive indices is given by

$$\Delta n \equiv n_+ - n_- = \frac{2\pi B}{n_0}(|E_-|^2 - |E_+|^2). \tag{4.2.24}$$

Note that this difference depends upon the value of the coefficient B but not the coefficient A. Since the left- and right-hand circular components propagate with different phase velocities, the polarization ellipse of the light will rotate as the beam propagates through the nonlinear medium; a similar effect occurs in the linear optics of optically active materials.

In order to determine the angle of rotation, we express the field amplitude as

$$\begin{aligned}\mathbf{E}(z) &= E_+\hat{\boldsymbol{\sigma}}_+ + E_-\hat{\boldsymbol{\sigma}}_- = A_+ e^{in_+\omega z/c}\hat{\boldsymbol{\sigma}}_+ + A_- e^{in_-\omega z/c}\hat{\boldsymbol{\sigma}}_- \\ &= (A_+ e^{i\Delta n\,\omega z/2c}\hat{\boldsymbol{\sigma}}_+ + A_- e^{-i\Delta n\,\omega z/2c}\hat{\boldsymbol{\sigma}}_-)e^{i[n_- + 1/2\,\Delta n]\omega z/c}.\end{aligned} \tag{4.2.25}$$

We now introduce the mean propagation constant $k_m = (n_- + \frac{1}{2}\Delta n)\omega/c$ and the angle

$$\theta = \tfrac{1}{2}\Delta n \frac{\omega}{c} z, \tag{4.2.26a}$$

in terms of which Eq. (4.2.25) becomes

$$\mathbf{E}(z) = (A_+\hat{\boldsymbol{\sigma}}_+ e^{i\theta} + A_-\hat{\boldsymbol{\sigma}}_- e^{-i\theta})e^{ik_m z}. \tag{4.2.26b}$$

As illustrated in Fig. 4.2.3, this equation describes a wave whose polarization ellipse is the same as that of the incident wave, but rotated through the angle θ (measured clockwise in the x−y plane, in conformity with the sign convention for rotation angles in optical activity). This conclusion can be demonstrated by

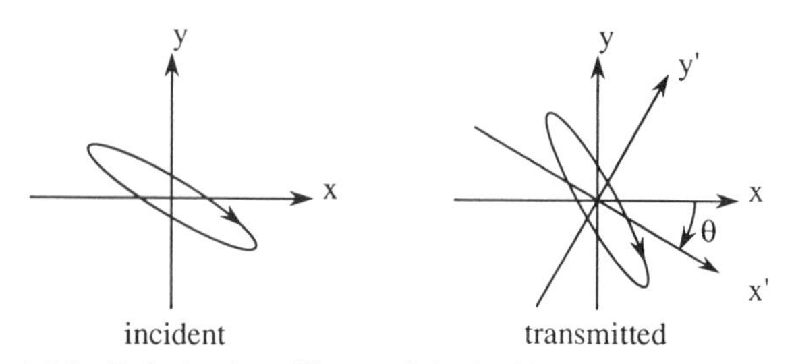

FIGURE 4.2.3 Polarization ellipses of the incident and transmitted waves.

noting that

$$\hat{\boldsymbol{\sigma}}_{\pm} e^{\pm i\theta} = \frac{\hat{\mathbf{x}}' \pm i\hat{\mathbf{y}}'}{\sqrt{2}}, \tag{4.2.27}$$

where $\hat{\mathbf{x}}'$ and $\hat{\mathbf{y}}'$ are polarization unit vectors in a new coordinate system rotated by angle θ with respect to the original coordinate system, that is,

$$x' = x\cos\theta - y\sin\theta, \tag{4.2.28a}$$

$$y' = x\sin\theta + y\cos\theta. \tag{4.2.28b}$$

Measurement of the rotation angle θ provides a sensitive method for determining the nonlinear coefficient B (see also Eqs. (4.2.24) and (4.2.26a)).

As mentioned above, there are two cases in which the polarization ellipse does not rotate. One case is that of circularly polarized light. In this case only one of the $\hat{\boldsymbol{\sigma}}_{\pm}$ components is present, and we see from Eq. (4.2.23) that the change in refractive index is given by

$$\delta n_{\text{circular}} = \frac{2\pi}{n_0} A|E|^2, \tag{4.2.29}$$

which clearly depends on the coefficient A but not upon the coefficient B. The other case in which there is no rotation is that of linearly polarized light. Since linearly polarized light is a combination of equal amounts of left- and right-hand circular components (i.e., $|E_-|^2 = |E_+|^2$), we see directly from Eq. (4.2.24) that the index difference Δn vanishes. If we let E denote the total field amplitude of the linearly polarized radiation, so that $|E|^2 = 2|E_+|^2 = 2|E_-|^2$, we find from Eq. (4.2.23) that for linearly polarized light the change in refractive index is given by

$$\delta n_{\text{linear}} = \frac{2\pi}{n_0}(A + \tfrac{1}{2}B)|E|^2. \tag{4.2.30}$$

Note that this change depends on the coefficients $A = 6\chi_{1122}$ and $B = 6\chi_{1221}$ as $A + \frac{1}{2}B$, which according to Eqs. (4.2.2) and (4.2.9) is equal to $3\chi_{1111}$. We see from Eqs. (4.2.29) and (4.2.30) that, for the usual case in which A and B have the same sign, linearly polarized light experiences a larger nonlinear change in refractive index than does circularly polarized light. In general the relative change in refractive index, $\delta n_{\text{linear}}/\delta n_{\text{circular}}$, is equal to $1 + B/2A$, which for the mechanisms described after Eq. (4.2.10) becomes

$$\frac{\delta n_{\text{linear}}}{\delta n_{\text{circular}}} = \begin{cases} 4 & \text{for molecular orientation,} \\ \frac{3}{2} & \text{for nonresonant electronic nonlinearities,} \\ 1 & \text{for electrostriction.} \end{cases} \tag{4.2.31}$$

4.3. Nonresonant Electronic Nonlinearities

Nonresonant electronic nonlinearities occur as the result of the nonlinear response of bound electrons on an applied optical field. This nonlinearity usually is not particularly large ($\chi^{(3)} \sim 10^{-14}$ esu is typical), but is of considerable importance because it is present in all dielectric materials. Furthermore, recent work has shown that certain organic nonlinear optical materials (such as polydiacetylene) can have nonresonant third-order susceptibilities as large as 10^{-9} esu due to the response of delocalized π electrons.

Nonresonant electronic nonlinearities are extremely fast, since they involve only virtual processes. The characteristic response time of this process is the time required for the electron cloud to become distorted in response to an applied optical field. This response time can be estimated as the orbital period of the electron in its motion about the nucleus, which according to the Bohr model of the atom is given by

$$\tau = 2\pi a_0/v,$$

where $a_0 = 0.5 \times 10^{-8}$ cm is the Bohr radius of the atom and $v \simeq c/137$ is a typical electronic velocity. We hence find that $\tau \simeq 10^{-16}$ s.

Classical, Anharmonic Oscillator Model of Electronic Nonlinearities

A simple model of electronic nonlinearities is the classical, anharmonic oscillator model described in Section 1.4. According to this model, one assumes that the potential well binding the electron to the atomic nucleus deviates from the parabolic potential of the usual Lorentz model. We approximate the actual potential well as

$$U(\mathbf{r}) = \tfrac{1}{2}m\omega_0^2|\mathbf{r}|^2 - \tfrac{1}{4}mb|\mathbf{r}|^4, \tag{4.3.1}$$

where b is a phenomenological nonlinear constant whose value is of the order of ω_0^2/d^2, where d is a typical atomic dimension. By solving the equation of motion for an electron in such a potential well, we obtained the expression (1.4.52) for the third-order susceptibility. When applied to the case of the nonlinear refractive index, this expression becomes

$$\chi_{ijkl}^{(3)}(\omega = \omega + \omega - \omega) = \frac{Nbe^4[\delta_{ij}\delta_{kl} + \delta_{ik}\delta_{jl} + \delta_{il}\delta_{jk}]}{3m^3D(\omega)^3D(-\omega)}. \tag{4.3.2}$$

where $D(\omega) = \omega_0^2 - \omega^2 - 2i\omega\gamma$. In the notation of Maker and Terhune (Eq. (4.2.10)), this result implies that

$$A = B = \frac{2Nbe^4}{m^3D(\omega)^3D(-\omega)}. \tag{4.3.3}$$

Hence, according to the classical, anharmonic oscillator model of electronic nonlinearities, A is equal to B for any value of the optical field frequency (whether resonant or nonresonant). For the case of far-off-resonant excitation (i.e., $\omega \ll \omega_0$), we can replace $D(\omega)$ by ω_0^2 in Eq. (4.3.2). If in addition we set b equal to ω_0^2/d^2, we find that

$$\chi^{(3)} \simeq \frac{Ne^4}{m^3 \omega_0^6 d^2}. \tag{4.3.4}$$

For the typical values $N = 4 \times 10^{22}$ cm^{-3}, $d = 3 \times 10^{-8}$ cm, and $\omega_0 = 7 \times 10^{15}$ rad/s, we find that $\chi^{(3)} \simeq 2 \times 10^{-14}$ esu.

Quantum-Mechanical Model of Nonresonant Electronic Nonlinearities

Let us now calculate the third-order susceptibility describing the nonlinear refractive index, using the laws of quantum mechanics. Since we are interested primarily in the case of nonresonant excitation, we make use of the expression for the nonlinear susceptibility in the form given by Eq. (3.2.33), that is,

$$\chi_{kjih}^{(3)}(\omega_\sigma, \omega_r, \omega_q, \omega_p) = \frac{N}{\hbar^3} \mathscr{P}_F \times \sum_{lmn} \left[\frac{\mu_{gn}^k \mu_{nm}^j \mu_{ml}^i \mu_{lg}^h}{(\omega_{ng} - \omega_\sigma)(\omega_{mg} - \omega_q - \omega_p)(\omega_{lg} - \omega_p)} \right], \tag{4.3.5}$$

where $\omega_\sigma = \omega_r + \omega_q + \omega_p$. We want to apply this expression to the case of the nonlinear refractive index, with the frequencies arranged as $\chi_{kjih}^{(3)}(\omega, \omega, \omega, -\omega) = \chi_{kjih}^{(3)}(\omega = \omega + \omega - \omega)$. However, Eq. (4.3.5) appears to have divergent contributions for this choice of frequencies, because the factor $\omega_{mg} - \omega_q - \omega_p$ in the denominator vanishes when the dummy index m is equal to g and when $\omega_p = -\omega_q = \pm\omega$. However, this divergence in fact exists in appearance only (Orr and Ward, 1971; Hanna, Yuratich, and Cotter, 1979); we can readily rearrange Eq. (4.3.5) into a form where no divergence appears. We first rewrite Eq. (4.3.5) as

$$\chi_{kjih}^{(3)}(\omega_\sigma, \omega_r, \omega_q, \omega_p) = \frac{N}{\hbar^3} \mathscr{P}_F \times \left[{\sum_{lmn}}' \frac{\mu_{gn}^k \mu_{nm}^j \mu_{ml}^i \mu_{lg}^h}{(\omega_{ng} - \omega_\sigma)(\omega_{mg} - \omega_q - \omega_p)(\omega_{lg} - \omega_p)} - \sum_{ln} \frac{\mu_{gn}^k \mu_{ng}^j \mu_{gl}^i \mu_{lg}^h}{(\omega_{ng} - \omega_\sigma)(\omega_q + \omega_p)(\omega_{lg} - \omega_p)} \right]. \tag{4.3.6}$$

Here the prime on the first summation indicates that the terms corresponding to $m = g$ are to be omitted from the summation over m; these terms are displayed explicitly in the second summation. The second summation, which appears to be divergent for $\omega_q = -\omega_p$, is now rearranged. We make use of the identity

$$\frac{1}{XY} = \frac{1}{(X+Y)Y} + \frac{1}{(X+Y)X}, \tag{4.3.7}$$

with $X = \omega_q + \omega_p$ and $Y = \omega_{lg} - \omega_p$, to express Eq. (4.3.6) as

$$\chi^{(3)}_{kjih}(\omega_\sigma, \omega_r, \omega_q, \omega_p) = \frac{N}{\hbar^3} \mathscr{P}_\mathrm{F} \times \left[\sum_{lmn}{}' \frac{\mu^k_{gn}\mu^j_{nm}\mu^i_{ml}\mu^h_{lg}}{(\omega_{ng} - \omega_\sigma)(\omega_{mg} - \omega_q - \omega_p)(\omega_{lg} - \omega_p)} - \sum_{ln} \frac{\mu^k_{gn}\mu^j_{ng}\mu^i_{gl}\mu^h_{lg}}{(\omega_{ng} - \omega_\sigma)(\omega_{lg} + \omega_q)(\omega_{lg} - \omega_p)} \right] \tag{4.3.8}$$

in addition to the contribution

$$\mathscr{P}_\mathrm{F} \sum_{ln} \frac{\mu^k_{gn}\mu^j_{ng}\mu^i_{gl}\mu^h_{lg}}{(\omega_{ng} - \omega_\sigma)(\omega_{lg} + \omega_q)(\omega_q + \omega_p)}. \tag{4.3.9}$$

However, this additional contribution vanishes, because for every term of the form

$$\frac{\mu^k_{gn}\mu^j_{ng}\mu^i_{gl}\mu^h_{lg}}{(\omega_{ng} - \omega_\sigma)(\omega_{lg} + \omega_q)(\omega_q + \omega_p)} \tag{4.3.10a}$$

that appears in Eq. (4.3.9), there is another term with the dummy summation indices n and l interchanged, with the pair $(-\omega_\sigma, k)$ interchanged with (ω_q, i), and with the pair (ω_p, h) interchanged with (ω_r, j); this term is of the form

$$\frac{\mu^i_{gl}\mu^h_{lg}\mu^k_{gn}\mu^j_{ng}}{(\omega_{lg} + \omega_q)(\omega_{ng} - \omega_\sigma)(\omega_r - \omega_\sigma)}. \tag{4.3.10b}$$

Since $\omega_\sigma = \omega_p + \omega_q + \omega_r$, it follows that $(\omega_q + \omega_p) = -(\omega_r - \omega_\sigma)$, and hence the expressions (4.3.10a) and (4.3.10b) are equal in magnitude but opposite in sign. The expression (4.3.8) for the nonlinear susceptibility is thus equivalent to Eq. (4.3.5), but is more useful for our present purposes because no apparent divergences are present.

We now specialize Eq. (4.3.8) to the case of the nonlinear refractive index with the choice of frequencies given by $\chi^{(3)}_{kjih}(\omega, \omega, \omega, -\omega)$. When we expand the permutation operator $\mathscr{P}_\mathrm{F}$, we find that each displayed term in Eq. (4.3.8) actually represents 24 terms. The resonance nature of each such term can be analyzed by means of diagrams of the sort shown in Fig. 3.2.3.* Rather than considering all 48 terms of the expanded version of Eq. (4.3.8), let us consider only the nearly resonant terms, which would be expected to make the largest contributions to $\chi^{(3)}$. One finds, after detailed analysis of Eq. (4.3.8), that the resonant contribution to the nonlinear susceptibility is given by

$$\begin{aligned}\chi^{(3)}_{kjih}(\omega,\omega,\omega,-\omega) &= \chi^{(3)}_{kjih}(\omega=\omega+\omega-\omega)\\ &= \frac{N}{6\hbar^3}\Bigg(\sum_{lmn}{}' [\mu^k_{gn}\mu^h_{nm}\mu^i_{ml}\mu^j_{lg} + \mu^k_{gn}\mu^h_{nm}\mu^j_{ml}\mu^i_{lg}\\ &\quad + \mu^h_{gn}\mu^k_{nm}\mu^i_{ml}\mu^j_{lg} + \mu^h_{gn}\mu^k_{nm}\mu^j_{ml}\mu^i_{lg}]\\ &\quad \times \frac{1}{(\omega_{ng}-\omega)(\omega_{mg}-2\omega)(\omega_{lg}-\omega)}\\ &\quad - \sum_{ln} [\mu^k_{gn}\mu^j_{ng}\mu^h_{gl}\mu^i_{lg} + \mu^k_{gn}\mu^i_{ng}\mu^h_{gl}\mu^j_{lg}\\ &\quad + \mu^h_{gn}\mu^i_{ng}\mu^k_{gl}\mu^j_{lg} + \mu^h_{gn}\mu^j_{ng}\mu^k_{gl}\mu^i_{lg}]\\ &\quad \times \frac{1}{(\omega_{ng}-\omega)(\omega_{lg}-\omega)(\omega_{lg}-\omega)}\Bigg].\end{aligned} \tag{4.3.11}$$

Here the first summation represents two-photon-resonant processes and the second summation represents one-photon-resonant processes, as illustrated in Fig. 4.3.1.

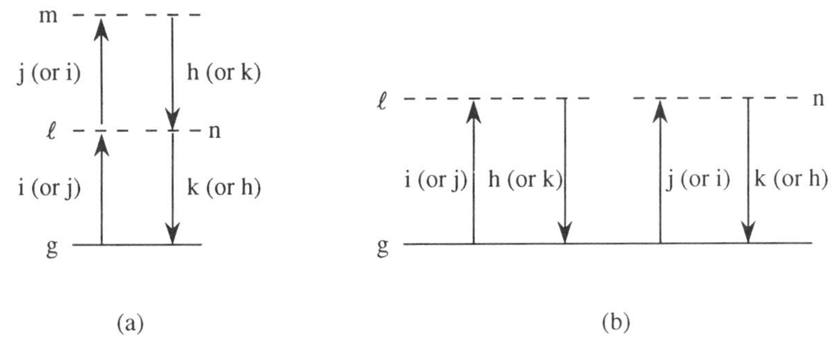

FIGURE 4.3.1 Resonance nature of the first (a) and second (b) summations of Eq. (4.3.11).

* Note, however, that Fig. 3.2.3 as drawn presupposes that the three input frequencies are all positive, whereas for the case of the nonlinear refractive index two of the input frequencies are positive and one is negative.

We can use Eq. (4.3.11) to obtain explicit expressions for the resonant contributions to the nonvanishing elements of the nonlinear susceptibility tensor for an isotropic medium. We find, for example, that $\chi_{1111}(\omega = \omega + \omega - \omega)$ is given by

$$\begin{aligned}\chi_{1111} = & \frac{2N}{3\hbar^3} \sum_{lmn}{}' \frac{\mu_{gn}^x \mu_{nm}^x \mu_{ml}^x \mu_{lg}^x}{(\omega_{ng} - \omega)(\omega_{mg} - 2\omega)(\omega_{lg} - \omega)} \\ & - \frac{2N}{3\hbar^3} \sum_{ln} \frac{\mu_{gn}^x \mu_{ng}^x \mu_{gl}^x \mu_{lg}^x}{(\omega_{ng} - \omega)(\omega_{lg} - \omega)(\omega_{lg} - \omega)}.\end{aligned} \tag{4.3.12}$$

Note that both one- and two-photon-resonant terms contribute to this expression. When ω is less than any resonance frequency of the material system, the two-photon contribution (the first term) tends to be positive. This contribution is positive because, in the presence of an applied optical field, there is some nonzero probability that the atom will reside in an excited state (state l or n as Fig. 4.3.1(a) is drawn). Since the (linear) polarizability of an atom in an excited state tends to be larger than that of an atom in the ground state, the effective polarizability of an atom is increased by the presence of an intense optical field; consequently this contribution to $\chi^{(3)}$ is positive. On the other hand, the one-photon contribution to χ_{1111} (the second term of Eq. (4.3.12)) is always negative when ω is less than any resonance frequency of the material system, because the product of matrix elements that appears in the numerator of this term is positive definite. We can understand this result from the point of view that the origin of one-photon-resonant contributions to the nonlinear susceptibility is saturation of the atomic response, which in the present case corresponds to a decrease of the positive linear susceptibility. We can also understand this result as a consequence of the ac Stark effect, which (as we shall see in Section 5.5) leads to an intensity-dependent increase in the separation of the lower and upper levels and consequently to a diminished optical response, as illustrated in Fig. 4.3.2.

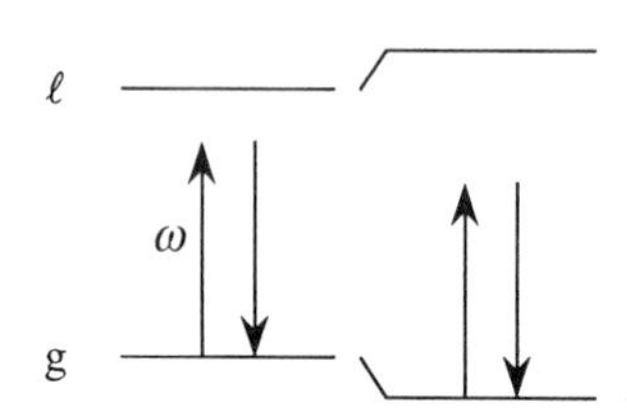

FIGURE 4.3.2 For $\omega < \omega_{lg}$ the ac Stark effect leads to an increase in the energy separation of the ground and excited states.

In a similar fashion, we find that the resonant contribution to χ_{1221} (or to $\frac{1}{6}B$ in the notation of Maker and Terhune) is given by

$$\chi_{1221} = \frac{2}{3}\frac{N}{\hbar^3}\sum_{lmn}{}' \frac{\mu_{gn}^x \mu_{nm}^x \mu_{ml}^y \mu_{lg}^y}{(\omega_{ng}-\omega)(\omega_{mg}-2\omega)(\omega_{lg}-\omega)}. \tag{4.3.13}$$

The one-photon-resonant terms do not contribute to χ_{1221}, since these terms involve the summation of the product of two matrix elements of the sort $\mu_{gl}^x \mu_{lg}^y$, and this contribution always vanishes.*

We also find that the resonant contribution to χ_{1122} (or to $\frac{1}{6}A$) is given by

$$\begin{aligned}\chi_{1122} &= \frac{N}{3\hbar^3}\sum_{lmn}{}' (\mu_{gn}^x \mu_{nm}^y \mu_{ml}^y \mu_{lg}^x + \mu_{gn}^x \mu_{nm}^y \mu_{ml}^x \mu_{lg}^y) \\ &\quad \times \frac{1}{(\omega_{ng}-\omega)(\omega_{mg}-2\omega)(\omega_{lg}-\omega)} \\ &\quad - \frac{N}{3\hbar^3}\sum_{ln} \frac{\mu_{gn}^x \mu_{ng}^x \mu_{gl}^y \mu_{lg}^y}{(\omega_{ng}-\omega)(\omega_{lg}-\omega)(\omega_{lg}-\omega)}.\end{aligned} \tag{4.3.14}$$

$\chi^{(3)}$ in the Low-Frequency Limit

In practice, we are often interested in determining the value of the third-order susceptibility under highly nonresonant conditions, that is, for the case in which the optical frequency is very much smaller than any resonance frequency of the atomic system. An example would be the nonlinear response of an insulating solid to visible radiation. In such cases, each of the terms in the expansion of the permutation operator in Eq. (4.3.8) makes an appreciable contribution to the nonlinear susceptibility, and no simplification such as those leading to Eqs. (4.3.11) through (4.3.14) is possible. It is an experimental fact that in the low-frequency limit both χ_{1122} and χ_{1221} (and consequently $\chi_{1111} = 2\chi_{1122} + \chi_{1221}$) are positive in sign. Also, the Kleinman symmetry condition becomes relevant under conditions of low-frequency excitation, which implies that χ_{1122} is equal to χ_{1221}, or that B is equal to A in the notation of Maker and Terhune.

We can use the results of the quantum-mechanical model to make an order-of-magnitude prediction of the size of the nonresonant third-order susceptibility. If we assume that the optical frequency ω is much smaller than all atomic resonance frequencies, we find from Eq. (4.3.5) that the nonresonant

* To see that this contribution vanishes, choose x to be the quantization axis. Then if μ_{gl}^x is nonzero, μ_{gl}^y must vanish, and vice versa.

TABLE 4.3.1 Nonlinear Optical Coefficients for Materials Showing Electronic Nonlinearities*

Material	χ_{1111} (esu)	n_2 (esu)
Diamond	15×10^{-14}	9×10^{-13}
Yttrium aluminum garnet	6×10^{-14}	3×10^{-13}
Sapphire	3×10^{-14}	2×10^{-13}
Borosilicate crown glass	2.5×10^{-14}	1.5×10^{-13}
Fused silica	2×10^{-14}	1.2×10^{-13}
CaF_2	1.6×10^{-14}	1×10^{-13}
LiF	1×10^{-14}	0.6×10^{-13}

* Values are obtained from optical frequency mixing experiments and hence do not include electrostrictive contributions, since electrostriction is a slow process that cannot respond at optical frequencies. The value of n_2 is calculated using $n_2 = 3\pi\chi_{1111}/n_0$. (Adapted from R. W. Hellwarth (1977), Tables 7.1 and 9.1.)

value of the nonlinear optical susceptibility is given by

$$\chi^{(3)} \simeq \frac{8N\mu^4}{\hbar^3\omega_0^3}, \tag{4.3.15}$$

where μ is a typical value of the dipole matrix element and ω_0 is a typical value of the atomic resonance frequency. It should be noted that while the predictions of the classical model (4.3.4) and the quantum-mechanical model (4.3.15) show different functional dependences on the displayed variables, the two expressions are equal if we identify d with the Bohr radius $a_0 = \hbar^2/me^2$, μ with the atomic unit of electric dipole moment $-ea_0$, and ω_0 with the Rydberg constant in angular frequency units, $\omega_0 = me^4/2\hbar^3$. Hence, the quantum-mechanical model also predicts that the third-order susceptibility is of the order of magnitude of 2×10^{-14} esu. The measured values of $\chi^{(3)}$ and n_2 for several materials that display nonresonant electronic nonlinearities are given in Table 4.3.1.

4.4. Nonlinearities Due to Molecular Orientation

Organic liquids that are composed of anisotropic molecules (i.e., molecules having an anisotropic polarizability tensor) typically possess a large value of n_2. The origin of this nonlinearity is the tendency of molecules to become aligned in the electric field of an applied optical wave. The optical wave then experiences a modified value of the refractive index because the average polarizability per molecule has been changed by the molecular alignment.

Consider, for example, the case of carbon disulfide (CS_2), which is illustrated in part (a) of Fig. 4.4.1. Carbon disulfide is a cigar-shaped molecule (i.e., a prolate spheroid), and consequently the polarizability α_3 experienced by an optical field that is parallel to its symmetry axis is larger than the polarizability α_1 experienced by a field that is perpendicular to its symmetry axis, that is,

$$\alpha_3 > \alpha_1. \tag{4.4.1}$$

Consider now what happens when such a molecule is subjected to a dc electric field, as shown in part (b) of the figure. Since α_3 is larger than α_1, the component of the induced dipole moment along the molecular axis will be disproportionately long. The induced dipole moment **p** thus will not be parallel to **E**, but will be offset from it in the direction of the symmetry axis. A torque

$$\boldsymbol{\tau} = \mathbf{p} \times \mathbf{E}. \tag{4.4.2}$$

will thus be exerted on the molecule. This torque is directed in such a manner as to twist the molecule into alignment with the applied electric field.

The tendency of the molecule to become aligned in the applied electric field is counteracted by thermal agitation, which tends to randomize the molecular orientation. We calculate the mean degree of molecular orientation through use of the Boltzmann factor. In order to do so, we first calculate the potential energy of the molecule in the applied electric field. If the applied field is changed by an amount $d\mathbf{E}$, the orientational potential energy is changed by the amount

$$dU = -\mathbf{p} \cdot d\mathbf{E} = -p_3\, dE_3 - p_1\, dE_1, \tag{4.4.3}$$

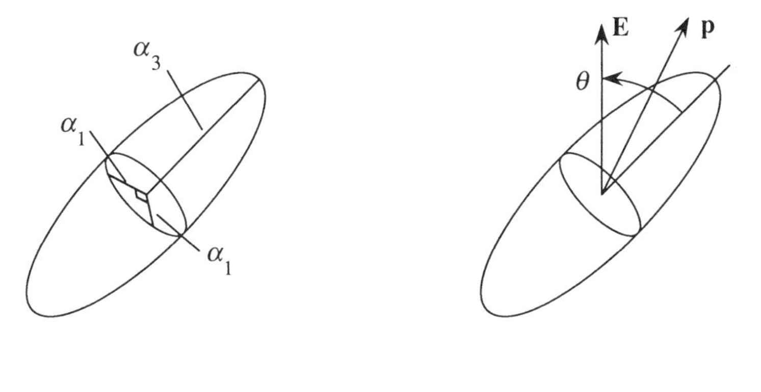

FIGURE 4.4.1 (a) A prolate spheroidal molecule such as carbon disulfide. (b) The dipole moment **p** induced by an electric field **E**.

where we have decomposed **E** into its components along the molecular axis (E_3) and perpendicular to the molecular axis (E_1). Since

$$p_3 = \alpha_3 E_3 \tag{4.4.4}$$

and

$$p_1 = \alpha_1 E_1, \tag{4.4.5}$$

we find that

$$dU = -\alpha_3 E_3\, dE_3 - \alpha_1 E_1\, dE_1, \tag{4.4.6}$$

which can be integrated to give

$$U = -\tfrac{1}{2}(\alpha_3 E_3^2 + \alpha_1 E_1^2). \tag{4.4.7}$$

If we now introduce the angle θ between **E** and the molecular axis (see Fig. 4.4.1b), we find that the orientational potential energy is given by

$$\begin{aligned} U &= -\tfrac{1}{2}[\alpha_3 E^2 \cos^2\theta + \alpha_1 E^2 \sin^2\theta] \\ &= -\tfrac{1}{2}\alpha_1 E^2 - \tfrac{1}{2}(\alpha_3 - \alpha_1)E^2 \cos^2\theta. \end{aligned} \tag{4.4.8}$$

Since $\alpha_3 - \alpha_1$ has been assumed to be positive, this result shows that the potential energy is lower when the molecular axis is parallel to **E** than when it is perpendicular to **E**, as illustrated in Fig. 4.4.2.

Our discussion thus far has assumed that the applied field is static. We now allow the field to vary in time at an optical frequency. For simplicity we assume that the light is linearly polarized; the general case of elliptical polarization is treated at the end of the present section. We thus replace **E** in Eq. (4.4.8) by the time-varying scalar quantity $\tilde{E}(t)$. The square of $\tilde{E}$ will contain frequency components near zero frequency and components at approximately twice the optical frequency ω. Since orientational relaxation times for molecules are typically of the order of a few picoseconds, the molecular orientation can respond to the frequency components near zero frequency but not to those

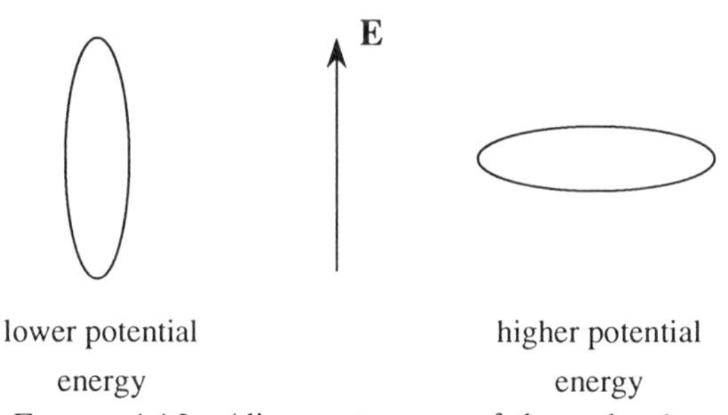

FIGURE 4.4.2 Alignment energy of the molecule.

near 2ω. We can thus formally replace E^2 in Eq. (4.4.8) by $\overline{\tilde{E}^2}$, where the bar denotes a time average over many cycles of the optical field.

We now calculate the intensity-dependent refractive index for such a medium. For simplicity, we first ignore local-field corrections, in which case the refractive index is given by

$$n^2 = 1 + 4\pi\chi = 1 + 4\pi N\langle\alpha\rangle, \tag{4.4.9}$$

where N is the number density of molecules and where $\langle\alpha\rangle$ denotes the expectation value of the molecular polarizability experienced by the incident radiation. To obtain an expression for $\langle\alpha\rangle$, we note that the mean orientational potential energy is given by $\langle U\rangle = -\frac{1}{2}|E|^2\langle\alpha\rangle$, which by comparison with the average of Eq. (4.4.8) shows that

$$\langle\alpha\rangle = \alpha_3\langle\cos^2\theta\rangle + \alpha_1\langle\sin^2\theta\rangle = \alpha_1 + (\alpha_3 - \alpha_1)\langle\cos^2\theta\rangle. \tag{4.4.10}$$

Here $\langle\cos^2\theta\rangle$ denotes the expectation value of $\cos^2\theta$ in thermal equilibrium and is given in terms of the Boltzmann distribution as

$$\langle\cos^2\theta\rangle = \frac{\int d\Omega \cos^2\theta \exp[-U(\theta)/kT]}{\int d\Omega \exp[-U(\theta)/kT]}, \tag{4.4.11}$$

where $\int d\Omega$ denotes an integration over all solid angles. For convenience, we introduce the intensity parameter

$$J = \tfrac{1}{2}(\alpha_3 - \alpha_1)\overline{\tilde{E}^2}/kT, \tag{4.4.12}$$

and let $d\Omega = 2\pi \sin\theta\, d\theta$. We then find that $\langle\cos^2\theta\rangle$ is given by

$$\langle\cos^2\theta\rangle = \frac{\int_0^\pi \cos^2\theta \exp(J\cos^2\theta)\sin\theta\, d\theta}{\int_0^\pi \exp(J\cos^2\theta)\sin\theta\, d\theta}. \tag{4.4.13}$$

Equations (4.4.9) through (4.4.13) can be used to determine the refractive index experienced by fields of arbitrary intensity $\overline{\tilde{E}^2}$.

Let us first calculate the refractive index experienced by a weak optical field, by taking the limit $J \to 0$. For this case we find that the average value of $\cos^2\theta$ is given by

$$\langle\cos^2\theta\rangle_0 = \frac{\int_0^\pi \cos^2\theta \sin\theta\, d\theta}{\int_0^\pi \sin\theta\, d\theta} = \frac{1}{3}, \tag{4.4.14}$$

and that according to Eq. (4.4.10) the mean polarizability is given by

$$\langle \alpha \rangle_0 = \tfrac{1}{3}\alpha_3 + \tfrac{2}{3}\alpha_1. \tag{4.4.15}$$

Using Eq. (4.4.9), we find that the refractive index is given by

$$n_0^2 = 1 + 4\pi N(\tfrac{1}{3}\alpha_3 + \tfrac{2}{3}\alpha_1). \tag{4.4.16}$$

Note that this result makes good physical sense: in the absence of interactions that tend to align the molecules, the mean polarizability is equal to one-third of that associated with the direction of the symmetry axis of the molecule plus two-thirds of that associated with directions perpendicular to this axis.

For the general case in which an intense optical field is applied, we find from Eqs. (4.4.9) and (4.4.10) that the refractive index is given by

$$n^2 = 1 + 4\pi N[\alpha_1 + (\alpha_3 - \alpha_1)\langle \cos^2\theta \rangle], \tag{4.4.17}$$

and hence by comparison with Eq. (4.4.16) that the square of the refractive index changes by the amount

$$\begin{aligned} n^2 - n_0^2 &= 4\pi N[\tfrac{1}{3}\alpha_1 + (\alpha_3 - \alpha_1)\langle \cos^2\theta \rangle - \tfrac{1}{3}\alpha_3] \\ &= 4\pi N(\alpha_3 - \alpha_1)(\langle \cos^2\theta \rangle - \tfrac{1}{3}). \end{aligned} \tag{4.4.18}$$

Since $n^2 - n_0^2$ is usually very much smaller than n_0^2, we can express the left-hand side of this equation as

$$n^2 - n_0^2 = (n - n_0)(n + n_0) \simeq 2n_0(n - n_0)$$

and hence find that the refractive index can be expressed as

$$n = n_0 + \delta n, \tag{4.4.19}$$

where the nonlinear change in refractive index is given by

$$\delta n \equiv n - n_0 = \frac{2\pi N}{n_0}(\alpha_3 - \alpha_1)(\langle \cos^2\theta \rangle - \tfrac{1}{3}). \tag{4.4.20}$$

The quantity $\langle \cos^2\theta \rangle$, given by Eq. (4.4.13), can be calculated in terms of a tabulated function (the Dawson integral). Figure 4.4.3 shows a plot of $\langle \cos^2\theta \rangle - \frac{1}{3}$ as a function of the intensity parameter $J = \frac{1}{2}(\alpha_3 - \alpha_1)\overline{\tilde{E}^2}/kT$.

In order to obtain an explicit formula for the change in refractive index, we expand the exponentials appearing in Eq. (4.4.13) and integrate the resulting expression term by term. We find that

$$\langle \cos^2\theta \rangle = \frac{1}{3} + \frac{4J}{45} + \frac{8J^2}{945} + \cdots. \tag{4.4.21}$$

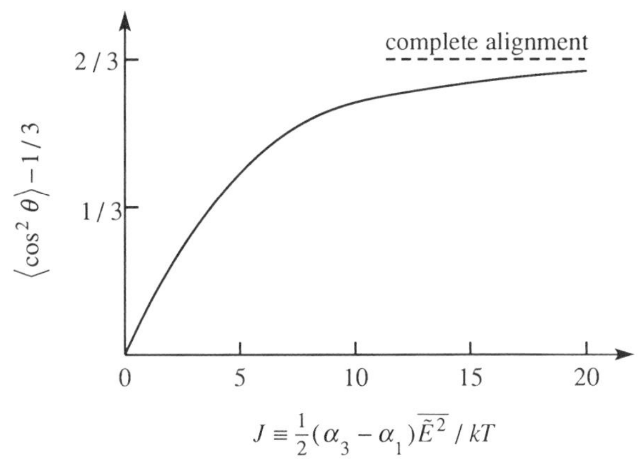

FIGURE 4.4.3 Variation of the quantity $(\langle\cos^2\theta\rangle - \frac{1}{3})$, which is proportional to the nonlinear change in refractive index δn, with the intensity parameter J. Note that for $J \lesssim 5$, δn increases nearly linearly with J.

Dropping all terms but the first two, we find from (4.4.20) that the change in the refractive index due to the nonlinear interaction is given by

$$\delta n = \frac{4\pi N}{2n_0}(\alpha_3 - \alpha_1)\frac{4J}{45} = \frac{4\pi}{45}\frac{N}{n_0}(\alpha_3 - \alpha_1)^2 \frac{\overline{\tilde{E}^2}}{kT}. \tag{4.4.22}$$

We can express this result as

$$\delta n = \bar{n}_2 \overline{\tilde{E}^2}, \tag{4.4.23}$$

where the second-order nonlinear refractive index is given by

$$\bar{n}_2 = \frac{4\pi N}{45 n_0}\frac{(\alpha_3 - \alpha_1)^2}{kT}. \tag{4.4.24}$$

Note that $\bar{n}_2$ is positive both for the case $\alpha_3 > \alpha_1$ (the case that we have been considering explicitly) and for the opposite case where $\alpha_3 < \alpha_1$. The reason for this behavior is that the torque experienced by the molecule is always directed in a manner that tends to align the molecule so that the light sees a *larger* value of the polarizability.

A more accurate prediction of the nonlinear refractive index is obtained by including the effects of local-field corrections. We begin with the Lorentz–Lorenz law (see also Eq. (3.8.8)).

$$\frac{n^2 - 1}{n^2 + 2} = \tfrac{4}{3}\pi N\langle\alpha\rangle \tag{4.4.25}$$

instead of the approximate relationship (4.4.9). By repeating the derivation leading to Eq. (4.4.24) with Eq. (4.4.9) replaced by Eq. (4.4.25) and with $\overline{\tilde{E}^2}$ replaced by the square of the Lorentz local field (see the discussion of Section 3.8), we find that the second-order nonlinear refractive index is given by

$$\bar{n}_2 = \frac{4\pi N}{45 n_0}\left(\frac{n_0^2+2}{3}\right)^4 \frac{(\alpha_3-\alpha_1)^2}{kT}. \tag{4.4.26}$$

Note that this result is consistent with the general prescription given in Section 3.8, which states that local-field effects can be included by multiplying the results obtained in the absence of local field corrections (that is, Eq. (4.4.24)) by the local-field correction factor $\mathscr{L}^{(3)} = [(n_0^2+2)/3]^4$ of Eq. (3.8.25).

Finally, we quote some numerical values appropriate to the case of carbon disulfide. The maximum possible value of δn is 0.58, and would correspond to a complete alignment of the molecules. The value $J = 1$ corresponds to a field strength of $E \simeq 3 \times 10^7$ V/cm. The value of $\bar{n}_2$ is hence equal to 1.3×10^{-11} esu.

Tensor Properties of $\chi^{(3)}$ for the Molecular Orientation Effect

Let us now consider the nonlinear response of a collection of anisotropic molecules to light of arbitrary polarization. Close *et al.* (1966) have shown that the mean polarizability in thermal equilibrium for a molecule whose three principal polarizabilities a, b, and c are distinct can be represented as

$$\langle\alpha_{ij}\rangle = \alpha\,\delta_{ij} + \gamma_{ij}, \tag{4.4.27}$$

where the linear contribution to the mean polarizability is given by

$$\alpha = \tfrac{1}{3}(a+b+c), \tag{4.4.28}$$

and where the lowest-order nonlinear correction term is given by

$$\gamma_{ij} = C\sum_{kl}(3\delta_{ik}\delta_{jl} - \delta_{ij}\delta_{kl})\overline{\tilde{E}_k^{\text{loc}}(t)\tilde{E}_l^{\text{loc}}(t)}. \tag{4.4.29}$$

Here the constant C is given by

$$C = \frac{(a-b)^2+(b-c)^2+(a-c)^2}{90kT}, \tag{4.4.30}$$

and $\tilde{E}^{\text{loc}}$ denotes the Lorentz local field. In the appendix to this section, we derive the result given by Eqs. (4.4.27) through (4.4.30) for the special case of an axially symmetric molecule; the derivation for the general case is left as a homework problem. Next, we use these results to determine the form of the third-order susceptibility tensor. We first ignore local-field corrections and replace $\tilde{E}_k^{\text{loc}}(t)$ by the macroscopic electric field $\tilde{E}_k(t)$, which we represent as

$$\tilde{E}_k(t) = E_k e^{-i\omega t} + \text{c.c.} \tag{4.4.31}$$

The electric-field-dependent factor appearing in Eq. (4.4.29) thus becomes

$$\overline{\tilde{E}_k^{\text{loc}}(t)\tilde{E}_l^{\text{loc}}(t)} = E_k E_l^* + E_k^* E_l. \tag{4.4.32}$$

Since we are ignoring local-field corrections, we can assume that the polarization is given by

$$P_i = \sum_j N\langle\alpha_{ij}\rangle E_j \tag{4.4.33}$$

and hence that the third-order contribution to the polarization is given by

$$P_i^{(3)} = N\sum_j \gamma_{ij} E_j. \tag{4.4.34}$$

By introducing the form for γ_{ij} given by Eqs. (4.4.29) and (4.4.32) into this expression, we find that

$$P_i^{(3)} = NC\sum_{jkl}(3\delta_{ik}\delta_{jl} - \delta_{ij}\delta_{kl})(E_k E_l^* + E_k^* E_l)E_j,$$

which can be written entirely in vector form as

$$\begin{aligned} \mathbf{P}^{(3)} &= NC[3(\mathbf{E}\cdot\mathbf{E}^*)\mathbf{E} + 3(\mathbf{E}\cdot\mathbf{E})\mathbf{E}^* - (\mathbf{E}\cdot\mathbf{E}^*)\mathbf{E} - (\mathbf{E}\cdot\mathbf{E}^*)\mathbf{E}] \\ &= NC[(\mathbf{E}\cdot\mathbf{E}^*)\mathbf{E} + 3(\mathbf{E}\cdot\mathbf{E})\mathbf{E}^*]. \end{aligned} \tag{4.4.35}$$

This result can be rewritten using the notation of Maker and Terhune (*cf.* Eq. (4.2.10)) as

$$P^{(3)} = A(\mathbf{E}\cdot\mathbf{E}^*)\mathbf{E} + \tfrac{1}{2}B(\mathbf{E}\cdot\mathbf{E})\mathbf{E}^*, \tag{4.4.36}$$

where the coefficients A and B are given by $B = 6A = 6NC$, which through use of the expression (4.4.30) for C becomes

$$B = 6A = N\left[\frac{(a-b)^2 + (b-c)^2 + (a-c)^2}{15kT}\right]. \tag{4.4.37}$$

This result shows that for the molecular orientation effect the ratio B/A is equal to 6, a result quoted earlier without proof (in Eq. (4.2.13a)). As in Eq. (4.4.26), local-field corrections can be included in the present formalism by replacing Eq. (4.4.37) by

$$B = 6A = \left(\frac{n_0^2 + 2}{3}\right)^4 N\left[\frac{(a-b)^2 + (b-c)^2 + (a-c)^2}{15kT}\right]. \tag{4.4.38}$$

Appendix to Section 4.4

The derivation that we presented above of the vector form of the nonlinear polarization due to the molecular orientation effect presupposed the validity of the starting equations (4.4.27) through (4.4.30). Here we derive these starting equations for the special case of an axially symmetric molecule. The derivation follows closely that of Owyoung (1971).

Consider an axially symmetric molecule whose polarizability tensor in the principal-axis coordinate system is described by

$$\boldsymbol{\alpha}^{\mathrm{P}} = \alpha_{ij}^{\mathrm{P}} = \begin{bmatrix} \alpha_1 & 0 & 0 \\ 0 & \alpha_1 & 0 \\ 0 & 0 & \alpha_3 \end{bmatrix}. \tag{4.4.39}$$

We need to express α_{ij}^{P} in a space-fixed (laboratory) coordinate system. The orientation of the molecule in this system can be described by the three Euler angles θ, ϕ, and ψ illustrated in Fig. 4.4.4. Here θ is the polar angle and ϕ is the azimuthal angle. The angle ψ specifies the rotation angle about the molecular α_3 axis; for the present case of a symmetric molecule ($\alpha_1 = \alpha_2$) this angle need not be specified.

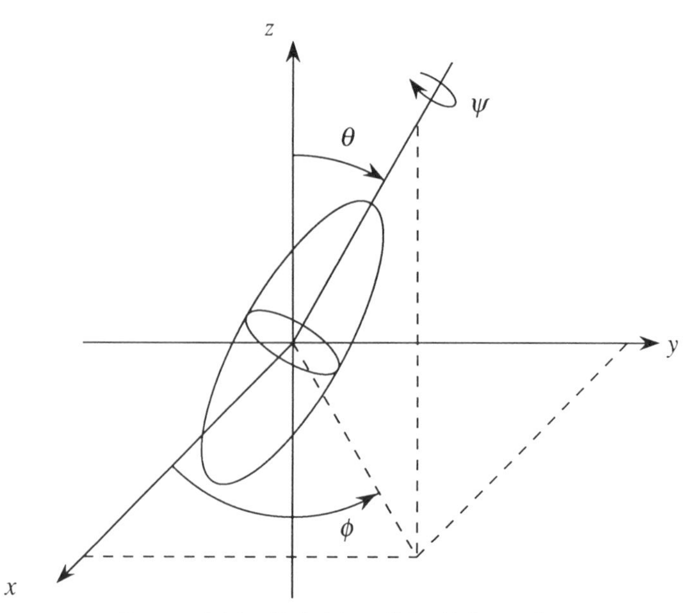

FIGURE 4.4.4 Definition of the Euler angles.

In the space-fixed coordinate system, the polarizability tensor is given by

$$\boldsymbol{\alpha}(\theta, \phi) = \mathbf{A}^T \boldsymbol{\alpha}^P \mathbf{A} \tag{4.4.40}$$

where $\mathbf{A}$ is the transformation matrix

$$= \begin{bmatrix} -\cos\psi\sin\phi - \cos\theta\cos\phi\sin\psi & \cos\psi\cos\phi - \cos\theta\sin\phi\sin\psi & \sin\psi\sin\theta \\ \sin\psi\sin\phi - \cos\theta\cos\phi\cos\psi & -\sin\psi\cos\phi - \cos\theta\sin\phi\cos\psi & \cos\psi\sin\theta \\ \sin\theta\cos\phi & \sin\theta\sin\phi & \cos\theta \end{bmatrix}, \tag{4.4.41}$$

and $\mathbf{A}^{\mathrm{T}}$ is its transpose. Through use of Eqs. (4.4.39) through (4.4.41) we find that

$$\phi) = \begin{bmatrix} \alpha_1 + (\alpha_3 - \alpha_1)\sin^2\theta\cos^2\phi & (\alpha_3 - \alpha_1)\sin^2\theta\sin\phi\cos\phi & (\alpha_3 - \alpha_1)\cos\theta\sin\theta\cos\phi \\ (\alpha_3 - \alpha_1)\sin^2\theta\sin\phi\cos\phi & \alpha_1 + (\alpha_3 - \alpha_1)\sin^2\theta\sin^2\phi & (\alpha_3 - \alpha_1)\cos\theta\sin\theta\sin\phi \\ (\alpha_3 - \alpha_1)\sin\theta\cos\theta\cos\phi & (\alpha_3 - \alpha_1)\sin\theta\cos\theta\sin\phi & \alpha_1 + (\alpha_3 - \alpha_1)\cos^2\theta \end{bmatrix}. \tag{4.4.42}$$

Note that this expression for the polarizability tensor in the laboratory coordinate system is independent of ψ, since $\boldsymbol{\alpha}$ is symmetric with respect to its principal axes 1 and 2.

We now calculate the mean polarizability in thermal equilibrium for an ensemble of such molecules in the presence of an applied electric field. The probability density that a given molecule will have its major axis oriented at angles (θ, ϕ) is given by

$$P(\theta, \phi) = \frac{\exp[-U(\theta, \phi)/kT]}{\int d\Omega \exp[-U(\theta, \phi)/kT]}, \tag{4.4.43}$$

where the orientational energy is given by

$$U(\theta, \phi) = -\frac{1}{2}\sum_{kl} \alpha_{kl}(\theta, \phi)\overline{\tilde{E}_k^{\mathrm{loc}}(t)\tilde{E}_l^{\mathrm{loc}}(t)}. \tag{4.4.44}$$

The ensemble-averaged polarizability is then given by

$$\langle \alpha_{ij} \rangle = \int d\Omega\, \alpha_{ij}(\theta, \phi) P(\theta, \phi). \tag{4.4.45}$$

We assume that the ratio $U(\theta, \phi)/kT$ is much smaller than unity, so that the exponentials can be approximated as

$$\exp\left[-\frac{U(\theta, \phi)}{kT}\right] = 1 - \frac{U(\theta, \phi)}{kT}. \tag{4.4.46}$$

Eqs. (4.4.42)–(4.4.45) can then be combined to give

$$\langle \alpha_{ij} \rangle = \frac{\displaystyle\int \alpha_{ij}(\theta,\phi)\,d\Omega + \sum_{kl} \frac{\overline{\tilde{E}_k^{\text{loc}}(t)\tilde{E}_l^{\text{loc}}(t)}}{2kT} \int \alpha_{ij}(\theta,\phi)\alpha_{kl}(\theta,\phi)\,d\Omega}{\displaystyle\int d\Omega + \sum_{kl} \frac{\overline{\tilde{E}_k^{\text{loc}}\tilde{E}_l^{\text{loc}}(t)}}{2kT} \int \sigma_{kl}(\theta,\phi)\,d\Omega}. \tag{4.4.47}$$

We note that $\int d\Omega = 4\pi$ and that the second term in the denominator is much smaller than the first. We thus expand the reciprocal of the denominator as a power series in the ratio of the second to first terms and find that to lowest order

$$\begin{aligned}\langle \alpha_{ij} \rangle = &\int \frac{d\Omega}{4\pi} \alpha_{ij}(\theta,\phi) + \sum_{kl} \frac{\overline{\tilde{E}_k^{\text{loc}}(t)\tilde{E}_l^{\text{loc}}(t)}}{2kT} \int \frac{d\Omega}{4\pi} \alpha_{ij}(\theta,\phi)\alpha_{kl}(\theta,\phi) \\ &- \sum_{kl} \frac{\overline{\tilde{E}_k^{\text{loc}}(t)\tilde{E}_l^{\text{loc}}(t)}}{2kT} \int \frac{d\Omega}{4\pi} \alpha_{ij}(\theta,\phi) \int \frac{d\Omega}{4\pi} \alpha_{kl}(\theta,\phi).\end{aligned} \tag{4.4.48}$$

The integrations can be performed explicitly. We let

$$\int d\Omega \rightarrow \int_0^{2\pi} d\phi \int_{-1}^{1} d(\cos\theta) \tag{4.4.49}$$

and find that

$$\int \frac{d\Omega}{4\pi} \alpha_{ij}(\theta,\phi) \equiv \langle \alpha_{ij} \rangle_0 = \alpha\,\delta_{ij}, \tag{4.4.50}$$

where

$$\alpha = \tfrac{2}{3}\alpha_1 + \tfrac{1}{3}\alpha_3, \tag{4.4.51}$$

and that

$$\begin{aligned}\int \frac{d\Omega}{4\pi} \alpha_{ij}(\theta,\phi)\alpha_{kl}(\theta,\phi) &\equiv \langle \alpha_{ij}\alpha_{kl} \rangle_0 \\ &= \begin{cases} \frac{4}{45}(\alpha_3-\alpha_1)^2 + \alpha^2 & \text{for} \quad i=j=k=l, \\ \alpha^2 - \frac{2}{45}(\alpha_3-\alpha_1)^2 & \text{for} \quad i=j\neq k=l, \\ \frac{1}{15}(\alpha_3-\alpha_1)^2 & \text{for} \quad i=k\neq j=l \\ & \text{or} \quad i=l\neq j=k. \end{cases}\end{aligned} \tag{4.4.52}$$

These results can be combined to represent the polarizability as

$$\langle a_{ij} \rangle = \langle \alpha_{ij} \rangle_0 + \sum_{kl} \frac{\overline{\tilde{E}_k^{\text{loc}}(t)\tilde{E}_l^{\text{loc}}(t)}}{2kT} (\langle \alpha_{ij}\alpha_{kl} \rangle_0 - \langle \alpha_{ij} \rangle_0 \langle \alpha_{kl} \rangle_0), \tag{4.4.53}$$

which can be written as

$$\langle \alpha_{ij} \rangle = \alpha \, \delta_{ij} + \gamma_{ij}, \tag{4.4.54}$$

where

$$\gamma_{ij} = \sum_{kl} \frac{\overline{\tilde{E}_k^{\text{loc}}(t) \tilde{E}_l^{\text{loc}}(t)}}{2kT} \tfrac{2}{45} (\alpha_3 - \alpha_1)^2 [\tfrac{3}{2}(\delta_{ik} \, \delta_{jl} + \delta_{il} \, \delta_{jk}) - \delta_{ij} \, \delta_{kl}]. \tag{4.4.55}$$

Note that since $\tilde{E}_k^{\text{loc}}(t)$ and $\tilde{E}_l^{\text{loc}}(t)$ appear in this last expression only as a symmetric product, we can replace the terms within the parentheses by $2\delta_{ik} \, \delta_{jl}$. We thus obtain the desired form

$$\gamma_{ij} = \sum_{kl} \frac{\overline{\tilde{E}_k^{\text{loc}}(t) \tilde{E}_l^{\text{loc}}(t)}}{45kT} (\alpha_3 - \alpha_1)^2 (3 \, \delta_{ik} \, \delta_{jl} - \delta_{ij} \, \delta_{kl}). \tag{4.4.56}$$

Note that γ_{ij} is traceless, that is, that $\sum_i \gamma_{ii} = 0$. Intuitively, we expect γ_{ij} to be traceless, since in applying an optical field to the medium we have not "added" any new polarizability to it; we have simply "rearranged" the polarizability that was initially present among the various tensor components.

Problems

1. Derive Eq. (4.2.2)

2. A 1-cm-long sample of carbon disulfide is illuminated by elliptically polarized light of intensity $I = 1 \text{ MW/cm}^2$. Determine how the angle through which the polarization ellipse is rotated depends upon the ellipticity of the light, and calculate numerically the maximum value of the rotation angle. Quantify the ellipticity in terms of the parameter δ $(-1 \leq \delta \leq 1)$ which defines the polarization unit vector through the relation

$$\hat{\boldsymbol{\epsilon}} = \frac{\hat{\mathbf{x}} + i \, \delta \hat{\mathbf{y}}}{(1 + \delta^2)^{1/2}}.$$

[Hint: The third-order susceptibility of carbon disulfide is due mainly to molecular orientation.]

3. Verify the statement made in the text that the first term in expression (4.3.12) is positive whenever ω is smaller than any resonance frequency of the atomic system.

4. Derive Eq. (4.4.27) for the general case in which a, b, and c are all distinct.

[This problem is extremely challenging.]

References

R. G. Brewer, J. R. Lifsitz, E. Garmire, R. Y. Chiao, and C. H. Townes, *Phys. Rev.* **166**, 326 (1968).

P. N. Butcher, *Nonlinear Optical Phenomena*, Ohio State University, 1965.

R. L. Carman, R. Y. Chiao, and P. L. Kelley, *Phys. Rev. Lett.* **17**, 1281 (1966).

R. Y. Chiao, and J. Godine, *Phys. Rev.* **185**, 430 (1969).

R. Y. Chiao, P. L. Kelley, and E. Garmire, *Phys. Rev. Lett.* **17**, 1158, (1966).

D. H. Close, C. R. Giuliano, R. W. Hellwarth, L. D. Hess, and F. J. McClung, *IEEE J. Quantum Electron.* **2**, 553 (1966).

D. C. Hanna, M. A. Yuratich, and D. Cotter, *Nonlinear Optics of Free Atoms and Molecules*, Springer-Verlag, Berlin, 1979.

R. W. Hellwarth, *Prog. Quantum Electron.* **5**, 1–68 (1977).

R. Landauer, *Phys. Lett.* **25A**, 416 (1967).

P. D. Maker and R. W. Terhune, *Phys. Rev.* **137**, A801 (1965).

P. D. Maker, R. W. Terhune, and C. M. Savage, *Phys. Rev. Lett.* **12**, 507 (1964).

B. J. Orr and J. F. Ward, *Molecular Phys.* **20**, 513 (1971).

A. Owyoung, *The Origins of the Nonlinear Refractive Indices of Liquids and Glasses*, Ph.D. dissertation, California Institute of Technology, 1971.

S. Saikan and M. Kiguchi, *Opt. Lett.* **7**, 555 (1982).

O. Svelto, in *Progress in Optics VII*, edited by E. Wolf, North Holland, 1974.

Chapter 5

Nonlinear Optics in the Two-Level Approximation

5.1. Introduction

Our treatment of nonlinear optics in the previous chapters has made use of power series expansions to relate the response of a material system to the strength of the applied optical field. In simple cases, this relation can be taken to be of the form

$$\tilde{P}(t) = \chi^{(1)}\tilde{E}(t) + \chi^{(2)}\tilde{E}(t)^2 + \chi^{(3)}\tilde{E}(t)^3 + \cdots. \tag{5.1.1}$$

However, there are circumstances under which such a power series expansion does not converge, and under such circumstances different methods must be employed to describe nonlinear optical effects. One example is that of a saturable absorber, where the absorption coefficient α is related to the intensity $I = nc|E|^2/2\pi$ of the applied optical field by the relation

$$\alpha = \frac{\alpha_0}{1 + I/I_s}, \tag{5.1.2}$$

where α_0 is the weak-field absorption coefficient and I_s is an optical constant called the saturation intensity. We can expand this equation in a power series to obtain

$$\alpha = \alpha_0\left[1 - \frac{I}{I_s} + \left(\frac{I}{I_s}\right)^2 - \left(\frac{I}{I_s}\right)^3 + \cdots\right]. \tag{5.1.3}$$

However, this series converges only for $I < I_s$, and hence only in this limit can saturable absorption be described by means of a power series of the sort given by Eq. (5.1.1).

It is primarily under conditions where a transition of the material system is resonantly excited that perturbation techniques fail to provide an adequate description of the response of the system to an applied optical field. However, under such conditions it is usually adequate to deal only with the two atomic levels that are resonantly connected by the optical field. The increased complexity entailed in describing the atomic system in a nonperturbative manner is thus compensated in part by the ability to make the two-level approximation. When only two levels are included in the theoretical analysis, there is no need to perform the sums over *all* atomic states that appear in the general quantum-mechanical expressions for $\chi^{(3)}$ given in Chapter 3.

For the most part, in the present chapter we will study how a monochromatic beam of frequency ω interacts with a collection of two-level atoms. The treatment is thus an extension of that of the previous chapter, which treated the interaction of a monochromatic beam with a nonlinear medium in terms of the third-order susceptibility $\chi^{(3)}(\omega = \omega + \omega - \omega)$. In addition, in the last two sections of the present chapter we generalize the treatment by studying non-degenerate four-wave mixing involving a collection of two-level atoms.

Even though the two-level approximation ignores many of the features present in real atomic systems, there is still an enormous richness in the physical processes that are described within the two-level approximation. Some of the processes that can occur and that are described in the present chapter include saturation effects, power broadening, Rabi oscillations, and optical Stark shifts. Parallel treatments of optical nonlinearities in two-level atoms can be found in the books of Allen and Eberly (1975) and Cohen-Tannoudji, Dupont-Roc, and Grynberg (1989).

5.2. Density Matrix Equations of Motion for a Two-Level Atom

We first consider the density matrix equations of motion for a two-level system in the absence of damping effects. Since damping mechanisms can be very different under different physical conditions, there is no *unique* way to include damping in the model. The present treatment thus serves as a starting point for the inclusion of damping by any mechanism.

The interaction we are treating is illustrated in Fig 5.2.1. The lower atomic level is denoted a and the upper level b. We represent the Hamiltonian for this system as

$$\hat{H} = \hat{H}_0 + \hat{V}(t), \tag{5.2.1}$$

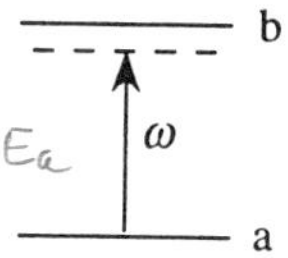

b

ω

a

FIGURE 5.2.1 Near-resonant excitation of a two-level atom.

where $\hat{H}_0$ denotes the atomic Hamiltonian and $\hat{V}(t)$ denotes the energy of interaction of the atom with the electromagnetic field. We denote the energies of the states a and b as

$$E_a = \hbar\omega_a \quad \text{and} \quad E_b = \hbar\omega_b. \tag{5.2.2}$$

The Hamiltonian $\hat{H}_0$ can thus be represented by the diagonal matrix whose elements are given by

$$H_{0,nm} = E_n \delta_{nm}. \tag{5.2.3}$$

We assume that the interaction energy can be adequately described in the electric dipole approximation, in which case the interaction Hamiltonian has the form

$$\hat{V}(t) = -\hat{\mu}\tilde{E}(t). \tag{5.2.4}$$

We also assume that the atomic wave functions corresponding to states a and b have definite parity so that the diagonal matrix elements of $\hat{\mu}$ vanish, that is, we assume that $\mu_{aa} = \mu_{bb} = 0$ and hence that

$$V_{aa} = V_{bb} = 0. \tag{5.2.5}$$

The only nonvanishing elements of $\tilde{V}$ are hence V_{ba} and V_{ab}, which are given explicitly by

$$V_{ba} = V_{ab}^* = -\mu_{ba}\tilde{E}(t). \tag{5.2.6}$$

We describe the state of this system in terms of the density matrix, which is given by

$$\hat{\rho} = \begin{bmatrix} \rho_{aa} & \rho_{ab} \\ \rho_{ba} & \rho_{bb} \end{bmatrix}, \tag{5.2.7}$$

where $\rho_{ba} = \rho_{ab}^*$. The time evolution of the density matrix is given in general by Eq. (3.3.21) as

$$\begin{aligned} \dot{\rho}_{nm} &= \frac{-i}{\hbar}[\hat{H}, \hat{\rho}]_{nm} = \frac{-i}{\hbar}[(\hat{H}\hat{\rho})_{nm} - (\hat{\rho}\hat{H})_{nm}] \\ &= \frac{-i}{\hbar}\sum_{\nu}(H_{n\nu}\rho_{\nu m} - \rho_{n\nu}H_{\nu m}). \end{aligned} \tag{5.2.8}$$

We now introduce the decomposition of the Hamiltonian into atomic and interaction parts (Eq. (5.2.1)) into this expression to obtain

$$\dot{\rho}_{nm} = -i\omega_{nm}\rho_{nm} - \frac{i}{\hbar}\sum_{\nu}(V_{n\nu}\rho_{\nu m} - \rho_{n\nu}V_{\nu m}), \tag{5.2.9}$$

where we have introduced the transition frequency $\omega_{nm} = (E_n - E_m)/\hbar$. For the case of the two-level atom, the indices n, m, and ν can take on the values a or b only, and the equations of motion for the density matrix elements are given explicitly as

$$\dot{\rho}_{ba} = -i\omega_{ba}\rho_{ba} + \frac{i}{\hbar}V_{ba}(\rho_{bb} - \rho_{aa}), \tag{5.2.10a}$$

$$\dot{\rho}_{bb} = \frac{-i}{\hbar}(V_{ba}\rho_{ab} - \rho_{ba}V_{ab}), \tag{5.2.10b}$$

$$\dot{\rho}_{aa} = \frac{-i}{\hbar}(V_{ab}\rho_{ba} - \rho_{ab}V_{ba}). \tag{5.2.10c}$$

It can be seen by inspection that

$$\dot{\rho}_{bb} + \dot{\rho}_{aa} = 0, \tag{5.2.11}$$

which shows that the total population $\rho_{bb} + \rho_{aa}$ is a conserved quantity. From the definition of the density matrix, we know that the diagonal elements of $\hat{\rho}$ represent probabilities of occupation, and hence that

$$\rho_{aa} + \rho_{bb} = 1. \tag{5.2.12}$$

No separate equation of motion is required for ρ_{ab}, because of the relation $\rho_{ab} = \rho_{ba}^*$.

Equations (5.2.10) constitute the density matrix equations of motion for a two-level atom in the absence of relaxation processes. These equations provide an adequate description of resonant nonlinear optical processes under conditions where relaxation processes can be neglected, such as excitation with short pulses whose duration is much less than the material relaxation times. We next see how these equations are modified in the presence of relaxation processes.

Closed Two-Level Atom

Let us first consider relaxation processes of the sort illustrated schematically in Fig. 5.2.2. We assume that the upper level b decays to the lower level at a

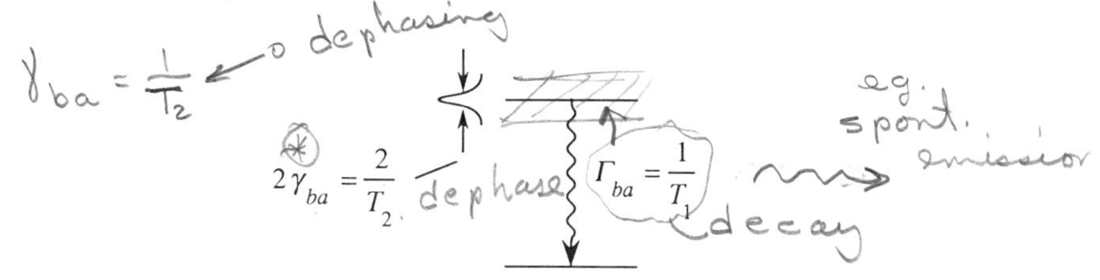

FIGURE 5.2.2 Relaxation processes of the closed two-level atom.

rate Γ_{ba} and therefore that the lifetime of the upper level is given by $T_1 = 1/\Gamma_{ba}$. Typically, the decay of the upper level would be due to spontaneous emission. This system is called closed, because any population that leaves the upper level enters the lower level. We also assume that the atomic dipole moment is dephased in the characteristic time T_2, leading to a transition linewidth (for weak applied fields) of characteristic width $\gamma_{ba} = 1/T_2$.*

We can describe these relaxation processes mathematically by adding decay terms phenomenologically to Eqs. (5.2.10); the modified equations are given by

$$\dot{\rho}_{ba} = -\left(i\omega_{ba} + \frac{1}{T_2}\right)\rho_{ba} + \frac{i}{\hbar}V_{ba}(\rho_{bb} - \rho_{aa}), \tag{5.2.13a}$$

$$\dot{\rho}_{bb} = \frac{-\rho_{bb}}{T_1} - \frac{i}{\hbar}(V_{ba}\rho_{ab} - \rho_{ba}V_{ab}), \tag{5.2.13b}$$

$$\dot{\rho}_{aa} = \frac{\rho_{bb}}{T_1} + \frac{i}{\hbar}(V_{ba}\rho_{ab} - \rho_{ba}V_{ab}). \tag{5.2.13c}$$

We can see by inspection that the condition

$$\dot{\rho}_{bb} + \dot{\rho}_{aa} = 0 \tag{5.2.14}$$

is still satisfied.

Since Eq. (5.2.13a) depends on the populations ρ_{bb} and ρ_{aa} only as the population difference, $\rho_{bb} - \rho_{aa}$, it is useful to consider the equation of motion satisfied by this difference. We subtract Eq. (5.2.13c) from Eq. (5.2.13b) to find that

$$\frac{d}{dt}(\rho_{bb} - \rho_{aa}) = \frac{-2\rho_{bb}}{T_1} - \frac{2i}{\hbar}(V_{ba}\rho_{ab} - \rho_{ba}V_{ab}). \tag{5.2.15}$$

The first term on the right-hand side can be rewritten using the relation

* In fact, as one can see from Eq. (5.3.25), the full width at half maximum in angular frequency units of the absorption line in the limit of weak fields is equal to $2\gamma_{ba}$.

$2\rho_{bb} = (\rho_{bb} - \rho_{aa}) + 1$ (which follows from Eq. (5.2.12)) to obtain

$$\frac{d}{dt}(\rho_{bb} - \rho_{aa}) = -\frac{(\rho_{bb} - \rho_{aa}) + 1}{T_1} - \frac{2i}{\hbar}(V_{ba}\rho_{ab} - \rho_{ba}V_{ab}). \tag{5.2.16}$$

This relation is often generalized by allowing the possibility that the population difference $(\rho_{bb} - \rho_{aa})^{\text{eq}}$ in thermal equilibrium can have some value other than -1, the value taken above by assuming that only downward spontaneous transitions could occur. This generalized version of Eq. (5.2.16) is given by

$$\frac{d}{dt}(\rho_{bb} - \rho_{aa}) = -\frac{(\rho_{bb} - \rho_{aa}) - (\rho_{bb} - \rho_{aa})^{\text{eq}}}{T_1} - \frac{2i}{\hbar}(V_{ba}\rho_{ab} - \rho_{ba}V_{ab}). \tag{5.2.17}$$

We therefore see that for a closed two-level system the density matrix equations of motion reduce to just two coupled equations, Eqs. (5.2.13a) and (5.2.17).

In order to justify the choice of relaxation terms used in Eqs. (5.2.13a) and (5.2.17), let us examine the nature of the solutions to these equations in the absence of an applied field, i.e., for $V_{ba} = 0$. The solution to Eq. (5.2.17) is

$$\begin{aligned}\rho_{bb}(t) - \rho_{aa}(t) &= (\rho_{bb} - \rho_{aa})^{\text{eq}} \\ &\quad + \{[\rho_{bb}(0) - \rho_{aa}(0)] - (\rho_{bb} - \rho_{aa})^{\text{eq}}\}e^{-t/T_1}.\end{aligned} \tag{5.2.18}$$

This equation shows that the population inversion $\rho_{bb}(t) - \rho_{aa}(t)$ relaxes from its initial value $\rho_{bb}(0) - \rho_{aa}(0)$ to its equilibrium value $(\rho_{bb} - \rho_{aa})^{\text{eq}}$ in a time of the order of T_1. For this reason, T_1 is called the population relaxation time.

Similarly, the solution to Eq. (5.2.13a) for the case $V_{ba} = 0$ is of the form

$$\rho_{ba}(t) = \rho_{ba}(0)e^{-(i\omega_{ba} + 1/T_2)t}. \tag{5.2.19}$$

We can interpret this result more directly by considering the expectation value of the induced dipole moment, which is given by

$$\begin{aligned}\langle\tilde{\mu}(t)\rangle &= \mu_{ab}\rho_{ba}(t) + \mu_{ba}\rho_{ab}(t) = \mu_{ab}\rho_{ba}(0)e^{-(i\omega_{ba} + 1/T_2)t} + \text{c.c.} \\ &= [\mu_{ab}\rho_{ba}(0)e^{-i\omega_{ba}t} + \text{c.c.}]e^{-t/T_2}.\end{aligned} \tag{5.2.20}$$

This result shows that, for an undriven atom, the dipole moment oscillates at frequency ω_{ba} and decays to zero in the characteristic time T_2, which is hence known as the dipole dephasing time.

For reasons that were discussed in relation to Eq. (3.3.25), T_1 and T_2 are related to the collisional dephasing rate γ_c by

$$\frac{1}{T_2} = \frac{1}{2T_1} + \gamma_c. \tag{5.2.21a}$$

For an atomic vapor, γ_c is usually described accurately by the formula

$$\gamma_c = C_s N + C_f N_f, \tag{5.2.21b}$$

where N is the number density of atoms having resonance frequency ω_{ba}, and N_f is the number density of any "foreign" atoms of a different atomic species having a different resonance frequency. The parameter C_s and C_f are coefficients describing self-broadening and foreign-gas broadening, respectively. As an example, for the resonance line (i.e., the 3s → 3p transition) of atomic sodium, T_1 is equal to 16 ns, $C_s = 1.50 \times 10^{-7}$ cm^3/s, and, for the case of foreign-gas broadening by collisions with argon atoms, $C_f = 2.53 \times 10^{-9}$ cm^3/s. The values of T_1, C_s, and C_f for other transitions are tabulated, for example, by Miles and Harris (1973).

Open Two-Level Atom

The open two-level atom is shown schematically in Fig. 5.2.3. Here the upper and lower levels are allowed to exchange population with associated reservoir levels. These levels might, for example, be magnetic sublevels or hyperfine levels associated with states a and b. The system is called open because the population that leaves the upper level does not necessarily enter the lower level. This model is often encountered in connection with laser theory, in which case the upper level or both levels are assumed to acquire population at some controllable pump rates, which we take to be λ_b and λ_a for levels b and a, respectively. In order to account for relaxation and pumping processes of the sort just described, the density matrix equations (5.2.10) are modified to become

$$\dot{\rho}_{ba} = -\left(i\omega_{ba} + \frac{1}{T_2}\right)\rho_{ba} + \frac{i}{\hbar} V_{ba}(\rho_{bb} - \rho_{aa}), \tag{5.2.22a}$$

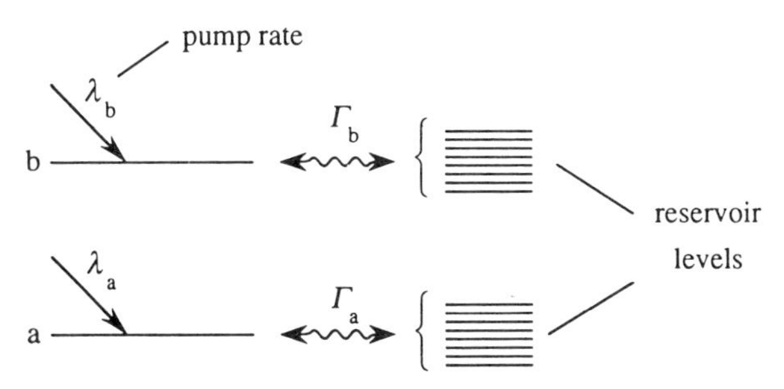

FIGURE 5.2.3 Relaxation processes for the open two-level atom.

$$\dot{\rho}_{bb} = \lambda_b - \Gamma_b(\rho_{bb} - \rho_{bb}^{\text{eq}}) - \frac{i}{\hbar}(V_{ba}\rho_{ab} - \rho_{ba}V_{ab}), \tag{5.2.22b}$$

$$\dot{\rho}_{aa} = \lambda_a - \Gamma_a(\rho_{aa} - \rho_{aa}^{\text{eq}}) + \frac{i}{\hbar}(V_{ba}\rho_{ab} - \rho_{ba}V_{ab}). \tag{5.2.22c}$$

Note that in this case the total population contained in the two levels a and b is not conserved, and that in general all three equations must be considered. The relaxation rates are related to the collisional dephasing rate γ_c and population rates Γ_b and Γ_a by

$$\frac{1}{T_2} = \tfrac{1}{2}(\Gamma_b + \Gamma_a) + \gamma_c.$$

Two-Level Atom with a Non-Radiatively-Coupled Third Level

The energy level scheme shown in Fig. 5.2.4 is often used to model a saturable absorber. Population spontaneously leaves the optically excited level b at a rate $\Gamma_{ba} + \Gamma_{bc}$, where Γ_{ba} is the rate of decay to the ground state a, and Γ_{bc} is the rate of decay to level c. Level c acts as a trap level; population decays from level c back to the ground state at a rate Γ_{ca}. In addition, any dipole moment associated with the transition between levels a and b is damped at a rate γ_{ba}. These relaxation processes are modeled by modifying Eqs. (5.2.10) to

$$\dot{\rho}_{ba} = -(i\omega_{ba} + \gamma_{ba})\rho_{ba} + \frac{i}{\hbar}V_{ba}(\rho_{bb} - \rho_{aa}), \tag{5.2.23a}$$

$$\dot{\rho}_{bb} = -(\Gamma_{ba} + \Gamma_{bc})\rho_{bb} - \frac{i}{\hbar}(V_{ba}\rho_{ab} - \rho_{ba}V_{ab}), \tag{5.2.23b}$$

$$\dot{\rho}_{cc} = \Gamma_{bc}\rho_{bb} - \Gamma_{ca}\rho_{cc}, \tag{5.2.23c}$$

$$\dot{\rho}_{aa} = \Gamma_{ba}\rho_{bb} + \Gamma_{ca}\rho_{cc} + \frac{i}{\hbar}(V_{ba}\rho_{ab} - \rho_{ba}V_{ab}). \tag{5.2.23d}$$

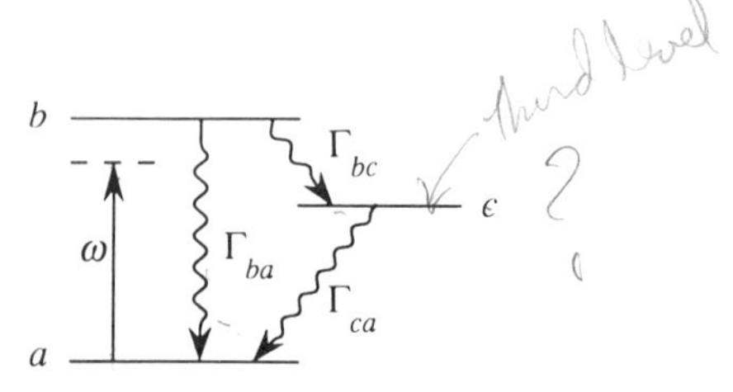

FIGURE 5.2.4 Relaxation processes for a two-level atom with a non-radiatively-coupled third level.

It can be seen by inspection that the population in the three levels is conserved, that is, that

$$\dot{\rho}_{aa} + \dot{\rho}_{bb} + \dot{\rho}_{cc} = 0.$$

5.3. Steady-State Response of a Two-Level Atom to a Monochromatic Field

We next examine the nature of the solution to the density matrix equations of motion for a two-level atom in the presence of a monochromatic, steady-state field. For definiteness, we treat the case of a closed two-level atom, although our results would be qualitatively similar for any of the models described above (see Problem 1 at the end of this chapter). For the closed two-level atomic system, the density matrix equations were shown above (Eqs. (5.2.13a) and (5.2.17)) to be of the form

$$\frac{d}{dt}\rho_{ba} = -\left(i\omega_{ba} + \frac{1}{T_2}\right)\rho_{ba} + \frac{i}{\hbar}V_{ba}(\rho_{bb} - \rho_{aa}), \tag{5.3.1}$$

$$\frac{d}{dt}(\rho_{bb} - \rho_{aa}) = -\frac{(\rho_{bb} - \rho_{aa}) - (\rho_{bb} - \rho_{aa})^{\text{eq}}}{T_1} - \frac{2i}{\hbar}(V_{ba}\rho_{ab} - \rho_{ba}V_{ab}). \tag{5.3.2}$$

In the electric dipole approximation, the interaction Hamiltonian for an applied field in the form of a monochromatic wave of frequency ω is given by

$$\hat{V} = -\hat{\mu}\tilde{E}(t) = -\hat{\mu}(Ee^{-i\omega t} + E^{*}e^{i\omega t}), \tag{5.3.3}$$

and the matrix elements of the interaction Hamiltonian are then given by

$$V_{ba} = -\mu_{ba}(Ee^{-i\omega t} + E^{*}e^{i\omega t}). \tag{5.3.4}$$

Equations (5.3.1) and (5.3.2) cannot be solved exactly for V_{ba} given by Eq. (5.3.4). However, they can be solved in an approximation known as the rotating wave approximation. We recall from the discussion of Eq. (5.2.20) that, in the absence of a driving field, ρ_{ba} tends to evolve in time as $\exp(-i\omega_{ba}t)$. For this reason, when ω is approximately equal to ω_{ba}, the part of V_{ba} that oscillates as $e^{-i\omega t}$ acts as a far more effective driving term for ρ_{ba} than does the part that oscillates as $e^{i\omega t}$. It is thus a good approximation to take V_{ba} not as Eq. (5.3.4) but instead as

$$V_{ba} = -\mu_{ba}Ee^{-i\omega t}. \tag{5.3.5}$$

This approximation is called the rotating wave approximation. Within this approximation, the density matrix equations of motion (5.3.1) and (5.3.2) become

$$\frac{d}{dt}\rho_{ba} = -\left(i\omega_{ba} + \frac{1}{T_2}\right)\rho_{ba} - \frac{i}{\hbar}\mu_{ba}Ee^{-i\omega t}(\rho_{bb} - \rho_{aa}), \tag{5.3.6}$$

$$\frac{d}{dt}(\rho_{bb} - \rho_{aa}) = -\frac{(\rho_{bb} - \rho_{aa}) - (\rho_{bb} - \rho_{aa})^{\text{eq}}}{T_1} + \frac{2i}{\hbar}(\mu_{ba}Ee^{-i\omega t}\rho_{ab} - \mu_{ab}E^*e^{i\omega t}\rho_{ba}). \tag{5.3.7}$$

Note that (in the rotating wave approximation) ρ_{ba} is driven only at nearly its resonance frequency ω_{ba}, and $\rho_{bb} - \rho_{aa}$ is driven only at nearly zero frequency, which is its natural frequency.

We next find the steady-state solution to Eqs. (5.3.6) and (5.3.7), that is, the solution that is valid long after the transients associated with the turn-on of the driving field have died out. We do so by introducing the slowly varying quantity σ_{ba}, defined by

$$\rho_{ba}(t) = \sigma_{ba}(t)e^{-i\omega t}. \tag{5.3.8}$$

Equations (5.3.6) and (5.3.7) then become

$$\frac{d}{dt}\sigma_{ba} = \left[i(\omega - \omega_{ba}) - \frac{1}{T_2}\right]\sigma_{ba} - \frac{i}{\hbar}\mu_{ba}E(\rho_{bb} - \rho_{aa}), \tag{5.3.9}$$

$$\frac{d}{dt}(\rho_{bb} - \rho_{aa}) = -\frac{(\rho_{bb} - \rho_{aa}) - (\rho_{bb} - \rho_{aa})^{\text{eq}}}{T_1} + \frac{2i}{\hbar}(\mu_{ba}E\sigma_{ab} - \mu_{ab}E^*\sigma_{ba}). \tag{5.3.10}$$

The steady-state solution can now be obtained by setting the left-hand sides of Eqs. (5.3.9) and (5.3.10) equal to zero. We thereby obtain two coupled equations, which we solve algebraically to obtain

$$\rho_{bb} - \rho_{aa} = \frac{(\rho_{bb} - \rho_{aa})^{\text{eq}}[1 + (\omega - \omega_{ba})^2T_2^2]}{1 + (\omega - \omega_{ba})^2T_2^2 + (4/\hbar^2)|\mu_{ba}|^2|E|^2T_1T_2}, \tag{5.3.11}$$

$$\rho_{ba} = \sigma_{ba}e^{-i\omega t} = \frac{\mu_{ba}Ee^{-i\omega t}(\rho_{bb} - \rho_{aa})}{\hbar(\omega - \omega_{ba} + i/T_2)}. \tag{5.3.12}$$

We now use this result to calculate the polarization (i.e., the dipole moment per unit volume), which is given in terms of the off-diagonal elements of the density matrix by

$$\tilde{P}(t) = N\langle\tilde{\mu}\rangle = N\,\mathrm{Tr}(\hat{\tilde{\rho}}\hat{\tilde{\mu}}) = N(\mu_{ab}\rho_{ba} + \mu_{ba}\rho_{ab}), \tag{5.3.13}$$

where N is the number density of atoms. We introduce the complex amplitude P of the polarization through the relation

$$\tilde{P}(t) = Pe^{-i\omega t} + \text{c.c.}, \tag{5.3.14}$$

and we define the susceptibility χ as the constant of proportionality relating P and E according to

$$P = \chi E. \tag{5.3.15}$$

We hence find from Eqs. (5.3.12) through (5.3.15) that the susceptibility is given by

$$\chi = \frac{N|\mu_{ba}|^2(\rho_{bb} - \rho_{aa})}{\hbar(\omega - \omega_{ba} + i/T_2)}, \tag{5.3.16}$$

where $\rho_{bb} - \rho_{aa}$ is given by Eq. (5.3.11). We introduce this expression for $\rho_{bb} - \rho_{aa}$ into Eq. (5.3.16) and rationalize the denominator to obtain the result

$$\chi = \frac{N(\rho_{bb} - \rho_{aa})^{\text{eq}}|\mu_{ba}|^2(\omega - \omega_{ba} - i/T_2)T_2^2/\hbar}{1 + (\omega - \omega_{ba})^2T_2^2 + (4/\hbar^2)|\mu_{ba}|^2|E|^2T_1T_2}. \tag{5.3.17}$$

Note that this expression gives the total susceptibility, including both its linear and nonlinear contributions.

We next introduce new notation to simplify this expression for the susceptibility. We introduce the quantity

$$\Omega = 2|\mu_{ba}||E|/\hbar, \tag{5.3.18}$$

which is known as the on-resonance Rabi frequency, and the quantity

$$\Delta = \omega - \omega_{ba}, \tag{5.3.19}$$

which is known as the detuning factor, so that the susceptibility can be expressed as

$$\chi = \left[N(\rho_{bb} - \rho_{aa})^{\text{eq}}|\mu_{ba}|^2\frac{T_2}{\hbar}\right]\frac{\Delta T_2 - i}{1 + \Delta^2T_2^2 + \Omega^2T_1T_2}. \tag{5.3.20}$$

Next, we express the combination of factors set off in square brackets in terms of the normal (i.e., linear) absorption coefficient of the material system, which is a directly measurable quantity. The absorption coefficient is given in general by

$$\alpha = 2\frac{\omega}{c}\,\text{Im}\,n = 2\frac{\omega}{c}\,\text{Im}[(1 + 4\pi\chi)^{1/2}], \tag{5.3.21a}$$

and whenever the condition $|\chi| \ll 1$ is valid, the absorption coefficient can be expressed as

$$\alpha = 4\pi \frac{\omega}{c} \operatorname{Im} \chi. \tag{5.3.21b}$$

If we let $\alpha_0(\Delta)$ denote the absorption coefficient experienced by a *weak* optical wave detuned from the atomic resonance by an amount Δ, we find by ignoring the contribution $\Omega^2 T_1 T_2$ to the denominator of Eq. (5.3.20) that $\alpha_0(\Delta)$ can be expressed as

$$\alpha_0(\Delta) = \frac{\alpha_0(0)}{1 + \Delta^2 T_2^2}, \tag{5.3.22a}$$

where the unsaturated, line-center absorption coefficient is given by

$$\alpha_0(0) = -\frac{4\pi\omega_{ba}}{c}\left[N(\rho_{bb} - \rho_{aa})^{\mathrm{eq}} |\mu_{ba}|^2 \frac{T_2}{\hbar}\right]. \tag{5.3.22b}$$

By introducing this last expression into Eq. (5.3.20), we find that the susceptibility can be expressed as

$$\chi = -\frac{\alpha_0(0)}{4\pi\omega_{ba}/c} \frac{\Delta T_2 - i}{1 + \Delta^2 T_2^2 + \Omega^2 T_1 T_2}. \tag{5.3.23}$$

In order to interpret this result, it is useful to express the susceptibility as $\chi = \chi' + i\chi''$ with its real and imaginary parts given by

$$\chi' = -\frac{\alpha_0(0)}{4\pi\omega_{ba}/c} \frac{1}{\sqrt{1 + \Omega^2 T_1 T_2}} \frac{\Delta T_2/\sqrt{1 + \Omega^2 T_1 T_2}}{1 + \Delta^2 T_2^2/(1 + \Omega^2 T_1 T_2)}, \tag{5.3.24a}$$

$$\chi'' = \frac{\alpha_0(0)}{4\pi\omega_{ba}/c} \left(\frac{1}{1 + \Omega^2 T_1 T_2}\right) \frac{1}{1 + \Delta^2 T_2^2/(1 + \Omega^2 T_1 T_2)}. \tag{5.3.24b}$$

We see from these expressions that, even in the presence of an intense laser field, χ' has a standard dispersive lineshape and χ'' has a Lorentzian lineshape. However, each of these lines has been broadened with respect to its weak-field width by the factor $(1 + \Omega^2 T_1 T_2)^{1/2}$. In particular, the width of the absorption line (full width at half maximum) is given by

$$\Delta\omega_{\mathrm{FWHM}} = \frac{2}{T_2}(1 + \Omega^2 T_1 T_2)^{1/2}. \tag{5.3.25}$$

The tendency of spectral lines to become broadened when measured using intense optical fields is known as power broadening. We also see (e.g., from Eq. 5.3.24b) that the line center value of χ'' (and consequently of the absorption coefficient α) is decreased with respect to its weak-field value by the factor $(1 + \Omega^2 T_1 T_2)^{-1}$. The tendency of the absorption to decrease when measured using intense optical fields is known as saturation. This behavior is illustrated in Fig. 5.3.1.

It is convenient to define, by means of the relation

$$\Omega^2 T_1 T_2 = \frac{|E|^2}{|E_s^{\circ}|^2}, \tag{5.3.26}$$

the quantity E_s°, which is known as the line-center saturation field strength. Through use of Eq. (5.3.18) we find that E_s° is given explicitly by

$$|E_s^{\circ}|^2 = \frac{\hbar^2}{4|\mu_{ba}|^2 T_1 T_2}. \tag{5.3.27}$$

The expression (5.3.23) for the susceptibility can be rewritten in terms of the saturation field strength as

$$\chi = \frac{-\alpha_0(0)}{4\pi\omega_{ba}/c} \frac{\Delta T_2 - i}{1 + \Delta^2 T_2^2 + |E|^2/|E_s^{\circ}|^2}. \tag{5.3.28}$$

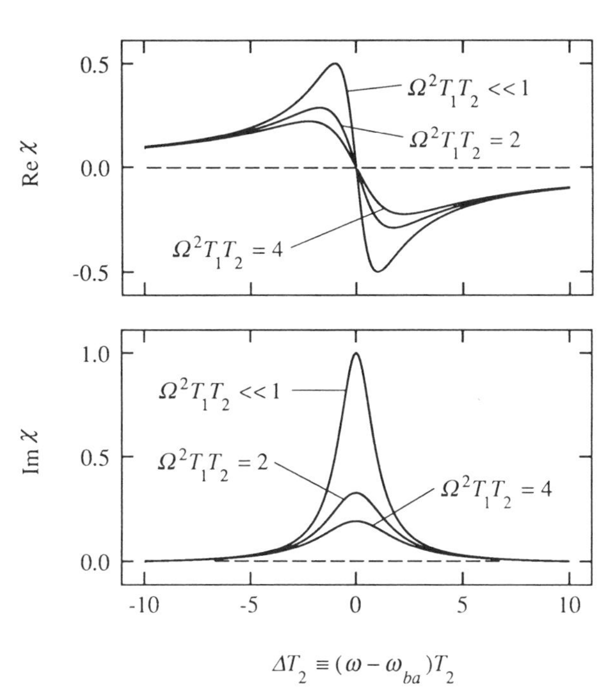

FIGURE 5.3.1 Real and imaginary parts of the susceptibility χ (in units of $\alpha_0 c/4\pi\omega_{ba}$) plotted as functions of the optical frequency ω for several values of the saturation parameter $\Omega^2 T_1 T_2$.

We see from this expression that the significance of E_s^o is that the absorption experienced by an optical wave tuned to line center (which is proportional to Im χ evaluated at $\Delta = 0$) drops to one-half its weak-field value when when the optical field has a strength of E_s^o. We can analogously define a saturation field strength for a wave of arbitrary detuning, which we denote E_s^Δ, by means of the relation

$$|E_s^\Delta|^2 = |E_s^o|^2(1 + \Delta^2 T_2^2). \tag{5.3.29}$$

We then see from Eq. (5.3.28) that Im χ drops to one-half its weak field value when a field of detuning Δ has a field strength of E_s^Δ.

It is also useful to define the saturation *intensity* for a wave at line center (assuming that $|n - 1| \ll 1$) as

$$I_s^o = \frac{c}{2\pi}|E_s^o|^2, \tag{5.3.30}$$

and the saturation intensity for a wave of arbitrary detuning as

$$I_s^\Delta = \frac{c}{2\pi}|E_s^\Delta|^2 = I_s^o(1 + \Delta^2 T_2^2). \tag{5.3.31}$$

In order to relate our present treatment of the nonlinear optical susceptibility to the perturbative treatment that we have used in the previous chapters, we next calculate the first- and third-order contributions to the susceptibility of a collection of two-level atoms. By performing a power series expansion of Eq. (5.3.28) in the quantity $|E|^2/|E_s^o|^2$ and retaining only the first and second terms, we find that the susceptibility can be approximated as

$$\chi \simeq \frac{-\alpha_0(0)}{4\pi\omega_{ba}/c}\left(\frac{\Delta T_2 - i}{1 + \Delta^2 T_2^2}\right)\left(1 - \frac{1}{1 + \Delta^2 T_2^2}\frac{|E|^2}{|E_s^o|^2}\right). \tag{5.3.32}$$

We now equate this expression with the usual power series expansion $\chi = \chi^{(1)} + 3\chi^{(3)}|E^2|$ (where $\chi^{(3)} \equiv \chi^{(3)}(\omega = \omega + \omega - \omega)$) to find that the first- and third-order susceptibilities are given by

$$\chi^{(1)} = \frac{-\alpha_0(0)}{4\pi\omega_{ba}/c}\,\frac{\Delta T_2 - i}{1 + \Delta^2 T_2^2}, \tag{5.3.33a}$$

$$\chi^{(3)} = \frac{\alpha_0(0)}{12\pi\omega_{ba}/c}\left[\frac{\Delta T_2 - i}{(1 + \Delta^2 T_2^2)^2}\right]\frac{1}{|E_s^o|^2}. \tag{5.3.33b}$$

For some purposes, it is useful to express the nonlinear susceptibility in terms of the line-center saturation intensity as

$$\chi^{(3)} = \frac{\alpha_0(0)}{24\pi^2\omega_{ba}/c^2}\left[\frac{\Delta T_2 - i}{(1 + \Delta^2 T_2^2)^2}\right]\frac{1}{I_s^o} \tag{5.3.34a}$$

or, through use of Eqs. (5.3.22a) and (5.3.31), in terms of the saturation intensity and absorption coefficient at the laser frequency as

$$\chi^{(3)} = \frac{\alpha_0(\Delta)}{24\pi^2\omega_{ba}/c^2} \frac{\Delta T_2 - i}{I_s^\Delta}. \tag{5.3.34b}$$

Note also that the third-order susceptibility can be related to the linear susceptibility by

$$\chi^{(3)} = \frac{-\chi^{(1)}}{3(1+\Delta^2 T_2^2)|E_s^\circ|^2} = \frac{-\chi^{(1)}}{3|E_s^\Delta|^2}. \tag{5.3.35}$$

Furthermore, through use of Eqs. (5.3.22b) and (5.3.27), the first- and third-order susceptibilities can be expressed in terms of microscopic quantities as

$$\chi^{(1)} = \left[N(\rho_{bb} - \rho_{aa})^{\text{eq}}|\mu_{ba}|^2 \frac{T_2}{\hbar}\right] \frac{\Delta T_2 - i}{1 + \Delta^2 T_2^2} \tag{5.3.36a}$$

$$\chi^{(3)} = -\tfrac{4}{3}N(\rho_{bb} - \rho_{aa})^{\text{eq}}|\mu_{ba}|^4 \frac{T_1 T_2^2}{\hbar^3} \frac{\Delta T_2 - i}{(1 + \Delta^2 T_2^2)^2}. \tag{5.3.36b}$$

In the limit $\Delta T_2 \gg 1$, the expression for $\chi^{(3)}$ reduces to

$$\chi^{(3)} = -\tfrac{4}{3}N(\rho_{bb} - \rho_{aa})^{\text{eq}}|\mu_{ba}|^4 \frac{1}{\hbar^3\Delta^3} \frac{T_1}{T_2}. \tag{5.3.37}$$

Let us consider the magnitudes of some of the physical quantities we have introduced in this section. Since the intensity of an optical wave with field strength E is given by $I = c|E|^2/2\pi$, the Rabi frequency of Eq. (5.3.18) can be expressed as

$$\Omega = \frac{2|\mu_{ba}|}{\hbar} \left(\frac{2\pi I}{c}\right)^{1/2}. \tag{5.3.38}$$

Assuming that $|\mu_{ba}| = 2.5ea_0 = 5.5 \times 10^{-18}$ esu (as it is for the 3s → 3p transition of atomic sodium) and that I is measured in W/cm^2, this relationship gives the numerical result

$$\Omega[\text{rad/sec}] = 2\pi(1 \times 10^9)\left(\frac{I[\text{W/cm}^2]}{127}\right)^{1/2}. \tag{5.3.39}$$

Hence, whenever the intensity I exceeds 127 W/cm^2, $\Omega/2\pi$ is greater than 1 GHz, which is a typical value of the Doppler-broadened linewidth of an atomic transition; such intensities are available from the focused output of even low-power, cw lasers.

The saturation intensity of an atomic transition can be quite small. Again using $|\mu_{ba}| = 5.5 \times 10^{-18}$ esu, and assuming that $T_1 = 16$ ns (the value for the

3p → 3s transition of atomic sodium) and that $T_2/T_1 = 2$ (the ratio for a radiatively broadened transition; see Eq. (5.2.21a)), we find from Eq. (5.3.30) that

$$I_s^0 = 5.27 \times 10^4 \frac{\text{erg}}{\text{cm}^2\,\text{s}} = 5.27 \frac{\text{mW}}{\text{cm}^2}. \tag{5.3.40}$$

Lastly, let us consider the magnitude of $\chi^{(3)}$ due to the near-resonant excitation of an atomic transition. We take the typical values $N = 10^{14}\ \text{cm}^{-3}$, $(\rho_{bb} - \rho_{aa})^{\text{eq}} = -1$, $\mu_{ba} = 5.5 \times 10^{-18}$ esu, $\Delta = \omega - \omega_{ba} = -2\pi c(1\ \text{cm}^{-1}) = -6\pi \times 10^{10}$ rad/s, and $T_2/T_1 = 2$, in which case we find from Eq. (5.3.37) that $\chi^{(3)} = 1.5 \times 10^{-8}$ esu. Note that this value is very much larger than the values of the nonresonant susceptibilities discussed in Chapter 4.

5.4. Optical Bloch Equations

In the previous two sections, we have treated the response of a two-level atom to an applied optical field by working directly with the density matrix equations of motion. We chose to work with the density matrix equations in order to establish a connection with the calculation of the second and third-order susceptibilities presented in Chapter 3. However, in theoretical quantum optics the response of a two-level atom is usually treated through use of the optical Bloch equations or through related theoretical formalisms. Although these various formalisms are equivalent in their predictions, the equations of motion look very different within different formalisms, and consequently different intuition regarding the nature of resonant optical nonlinearities is obtained. In this section, we review several of these formalisms.

We have seen above (Eqs. (5.3.9) and (5.3.10)) that the density matrix equations describing the interaction of a closed two-level atomic system with the optical field

$$\tilde{E}(t) = E(t)e^{-i\omega t} + \text{c.c.} \tag{5.4.1}$$

can be written in the rotating wave approximation as

$$\frac{d}{dt}\sigma_{ba} = \left[i(\omega - \omega_{ba}) - \frac{1}{T_2}\right]\sigma_{ba} - \frac{i}{\hbar}\mu_{ba}E(\rho_{bb} - \rho_{aa}) \tag{5.4.2a}$$

$$\frac{d}{dt}(\rho_{bb} - \rho_{aa}) = -\frac{(\rho_{bb} - \rho_{aa}) - (\rho_{bb} - \rho_{aa})^{\text{eq}}}{T_1} + \frac{2i}{\hbar}(\mu_{ba}E\sigma_{ab} - \mu_{ab}E^*\sigma_{ba}), \tag{5.4.2b}$$

where the slowly varying, off-diagonal density matrix component $\sigma_{ba}(t)$ is defined by

$$\rho_{ba}(t) = \sigma_{ba}(t)e^{-i\omega t}. \tag{5.4.3}$$

The form of Eqs. (5.4.2) can be greatly simplified by introducing the following quantities:

1. The population inversions

$$w = \rho_{bb} - \rho_{aa} \quad \text{and} \quad w^{\text{eq}} = (\rho_{bb} - \rho_{aa})^{\text{eq}}, \tag{5.4.4a}$$

2. The detuning of the optical field from resonance,*

$$\Delta = \omega - \omega_{ba}, \tag{5.4.4b}$$

and

3. The atom–field coupling constant

$$\kappa = 2\mu_{ba}/\hbar.$$

We also drop the subscripts on σ_{ba} for compactness. The density matrix equations of motion (5.4.2) then take the simpler form

$$\frac{d}{dt}\sigma = \left(i\Delta - \frac{1}{T_2}\right)\sigma - \tfrac{1}{2}i\kappa E w, \tag{5.4.5a}$$

$$\frac{d}{dt}w = -\frac{w - w^{\text{eq}}}{T_1} + i(\kappa E\sigma^* - \kappa^* E^* \sigma). \tag{5.4.5b}$$

It is instructive to consider the equation of motion satisfied by the complex amplitude of the induced dipole moment. The expectation value of the induced dipole moment is given by

$$\langle\hat{\tilde{\mu}}\rangle = \rho_{ba}\mu_{ab} + \rho_{ab}\mu_{ba} = \sigma_{ba}\mu_{ab}e^{-i\omega t} + \sigma_{ab}\mu_{ba}e^{i\omega t}. \tag{5.4.6}$$

If we define the complex amplitude p of the dipole moment $\langle\hat{\tilde{\mu}}\rangle$ through the relation

$$\langle\hat{\tilde{\mu}}\rangle = pe^{-i\omega t} + \text{c.c.}, \tag{5.4.7}$$

we find by comparison with Eq. (5.4.6) that

$$p = \sigma_{ba}\mu_{ab}. \tag{5.4.8}$$

* Note that some authors use the opposite sign convention for Δ.

Equations (5.4.5) can hence be rewritten in terms of the dipole amplitude p as

$$\frac{dp}{dt} = \left(i\Delta - \frac{1}{T_2}\right)p - \frac{\hbar}{4}i|\kappa|^2 Ew, \tag{5.4.9a}$$

$$\frac{dw}{dt} = -\frac{w - w^{\text{eq}}}{T_1} - \frac{4}{\hbar}\,\text{Im}(Ep^*). \tag{5.4.9b}$$

These equations illustrate the nature of the coupling between the atom and the optical field. Note that they are linear in the atomic variables p and w and in the applied field amplitude E. However, the coupling is parametric: the dipole moment p is driven by a term that depends on the product of E with the inversion w, and likewise the inversion is driven by a term that depends on the product of E with p.

For those cases in which the field amplitude E can be taken to be a real quantity, the density matrix equations (5.4.5) can be simplified in a different way. We assume that the phase convention for describing the atomic energy eigenstates has been chosen such that μ_{ba} and hence κ are real quantities. It is then useful to express the density matrix element σ in terms of two real quantities u and v as

$$\sigma = \tfrac{1}{2}(u - iv). \tag{5.4.10}$$

The factor of one-half and the minus sign are used here to conform with convention (Allen and Eberly, 1975). This definition is introduced into Eq. (5.4.5a), which becomes

$$\frac{1}{2}\frac{d}{dt}(u - iv) = \frac{1}{2}\left(i\Delta - \frac{1}{T_2}\right)(u - iv) - \tfrac{1}{2}i\kappa Ew.$$

This equation can be separated into its real and imaginary parts as

$$\frac{d}{dt}u = \Delta v - \frac{u}{T_2}, \tag{5.4.11a}$$

$$\frac{d}{dt}v = -\Delta u - \frac{v}{T_2} + \kappa Ew. \tag{5.4.11b}$$

Similarly, Eq. (5.4.5b) becomes

$$\frac{d}{dt}w = -\frac{w - w^{\text{eq}}}{T_1} - \kappa Ev. \tag{5.4.11c}$$

The set (5.4.11) is known as the optical Bloch equations.

We next show that in the absence of relaxation processes (i.e., in the limit T_1, $T_2 \to \infty$) the variables u, v, and w obey the conservation law

$$u^2 + v^2 + w^2 = 1. \tag{5.4.12}$$

First, we note that the time derivative of $u^2 + v^2 + w^2$ vanishes:

$$\begin{aligned} \frac{d}{dt}(u^2 + v^2 + w^2) &= 2u\frac{du}{dt} + 2v\frac{dv}{dt} + 2w\frac{dw}{dt} \\ &= 2u\,\Delta v - 2v\,\Delta u + 2v\kappa E w - 2w\kappa E v \\ &= 0, \end{aligned} \tag{5.4.13}$$

where we have used Eqs. (5.4.11) in obtaining expressions for the time derivatives. We hence see that $u^2 + v^2 + w^2$ is a constant. Next, we note that before the optical field is applied the atom must be in its ground state and hence that $w = -1$ and $u = v = 0$ (as there can be no probability amplitude to be in the upper level). In this case we see that $u^2 + v^2 + w^2$ is equal to 1, but since the quantity $u^2 + v^2 + w^2$ is conserved, it must have this value at all times. We also note that since all of the damping terms in Eqs. (5.4.11) have negative signs associated with them, it must generally be true that

$$u^2 + v^2 + w^2 \leq 1. \tag{5.4.14}$$

Harmonic Oscillator Form of the Density Matrix Equations

Still different intuition regarding the nature of resonant optical nonlinearities can be obtained by considering the equation of motion satisfied by the expectation value of the dipole moment induced by the applied field (rather than considering the equation satisfied by its complex amplitude). This quantity is given by

$$\tilde{M} \equiv \langle \hat{\tilde{\mu}} \rangle = \rho_{ba}\mu_{ab} + \text{c.c.} \tag{5.4.15}$$

For simplicity of notation, we have introduced the new symbol $\tilde{M}$ rather than continuing to use $\langle \hat{\tilde{\mu}} \rangle$. Note that $\tilde{M}$ is a real quantity that oscillates at an optical frequency.

We take the density matrix equations of motion in the form

$$\dot{\rho}_{ba} = -\left(i\omega_{ba} + \frac{1}{T_2}\right)\rho_{ba} - \frac{i}{\hbar}\mu_{ba}\tilde{E}w, \tag{5.4.16a}$$

$$\dot{w} = -\frac{w - w^{\text{eq}}}{T_1} - \frac{4\tilde{E}}{\hbar}\,\text{Im}(\mu_{ab}\rho_{ba}), \tag{5.4.16b}$$

where the dot denotes a time derivative. These equations follow from Eqs. (5.2.6), (5.2.13a), (5.2.17) and the definition $w = \rho_{bb} - \rho_{aa}$. Here $\tilde{E}$ is the real, time-varying optical field; note that we have not made the rotating wave approximation. We find by direct time differentiation of Eq. (5.4.15) and subsequent use of Eq. (5.4.16a) that $\dot{\tilde{M}}$ is given by

$$\begin{aligned}\dot{\tilde{M}} &= \dot{\rho}_{ba}\mu_{ab} + \text{c.c.} \\ &= -\left(i\omega_{ba} + \frac{1}{T_2}\right)\rho_{ba}\mu_{ab} - \frac{i}{\hbar}|\mu_{ba}|^2\tilde{E}w + \text{c.c.} \\ &= -\left(i\omega_{ba} + \frac{1}{T_2}\right)\rho_{ba}\mu_{ab} + \text{c.c.}\end{aligned} \tag{5.4.17}$$

We have dropped the second term in the second-to-last form because it is imaginary and disappears when added to its complex conjugate. Next, we calculate $\ddot{\tilde{M}}$ by taking the time derivative of Eq. (5.4.17) and introducing expression (5.4.16a) for $\dot{\rho}_{ba}$:

$$\begin{aligned}\ddot{\tilde{M}} &= -\left(i\omega_{ba} + \frac{1}{T_2}\right)\dot{\rho}_{ba}\mu_{ab} + \text{c.c.} \\ &= \left(i\omega_{ba} + \frac{1}{T_2}\right)^2\rho_{ba}\mu_{ab} + \frac{i}{\hbar}\left(i\omega_{ba} + \frac{1}{T_2}\right)|\mu_{ba}|^2\tilde{E}w + \text{c.c.},\end{aligned}$$

or

$$\ddot{\tilde{M}} = \left(-\omega_{ba}^2 + \frac{2i\omega_{ba}}{T_2} + \frac{1}{T_2^2}\right)\rho_{ba}\mu_{ab} - \frac{\omega_{ba}}{\hbar}|\mu_{ba}|^2\tilde{E}w + \text{c.c.} \tag{5.4.18}$$

If we now we introduce Eqs. (5.4.15) and (5.4.17) into this expression, we find that $\tilde{M}$ obeys the equation

$$\ddot{\tilde{M}} + \frac{2}{T_2}\dot{\tilde{M}} + \omega_{ba}^2\tilde{M} = \frac{-\tilde{M}}{T_2^2} - \frac{2\omega_{ba}}{\hbar}|\mu_{ba}|^2\tilde{E}w. \tag{5.4.19}$$

Since ω_{ba}^2 is much larger than $1/T_2^2$ in all physically realistic circumstances, we can drop the first term on the right-hand side of this expression to obtain the result

$$\ddot{\tilde{M}} + \frac{2}{T_2}\dot{\tilde{M}} + \omega_{ba}^2\tilde{M} = -\frac{2\omega_{ba}}{\hbar}|\mu_{ba}|^2\tilde{E}w. \tag{5.4.20}$$

This is the equation of a damped, driven harmonic oscillator. Note that the driving term is proportional to the product of the applied field strength $\tilde{E}(t)$ with the inversion w.

We next consider the equation of motion satisfied by the inversion w. In order to simplify Eq. (5.4.16b), we need an explicit expression for $\mathrm{Im}(\rho_{ba}\mu_{ab})$. To find such an expression, we rewrite Eq. (5.4.17) as

$$\begin{aligned}\dot{\tilde{M}} &= -\left(i\omega_{ba} + \frac{1}{T_2}\right)\rho_{ba}\mu_{ab} + \text{c.c.} \\ &= -i\omega_{ba}(\rho_{ba}\mu_{ab} - \text{c.c.}) - \frac{1}{T_2}(\rho_{ba}\mu_{ab} + \text{c.c.}) \\ &= 2\omega_{ba}\,\mathrm{Im}(\rho_{ba}\mu_{ab}) - \frac{\tilde{M}}{T_2},\end{aligned} \tag{5.4.21}$$

which shows that

$$\mathrm{Im}(\rho_{ba}\mu_{ab}) = \frac{1}{2\omega_{ba}}\left(\dot{\tilde{M}} + \frac{\tilde{M}}{T_2}\right). \tag{5.4.22}$$

This result is now introduced into Eq. (5.4.16b), which becomes

$$\dot{w} = -\frac{w - w^{\text{eq}}}{T_1} - \frac{2\tilde{E}}{\hbar\omega_{ba}}\left(\dot{\tilde{M}} + \frac{\tilde{M}}{T_2}\right). \tag{5.4.23}$$

Since $\dot{\tilde{M}}$ oscillates at an optical frequency (which is much larger than $1/T_2$), the term $\tilde{M}/T_2$ can be omitted, yielding the result

$$\dot{w} = -\frac{w - w^{\text{eq}}}{T_1} - \frac{2}{\hbar\omega_{ba}}\tilde{E}\dot{\tilde{M}}. \tag{5.4.24}$$

We see that the inversion w is driven by the product of $\tilde{E}$ with $\dot{\tilde{M}}$, which is proportional to the part of $\tilde{M}$ that is 90 degrees out of phase with $\tilde{E}$. We also see that w relaxes to its equilibrium value w^{eq} (which is typically equal to -1) in a time of the order of T_1.

Equations (5.4.20) and (5.4.24) provide a description of the two-level atomic system. Note that each equation is linear in the atomic variables $\tilde{M}$ and w. The origin of the nonlinear response of atomic systems lies in the fact that the coupling to the optical field depends parametrically on the atomic variables. A linear harmonic oscillator, for example, would be described by Eq. (5.4.20) with the inversion w held fixed at the value -1. The fact that, for an atom, the coupling depends on the inversion w, whose value depends on the applied field strength as described by Eq. (5.4.24), leads to nonlinearities.

Adiabatic Following Limit

The treatment of Section 5.3 considered the steady-state response of a two-level atom to a cw laser field. The adiabatic following limit (Grischkowsky,

1970) is another limit in which it is relatively easy to obtain solutions to the density matrix equations of motion. The nature of the adiabatic following approximation is as follows: We assume that the optical field is in the form of a pulse whose length τ_p obeys the condition

$$\tau_p \ll T_1, T_2; \tag{5.4.25}$$

we thus assume that essentially no relaxation occurs during the extent of the optical pulse. In addition, we assume that the laser is detuned sufficiently far from resonance that

$$|\omega - \omega_{ba}| \gg T_2^{-1}, \tau_p^{-1}, \mu_{ba}E/\hbar, \tag{5.4.26}$$

that is, we assume that the detuning is greater than the transition linewidth, that no Fourier component of the pulse extends to the transition frequency, and that the transition is not power-broadened into resonance with the pulse. These conditions ensure that no appreciable population is excited to the upper level by the laser pulse.

To simplify the following analysis, we introduce the (complex) Rabi frequency

$$\Omega(t) = 2\mu_{ba}E(t)/\hbar \tag{5.4.27}$$

where $E(t)$ gives the time evolution of the pulse envelope. The density matrix equations of motion (5.4.5) then become, in the limit $T_1 \to \infty$, $T_2 \to \infty$,

$$\frac{d\sigma}{dt} = i\,\Delta\sigma - \tfrac{1}{2}i\Omega w, \tag{5.4.28a}$$

$$\frac{dw}{dt} = -i(\Omega^*\sigma - \Omega\sigma^*). \tag{5.4.28b}$$

We note that the quantity $w^2 + 4\sigma\sigma^*$ is a constant of the motion whose value is given by

$$w^2(t) + 4|\sigma(t)|^2 = 1. \tag{5.4.29}$$

This conclusion is verified by means of a derivation analogous to that leading to Eq. (5.4.12).

We now make the adiabatic following approximation, that is, we assume that for all times the atomic response is nearly in steady state with the applied field. We thus set $d\sigma/dt$ and dw/dt equal to zero in Eqs. (5.4.28a) and (5.4.28b). The simultaneous solution of these equations (which in fact is just the solution to (5.4.28a)) is given by

$$\sigma(t) = \frac{w(t)\Omega(t)}{2\Delta}. \tag{5.4.30}$$

Since $w(t)$ is a real quantity, this result shows that $\sigma(t)$ is always in phase with the driving field $\Omega(t)$. We now combine Eqs. (5.4.29) and (5.4.30) to obtain the equation

$$w(t)^2 + \frac{w(t)^2|\Omega|^2}{\Delta^2} = 1, \tag{5.4.31}$$

which can be solved for $w(t)$ to obtain

$$w(t) = \frac{-|\Delta|}{\sqrt{\Delta^2 + |\Omega(t)|^2}}. \tag{5.4.32}$$

This expression can now be substituted back into Eq. (5.4.30) to obtain the result

$$\sigma(t) = -\frac{\Delta}{|\Delta|}\frac{\frac{1}{2}\Omega(t)}{\sqrt{\Delta^2 + |\Omega(t)|^2}}. \tag{5.4.33}$$

We now use these results to deduce the value of the nonlinear susceptibility. As in Eqs. (5.3.11) through (5.3.17), the polarization P is related to $\sigma(t)$ (recall that $\sigma = \sigma_{ba}$) through

$$P = N\mu_{ab}\sigma, \tag{5.4.34}$$

which through use of Eq. (5.4.33) becomes

$$P = -\frac{\Delta}{|\Delta|}\frac{\frac{1}{2}N\mu_{ab}\Omega(t)}{\sqrt{\Delta^2 + |\Omega(t)|^2}}. \tag{5.4.35}$$

Our derivation has assumed that the condition $|\Delta| \gg |\Omega|$ is valid. We can thus expand Eq. (5.4.35) in a power series in the small quantity $|\Omega|/\Delta$ to obtain

$$P = \frac{\Delta\Omega}{\Delta^2}\frac{-\frac{1}{2}N\mu_{ab}}{(1 + |\Omega|^2/\Delta^2)^{1/2}} = -\frac{\Delta\Omega}{\Delta^2}\frac{1}{2}N\mu_{ab}\left(1 - \frac{1}{2}\frac{|\Omega|^2}{\Delta^2} + \cdots\right). \tag{5.4.36}$$

The contribution to P that is third-order in the applied field is thus given by

$$\frac{|\Omega|^2\Omega\,\Delta N\mu_{ab}}{4\,\Delta^4} = \frac{2N|\mu_{ab}|^4}{\hbar^3\,\Delta^3}|E|^2E, \tag{5.4.37}$$

where, in obtaining the second form, we have used the fact that $\Omega = 2\mu_{ba}E/\hbar$. By convention, the coefficient of $|E|^2E$ is $3\chi^{(3)}$, and hence we find that

$$\chi^{(3)} = \frac{2N|\mu_{ba}|^4}{3\hbar^3\,\Delta^3}. \tag{5.4.38}$$

Note that this prediction is identical to that of the steady-state theory

(Eq. (5.3.37)) in the limit $\Delta T_2 \gg 1$ for the case of a radiatively broadened transition (i.e., $T_2/T_1 = 2$) for which $(\rho_{bb} - \rho_{aa})^{\text{eq}} = -1$.

5.5. Rabi Oscillations and Dressed Atomic States

In this section we consider the response of a two-level atom to an optical field sufficiently intense to remove a significant fraction of the population from the atomic ground state. One might think that the only consequence of a field this intense would be to lower the overall response of the atom. Such is not the case, however. Stark shifts induced by the laser field profoundly modify the energy-level structure of the atom, leading to new resonances in the optical susceptibility. In the present section, we explore some of the processes that occur in the presence of a strong driving field.

Rabi Solution of the Schrödinger Equation

Let us consider the solution to the Schrödinger equation for a two-level atom in the presence of an intense optical field.* We describe the state of the system in terms of the atomic wave function ψ, which obeys the Schrödinger equation

$$i\hbar \frac{\partial \psi}{\partial t} = \hat{H}\psi \tag{5.5.1}$$

with the Hamiltonian operator $\hat{H}$ given by

$$\hat{H} = \hat{H}_0 + \hat{V}(t). \tag{5.5.2}$$

Here $\hat{H}_0$ represents the Hamiltonian of a free atom, and $\hat{V}(t)$ represents the energy of interaction with the applied field. In the electric dipole approximation, $\hat{V}(t)$ is given by

$$\hat{V}(t) = -\hat{\mu}\tilde{E}(t), \tag{5.5.3}$$

where the dipole moment operator is given by $\hat{\mu} = -e\hat{r}$.

We assume that the applied field is given by $\tilde{E}(t) = Ee^{-i\omega t} + \text{c.c.}$ with E constant, and that the field is nearly resonant with an allowed transition between the atomic ground state a and some other level b, as shown in Fig. 5.2.1. Since the effect of the interaction is to mix states a and b, the atomic wave

* See also Sargent *et al.* (1974), p. 26, or Dicke and Wittke (1960), p. 203.

function in the presence of the applied field can be represented as

$$\psi(\mathbf{r}, t) = C_a(t)u_a(\mathbf{r})e^{-i\omega_a t} + C_b(t)u_b(\mathbf{r})e^{-i\omega_b t}. \quad (5.5.4)$$

Here $u_a(\mathbf{r})e^{-i\omega_a t}$ represents the wave function of the atomic ground state a, and $u_b(\mathbf{r})e^{-i\omega_b t}$ represents the wave function of the excited state b. We assume that these wave functions are orthonormal in the sense that

$$\int d^3r\, u_i^*(\mathbf{r})u_j(\mathbf{r}) = \delta_{ij}. \quad (5.5.5)$$

The quantities $C_a(t)$ and $C_b(t)$ that appear in Eq. (5.5.4) can be interpreted as the probability amplitudes that at time t the atom is in state a or state b, respectively.

We next derive the equations of motion for $C_a(t)$ and $C_b(t)$, using methods analogous to those used in Section 3.2. By introducing Eq. (5.5.4) into the Schrödinger equation (5.5.1), multiplying the resulting equation by u_a^*, and integrating this equation over all space, we find that

$$\dot{C}_a = \frac{1}{i\hbar} C_b V_{ab} e^{-i\omega_{ba} t}, \quad (5.5.6)$$

where we have introduced the resonance frequency $\omega_{ba} = \omega_b - \omega_a$ and the interaction matrix element

$$V_{ab} = V_{ba}^* = \int d^3r\, u_a^* \hat{V} u_b. \quad (5.5.7)$$

Similarly, by multiplying instead by u_b^* and again integrating over all space, we find that

$$\dot{C}_b = \frac{1}{i\hbar} C_a V_{ba} e^{i\omega_{ba} t}. \quad (5.5.8)$$

We now explicitly introduce the form of the interaction Hamiltonian and represent the interaction matrix elements as

$$V_{ab}^* = V_{ba} = -\mu_{ba}\tilde{E}(t) = -\mu_{ba}(Ee^{-i\omega t} + E^* e^{i\omega t}). \quad (5.5.9)$$

Equations (5.5.6) and (5.5.8) then become

$$\dot{C}_a = \frac{-\mu_{ab}}{i\hbar} C_b (E^* e^{-i(\omega_{ba} - \omega)t} + Ee^{-i(\omega_{ba} + \omega)t}) \quad (5.5.10a)$$

and

$$\dot{C}_b = \frac{-\mu_{ba}}{i\hbar} C_a (Ee^{i(\omega_{ba} - \omega)t} + E^* e^{+i(\omega_{ba} + \omega)t}). \quad (5.5.10b)$$

We next make the rotating wave approximation, that is, we drop the rapidly oscillating second terms in these equations and retain only the first terms.* We also introduce the detuning factor

$$\Delta = \omega - \omega_{ba}. \tag{5.5.11}$$

The coupled equations (5.5.10) then reduce to the set

$$\dot{C}_a = i\frac{\mu_{ab}E^*}{\hbar}C_b e^{i\Delta t}, \tag{5.5.12a}$$

$$\dot{C}_b = i\frac{\mu_{ba}E}{\hbar}C_a e^{-i\Delta t}. \tag{5.5.12b}$$

This set of equations can be readily solved by adopting a trial solution of the form

$$C_a = Ke^{-i\lambda t}. \tag{5.5.13}$$

This expression is introduced into Eq. (5.5.12a), which shows that C_b must be of the form

$$C_b = \frac{-\hbar\lambda K}{\mu_{ab}E^*}e^{-i(\lambda+\Delta)t}. \tag{5.5.14}$$

This form for C_b and the trial solution (5.5.13) for C_a are now introduced into Eq. (5.5.12b), which shows that the characteristic frequency λ must obey the equation

$$\lambda(\lambda + \Delta) = \frac{|\mu_{ba}|^2|E|^2}{\hbar^2}. \tag{5.5.15}$$

The solutions of this equation are given by

$$\lambda_\pm = -\tfrac{1}{2}\Delta \pm \tfrac{1}{2}\Omega', \tag{5.5.16}$$

where we have introduced the generalized (or detuned) Rabi frequency

$$\Omega' = (|\Omega|^2 + \Delta^2)^{1/2} \tag{5.5.17}$$

and where, as before, $\Omega = 2\mu_{ba}E/\hbar$ denotes the complex Rabi frequency. The general solution to Eqs. (5.5.12) for $C_a(t)$ can thus be expressed as

$$C_a(t) = e^{(1/2)i\Delta t}(A_+e^{-(1/2)i\Omega' t} + A_-e^{(1/2)i\Omega' t}), \tag{5.5.18a}$$

where A_+ and A_- are constants of integration whose values depend on the

* See also the discussion preceding Eq. (5.3.5).

initial conditions. The corresponding expression for $C_b(t)$ is obtained by introducing this result into Eq. (5.5.12a):

$$\begin{aligned} C_b(t) &= \frac{-i\hbar\dot{C}_a}{\mu_{ab}E^*} e^{-i\Delta t} \\ &= e^{-(1/2)i\Delta t}\left(\frac{\Delta - \Omega'}{\Omega^*} A_+ e^{-(1/2)i\Omega' t} + \frac{\Delta + \Omega'}{\Omega^*} A_- e^{(1/2)i\Omega' t}\right). \end{aligned} \tag{5.5.18b}$$

Equations (5.5.18) give the general solution to Eqs. (5.5.12). Next, we find the specific solution for two different sets of initial conditions.

Solution for an Atom Initially in the Ground State

One realistic set of initial conditions corresponds to an atom known to be in the ground state at time $t = 0$, so that

$$C_a(0) = 1 \quad \text{and} \quad C_b(0) = 0. \tag{5.5.19}$$

Equation (5.5.18a) evaluated at $t = 0$ then shows that

$$A_+ + A_- = 1, \tag{5.5.20}$$

while Eq. (5.5.18b) evaluated at $t = 0$ shows that

$$(\Delta - \Omega')A_+ + (\Delta + \Omega')A_- = 0. \tag{5.5.21}$$

These equations are solved algebraically to find that

$$A_+ = 1 - A_- = \frac{\Omega' + \Delta}{2\Omega'}. \tag{5.5.22}$$

The probability amplitudes $C_a(t)$ and $C_b(t)$ are now determined by introducing these expressions for A_+ and A_- into Eqs. (5.5.18), to obtain

$$\begin{aligned} C_a(t) &= e^{(1/2)i\Delta t}\left[\left(\frac{\Omega' + \Delta}{2\Omega'}\right) e^{-(1/2)i\Omega' t} + \left(\frac{\Omega' - \Delta}{2\Omega'}\right) e^{(1/2)i\Omega' t}\right] \\ &= e^{(1/2)i\Delta t}\left[\cos(\tfrac{1}{2}\Omega' t) - \frac{i\Delta}{\Omega'}\sin(\tfrac{1}{2}\Omega' t)\right] \end{aligned} \tag{5.5.23}$$

and

$$\begin{aligned} C_b(t) &= e^{-(1/2)i\Delta t}\left(\frac{-\Omega}{2\Omega'} e^{-(1/2)i\Omega' t} + \frac{\Omega}{2\Omega'} e^{(1/2)i\Omega' t}\right) \\ &= ie^{-(1/2)i\Delta t}\left[\frac{\Omega}{\Omega'}\sin(\tfrac{1}{2}\Omega' t)\right]. \end{aligned} \tag{5.5.24}$$

The probability that the atom is in level a at time t is hence given by

$$|C_a|^2 = \cos^2(\tfrac{1}{2}\Omega' t) + \frac{\Delta^2}{\Omega'^2}\sin^2(\tfrac{1}{2}\Omega' t), \tag{5.5.25}$$

while the probability of being in level b is given by

$$|C_b|^2 = \frac{|\Omega|^2}{\Omega'^2}\sin^2(\tfrac{1}{2}\Omega' t). \tag{5.5.26}$$

Note that (since $\Omega'^2 = |\Omega|^2 + \Delta^2$)

$$|C_a|^2 + |C_b|^2 = 1, \tag{5.5.27}$$

which shows that probability is conserved.

For the case of exact resonance ($\Delta = 0$), Eqs. (5.5.25) and (5.5.26) reduce to

$$|C_a|^2 = \cos^2(\tfrac{1}{2}|\Omega| t), \tag{5.5.28a}$$

$$|C_b|^2 = \sin^2(\tfrac{1}{2}|\Omega| t), \tag{5.5.28b}$$

and the probabilities oscillate between zero and one in the simple manner illustrated in Fig. 5.5.1. Note that, since the probability amplitude C_a oscillates at angular frequency $|\Omega|/2$, the probability $|C_a|^2$ oscillates at angular frequency $|\Omega|$, that is, at the Rabi frequency. As the detuning Δ is increased, the angular frequency at which the population oscillates increases, since the generalized Rabi frequency is given by $\Omega' = [|\Omega|^2 + \Delta^2]^{1/2}$, but the amplitude of the oscillation decreases, as shown in Fig. 5.5.2.

Next we calculate the expectation value of the atomic dipole moment for an atom known to be in the atomic ground state at time $t = 0$. This quantity is given by

$$\langle \tilde{\mu} \rangle = \langle \psi | \hat{\mu} | \psi \rangle, \tag{5.5.29}$$

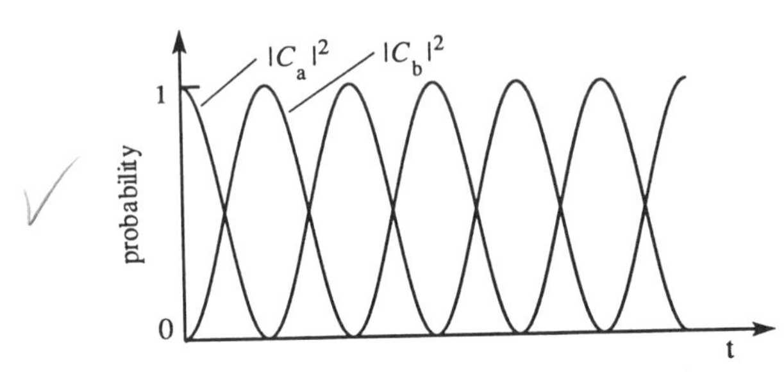

FIGURE 5.5.1 Rabi oscillations of the populations in the ground ($|C_a|^2$) and excited ($|C_b|^2$) states for the case of exact resonance ($\Delta = 0$).

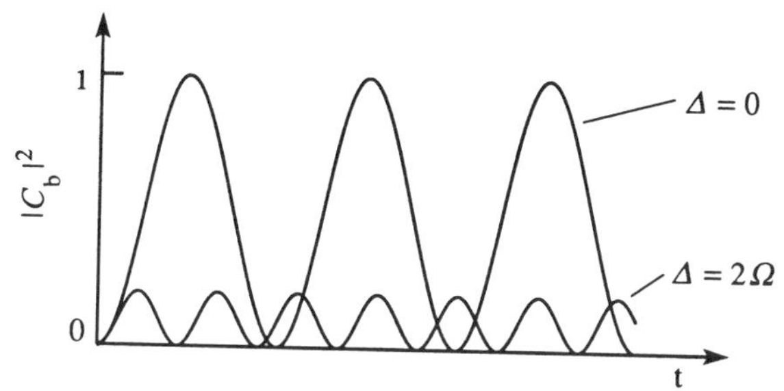

FIGURE 5.5.2 Rabi oscillations of the excited state population for two values of $\Delta \equiv \omega - \omega_{ba}$.

where $\psi(\mathbf{r}, t)$ is given by Eq. (5.5.4). We assume as before that $\langle a|\hat{\mu}|a\rangle = \langle b|\hat{\mu}|b\rangle = 0$, and we denote the nonvanishing matrix elements of $\hat{\mu}$ by

$$\mu_{ab} = \langle a|\hat{\mu}|b\rangle = \langle b|\hat{\mu}|a\rangle^* = \mu_{ba}^*. \tag{5.5.30}$$

We thus find that the induced dipole moment is given by

$$\langle\tilde{\mu}\rangle = C_a^* C_b \mu_{ab} e^{-i\omega_{ba}t} + \text{c.c.} \tag{5.5.31}$$

or, introducing Eqs. (5.5.23) and (5.5.24) for C_a and C_b, by

$$\langle\tilde{\mu}\rangle = \mu_{ab}\frac{\Omega}{\Omega'}\left[\frac{-\Delta}{2\Omega'}e^{-i\omega t} + \frac{1}{4}\left(\frac{\Delta}{\Omega'} - 1\right)e^{-i(\omega+\Omega')t} + \frac{1}{4}\left(\frac{\Delta}{\Omega'} + 1\right)e^{-i(\omega-\Omega')t}\right] + \text{c.c.} \tag{5.5.32}$$

This result shows that the atomic dipole oscillates not only at the driving frequency ω but also at the Rabi sideband frequencies $\omega + \Omega'$ and $\omega - \Omega'$. We can understand the origin of this effect by considering the frequencies that are present in the atomic wave function. We recall that the wave function is given by Eq. (5.5.4), where (according to Eqs. (5.5.23) and (5.5.24)) $C_a(t)$ contains frequencies $-\frac{1}{2}(\Delta \pm \Omega')$ and $C_b(t)$ contains frequencies $\frac{1}{2}(\Delta \pm \Omega')$. Figure 5.5.3 shows graphically the frequencies that are present in the atomic wave function. Note that the frequencies at which the atomic dipole oscillates correspond to differences of the various frequency components of the wave function.

Dressed States

Also see notes

Another important solution to the Schrödinger equation for a two-level atom is that corresponding to the dressed atomic states (Autler and Townes, 1955;

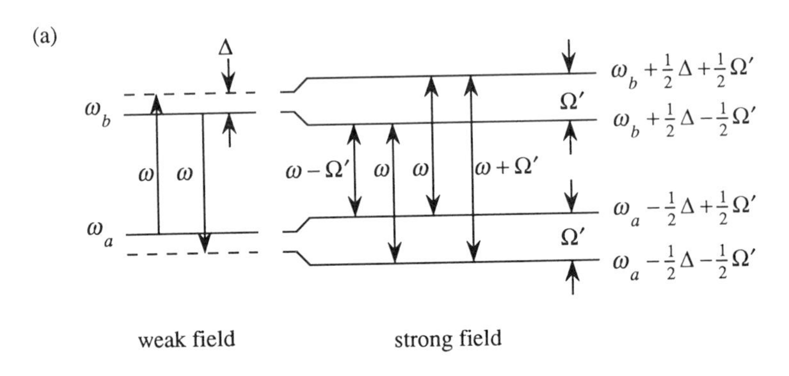

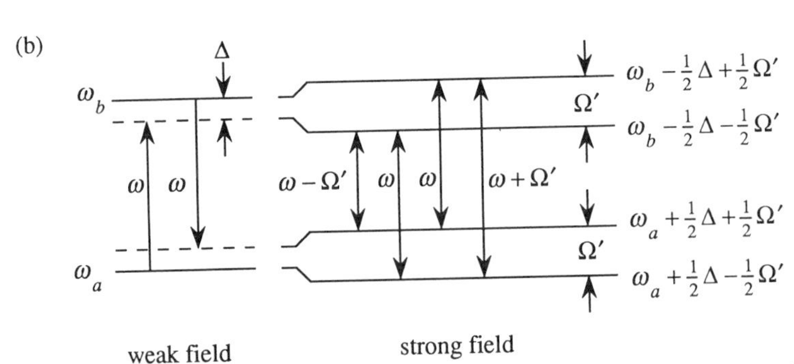

FIGURE 5.5.3 Frequency spectrum of the atomic wave function given by Eq. (5.5.6) (with $C_a(t)$ and $C_b(t)$ given by Eqs. (5.5.28) and (5.5.29)) for the case of (a) positive detuning ($\Delta > 0$) and (b) negative detuning ($\Delta < 0$).

Cohen-Tannoudji and Reynaud, 1977). The characteristic feature of these states is that the probability to be in atomic level a (or b) is constant in time. As a consequence, the probability amplitudes $C_a(t)$ and $C_b(t)$ can depend on time only in terms of exponential phase factors. However, $C_a(t)$ and $C_b(t)$ are given in general by Eqs. (5.5.18).

There are two ways in which this solution can lead to time-independent probabilities of occupancy for levels a and b. One such solution, which we designate as ψ_+, corresponds to the case in which the integration constants A_+ and A_- have the values

$$A_+ = 1, \quad A_- = 0 \qquad (\text{for } \psi_+); \tag{5.5.33a}$$

the other solution, which we designate as ψ_-, corresponds to the case in which

$$A_+ = 0, \quad A_- = 1 \qquad (\text{for } \psi_-). \tag{5.5.33b}$$

Explicitly, the atomic wave function corresponding to each of these solutions is given, through use of Eqs. (5.5.4), (5.5.18), and (5.5.33), as

$$\begin{aligned}\psi_\pm = N_\pm\{u_a(\mathbf{r})\exp[-i(\omega_a - \tfrac{1}{2}\Delta \pm \tfrac{1}{2}\Omega')t] \\ + \frac{\Delta \mp \Omega'}{\Omega^*} u_b(\mathbf{r})\exp[-i(\omega_b + \tfrac{1}{2}\Delta \pm \tfrac{1}{2}\Omega')t]\},\end{aligned} \tag{5.5.34}$$

where $N_\pm$ is a normalization constant. The value of this constant is determined by requiring that

$$\int |\psi_\pm|^2 \, d^3r = 1. \tag{5.5.35}$$

By introducing Eq. (5.3.34) into this expression and performing the integrations, we find that

$$|N_\pm|^2\left[1 + \frac{(\Delta \mp \Omega')^2}{|\Omega|^2}\right] = 1. \tag{5.5.36}$$

For future convenience, we choose the phases of $N_\pm$ so that $N_\pm$ are given by

$$N_\pm = \frac{\Omega^*}{\Omega'}\left[\frac{\Omega'}{2(\Omega' \mp \Delta)}\right]^{1/2} \tag{5.5.37}$$

The normalized dressed-state wave functions are hence given by

$$\begin{aligned}\psi_\pm = \frac{\Omega^*}{\Omega'}\left[\frac{\Omega'}{2(\Omega' \mp \Delta)}\right]^{1/2} u_a(\mathbf{r})\exp[-i(\omega_a - \tfrac{1}{2}\Delta \pm \tfrac{1}{2}\Omega')t] \\ \mp \left[\frac{\Omega' \mp \Delta}{2\Omega'}\right]^{1/2} u_b(\mathbf{r})\exp[-i(\omega_b + \tfrac{1}{2}\Delta \pm \tfrac{1}{2}\Omega')t].\end{aligned} \tag{5.5.38}$$

We next examine some of the properties of the dressed states. The probability amplitude for an atom in the dressed state $\psi_\pm$ to be in the atomic level a is given by

$$\langle a|\psi_\pm\rangle = \frac{\Omega^*}{\Omega'}\left[\frac{\Omega'}{2(\Omega' \mp \Delta)}\right]^{1/2} \exp[-i(\omega_a - \tfrac{1}{2}\Delta \pm \tfrac{1}{2}\Omega')t], \tag{5.5.39}$$

and hence the probability of finding the atom in the state a is given by

$$|\langle a|\psi_\pm\rangle|^2 = \frac{|\Omega|^2}{\Omega'^2}\,\frac{\Omega'}{2(\Omega' \mp \Delta)} = \frac{|\Omega|^2}{2\Omega'(\Omega' \mp \Delta)}. \tag{5.5.40}$$

Similarly, the probability amplitude of finding the atom in state b is given by

$$\langle b|\psi_\pm\rangle = \mp\left(\frac{\Omega' \mp \Delta}{2\Omega'}\right)^{1/2} \exp[-i(\omega_b + \tfrac{1}{2}\Delta \pm \tfrac{1}{2}\Omega')t], \tag{5.5.41}$$

and hence the probability of finding the atom in the state b is given by

$$|\langle b|\psi_{\pm}\rangle|^2 = \frac{\Omega' \pm \Delta}{2\Omega'}. \tag{5.5.42}$$

Note that these probabilities of occupancy are indeed constant in time; in this sense the dressed states constitute the stationary states of the coupled atom–field system.

The dressed states $\psi_{\pm}$ are solutions of Schrödinger's equation in the presence of the total Hamiltonian $\hat{H} = \hat{H}_0 + \hat{V}(t)$. Thus, if the system is known to be in state ψ_+ (or ψ_-) at the time $t = 0$, the system will remain in this state, even though the system is subject to the interaction Hamiltonian $\hat{V}$. They are stationary states in the sense mentioned above, that the probability of finding the atom in either of the atomic states a or b is constant in time. Although the states $\psi_{\pm}$ are stationary states, they are not energy eigenstates, because the Hamiltonian $\hat{H}$ depends explicitly on time.

It is easy to demonstrate that the dressed states are orthogonal; that is, that

$$\langle\psi_+|\psi_-\rangle = 0. \tag{5.5.43}$$

The expectation value of the induced dipole moment for an atom in a dressed state is given by

$$\langle\psi_{\pm}|\hat{\mu}|\psi_{\pm}\rangle = \pm\frac{\Omega}{2\Omega'}\mu_{ab}e^{-i\omega t} + \text{c.c.} \tag{5.5.44}$$

Thus the induced dipole moment of an atom in a dressed state oscillates only at the driving frequency. However, the dipole transition moment between the dressed states is nonzero:

$$\begin{aligned}\langle\psi_{\pm}|\hat{\mu}|\psi_{\mp}\rangle &= \pm\,\mu_{ab}\frac{\Omega}{2\Omega'}\left(\frac{\Omega' \pm \Delta}{\Omega' \mp \Delta}\right)^{1/2} e^{-i(\omega \mp \Omega')t} \\ &\mp\,\mu_{ba}\frac{\Omega^*}{2\Omega'}\left(\frac{\Omega' \mp \Delta}{\Omega' \pm \Delta}\right)^{1/2} e^{i(\omega \mp \Omega')t}.\end{aligned} \tag{5.5.45}$$

The properties of the dressed states are summarized in the frequency level diagram shown for the case of positive Δ in Fig. 5.5.4(a) and for the case of negative Δ in Fig. 5.5.4(b).

Next, we consider the limiting form of the dressed states for the case of a weak applied field, i.e., for $|\Omega| \ll |\Delta|$. In this limit, we can approximate the generalized Rabi frequency Ω' as

$$\begin{aligned}\Omega' &= (|\Omega|^2 + \Delta^2)^{1/2} = |\Delta|\left(1 + \frac{|\Omega|^2}{\Delta^2}\right)^{1/2} \\ &\simeq |\Delta|\left(1 + \frac{1}{2}\frac{|\Omega|^2}{\Delta^2}\right).\end{aligned} \tag{5.5.46}$$

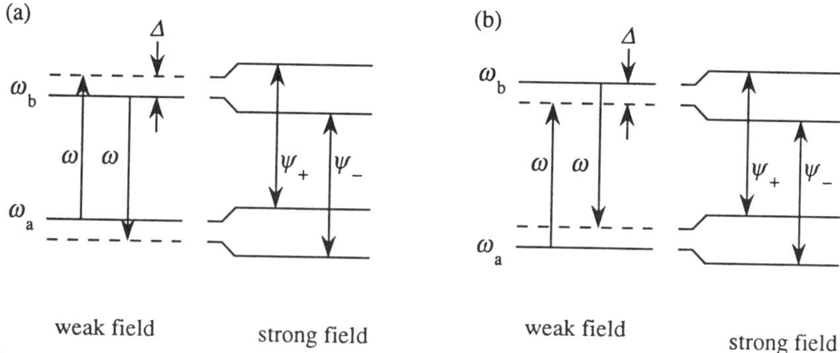

FIGURE 5.5.4 The dressed atomic states ψ_+ and ψ_- for Δ positive (a) and negative (b).

Using this result, we can approximate the dressed-state wave functions of Eq. (5.5.38) for the case of positive Δ as

$$\psi_+ = \frac{\Omega^*}{|\Omega|} u_a e^{-i\omega_a t} - \frac{1}{2}\frac{|\Omega|}{\Delta} u_b e^{-i(\omega_b + \Delta)t}, \tag{5.5.47a}$$

$$\psi_- = \frac{\Omega^*}{2\Delta} u_a e^{-i(\omega_a - \Delta)t} + u_b e^{-i\omega_b t}. \tag{5.5.47b}$$

We note that in this limit ψ_+ is primarily ψ_a and ψ_- is primarily ψ_b. The smaller contribution to ψ_+ can be identified with the virtual level induced by the transition. For the case of negative Δ, we obtain

$$\psi_+ = -\frac{\Omega^*}{2\Delta} u_a e^{-i(\omega_a - \Delta)t} - u_b e^{-i\omega_b t}, \tag{5.5.48a}$$

$$\psi_- = \frac{\Omega^*}{|\Omega|} u_a e^{-i\omega_a t} - \frac{|\Omega|}{2\Delta} u_b e^{-i(\omega_b + \Delta)t}. \tag{5.5.48b}$$

Now ψ_+ is primarily ψ_b, and ψ_- is primarily ψ_a. These results are illustrated in Fig. 5.5.5. Note that these results have been anticipated in drawing certain of

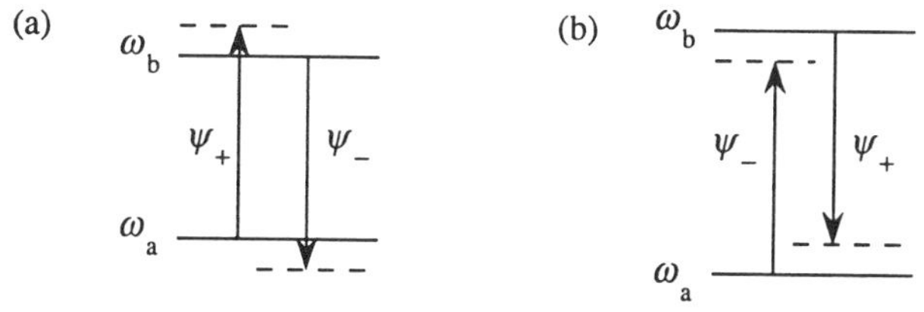

FIGURE 5.5.5 The weak-field limit of the dressed states ψ_+ and ψ_- for the case of (a) positive and (b) negative detuning Δ.

the levels as dashed lines in the weak-field limit of the diagrams shown in Figs. 5.5.3 and 5.5.4.

Inclusion of Relaxation Phenomena

In the absence of damping phenomena, it is adequate to treat the response of a two-level atom to an applied optical field by solving Schrödinger's equation for the time evolution of the wave function. We have seen that under such circumstances the population inversion oscillates at the generalized Rabi frequency $\Omega' = (\Omega^2 + \Delta^2)^{1/2}$. If damping effects are present, we expect that these Rabi oscillations will eventually become damped out and that the population difference will approach some steady-state value. In order to treat this behavior, we need to solve the density matrix equations of motion with the inclusion of damping effects. We take the density matrix equations in the form

$$\dot{p} = \left(i\Delta - \frac{1}{T_2}\right)p - \frac{i}{\hbar}|\mu|^2 E w, \tag{5.5.49a}$$

$$\dot{w} = -\frac{w+1}{T_1} - \frac{2i}{\hbar}(pE^* - p^*E), \tag{5.5.49b}$$

and we assume that at $t = 0$ the atom is in its ground state, that is, that

$$p(0) = 0, \qquad w(0) = -1, \tag{5.5.50}$$

and that the field $\tilde{E}(t)$ is turned on at $t = 0$ and oscillates harmonically thereafter (i.e., $E = 0$ for $t < 0$, $E =$ constant for $t \geq 0$).

Equations (5.5.49) can be solved in general under the conditions given above (see Problem 4 at the end of this chapter). For the special case in which $T_1 = T_2$, the form of the solution is considerably simpler than in the general case. The solution to Eqs. (5.5.49) for the population inversion for this special case is given by

$$w(t) = w_0 - (1 + w_0)\cos\Omega' t\, e^{-t/T_2}\left[\cos\Omega' t + \frac{1}{\Omega' T_2}\sin\Omega' t\right], \tag{5.5.51a}$$

where

$$w_0 = \frac{-(1 + \Delta^2 T_2^2)}{1 + \Delta^2 T_2^2 + \Omega^2 T_1 T_2}. \tag{5.5.51b}$$

The nature of this solution is shown in Fig. 5.5.6. Note that the Rabi oscillations are damped out in a time of the order of T_2. Once the Rabi oscillations have damped out, the system enters one of the dressed states of the coupled atom–field system.

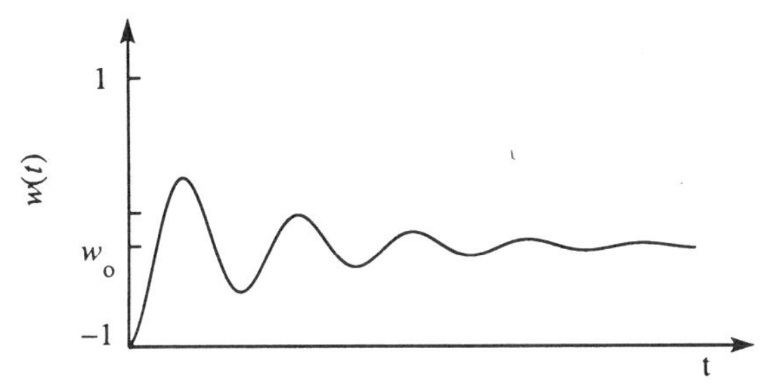

FIGURE 5.5.6 Damped Rabi oscillation.

In summary, we have just seen that, in the absence of damping effects, the population inversion of a strongly driven two-level atom oscillates at the generalized Rabi frequency $\Omega' = (\Omega^2 + \Delta^2)^{1/2}$ and that consequently the induced dipole moment oscillates at the applied frequency ω and also at the Rabi sideband frequencies $\omega \pm \Omega'$. However, we have also seen that, in the presence of dephasing processes, the Rabi oscillations die out in a characteristic time given by the dipole dephasing time T_2. Hence, Rabi oscillations are not present in the steady state.

In the next section, we explore the nature of the response of the atom to a strong field at frequency ω and a weak field at frequency $\omega + \delta$. If the frequency difference δ (or its negative $-\delta$) between these two fields is nearly equal to the generalized Rabi frequency Ω', the beat frequency between the two applied fields can act as a source term to drive the Rabi oscillation. We shall find that, in the presence of such a field, the population difference oscillates at the beat frequency δ, and that the induced dipole moment contains the frequency components ω and $\omega \pm \delta$.

5.6. Optical Wave Mixing in Two-Level Systems

In the present section we consider the response of a collection of two-level atoms to the simultaneous presence of a strong optical field (which we call the pump field) and one or more weak optical fields (which we call probe fields). These latter fields are considered weak in the sense that they alone cannot saturate the response of the atomic system.

An example of such an occurrence is saturation spectroscopy, using a setup of the sort shown in Fig. 5.6.1. In such an experiment, one determines how the response of the medium to the probe wave is modified by the presence of the pump wave. Typically, one might measure the transmission of

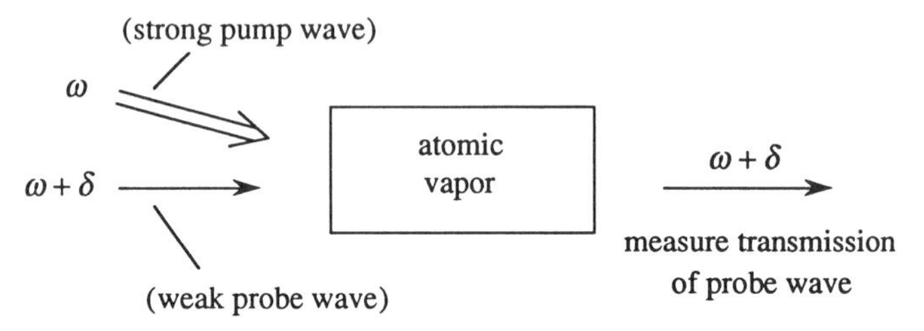

FIGURE 5.6.1 Saturation spectroscopy setup.

the probe wave as a function of the frequency ω and intensity of the pump wave and of the frequency detuning δ between the pump and probe waves. The results of such experiments can be used to obtain information regarding the dipole transition moments and the relaxation times T_1 and T_2.

Another example of the interactions considered in this section is the multiwave mixing experiment shown in part (a) of Fig. 5.6.2. Here the pump wave at frequency ω and the probe wave at frequency $\omega + \delta$ are copropagating (or nearly copropagating) through the medium. For this geometry, the four-wave mixing process shown in part (b) of the figure becomes phase-matched (or nearly phase-matched), and this process leads to the generation of the symmetric sideband at frequency $\omega - \delta$.

At low intensities of the pump laser, the response of the atomic system at the frequencies $\omega + \delta$ and $\omega - \delta$ can be calculated using perturbation theory of the sort developed in Chapter 3. In this limit, one finds that the absorption

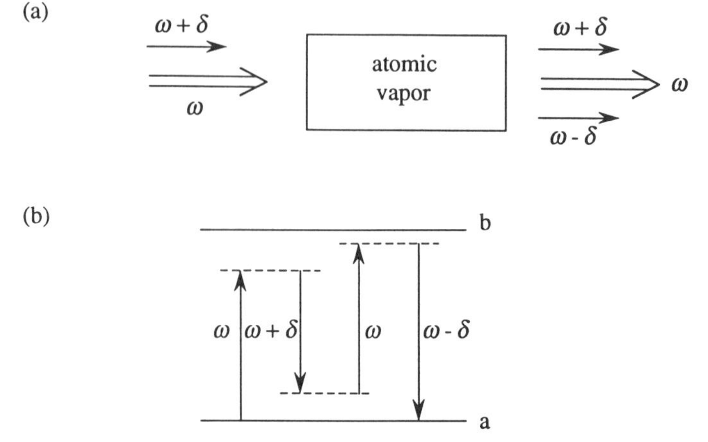

FIGURE 5.6.2 (a) Forward four-wave mixing. (b) Energy level description of the four-wave mixing process, drawn for clarity for the case in which δ is negative.

(and dispersion) experienced by the probe wave in the geometry of Fig. 5.6.1 is somewhat reduced by the presence of the pump wave. One also finds that, for the geometry of Fig. 5.6.2, the intensity of the generated sideband at frequency $\omega - \delta$ increases quadratically as the pump intensity is increased.

In this section we show that the character of these nonlinear processes is profoundly modified when the intensity of the pump laser is increased to the extent that perturbation theory is not sufficient to describe the interaction. These higher-order processes become important when the Rabi frequency Ω associated with the pump field is greater than both the detuning Δ of the pump wave from the atomic resonance and the transition linewidth $1/T_2$. Under this condition, the atomic energy levels are strongly modified by the pump field, leading to new resonances in the absorptive and mixing responses. In particular, we shall find that these new resonances can be excited when the pump–probe detuning δ is equal to $\pm\Omega'$, where Ω' is the generalized Rabi frequency.

Solution of the Density Matrix Equations for a Two-Level Atom in the Presence of Pump and Probe Fields

We have seen in Section 5.4 that the dynamical behavior of a two-level atom in the presence of the optical field

$$\tilde{E}(t) = Ee^{-i\omega t} + \text{c.c.} \tag{5.6.1}$$

can be described in terms of equations of motion for the population inversion $w = \rho_{bb} - \rho_{aa}$ and the complex dipole amplitude $p = \mu_{ab}\sigma_{ba}$, which is related tc the expectation value $\tilde{p}(t)$ of the atomic dipole moment by

$$\tilde{p}(t) = pe^{-i\omega t} + \text{c.c.} \tag{5.6.2}$$

The equations of motion for p and w are given explicitly by

$$\frac{dp}{dt} = \left(i\Delta - \frac{1}{T_2}\right)p - \frac{i}{\hbar}|\mu_{ba}|^2 Ew, \tag{5.6.3}$$

$$\frac{dw}{dt} = -\frac{w - w^{\text{eq}}}{T_1} + \frac{4}{\hbar}\operatorname{Im}(pE^*), \tag{5.6.4}$$

where $\Delta = \omega - \omega_{ba}$. For the problem at hand, we represent the amplitude of the applied optical field as

$$E = E_0 + E_1 e^{-i\delta t}, \tag{5.6.5}$$

where we assume that $|E_1| \ll |E_0|$. By introducing Eq. (5.6.5) into Eq. (5.6.1), we find that the electric field can alternatively be expressed as

$$\tilde{E}(t) = E_0 e^{-i\omega t} + E_1 e^{-i(\omega+\delta)t} + \text{c.c.}; \tag{5.6.6}$$

hence E_0 and E_1 represent the complex amplitudes of the pump and probe waves, respectively.

Equations (5.6.3) and (5.6.4) cannot readily be solved exactly for the field given in Eq. (5.6.5). Instead, our strategy will be to find a solution that is exact for an applied strong field E_0 and is correct to lowest order in the amplitude E_1 of the weak field. We hence require that the steady-state solution of Eq. (5.6.3) and (5.6.4) be of the form

$$p = p_0 + p_1 e^{-i\delta t} + p_{-1} e^{i\delta t} \tag{5.6.7}$$

and

$$w = w_0 + w_1 e^{-i\delta t} + w_{-1} e^{i\delta t}, \tag{5.6.8}$$

where p_0 and w_0 denote the solution for the case in which only the pump field E_0 is present, and where the other terms are assumed to be small in the sense that

$$|p_1|, |p_{-1}| \ll |p_0|, \qquad |w_1|, |w_{-1}| \ll |w_0|. \tag{5.6.9}$$

Note that, to lowest order in the amplitude E_1 of the probe field, 0 and $\pm\delta$ are the only frequencies that can be present in the solution of Eqs. (5.6.3) and (5.6.4). Note also that, in order for $w(t)$ to be a real quantity, w_{-1} must be equal to w_1^*. Hence $w(t)$ is of the form $w(t) = w_0 + 2|w_1|\cos(\delta t + \phi)$, where ϕ is the phase of w. Thus, in the simultaneous presence of pump and probe fields, the population difference oscillates harmonically at the pump–probe frequency difference, and w_1 represents the complex amplitude of the population oscillation.

We now introduce the trial solution (5.6.7) and (5.6.8) into the density matrix equations (5.6.3) and (5.6.4) and equate terms with the same time dependence. In accordance with our perturbation assumptions, we drop any term that contains the product of more than one small quantity. Then, for example, the zero-frequency part of the equation of motion for the dipole amplitude, Eq. (5.6.3), becomes

$$0 = \left(i\Delta - \frac{1}{T_2}\right)p_0 - \frac{i}{\hbar}|\mu_{ba}|^2 E_0 w_0,$$

whose solution is

$$p_0 = \frac{\hbar^{-1}|\mu_{ba}|^2 E_0 w_0}{\Delta + i/T_2}. \tag{5.6.10}$$

Likewise, the part of Eq. (5.6.3) oscillating as $e^{-i\delta t}$ is

$$-i\delta p_1 = \left(i\Delta - \frac{1}{T_2}\right)p_1 - \frac{i}{\hbar}|\mu_{ba}|^2(E_0 w_1 + E_1 w_0),$$

which can be solved to obtain

$$p_1 = \frac{\hbar^{-1}|\mu_{ba}|^2(E_0 w_1 + E_1 w_0)}{(\Delta + \delta) + i/T_2}; \tag{5.6.11}$$

and the part of Eq. (5.6.3) oscillating as $e^{i\delta t}$ is

$$i\delta p_{-1} = \left(i\Delta - \frac{1}{T_2}\right)p_{-1} - \frac{i}{\hbar}|\mu_{ba}|^2(E_0 w_{-1}),$$

which can be solved to obtain

$$p_{-1} = \frac{\hbar^{-1}|\mu_{ba}|^2 E_0 w_{-1}}{(\Delta - \delta) + i/T_2}. \tag{5.6.12}$$

Next, we consider the solution of the inversion equation (5.6.4). We introduce the trial solution (5.6.7) and (5.6.8) into this equation. The zero-frequency part of the resulting expression is

$$0 = -\frac{w_0 - e^{\text{eq}}}{T_1} + \frac{4}{\hbar}\operatorname{Im}(p_0 E_0^*). \tag{5.6.13}$$

We now introduce the expression (5.6.10) for p_0 into this expression to obtain

$$\frac{w_0 - w^{\text{eq}}}{T_1} = \Omega^2 w_0 \operatorname{Im}\left(\frac{\Delta - i/T_2}{\Delta^2 + 1/T_2^2}\right) = \frac{-\Omega^2 w_0/T_2}{\Delta^2 + 1/T_2^2}, \tag{5.6.14}$$

where we have introduced the on-resonance Rabi frequency $\Omega = 2|\mu E|/\hbar$. We now solve Eq. (5.6.14) algebraically for w_0 to obtain

$$w_0 = \frac{w^{\text{eq}}(1 + \Delta^2 T_2^2)}{1 + \Delta^2 T_2^2 + \Omega^2 T_1 T_2}. \tag{5.6.15}$$

We next consider the oscillating part of Eq. (5.6.4). The part of $\operatorname{Im}(pE^*)$ oscillating at frequencies $\pm\delta$ is given by

$$\begin{aligned}\operatorname{Im}(pE^*) &= \operatorname{Im}(p_0 E_1^* e^{i\delta t} + p_1 E_0^* e^{-i\delta t} + p_{-1} E_0^* e^{i\delta t}) \\ &= \frac{1}{2i}(p_0 E_1^* e^{i\delta t} + p_1 E_0^* e^{-i\delta t} + p_{-1} E_0^* e^{i\delta t} \\ &\quad - p_0^* E_1 e^{-i\delta t} - p_1^* E_0 e^{i\delta t} - p_{-1}^* E_0 e^{-i\delta t}),\end{aligned} \tag{5.6.16}$$

where in obtaining the second form we have used the identity $\operatorname{Im} z = (z - z^*)/2i$. We now introduce this result into Eq. (5.6.4). The part of the resulting expression which varies as $e^{-i\delta t}$ is

$$-i\delta w_1 = \frac{-w_1}{T_1} - \frac{2i}{\hbar}(p_1 E_0^* - p_0^* E_1 - p_{-1}^* E_0).$$

This expression is solved for w_1 to obtain

$$w_1 = \frac{2\hbar^{-1}(p_1 E_0^* - p_0^* E_1 - p_{-1}^* E_0)}{\delta + i/T_1}. \tag{5.6.17}$$

We similarly find from the part of Eq. (5.6.4) oscillating as $e^{i\delta t}$ that

$$w_{-1} = \frac{2\hbar^{-1}(p_1^* E_0 - p_0 E_1^* - p_{-1} E_0^*)}{\delta - i/T_1}. \tag{5.6.18}$$

Note that $w_{-1} = w_1^*$, as required from the condition that $w(t)$ as given by Eq. (5.6.8) be real.

At this point we have a set of six coupled equations [(5.6.10), (5.6.11), (5.6.12), (5.6.15), (5.6.17), (5.6.18)] for the quantities $p_0, p_1, p_{-1}, w_0, w_1, w_{-1}$. However, we note that w_0 is given by Eq. (5.6.15) in terms of known quantities. Our strategy will thus be to solve next for w_1, since the other unknown quantities are simply related to w_0 and w_1. We thus introduce the expressions for p_1, p_0, and p_{-1} into Eq. (5.6.17), which becomes

$$\left(\delta + \frac{i}{T_1}\right) w_1 = \left(\frac{|E_0|^2 w_1}{\Delta + \delta + i/T_2} + \frac{E_1 E_0^* w_0}{\Delta + \delta + i/T_2} - \frac{E_1 E_0^* w_0}{\Delta - i/T_2} - \frac{|E_0|^2 w_1}{\Delta - \delta - i/T_2}\right) \times \frac{2|\mu_{ba}|^2}{\hbar^2}.$$

This equation is now solved algebraically for w_1, yielding

$$w_1 = -\frac{2 w_0 |\mu_{ba}|^2 E_1 E_0^* \hbar^{-2} \dfrac{(\delta - \Delta + i/T_2)(\delta + 2i/T_2)}{\Delta - i/T_2}}{\left(\delta + \dfrac{i}{T_1}\right)\left(\delta - \Delta + \dfrac{i}{T_2}\right)\left(\Delta + \delta + \dfrac{i}{T_2}\right) - \Omega^2\left(\delta + \dfrac{i}{T_2}\right)}. \tag{5.6.19}$$

The combination of terms that appears in the denominator of this expression appears repeatedly in the subsequent equations. For convenience we define the quantity

$$D(\delta) = \left(\delta + \frac{i}{T_1}\right)\left(\delta - \Delta + \frac{i}{T_2}\right)\left(\delta + \Delta + \frac{i}{T_2}\right) - \Omega^2\left(\delta + \frac{i}{T_2}\right), \tag{5.6.20}$$

so that Eq. (5.5.19) can be written as

$$w_1 = -2w_0|\mu_{ba}|^2 E_1 E_0^* \hbar^{-2} \frac{(\delta - \Delta + i/T_2)(\delta + 2i/T_2)}{(\Delta - i/T_2)D(\delta)}. \tag{5.6.21}$$

Note that w_1 (and consequently p_1 and p_{-1}) shows a resonance whenever the pump wave is tuned to line center so that $\Delta = 0$, or whenever a zero occurs in the function $D(\delta)$. We thus examine the resonant nature of the function $D(\delta)$. We first consider the limit $\Omega^2 \to 0$, that is, the $\chi^{(3)}$ perturbation theory limit. In this limit $D(\delta)$ is automatically factored into the product of three terms as

$$D(\delta) = \left(\delta + \frac{i}{T_1}\right)\left(\delta - \Delta + \frac{i}{T_2}\right)\left(\Delta + \delta + \frac{i}{T_2}\right), \tag{5.6.22}$$

and we see by inspection that zeros of $D(\delta)$ occur near

$$\delta = 0, \pm\Delta. \tag{5.6.23}$$

The positions of these frequencies are indicated in part (a) of Fig. 5.6.3. However, inspection of Eq. (5.6.21) shows that no resonance in w_1 occurs at $\delta = \Delta$, because the factor $\delta - \Delta + i/T_2$ in the numerator exactly cancels the same factor in the denominator. However, a resonance occurs *near* $\delta = \Delta$ when the term containing Ω^2 in (5.6.20) is not ignored. $\chi^{(5)}$ is the lowest-order contribution to this resonance.

(a)

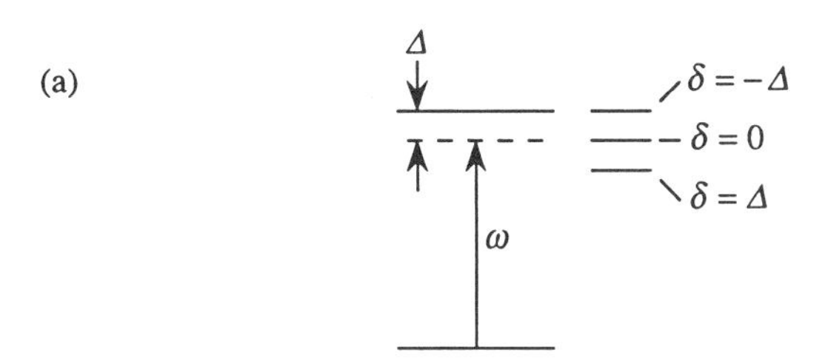

(b)

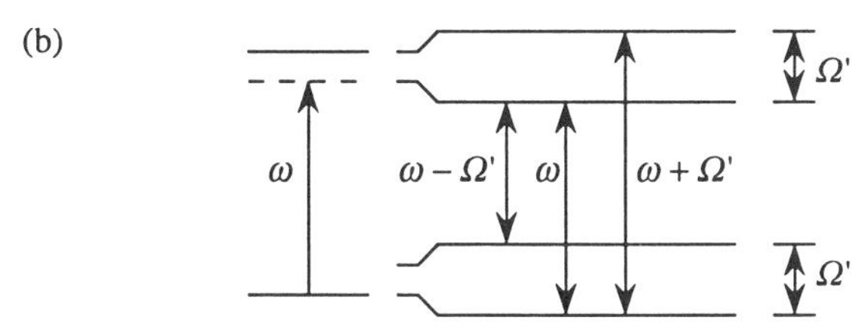

FIGURE 5.6.3 Resonances in the response of a two-level atom to pump and probe fields, as given by the function $D(\delta)$, (a) in the limit $\Omega^2 \to 0$, and (b) in the general case.

In the general case in which Ω^2 is not small, the full form of Eq. (5.6.20) must be used. In order to determine its resonant structure, we write $D(\delta)$ in terms of its real and imaginary parts as

$$D(\delta) = \delta\left(\delta^2 - \Omega'^2 - \frac{1}{T_2^2} - \frac{2}{T_1 T_2}\right) + i\left(\frac{\delta^2 - \Delta^2}{T_1} + \frac{2\delta^2}{T_2} - \frac{\Omega^2}{T_2} - \frac{1}{T_1 T_2^2}\right), \tag{5.6.24}$$

where we have introduced the detuned Rabi frequency $\Omega' = (\Omega^2 + \Delta^2)^{1/2}$. We see by inspection that the real part of D vanishes for

$$\delta = 0, \qquad \delta = \pm\left(\Omega'^2 + \frac{1}{T_2^2} + \frac{2}{T_1 T_2}\right)^{1/2}. \tag{5.6.25}$$

If we now assume that $\Omega' T_2$ is much greater than unity, these three resonances will be well separated, and we can describe their properties separately. In this limit, the function $D(\delta)$ becomes

$$D(\delta) = \delta(\delta^2 - \Omega'^2) + i\left(\frac{\delta^2 - \Delta^2}{T_1} + \frac{2\delta^2 - \Omega^2}{T_2}\right), \tag{5.6.26}$$

and the three resonances occur at

$$\delta = 0, \ \pm\Omega'. \tag{5.6.27}$$

Near the resonance at $\delta = 0$, $D(\delta)$ can be approximated as

$$D(\delta) = -\Omega'^2(\delta + i\Gamma_0), \tag{5.6.28a}$$

where

$$\Gamma_0 = \frac{\Delta^2/T_1 + \Omega^2/T_2}{\Delta^2 + \Omega^2} \tag{5.6.28b}$$

represents the width of this resonance. Likewise, near the resonances at $\delta = \mp\Omega'$, $D(\delta)$ can be approximated as

$$D(\delta) = 2\Omega'^2[(\delta \pm \Omega') + i\Gamma_\pm], \tag{5.6.29a}$$

where

$$\Gamma_\pm = \frac{\Omega^2/T_1 + (2\Delta^2 + \Omega^2)/T_2}{2(\Omega^2 + \Delta^2)} \tag{5.6.29b}$$

represents the width of these resonances. Note that the positions of these resonances can be understood in terms of the energies of the dressed atomic states, as illustrated in Fig. 5.6.3(b). Note also that, for the case of weak optical excitation (i.e., for $\Omega^2 \ll \Delta^2$), Γ_0 approaches the population decay rate $1/T_1$, and $\Gamma_\pm$ approach the dipole dephasing rate $1/T_2$. In the limit of strong optical excitation (i.e., for $\Omega^2 \gg \Delta^2$), Γ_0 approaches the limit $1/T_2$ and $\Gamma_\pm$ approach the limit $\frac{1}{2}(1/T_1 + 1/T_2)$.

We next calculate the response of the atomic dipole at the sideband frequencies $\pm\delta$. We introduce the expression (5.6.19) for w_1 into Eq. (5.6.11) for p_1 and obtain

$$p_1 = \frac{\hbar^{-1}|\mu_{ba}|^2 w_0 E_1}{\Delta + \delta + i/T_2} \times \left[1 - \frac{\frac{1}{2}\Omega^2(\delta - \Delta + i/T_2)(\delta + 2i/T_2)/(\Delta - i/T_2)}{(\delta + i/T_1)(\delta - \Delta + i/T_2)(\Delta + \delta + i/T_2) - \Omega^2(\delta + i/T_2)}\right]. \tag{5.6.30}$$

Written in this form, we see that the response at the probe frequency $\omega + \delta$ can be considered to be the sum of two contributions. The first is the result of the dc part of the population difference w. The second is the result of population oscillations. The first term is resonant only at $\delta = -\Delta$, whereas the second term contains the additional resonances associated with the function $D(\delta)$. Sargent (1978) has pointed out that the second term obeys the relation

$$\int_{-\infty}^{\infty} \frac{-\frac{1}{2}\Omega^2 \dfrac{(\delta - \Delta + i/T_2)(\delta + 2i/T_2)}{\Delta - i/T_2}}{\left(\delta + \dfrac{i}{T_1}\right)\left(\delta - \Delta + \dfrac{i}{T_2}\right)\left(\Delta + \delta + \dfrac{i}{T_2}\right) - \Omega^2\left(\delta + \dfrac{i}{T_2}\right)} d\delta = 0. \tag{5.6.31}$$

Thus, the second term, which results from population oscillations, does not modify the integrated absorption of the atom in the presence of a pump field; it simply leads to a spectral redistribution of probe-wave absorption.

A certain simplification of Eq. (5.6.30) can be obtained by combining the two terms algebraically so that p_1 can be expressed as

$$p_1 = \frac{\hbar^{-1}|\mu_{ba}|^2 w_0 E_1}{D(\delta)}\left[\left(\delta + \frac{i}{T_1}\right)\left(\delta - \Delta + \frac{i}{T_2}\right) - \tfrac{1}{2}\Omega^2 \frac{\delta}{\Delta - i/T_2}\right]. \tag{5.6.32}$$

Finally, we calculate the response at the sideband opposite to the applied probe wave through use of Eqs. (5.6.12) and (5.6.21) and the fact that $w_{-1} = w_1^*$, as noted in the discussion following Eq. (5.6.18). We obtain the result

$$p_{-1} = \frac{2w_0|\mu_{ba}|^4 E_0^2 E_1^* \dfrac{(\delta - \Delta - i/T_2)(-\delta + 2i/T_2)}{\Delta + i/T_2}}{\hbar^3\left(\Delta - \delta + \dfrac{i}{T_2}\right)D^*(\delta)}. \tag{5.6.33}$$

Nonlinear Susceptibility and Coupled-Amplitude Equations

Let us now use these results to determine the forms of the nonlinear polarization and the nonlinear susceptibility. Since p_1 is the complex amplitude of the dipole moment at frequency $\omega + \delta$ induced by a probe wave at this frequency, the polarization at this frequency is $P(\omega + \delta) = Np_1$. If we set $P(\omega + \delta)$ equal to $\chi_{\text{eff}}^{(1)}(\omega + \delta)E_1$, we find that $\chi_{\text{eff}}^{(1)}(\omega + \delta) = Np_1/E_1$, or through use of Eq. (5.6.32) that

$$\chi_{\text{eff}}^{(1)}(\omega + \delta) = \frac{N|\mu_{ba}|^2 w_0}{\hbar D(\delta)}\left[\left(\delta + \frac{i}{T_1}\right)\left(\delta - \Delta + \frac{i}{T_2}\right) - \tfrac{1}{2}\Omega^2 \frac{\delta}{\Delta - i/T_2}\right]. \tag{5.6.34}$$

We have called this quantity an effective linear susceptibility because it depends on the intensity of the pump wave. Similarly, the part of the nonlinear polarization oscillating at frequency $\omega - \delta$ is given by $P(\omega - \delta) = Np_{-1}$. If we set this quantity equal to $3\chi_{\text{eff}}^{(3)}[\omega - \delta = \omega + \omega - (\omega + \delta)]E_0^2 E_1^*$, we find through use of Eq. (5.6.33) that

$$\chi_{\text{eff}}^{(3)}[\omega - \delta = \omega + \omega - (\omega + \delta)] = \frac{2Nw_0|\mu_{ba}|^4 \dfrac{(\delta - \Delta - i/T_2)(-\delta + 2i/T_2)}{\Delta + i/T_2}}{3\hbar^3\left(\Delta - \delta + \dfrac{i}{T_2}\right)D^*(\delta)}. \tag{5.6.35}$$

We have called this quantity an effective third-order susceptibility, because it too depends on the laser intensity.

The calculation just presented has assumed that E_1 (the field at frequency $\omega + \delta$) is the only weak wave that is present. However, for the geometry of Fig. 5.5.2, a weak wave at frequency $\omega - \delta$ is generated by the interaction, and the response of the medium to this wave must also be taken into consideration. If we let E_{-1} denote the complex amplitude of this new wave, we find that we can represent the total response of the medium through the equations

$$P(\omega + \delta) = \chi_{\text{eff}}^{(1)}(\omega + \delta)E_1 + 3\chi_{\text{eff}}^{(3)}[\omega + \delta = \omega + \omega - (\omega - \delta)]E_0^2 E_{-1}^*, \tag{5.6.36a}$$

$$P(\omega - \delta) = \chi_{\text{eff}}^{(1)}(\omega - \delta)E_{-1} + 3\chi_{\text{eff}}^{(3)}[\omega - \delta = \omega + \omega - (\omega + \delta)]E_0^2 E_1^*. \tag{5.6.36b}$$

Formulas for the new quantities $\chi_{\text{eff}}^{(1)}(\omega - \delta)$ and $\chi_{\text{eff}}^{(3)}[\omega + \delta = \omega + \omega - (\omega - \delta)]$ can be obtained by formally replacing δ by $-\delta$ in Eqs. (5.6.34) and (5.6.35).

The nonlinear response of the medium as described by Eqs. (5.6.36) will of course influence the propagation of the weak waves at frequencies $\omega \pm \delta$. We can describe the propagation of these waves by means of coupled-amplitude equations that we derive using methods described in Chapter 2. We introduce the slowly varying amplitudes $A_{\pm 1}$ of the weak waves by means of the equation

$$E_{\pm 1} = A_{\pm 1} e^{ik_{\pm 1} z}, \tag{5.6.37a}$$

where the propagation constant is given by

$$k_{\pm 1} = n_{\pm 1}(\omega \pm \delta)/c. \tag{5.6.37b}$$

Here $n_{\pm 1}$ is the real part of the refractive index experienced by each of the sidebands, and is given by

$$n_{\pm 1}^2 = 1 + 4\pi \operatorname{Re} \chi_{\text{eff}}^{(1)}(\omega \pm \delta). \tag{5.6.37c}$$

We now introduce the nonlinear polarization of Eqs. (5.6.36) and the field decomposition of Eq. (5.6.37) into the wave equation in the form of Eq. (2.1.21), and assume the validity of the slowly-varying-amplitude approximation. We find that the slowly varying amplitudes must obey the set of coupled equations

$$\frac{dA_1}{dz} = -\alpha_1 A_1 + \kappa_1 A_{-1}^* e^{i\Delta k z} \tag{5.6.38a}$$

$$\frac{dA_{-1}}{dz} = -\alpha_{-1} A_{-1} + \kappa_{-1} A_1^* e^{i\Delta k z} \tag{5.6.38b}$$

where we have introduced the nonlinear absorption coefficients

$$\alpha_{\pm 1} = -2\pi \frac{\omega \pm \delta}{n_{\pm 1} c} \operatorname{Im} \chi_{\text{eff}}^{(1)}(\omega \pm \delta), \tag{5.6.39a}$$

the nonlinear coupling coefficients

$$\kappa_{\pm 1} = -6\pi i \frac{\omega \pm \delta}{n_{\pm 1} c} \chi_{\text{eff}}^{(3)}[\omega \pm \delta = \omega + \omega - (\omega \mp \delta)] A_0^2, \tag{5.6.39b}$$

and the wave vector mismatch

$$\Delta k = 2k_0 - k_1 - k_{-1}, \tag{5.6.39c}$$

where k_0 is the magnitude of the wave vector of the pump wave.*

* We have arbitrarily placed the real part of $\chi_{\text{eff}}^{(1)}$ in $n_{\pm 1}$ and the imaginary part in $\alpha_{\pm 1}$. We could equivalently have placed all of $\chi_{\text{eff}}^{(1)}$ in a complex absorption coefficient $\alpha_{\pm 1}$ and set Δk equal to zero, or could have placed all of $\chi_{\text{eff}}^{(1)}$ in a complex refractive index $n_{\pm 1}$ and set $\alpha_{\pm 1}$ equal to zero. We have chosen the present convention because it illustrates most clearly the separate effects of absorption and of wave vector mismatch.

The coupled wave equations given by Eqs. (5.6.38) can be solved explicitly for arbitrary boundary conditions. We shall not present the solution here; it is formally equivalent to the solution presented in Chapter 9 to the equations describing Stokes–anti-Stokes coupling in stimulated Raman scattering. The nature of the solution to Eqs. (5.6.38) for the case of a two-level atomic system has been described in detail by Boyd *et al.* (1981). These authors find that significant amplification of the A_1 and A_{-1} waves can occur in the near-forward direction due to the four-wave mixing processes described by Eqs. (5.6.38). They also find that the gain is particularly large when the detuning δ (or its negative $-\delta$) is approximately equal to the generalized Rabi frequency Ω'.

Let us next consider the nature of the solution of Eqs. (5.6.38) for the special case of the geometry shown in Fig. 5.6.1. For this geometry, due to the large angle θ between the pump and probe beams, the magnitude Δk of the wave vector mismatch is very large, and as a result the coupled-amplitude equations (5.6.38) decouple into the two equations

$$\frac{dA_1}{dz} = -\alpha_1 A_1, \qquad \frac{dA_{-1}}{dz} = -\alpha_{-1} A_{-1}. \tag{5.6.40}$$

Recall that $\alpha_{\pm 1}$ denotes the absorption coefficient experienced by the probe wave at frequency $\omega \pm \delta$, and that $\alpha_{\pm 1}$ depends on the probe–pump detuning δ, on the detuning Δ of the pump wave from the atomic resonance, and on the intensity I of the pump wave.

The dependence of α_1 on the probe–pump detuning δ is illustrated for one representative case in part (a) of Fig. 5.6.4. We see that three features appear in the probe absorption spectrum. One of these features is centered on the laser frequency, and the other two occur at the *Rabi sidebands* of the laser frequency, that is, they occur at frequencies detuned from the laser frequency by the generalized Rabi frequency $\Omega' = (\Omega^2 + \Delta^2)^{1/2}$ associated with the atomic response. Note that α_1 can become negative for two of these features; the gain associated with these features was predicted by Mollow (1972) and has been observed experimentally by Wu *et al.* (1977) and by Gruneisen *et al.* (1988, 1989). The gain feature that occurs near $\delta = 0$ can be considered to be a form of stimulated Rayleigh scattering (see also Chapter 9). The gain associated with these features has been utilized to construct optical parametric oscillators (Grandclement *et al.*, 1987).

Part (b) of Fig. 5.6.4 shows the origin of each of the features shown in part (a). The leftmost portion of this figure shows how the dressed states of the atom are related to the unperturbed atomic energy states. The next diagram, labeled TP, shows the origin of the three-photon resonance. Here the atom

(a)

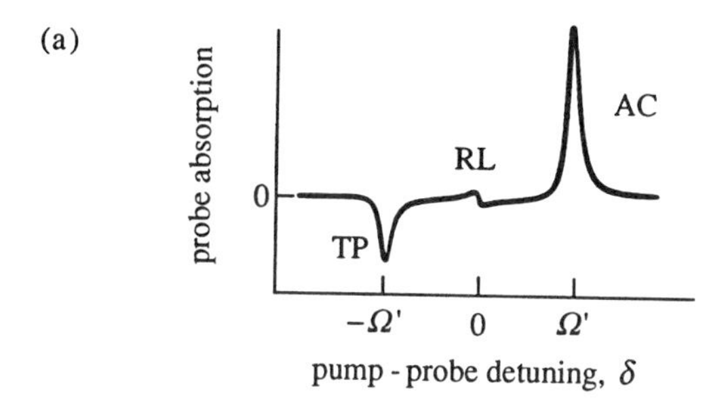

(b)

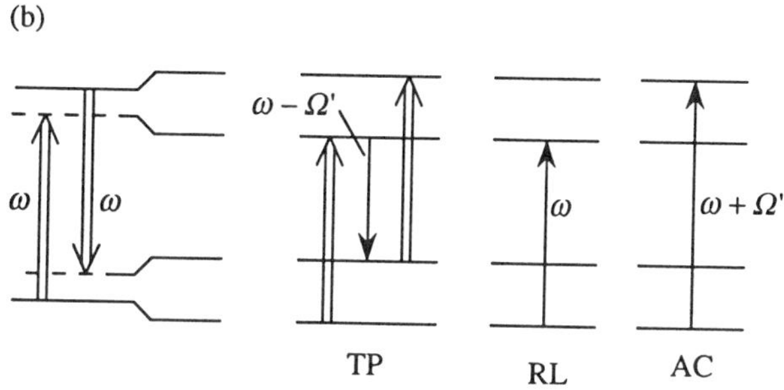

FIGURE 5.6.4 (a) Absorption spectrum of a probe wave in the presence of a strong pump wave for the case $\Delta T_2 = -3$, $\Omega T_2 = 8$, and $T_2/T_1 = 2$. (b) Each of the features in the spectrum shown in part (a) is identified by the corresponding transition between dressed states of the atom. TP denotes the three-photon resonance, RL denotes the Rayleigh resonance, and AC denotes the ac-Stark-shifted atomic resonance.

makes a transition from the lowest dressed level to the highest dressed level by the simultaneous absorption of two pump photons and the emission of a photon at the Rabi sideband frequency $\omega - \Omega'$. This process can amplify a wave at the Rabi sideband frequency, as indicated by the region of negative absorption labeled TP in part (a). The third diagram of part (b), labeled RL, shows the origin of the stimulated Rayleigh resonance. The Rayleigh resonance corresponds to a transition from the lower level of the lower doublet to the lower level of the upper doublet (as illustrated) or from the upper level of the lower doublet to the upper level of the upper doublet. Each of these transitions is centered on the frequency of the pump laser. The final diagram of part (b) of the figure, labeled AC, corresponds to the usual absorptive resonance of the atom as modified by the ac Stark effect. For the sign of the detuning used in the diagram, the atomic absorption is shifted to higher frequencies. Note that this last feature can lead only to absorption, whereas the first two features can lead to amplification.

Problems

1. Determine how the saturated absorption of an atomic transition depends on the intensity of the incident (monochromatic) laser field for the case of an open two-level atom and for a two-level atom with a non-radiatively-coupled intermediate level, and compare these results with those derived in Section 5.3 for a closed two-level atom.

2. $\chi^{(3)}$ *for an impurity-doped solid.* One is often interested in determining the third-order susceptibility of a collection of two-level atoms contained in a medium of constant (i.e., wavelength-independent and non-intensity-dependent) refractive index n_0. Show that the third-order susceptibility of such a system is given by Eq. (5.3.36) in the form shown, or by Eq. (5.3.33b) with a factor of n_0 introduced in the numerator, or by Eq. (5.3.34a) or (5.3.34b) with a factor of n_0^2 introduced in the numerator. In cases in which I_s^0, I_s^Δ, or $\alpha_0(\Delta)$ appears in the expression, it is to be understood that the expressions (5.3.30) and (5.3.31) for I_s^0 and I_s^Δ should each be multiplied by a factor of n_0 and the expression (5.3.22b) for $\alpha_0(0)$ should be divided by a factor of n_0.

3. Verify Eq. (5.5.43).

4. The intent of this problem is to determine the influence of T_1- and T_2-type relaxation processes on Rabi oscillations of the sort predicted by the solution to the Schrödinger equation for an atom in the presence of an intense, near-resonant driving field. In particular, you are to solve the Bloch equation in the form of Eqs. (5.5.49) for the time evolution of an atom known to be in the ground state at time $t = 0$ and subjected to a field $Ee^{-i\omega t}$ + c.c. that is turned on at time $t = 0$. In addition, sketch the behavior of w and of p as functions of time.

   [Hint: At a certain point in the calculation, the mathematical complexity will be markedly reduced by assuming that $T_1 = T_2$. Make this simplification only when it becomes necessary.]

5. Consider the question of estimating the response time of nonresonant electronic nonlinearities of the sort described in Section 4.3. Student A argues that it is well known that the response time under such conditions is of the order of the reciprocal of the detuning of the laser field from the nearest atomic resonance. Student B argues that only relaxation processes can allow a system to enter the steady state, and that consequently the response time is of the order of the longer of T_1 and T_2, that is, is of the order of T_1. Who is right, and in what sense is each of them correct?

[Hint: Consider how the graph shown in Fig. 5.5.6 and the analogous graph of $p(t)$ would look in the limit of $\Delta \gg \Omega$, $\Delta T_2 \gg 1$.]
[Partial answer: The nonlinearity turns on in a time Δ^{-1} but does not reach its steady-state value until a time of the order of T_1.]

6. Verify Eq. (5.6.31).

References

L. D. Allen and J. H. Eberly, *Optical Resonance and Two-Level Atoms*, Wiley, New York, 1975.
S. H. Autler and C. H. Townes, *Phys. Rev.* **100**, 703 (1955).
R. W. Boyd, M. G. Raymer, P. Narum, and D. J. Harter, *Phys. Rev. A* **24**, 411 (1981).
C. Cohen-Tannoudji and S. Reynaud, *J. Phys. B* **10**, 345 (1977); **10**, 365 (1977); **10**, 2311 (1977).
C. Cohen-Tannoudji, J. Dupont-Roc, and G. Grynberg, *Photons and Atoms*, Wiley, New York, 1989; *Atom–Photon Interactions*, Wiley, New York, 1991.
R. H. Dicke and J. P. Wittke, *Introduction to Quantum Mechanics*, Addison-Wesley, Reading, Mass., 1960.
D. Grandclement, G. Grynberg, and M. Pinard, *Phys. Rev. Lett.*, **59**, 44 (1987); see also D. Grandclement, D. Pinard, and G. Grynberg, *IEEE J. Quantum Electron.* **25**, 580 (1989).
D. Grischkowsky, *Phys. Rev. Lett.* **24**, 866 (1970); see also D. Grischkowsky and J. A. Armstrong, *Phys. Rev. A* **6**, 1566 (1972); D. Grischkowsky, *Phys. Rev. A* **7**, 2096 (1973); D. Grischkowsky, E. Courtens, and J. A. Armstrong, *Phys. Rev. Lett.* **31**, 422 (1973).
M. T. Gruneisen, K. R. MacDonald, and R. W. Boyd, *J. Opt. Soc. Am. B* **5**, 123 (1988); M. T. Gruneisen, K. R. MacDonald, A. L. Gaeta, R. W. Boyd, and D. J. Harter, *Phys. Rev. A* **40**, 3464 (1989).
R. B. Miles and S. E. Harris, *IEEE J. Quantum Electron.* **9** 470 (1973).
B. R. Mollow, *Phys. Rev. A* **5**, 2217 (1972).
M. Sargent III, *Phys. Rep.* **43,** 223, (1978).
M. Sargent III, M. O. Scully, and W. E. Lamb, Jr., *Laser Physics*, Addison-Wesley, Reading, Mass., 1974.
F. Y. Wu, S. Ezekiel, M. Ducloy, and B. R. Mollow, *Phys. Rev. Lett.* **38**, 1077 (1977).

Chapter 6

Processes Resulting from the Intensity-Dependent Refractive Index

In this chapter, we explore several different processes that occur as a result of the nonlinear refractive index.

6.1. Optical Phase Conjugation

Optical phase conjugation is a process that can be used to remove the effects of aberrations from certain types of optical systems.

The nature of the phase conjugation process is illustrated in Fig. 6.1.1. Part (a) of the figure shows an optical wave falling at normal incidence onto an ordinary metallic mirror. We see that the most advanced portion of the incident wavefront remains the most advanced after reflection has occurred. Part (b) of the figure shows the same wavefront falling onto a phase-conjugate mirror. In this case the most advanced portion turns into the most retarded portion in the reflection process. For this reason, optical phase conjugation is sometimes referred to as wavefront reversal. Note, however, that the wavefront is reversed only with respect to normal geometrical reflection; in fact the generated wavefront exactly replicates the incident wavefront but propagates in the opposite direction. For this reason, optical phase conjugation is sometimes referred to as the generation of a time-reversed wavefront.

The reason why the process illustrated in part (b) of Fig. 6.1.1 is called phase conjugation can be understood by introducing a mathematical description of the process. We represent the wave incident on the phase-conjugate mirror

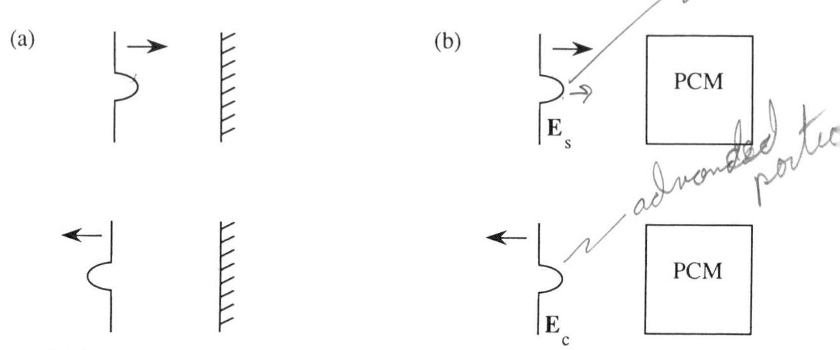

FIGURE 6.1.1 Reflection from (a) an ordinary mirror and (b) a phase-conjugate mirror.

(called the signal wave) as

$$\tilde{\mathbf{E}}_s(\mathbf{r}, t) = \mathbf{E}_s e^{-i\omega t} + \text{c.c.} \tag{6.1.1}$$

When illuminated by such a wave, a phase-conjugate mirror produces a reflected wave, called the phase-conjugate wave, described by

$$\tilde{\mathbf{E}}_c(\mathbf{r}, t) = r\mathbf{E}_s^* e^{-i\omega t} + \text{c.c.}, \tag{6.1.2}$$

where r represents the amplitude reflection coefficient of the mirror. In order to determine the significance of replacing $\mathbf{E}_s$ by $\mathbf{E}_s^*$ in the reflection process, it is useful to represent $\mathbf{E}_s$ as the product

$$\mathbf{E}_s = \hat{\boldsymbol{\epsilon}}_s A_s e^{i\mathbf{k}_s \cdot \mathbf{r}}, \tag{6.1.3}$$

where $\hat{\boldsymbol{\epsilon}}_s$ represents the polarization unit vector, A_s the slowly varying field amplitude, and $\mathbf{k}_s$ the mean wave vector of the incident light. The complex conjugate of Eq. (6.1.3) is given explicitly by

$$\mathbf{E}_s^* = \hat{\boldsymbol{\epsilon}}_s^* A_s^* e^{-i\mathbf{k}_s \cdot \mathbf{r}}. \tag{6.1.4}$$

We thus see that the action of an ideal phase-conjugate mirror is threefold:

1. The complex polarization unit vector of the incident radiation is replaced by its complex conjugate. For example, right-hand circular light remains right-hand circular in reflection from a phase-conjugate mirror rather than being converted into left-hand circular light, as is the case in reflection at normal incidence from a metallic mirror.
2. A_s is replaced by A_s^*, implying that the wavefront is reversed in the sense illustrated in Fig. 6.1.1(b).
3. $\mathbf{k}_s$ is replaced by $-\mathbf{k}_s$, showing that the incident wave is reflected back into its direction of incidence. From the point of view of ray optics, this

result shows that each ray of the incident beam is precisely reflected back onto itself.

Note that Eqs. (6.1.1) through (6.1.4) imply that

$$\tilde{\mathbf{E}}_c(\mathbf{r}, t) = r\tilde{\mathbf{E}}_s(\mathbf{r}, -t). \tag{6.1.5}$$

This result shows that the phase conjugation process can be thought of as the generation of a time-reversed wavefront.

It is important to note that the description given here refers to an *ideal* phase-conjugate mirror. Many physical devices that are ordinarily known as phase-conjugate mirrors are imperfect either in the sense that they do not possess all three properties listed above or in the sense that they possess these properties only approximately. For example, many phase-conjugate mirrors are highly imperfect in their polarization properties, even though they are nearly perfect in their ability to perform wavefront reversal.

Aberration Correction by Phase Conjugation

The process of phase conjugation is able to remove the effects of aberrations under conditions where a beam of light passes twice in opposite directions through an aberrating medium. The reason why optical phase conjugation leads to aberration correction is illustrated in Fig. 6.1.2. Here an initially plane wavefront propagates through an aberrating medium. The aberration may be due to turbulence in the earth's atmosphere, inhomogeneities in the refractive index of a piece of glass, or a poorly designed optical system. The wavefront of the light leaving the medium therefore becomes distorted in the manner shown schematically in the figure. If this aberrated wavefront is now allowed to fall onto a phase-conjugate mirror, a conjugate wavefront will be generated, and

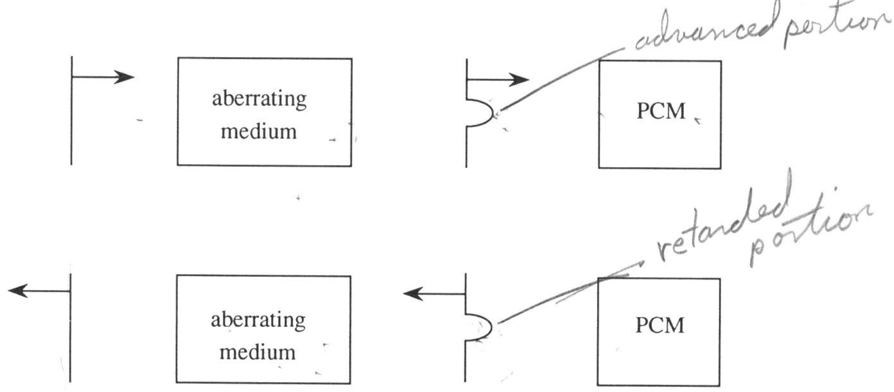

FIGURE 6.1.2 Aberration correction by optical phase conjugation.

the sense of the wavefront distortion will be inverted in this reflected wave. As a result, when this wavefront passes through the aberrating medium again, an undistorted output wave will emerge.

Let us now see how to demonstrate mathematically that optical phase conjugation leads to aberration correction. (Our treatment here is similar to that of A. Yariv and R. A. Fisher in Fisher, 1983.) We consider a wave $\tilde{E}(\mathbf{r}, t)$ propagating through a lossless material of nonuniform refractive index $n(\mathbf{r}) = [\epsilon(\mathbf{r})]^{1/2}$, as shown in Fig. 6.1.3.

We assume that the spatial variation of $\epsilon(\mathbf{r})$ occurs on a scale that is much larger than an optical wavelength. The optical field in this region must obey the wave equation, which we write in the form

$$\nabla^2 \tilde{E} - \frac{\epsilon(\mathbf{r})}{c^2} \frac{\partial^2 \tilde{E}}{\partial t^2} = 0. \tag{6.1.6}$$

We represent the field propagating to the right through this region as

$$\tilde{E}(\mathbf{r}, t) = A(\mathbf{r}) e^{i(kz - \omega t)} + \text{c.c.}, \tag{6.1.7}$$

where the field amplitude $A(\mathbf{r})$ is assumed to be a slowly varying function of $\mathbf{r}$. Since we have singled out the z direction as the mean direction of propagation, it is convenient to express the Laplacian operator which appears in Eq. (6.1.6) as

$$\nabla^2 = \frac{\partial^2}{\partial z^2} + \nabla_T^2, \tag{6.1.8}$$

where $\nabla_T^2 = \partial^2/\partial x^2 + \partial^2/\partial y^2$ is called the transverse Laplacian. Equations (6.1.7) and (6.1.8) are now introduced into Eq. (6.1.6), which becomes

$$\nabla_T^2 A + \left[\frac{\omega^2 \epsilon(\mathbf{r})}{c^2} - k^2 \right] A + 2ik \frac{\partial A}{\partial z} = 0. \tag{6.1.9}$$

In writing this equation in the form shown, we have omitted the term $\partial^2 A/\partial z^2$ because $A(\mathbf{r})$ has been assumed to be slowly varying.

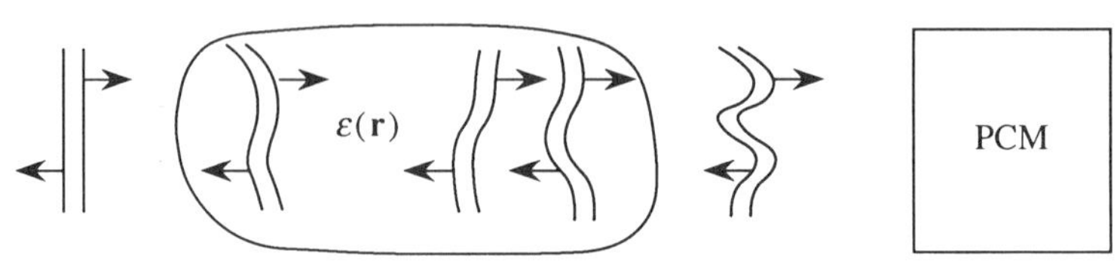

FIGURE 6.1.3 Conjugate waves propagating through an inhomogeneous optical material.

Since this equation is generally valid, so is its complex conjugate, which is given explicitly by

$$\nabla_T^2 A^* + \left[\frac{\omega^2 \epsilon(\mathbf{r})}{c^2} - k^2\right] A^* - 2ik\frac{\partial A^*}{\partial z} = 0. \tag{6.1.10}$$

However, this equation describes the wave

$$\tilde{E}_c(\mathbf{r}, t) = A^*(\mathbf{r}) e^{i(-kz - \omega t)} + \text{c.c.}, \tag{6.1.11}$$

which is a wave propagating in the negative z direction whose complex amplitude is *everywhere* the complex conjugate of the forward-going wave. This proof shows that if the phase-conjugate mirror can generate a backward-going wave whose amplitude is the complex conjugate of that of the forward-going wave at any one plane (say the input face of the mirror), then the field amplitude of the backward-going wave will be the complex conjugate of that of the forward-going wave at *all* points in front of the mirror. In particular, if the forward-going wave is a plane wave before entering the aberrating medium, then the backward-going (i.e., conjugate) wave emerging from the aberrating medium will also be a plane wave. A physical process that can produce such a conjugate wavefront is described in the next subsection.

Phase Conjugation by Degenerate Four-Wave Mixing

It has been shown by Hellwarth (1977) and by Yariv and Pepper (1977) that the phase conjugate of an incident wave can be created by the process of degenerate four-wave mixing (DFWM) using the geometry shown in Fig. 6.1.4. This four-wave mixing process is degenerate in the sense that all four interacting waves have the same frequency. In this process, a lossless nonlinear medium characterized by a third-order nonlinear susceptibility $\chi^{(3)}$ is illuminated by two strong counterpropagating pump waves E_1 and E_2 and by a signal wave E_3. The pump waves are usually taken to be plane waves, although in principle they can possess any wavefront structure as long as their amplitudes are complex conjugates of one another. The signal wave is allowed to

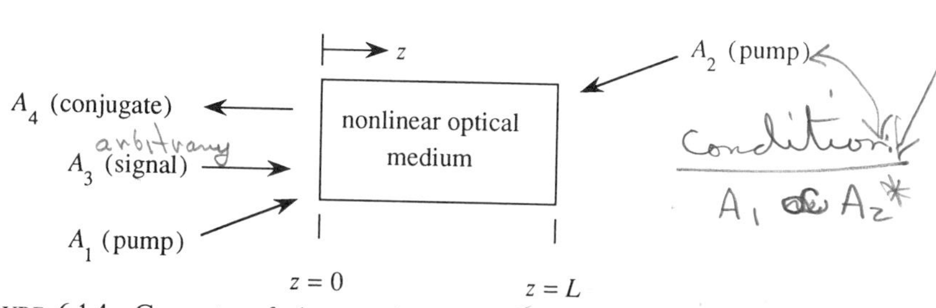

FIGURE 6.1.4 Geometry of phase conjugation by degenerate four-wave mixing.

have an arbitrary wavefront. In this section we show that, as a result of the nonlinear coupling between these waves, a new wave E_4 is created that is the phase conjugate of E_3. We also derive an expression (Eq. (6.1.37b)) that describes the efficiency with which the conjugate wave is generated.

Since the mathematical development that follows is somewhat involved, it is useful to consider first in simple terms why the interaction illustrated in Fig. 6.1.4 leads to the generation of a conjugate wavefront. We represent the four interacting waves by

$$\begin{aligned}\tilde{E}_i(\mathbf{r}, t) &= E_i(\mathbf{r})e^{-i\omega t} + \text{c.c.}\\ &= A_i(\mathbf{r})e^{i(\mathbf{k}_i \cdot \mathbf{r} - \omega t)} + \text{c.c.}\end{aligned} \tag{6.1.12}$$

for $i = 1, 2, 3, 4$, where the A_i are slowly varying quantities. The nonlinear polarization produced within the medium by the three input waves will have, in addition to a large number of other terms, a term of the form

$$P^{\text{NL}} = 6\chi^{(3)}E_1 E_2 E_3^* = 6\chi^{(3)}A_1 A_2 A_3^* e^{i(\mathbf{k}_1 + \mathbf{k}_2 - \mathbf{k}_3)\cdot\mathbf{r}}. \tag{6.1.13}$$

Since we have assumed that the pump waves E_1 and E_2 are counterpropagating, their wave vectors are related by

$$\mathbf{k}_1 + \mathbf{k}_2 = 0, \tag{6.1.14}$$

and hence Eq. (6.1.13) becomes

$$P^{\text{NL}} = 6\chi^{(3)}A_1 A_2 A_3^* e^{-i\mathbf{k}_3\cdot\mathbf{r}}. \tag{6.1.15}$$

We see that this contribution to the nonlinear polarization has a spatial dependence that allows it to act as a phase-matched source term for a conjugate wave (E_4) having wave vector $-\mathbf{k}_3$, and hence we see that the wave vectors of the signal and conjugate waves are related by

$$\mathbf{k}_3 + \mathbf{k}_4 = 0. \tag{6.1.16}$$

The field amplitude of the wave generated by the nonlinear polarization of Eq. (6.1.15) will be proportional to $A_1 A_2 A_3^*$. This wave will be the phase conjugate of A_3 whenever the phase of the product $A_1 A_2$ is spatially invariant, either because A_1 and A_2 both represent plane waves and hence are each constant or because A_1 and A_2 are phase conjugates of one another (because if A_2 is proportional to A_1^*, then $A_1 A_2$ will be proportional to the real quantity $|A_1|^2$).

We can also understand the interaction shown in Fig. 6.1.4 from the following point of view. The incoming signal wave of amplitude A_3 interferes with one of the pump waves (e.g., the forward-going pump wave of amplitude A_1) to form a spatially varying intensity distribution. Due to the nonlinear response of the medium, a refractive index variation accompanies this interference pattern. This variation acts as a volume diffraction grating, which scatters the other pump wave to form the outgoing conjugate wave of amplitude A_4.

Let us now treat the degenerate four-wave mixing process more rigorously. The total field amplitude within the nonlinear medium is given by

$$E = E_1 + E_2 + E_3 + E_4. \tag{6.1.17}$$

This field produces a nonlinear polarization within the medium, which is given by

$$P = 3\chi^{(3)}(\omega = \omega + \omega - \omega)E^2E^*. \tag{6.1.18}$$

The product E^2E^* that appears on the right-hand side of this equation contains a large number of terms with different spatial dependences. Those terms with spatial dependence of the form

$$e^{i\mathbf{k}_i \cdot \mathbf{r}} \quad \text{for} \quad i = 1, 2, 3, 4 \tag{6.1.19}$$

are particularly important because they can act as phase-matched source terms for one of the four interacting waves. The polarization amplitudes associated with these phase-matched contributions are as follows:

$$\begin{aligned}
P_1 &= 3\chi^{(3)}[E_1^2E_1^* + 2E_1E_2E_2^* + 2E_1E_3E_3^* + 2E_1E_4E_4^* + 2E_3E_4E_2^*],\\
P_2 &= 3\chi^{(3)}[E_2^2E_2^* + 2E_2E_1E_1^* + 2E_2E_3E_3^* + 2E_2E_4E_4^* + 2E_3E_4E_1^*],\\
P_3 &= 3\chi^{(3)}[E_3^2E_3^* + 2E_3E_1E_1^* + 2E_3E_2E_2^* + 2E_3E_4E_4^* + 2E_1E_2E_4^*],\\
P_4 &= 3\chi^{(3)}[E_4^2E_4^* + 2E_4E_1E_1^* + 2E_4E_2E_2^* + 2E_4E_3E_3^* + 2E_1E_2E_3^*].
\end{aligned} \tag{6.1.20}$$

We next assume that the fields E_3 and E_4 are much weaker than the pump fields E_1 and E_2. In the above expressions we therefore drop those terms that contain more than one weak-field amplitude. We hence obtain

$$\begin{aligned}
P_1 &= 3\chi^{(3)}[E_1^2E_1^* + 2E_1E_2E_2^*],\\
P_2 &= 3\chi^{(3)}[E_2^2E_2^* + 2E_2E_1E_1^*],\\
P_3 &= 3\chi^{(3)}[2E_3E_1E_1^* + 2E_3E_2E_2^* + 2E_1E_2E_4^*],\\
P_4 &= 3\chi^{(3)}[2E_4E_1E_1^* + 2E_4E_2E_2^* + 2E_1E_2E_3^*].
\end{aligned} \tag{6.1.21}$$

Note that, at the present level of approximation, the E_3 and E_4 fields are each driven by a polarization that depends on the amplitudes of all of the fields, but that the polarizations driving the E_1 and E_2 fields depend only upon E_1 and E_2 themselves. We thus first consider the problem of calculating the spatial evolution of the pump field amplitudes E_1 and E_2. We can then later use these known amplitudes when we calculate the spatial evolution of the signal and conjugate waves.

We assume that each of the interacting waves obeys the wave equation in the form

$$\nabla^2 \tilde{E}_i - \frac{\epsilon}{c^2} \frac{\partial^2 \tilde{E}_i}{\partial t^2} = \frac{4\pi}{c^2} \frac{\partial^2}{\partial t^2} \tilde{P}_i. \tag{6.1.22}$$

We now introduce Eqs. (6.1.12) and (6.1.21) into this equation and make the slowly-varying-amplitude approximation. Also, we let z' be the spatial coordinate measured in the direction of propagation of the E_1 field, and we assume for simplicity that the pump waves have plane wavefronts. We then find that the pump field A_1 must obey the equation

$$\left[\left(-k_1^2 + 2ik_1 \frac{d}{dz'} + \frac{\epsilon\omega^2}{c^2}\right) A_1\right] e^{i(k_1 z' - \omega t)}$$
$$= -\frac{4\pi}{c^2} \omega^2 3\chi^{(3)} [|A_1|^2 + 2|A_2|^2] A_1 e^{i(k_1 z' - \omega t)},$$

which, after simplification, becomes

$$\frac{dA_1}{dz'} = \frac{6\pi i\omega}{nc} \chi^{(3)} [|A_1|^2 + 2|A_2|^2] A_1 \equiv i\kappa_1 A_1. \tag{6.1.23a}$$

We similarly find that the backward-going pump wave is described by the equation

$$\frac{dA_2}{dz'} = \frac{-6\pi i\omega}{nc} \chi^{(3)} [|A_2|^2 + 2|A_1|^2] A_2 \equiv -i\kappa_2 A_2. \tag{6.1.23b}$$

Since κ_1 and κ_2 are real quantities, these equations show that A_1 and A_2 each undergo phase shifts as they propagate through the nonlinear medium. The phase shift experienced by each wave depends both on its own intensity and on that of the other wave. Note that each wave shifts the phase of the other wave by twice as much as it shifts its own phase, in consistency with the general result described in the discussion following Eq. (4.1.14). These phase shifts can induce a phase mismatch into the process that generates the phase-conjugate signal. Note that since only the phases (and not the amplitudes) of the pump waves are affected by the nonlinear coupling, the quantities $|A_1|^2$ and $|A_2|^2$ are spatially invariant, and hence the quantities κ_1 and κ_2 that appear in Eqs. (6.1.23) are in fact constants. These equations can therefore be solved directly to obtain

$$A_1(z') = A_1(0)e^{i\kappa_1 z'}, \tag{6.1.24a}$$

$$A_2(z') = A_2(0)e^{-i\kappa_2 z'}. \tag{6.1.24b}$$

The product $A_1 A_2$ that appears in the expression (6.1.15) for the nonlinear polarization responsible for producing the phase-conjugate wave therefore varies spatially as

$$A_1(z')A_2(z') = A_1(0)A_2(0)e^{i(\kappa_1 - \kappa_2)z'}; \tag{6.1.25}$$

the factor $e^{i(\kappa_1 - \kappa_2)z'}$ shows the effect of wave vector mismatch. If the two pump beams have equal intensities, so that $\kappa_1 = \kappa_2$, the product $A_1 A_2$ becomes spatially invariant, so that

$$A_1(z')A_2(z') = A_1(0)A_2(0), \tag{6.1.26}$$

and in this case the interaction is perfectly phase-matched. We will henceforth assume that the pump intensities are equal.

We next consider the coupled-amplitude equations describing the signal and conjugate fields, $\tilde{E}_3$ and $\tilde{E}_4$. We assume for simplicity that the incident signal wave has plane wavefronts. This is actually not a restrictive assumption, because an arbitrary signal field can be decomposed into plane wave components, each of which will couple to a plane wave component of the conjugate field $\tilde{E}_4$. Under this assumption, the wave equation (6.1.22) applied to the signal and conjugate fields leads to the coupled-amplitude equations

$$\frac{dA_3}{dz} = \frac{12\pi i\omega}{nc}\chi^{(3)}[(|A_1|^2 + |A_2|^2)A_3 + A_1 A_2 A_4^*], \tag{6.1.27a}$$

$$\frac{dA_4}{dz} = -\frac{12\pi i\omega}{nc}\chi^{(3)}[(|A_1|^2 + |A_2|^2)A_4 + A_1 A_2 A_3^*]. \tag{6.1.27b}$$

For convenience, we write these equations as

$$\frac{dA_3}{dz} = i\kappa_3 A_3 + i\kappa A_4^*, \tag{6.1.28a}$$

$$\frac{dA_4}{dz} = -i\kappa_3 A_4 - i\kappa A_3^*, \tag{6.1.28b}$$

where

$$\kappa_3 = \frac{12\pi\omega}{nc}\chi^{(3)}(|A_1|^2 + |A_2|^2), \tag{6.1.29a}$$

$$\kappa = \frac{12\pi\omega}{nc}\chi^{(3)}A_1 A_2. \tag{6.1.29b}$$

The set of equations (6.1.28) can be simplified through a change of variables. We let

$$A_3 = A_3' e^{i\kappa_3 z}, \tag{6.1.30a}$$

$$A_4 = A_4' e^{-i\kappa_3 z}. \tag{6.1.30b}$$

Note that the primed and unprimed variables coincide at the input face of the interaction region, that is, at the plane $z = 0$. We introduce these relations into Eq. (6.1.28a), which becomes

$$i\kappa_3 A_3' e^{i\kappa_3 z} + \frac{dA_3'}{dz} e^{i\kappa_3 z} = i\kappa_3 A_3' e^{i\kappa_3 z} + i\kappa A_4'^* e^{i\kappa_3 z},$$

or

$$\frac{dA_3'}{dz} = i\kappa A_4'^*. \tag{6.1.31a}$$

We similarly find that Eq. (6.1.28b) becomes

$$\frac{dA_4'}{dz} = -i\kappa A_3'^*. \tag{6.1.31b}$$

This set of equations shows why degenerate four-wave mixing leads to phase conjugation: The generated field A_4' is driven only by the complex conjugate of the input field amplitude. We note that this set of equations is *formally* identical to the set that we would have obtained if we had taken the driving polarizations of Eq. (6.1.21) to be simply

$$P_1 = P_2 = 0, \qquad P_3 = 6\chi^{(3)} E_1 E_2 E_4^*, \qquad P_4 = 6\chi^{(3)} E_1 E_2 E_3^*, \tag{6.1.32}$$

that is, if we had ignored the modification of the pump waves due to the nonlinear interaction.

Next, we solve the set of equations (6.1.31). We take the derivative of Eq. (6.1.31b) with respect to z and introduce Eq. (6.1.31a) to obtain*

$$\frac{d^2 A_4'}{dz^2} = -i\kappa \frac{dA_3'^*}{dz} = -|\kappa|^2 A_4',$$

or

$$\frac{d^2 A_4'}{dz^2} + |\kappa|^2 A_4' = 0. \tag{6.1.33}$$

* We are assuming throughout this discussion that $\chi^{(3)}$ and hence κ are real; we have written the equation in the form shown for generality and for consistency with other cases where κ is complex.

This result shows that the spatial dependence of A_4' must be of the form

$$A_4'(z) = B \sin|\kappa| z + C \cos|\kappa| z. \tag{6.1.34}$$

In order to determine the constants B and C, we must specify the boundary conditions for each of the two weak waves. In particular, we assume that $A_3'^*(0)$ and $A_4'(L)$ are specified. In this case, the solution of Eq. (6.1.33) is

$$A_3'^*(z) = -\frac{i|\kappa| \sin|\kappa| z}{\kappa \cos|\kappa| L} A_4'(L) + \frac{\cos[|\kappa|(z-L)]}{\cos|\kappa| L} A_3'^*(0), \tag{6.1.35a}$$

$$A_4'(z) = \frac{\cos|\kappa| z}{\cos|\kappa| L} A_4'(L) - \frac{i\kappa}{|\kappa|} \frac{\sin[|\kappa|(z-L)]}{\cos|\kappa| L} A_3'^*(0). \tag{6.1.35b}$$

However, for the case of four-wave mixing for optical phase conjugation, we can usually assume that there is no conjugate wave injected into the medium at $z = L$, that is, we can assume that

$$A_4'(L) = 0. \tag{6.1.36}$$

Furthermore we are usually interested only in the output values of the two interacting fields. These output field amplitudes are then given by

$$A_3'^*(L) = \frac{A_3'^*(0)}{\cos|\kappa| L}, \tag{6.1.37a}$$

$$A_4'(0) = \frac{i\kappa}{|\kappa|} (\tan|\kappa| L) A_3'^*(0). \tag{6.1.37b}$$

Note that the transmitted signal wave $A_3'^*(L)$ is always more intense than the incident wave. Note also that the output conjugate wave $A_4(0)$ can have any intensity ranging from zero to infinity, the actual value depending on the particular value of $|\kappa| L$. The reflectivity of a phase-conjugate mirror based on degenerate four-wave mixing can exceed 100% because the mirror is actively pumped by externally applied waves, which can supply energy.

From the point of view of energetics, we can describe the process of degenerate four-wave mixing as a process in which one photon from each of the pump waves is annihilated and one photon is added to each of the signal and conjugate waves, as shown in Fig. 6.1.5. Hence, the conjugate wave A_4 is created, and the signal wave A_3 is amplified. The degenerate four-wave mixing process with counterpropagating pump waves is automatically phase-matched (when the two pump waves have equal intensity or whenever we can ignore the nonlinear phase shifts experienced by each wave). We see that this is true because no phase-mismatch terms of the sort $e^{\pm i \Delta k z}$ appear on the right-hand sides of Eqs. (6.1.31).

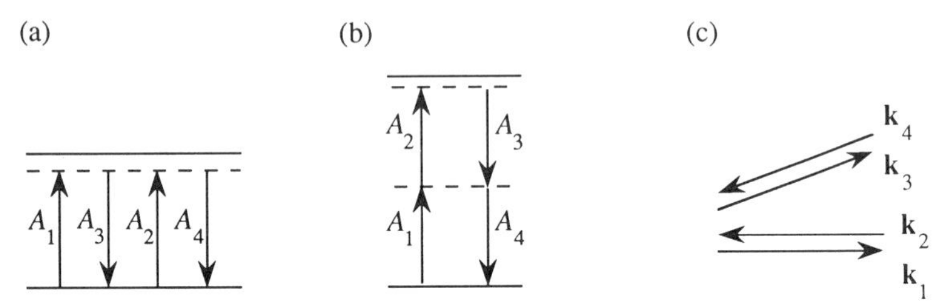

FIGURE 6.1.5 Parts (a) and (b) are energy level diagrams describing two different interactions that can lead to phase conjugation by degenerate four-wave mixing. In either case, the interaction involves the simultaneous annihilation of two pump photons with the creation of signal and conjugate photons. Diagram (a) describes the dominant interaction if the applied field frequency is nearly resonant with a one-photon transition of the material system, whereas (b) describes the dominant interaction under conditions of two-photon resonant excitation. Part (c) shows the wave vectors of the four interacting waves. Since $\mathbf{k}_1 + \mathbf{k}_2 - \mathbf{k}_3 - \mathbf{k}_4 = 0$, the process is perfectly phase-matched.

The fact that degenerate four-wave mixing in the phase conjugation geometry is automatically phase-matched has a very simple physical interpretation. Since this process entails the annihilation of two pump photons and the creation of a signal and conjugate photon, the total input energy is $2\hbar\omega$ and the total input momentum is $\hbar(\mathbf{k}_1 + \mathbf{k}_2) = 0$, whereas the total output energy is $2\hbar\omega$ and the total output momentum is $\hbar(\mathbf{k}_3 + \mathbf{k}_4) = 0$. If the two pump beams are not exactly counterpropagating, then $\hbar(\mathbf{k}_1 + \mathbf{k}_2)$ does not vanish and the phase-matching condition is not automatically satisfied.

The first experimental demonstration of phase conjugation by degenerate four-wave mixing was performed by Bloom and Bjorklund (1977). Their experimental setup is shown in Fig. 6.1.6. They observed that the presence of the aberrating glass did not lower the resolution of the system when the mirror was aligned to retroreflect the pump laser beam onto itself. However, when this mirror was partially misaligned, the return beam passed through a different portion of the aberrating glass and the resolution of the system was degraded.

Polarization Properties of Phase Conjugation

Our discussion thus far has treated phase conjugation in the scalar approximation and has shown that phase conjugation can be used to remove the effects of wavefront aberrations. It is often desirable that phase conjugation be able to remove the effects of polarization distortions as well. An example

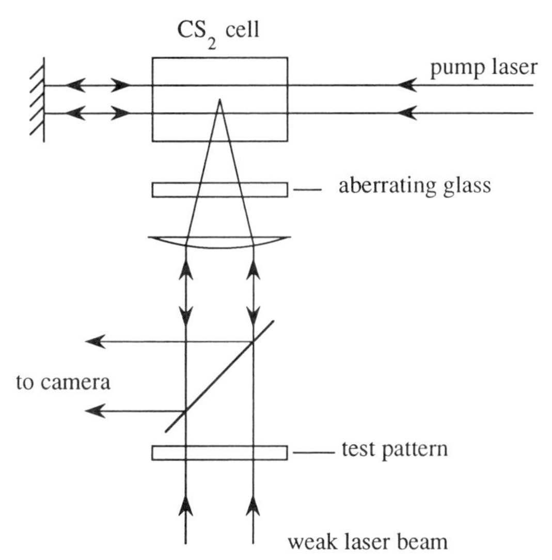

FIGURE 6.1.6 Experimental setup for studying phase conjugation by degenerate four-wave mixing.

is shown in Fig. 6.1.7. Here a beam of light that initially is linearly polarized passes through a stressed optical component. As a result of stress-induced birefringence, the state of polarization of the beam becomes distorted nonuniformly over the cross section of the beam. This beam then falls onto a phase-conjugate mirror. If this mirror is ideal in the sense that the polarization unit vector $\hat{\epsilon}$ of the incident light is replaced by its complex conjugate in the reflected beam, the effects of the polarization distortion will be removed in the second pass through the stressed optical component, and the beam will be returned to its initial state of linear polarization. A phase-conjugate mirror

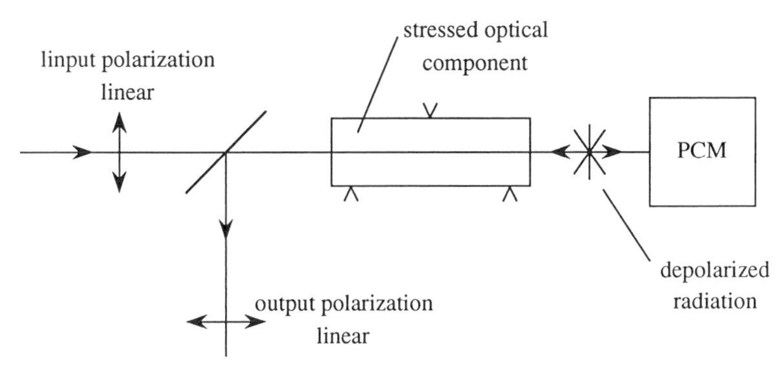

FIGURE 6.1.7 Polarization properties of phase conjugation.

that produces a reflected beam that is both a wavefront conjugate and a polarization conjugate is often called a vector phase-conjugate mirror.

In order to describe the polarization properties of the degenerate four-wave mixing process, we consider the geometry shown in Fig. 6.1.8, where **F**, **B**, and **S** denote the amplitudes of the forward- and backward-going pump waves and of the signal wave, respectively. The total applied field is thus given by

$$\mathbf{E} = \mathbf{F} + \mathbf{B} + \mathbf{S}. \tag{6.1.38}$$

We assume that the angle θ between the signal and forward-going pump wave is much smaller than unity, so that only the x and y components of the incident fields have appreciable amplitudes. We also assume that the nonlinear optical material is isotropic, so that the third-order nonlinear optical susceptibility is given by Eq. (4.2.5) as

$$\chi_{ijkl}(\omega = \omega + \omega - \omega) = \chi_{1122}(\delta_{ij}\,\delta_{kl} + \delta_{ik}\,\delta_{jl}) + \chi_{1221}\delta_{il}\,\delta_{jk}, \tag{6.1.39}$$

and so that the nonlinear polarization can be expressed as

$$\begin{aligned} \mathbf{P} &= 6\chi_{1122}(\mathbf{E}\cdot\mathbf{E}^*)\mathbf{E} + 3\chi_{1221}(\mathbf{E}\cdot\mathbf{E})\mathbf{E}^* \\ &\equiv A(\mathbf{E}\cdot\mathbf{E}^*)\mathbf{E} + \tfrac{1}{2}B(\mathbf{E}\cdot\mathbf{E})\mathbf{E}^*. \end{aligned} \tag{6.1.40}$$

If we now introduce Eq. (6.1.38) into Eq. (6.1.40), we find that the phase-matched contribution to the nonlinear polarization that acts as a source for the conjugate wave is given by

$$\begin{bmatrix} P_x \\ P_y \end{bmatrix} = 6 \begin{bmatrix} \chi_{1111}B_xF_x + \chi_{1221}B_yF_y & \chi_{1122}(B_xF_y + B_yF_x) \\ \chi_{1122}(B_yF_x + B_xF_y) & \chi_{1111}B_yF_y + \chi_{1221}B_xF_x \end{bmatrix} \begin{bmatrix} S_x^* \\ S_y^* \end{bmatrix}, \tag{6.1.41}$$

where $\chi_{1111} = 2\chi_{1122} + \chi_{1221}$. The polarization properties of the phase conjugation process will be ideal (i.e., vector phase conjugation will be obtained) whenever the two-by-two transfer matrix of Eq. (6.1.41) is a multiple of the identity matrix. Under these conditions, both cartesian components of the incident field are reflected with equal efficiency and no coupling between orthogonal components occurs.

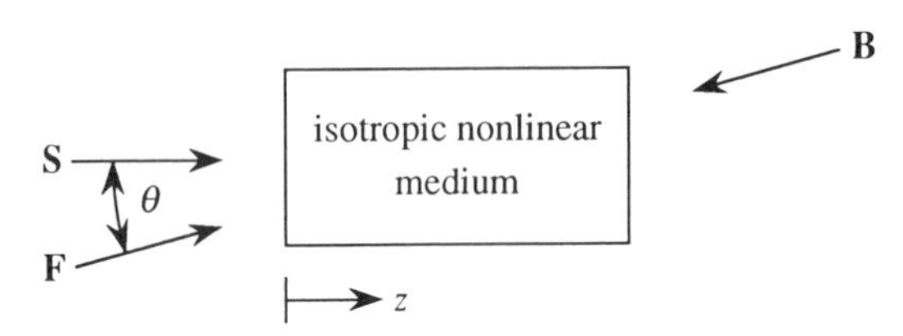

FIGURE 6.1.8 Geometry of vector phase conjugation.

There are two different ways in which the matrix in Eq. (6.1.41) can reduce to a multiple of the identity matrix. One way is for $A = 6\chi_{1122}$ to vanish identically. In this case Eq. (6.1.41) becomes

$$\begin{aligned}\begin{bmatrix} P_x \\ P_y \end{bmatrix} &= 6\chi_{1221}(B_x F_x + B_y F_y)\begin{bmatrix} 1 & 0 \\ 0 & 1 \end{bmatrix}\begin{bmatrix} S_x^* \\ S_y^* \end{bmatrix} \\ &= 6\chi_{1221}(B_x F_x + B_y F_y)\begin{bmatrix} S_x^* \\ S_y^* \end{bmatrix},\end{aligned} \tag{6.1.42}$$

and hence the nonlinear polarization is proportional to the complex conjugate of the signal amplitude for *any* choice of the polarization vectors of the pump waves. This result can be understood directly in terms of Eq. (6.1.40), which shows that **P** has the vector character of **E*** whenever χ_{1122} vanishes. However, χ_{1122} (or A) vanishes identically only under very unusual circumstances. The only known case for this condition to occur is that of degenerate four-wave mixing in an atomic vapor utilizing a two-photon resonance between certain atomic states. This situation has been analyzed by Grynberg (1984) and studied experimentally by Malcuit *et al.* (1988). The analysis can be described most simply for the case of a transition between two S states of an atom with zero electron spin. The four-wave mixing process can then be described graphically by the diagram shown in Fig. 6.1.9. Since the lower and upper level each possess zero angular momentum, the sum of the angular momenta of the signal and conjugate photons must be zero, and this condition implies that the polarization unit vectors of the two waves must be related by complex conjugation.

For most physical mechanisms giving rise to optical nonlinearities, the coefficient A does not vanish. (Recall that for molecular orientation $B/A = 6$, for electrostriction $B/A = 0$, and for nonresonant electronic nonlinearities $B/A = 1$.) For the general case in which A is not equal to 0, vector phase conjugation in the geometry in Fig. 6.1.8 can be obtained only when the pump

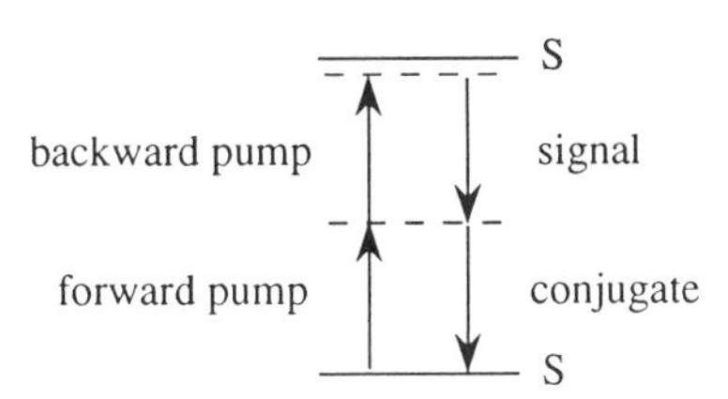

FIGURE 6.1.9 Phase conjugation by degenerate four-wave mixing using a two-photon transition.

waves are circularly polarized and counterrotating. By counterrotating, we mean that if the forward-going wave is described by

$$\tilde{\mathbf{F}}(z,t) = F\frac{\hat{\mathbf{x}} - i\hat{\mathbf{y}}}{\sqrt{2}} e^{i(kz-\omega t)} + \text{c.c.}, \tag{6.1.43a}$$

then the backward-going wave is described by

$$\tilde{\mathbf{B}}(z,t) = B\frac{\hat{\mathbf{x}} + i\hat{\mathbf{y}}}{\sqrt{2}} e^{i(-kz-\omega t)} + \text{c.c.} \tag{6.1.43b}$$

These waves are counterrotating in the sense that, for any fixed value of z, $\tilde{\mathbf{F}}$ rotates clockwise in time in the x–y plane and $\tilde{\mathbf{B}}$ rotates counterclockwise in time. However, both waves are right-hand circularly polarized, since, by convention, the handedness of a wave is the sense of rotation as determined when looking into the beam.

In the notation of Eq. (6.1.41), the amplitudes of the fields described by Eqs. (6.1.43) are given by

$$\begin{aligned} F_x &= \frac{F}{\sqrt{2}} e^{ikz}, \qquad & F_y &= -i\frac{F}{\sqrt{2}} e^{ikz}, \\ B_x &= \frac{B}{\sqrt{2}} e^{-ikz}, \qquad & B_y &= i\frac{B}{\sqrt{2}} e^{-ikz}, \end{aligned} \tag{6.1.44}$$

and hence Eq. (6.1.41) becomes

$$\begin{bmatrix} P_x \\ P_y \end{bmatrix} = 3FB(\chi_{1111} + \chi_{1221}) \begin{bmatrix} 1 & 0 \\ 0 & 1 \end{bmatrix} \begin{bmatrix} S_x^* \\ S_y^* \end{bmatrix}. \tag{6.1.45}$$

We see that the transfer matrix is again a multiple of the identity matrix and hence that the nonlinear polarization vector is proportional to the complex conjugate of the signal field vector. The fact that degenerate four-wave mixing excited by counterrotating pump waves leads to vector phase conjugation was predicted theoretically by Zel'dovich and Shkunov (1979) and was verified experimentally by Martin *et al.* (1980).

The reason why degenerate four-wave mixing with counterrotating pump waves leads to vector phase conjugation can be understood in terms of the conservation of linear and angular momentum. As described above, phase conjugation can be visualized as a process in which one photon from each pump wave is annihilated and a signal and conjugate photon are simultaneously created. Since the pump waves are counterpropagating and counterrotating, the total linear and angular momenta of the two input photons must vanish. Then conservation of linear and angular momentum requires that the conjugate wave must be emitted in a direction opposite to the direction of propagation of the signal wave and that its polarization vector must rotate in a sense opposite to that of the signal wave.

6.2. Self-Focusing of Light

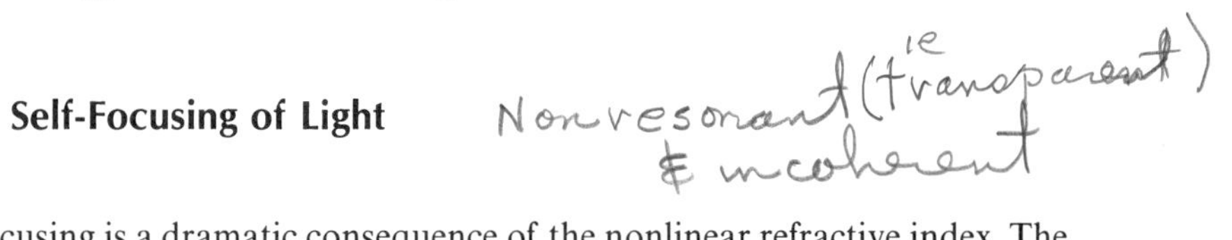

Self-focusing is a dramatic consequence of the nonlinear refractive index. The nature of the self-focusing process is illustrated schematically in Fig. 6.2.1. Here a laser beam with a gaussian transverse intensity distribution is incident upon an optical medium whose refractive index is given by

$$n = n_0 + n_2 I. \tag{6.2.1}$$

We assume that the nonlinear refractive index coefficient n_2 is positive. As a result of this nonlinear response, the refractive index of the material is larger at the center of the laser beam than at its periphery, with the result that the

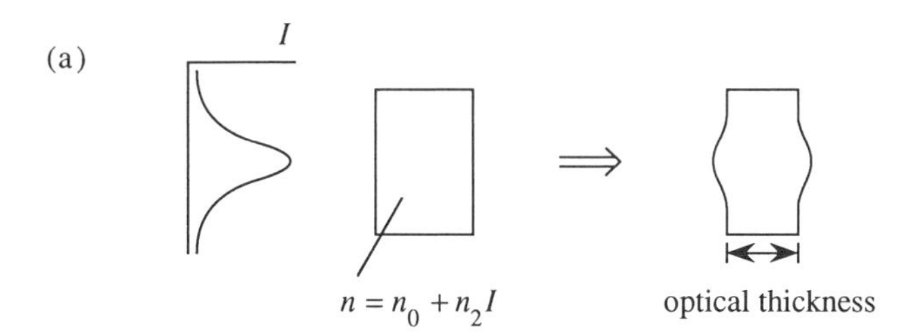

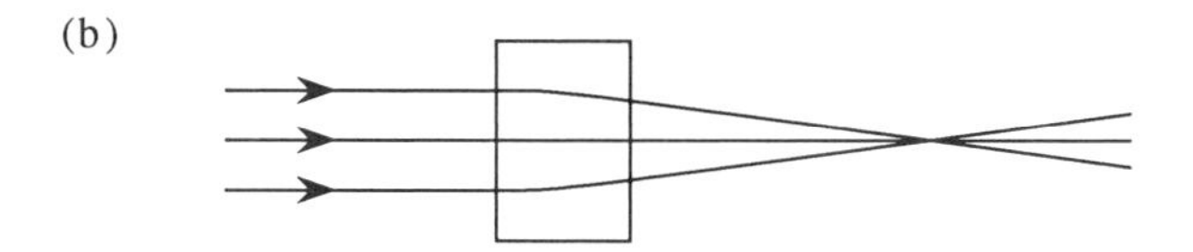

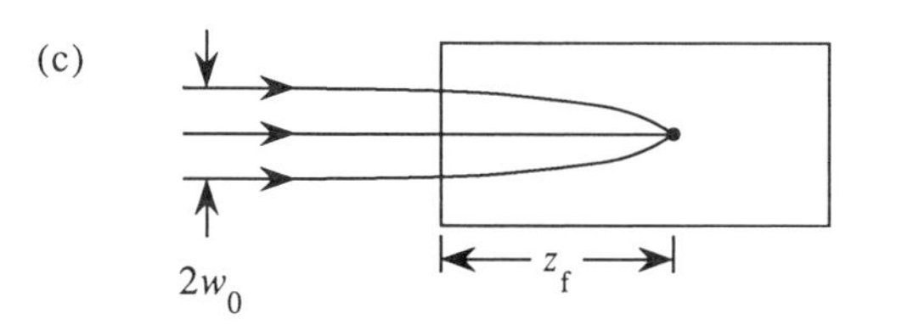

FIGURE 6.2.1 Self-focusing of light.

medium is in effect turned into a positive lens. The incident laser beam will tend to be brought to a focus by the action of this lens. If the medium is short enough, this focus will occur outside of the medium, as shown in the ray diagram of part (b) of the figure. However, if the medium is sufficiently long (or if the beam intensity is sufficiently large) the focus will occur within the nonlinear medium (part (c) of the figure), often leading to catastrophic damage of the material.

Self-focusing can occur only if the power of the laser beam is sufficiently large. If the power of the beam is too small, the tendency of the beam to contract due to the nonlinear change in refractive index will be too small to counteract the tendency of the beam to spread due to diffraction effects.

Self-Trapping of Light

When the tendency of the beam to spread due to diffraction is precisely compensated by the tendency of the beam to contract due to self-focusing, a phenomenon known as self-trapping may occur (Fig. 6.2.2). Here the beam maintains a small diameter d over a distance much longer than the usual longitudinal extent (approximately d^2/λ) of the focal region of a beam of characteristic transverse dimension d. One can think of the self-trapping process as the propagation of a light wave through a dielectric waveguide created within the material by the light itself by means of the nonlinearity of the refractive index. The self-trapping process is usually unstable, and small perturbations in the beam diameter will lead either to rapid spreading of the beam due to diffraction or to catastrophic collapse due to self-focusing. However, the process of self-trapping has been observed very cleanly in the experiments of Bjorkholm and Ashkin (1974) utilizing the nonlinear response of an atomic sodium vapor.

We next consider a simple model that predicts when self-trapping of light can occur. We make the simplifying assumption that the laser beam has a flat-top intensity distribution, as shown in part (a) of Fig. 6.2.3. The refractive index distribution within the nonlinear medium then has the form shown in part (b) of the figure, which shows a cut through the medium that includes the symmetry axis of the laser beam. Here the refractive index of the bulk of the

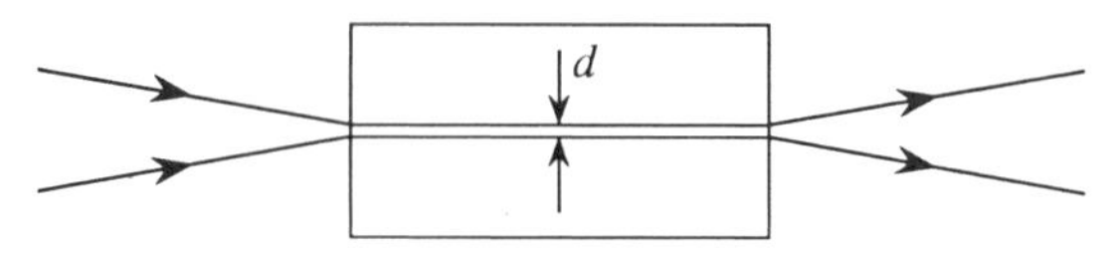

FIGURE 6.2.2 Self-trapping of light.

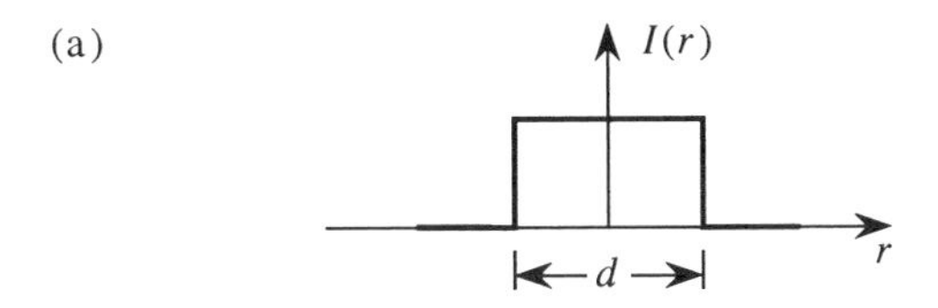

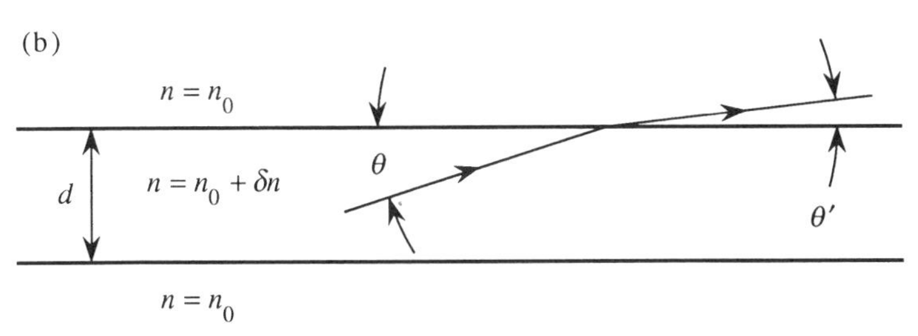

FIGURE 6.2.3 (a) Radial intensity distribution of the laser beam. (b) A ray of light incident on the boundary formed by the edge of the laser beam.

material is denoted by n_0 and the refractive index of that part of the medium exposed to the laser beam is denoted by $n_0 + \delta n$, where δn is the nonlinear contribution to the refractive index. Also shown in part (b) of the figure is a ray of light incident on the boundary between the two regions. It is one ray of the bundle that makes up the laser beam. This ray will remain trapped within the laser beam if it undergoes total internal reflection at the boundary between the two regions. Total internal reflection occurs if θ is less than the critical angle θ_0 for total internal reflection, which is given by the equation

$$\cos\theta_0 = \frac{n_0}{n_0 + \delta n}. \tag{6.2.2}$$

Since δn is very much smaller than n_0 for most nonlinear optical materials, and consequently θ_0 is much smaller than unity, Eq. (6.2.2) can be approximated by

$$1 - \tfrac{1}{2}\theta_0^2 = 1 - \frac{\delta n}{n_0},$$

which shows that the critical angle is related to the nonlinear change in refractive index by

$$\theta_0 = (2\,\delta n/n_0)^{1/2}. \tag{6.2.3}$$

A laser beam of diameter d will contain rays within a cone whose maximum angular extent is of the order of magnitude of characteristic diffraction angle

$$\theta_{\mathrm{d}} = \frac{0.61\lambda}{n_0 d}, \tag{6.2.4}$$

where λ is the wavelength of the light in vacuum. We expect that self-trapping will occur if total internal reflection occurs for all of the rays contained within the beam, that is, if $\theta_{\mathrm{d}} = \theta_0$. By comparing Eqs. (6.2.3) and (6.2.4), we see that self-trapping will occur if

$$\delta n = \tfrac{1}{2} n_0 (0.61\lambda / d n_0)^2, \tag{6.2.5a}$$

or equivalently, if

$$d = 0.61\lambda (2 n_0\, \delta n)^{-1/2}. \tag{6.2.5b}$$

If we now use Eq. (6.2.1) to replace δn by $n_2 I$, we see that the diameter of a self-trapped filament is related to the intensity of the light within the filament by

$$d = 0.61\lambda (2 n_0 n_2 I)^{-1/2}. \tag{6.2.6}$$

The power contained in a filament whose diameter is given by Eq. (6.2.6) is given by

$$P_{\mathrm{cr}} = \frac{\pi}{4} d^2 I = \frac{\pi (0.61)^2 \lambda^2}{8 n_0 n_2}. \tag{6.2.7}$$

Note that the predicted power is independent of the beam diameter. Our simple model hence predicts that, for a given material and a given laser wavelength, the power contained in a self-trapped filament has a unique value, even though the diameter d of the filament is not uniquely determined. The significance of this value of the power, called the critical power P_{cr}, is that self-focusing will occur if the power P of a laser beam is greater than P_{cr}, whereas self-focusing cannot occur if P is less than P_{cr}. Only when P is precisely equal to P_{cr} is self-trapping possible. Note that the *power*, not the *intensity*, of the laser beam is crucial in determining whether self-focusing will occur.

When the power P exceeds the critical power P_{cr} and self-focusing does occur, the beam will usually break up into several filaments, each of which contains power P_{cr}. The theory of filament formation has been described by Bespalov and Talanov (1966).

It is instructive to determine the numerical value of P_{cr}. For carbon disulfide (CS_2), n_2 for linearly polarized light is equal to $2.6 \times 10^{-14}\ \mathrm{cm^2/W}$, n_0 is equal to 1.7, and P_{cr} at a wavelength of 1 μm is equal to 33 kW. For typical crystals and glasses, n_2 is in the range 5×10^{-16} to $5 \times 10^{-15}\ \mathrm{cm^2/W}$ and P_{cr} is in the range 0.2 to 2 MW.

Simple Model of Self-Focusing

Let us now assume that the power P contained in the laser beam is very much larger that the critical power P_{cr}, so that the action of self-focusing overwhelms that of diffraction. We assume that the laser beam has a gaussian transverse intensity distribution, and that the beam waist is located at the entrance face of the nonlinear optical medium. We use Fermat's principle to estimate the distance z_f from the input face to the position of the focus created by self-focusing (see Fig. 6.2.1(c)). Our treatment is similar to that of Svelto (1974).

Fermat's principle tells us that the optical path length between the entrance face and the self-focus is the same for the central ray and for a ray at the periphery of the input beam. For the central ray, the physical path length is z_f, and we can estimate the refractive index experienced by this ray as $n_0 + \delta n$, where we take δn to be the nonlinear contribution to the refractive index at the center of the beam at the input face. For the peripheral ray, we estimate the physical path length as $(z_f^2 + w_0^2)^{1/2}$ and the mean refractive index as $n_0 + \frac{1}{2}\delta n$. The factor of $\frac{1}{2}$ accounts for the fact that the nonlinear contribution to the refractive index will be smaller than for the central ray.

By equating the two optical path lengths we find that

$$z_f(n_0 + \delta n) = (z_f^2 + w_0^2)^{1/2}(n_0 + \tfrac{1}{2}\delta n). \tag{6.2.8}$$

We approximate the right-hand side of this equation as

$$z_f[1 + \tfrac{1}{2}(w_0/z_f)^2 + \cdots](n_0 + \tfrac{1}{2}\delta n) \simeq z_f n_0 + \tfrac{1}{2}z_f\,\delta n + \tfrac{1}{2}z_f n_0(w_0/z_f)^2,$$

so that Eq. (6.2.8) becomes approximately

$$z_f n_0 + z_f\,\delta n = z_f n_0 + \tfrac{1}{2}z_f\,\delta n + \tfrac{1}{2}z_f n_0(w_0/z_f)^2,$$

whose solution for z_f is

$$z_f = w_0(n_0/\delta n)^{1/2}. \tag{6.2.9}$$

This result can be expressed in terms of the critical angle θ_0 for total internal reflection, which is given by Eq. (6.2.3), as

$$z_f = \sqrt{2}\,\frac{w_0}{\theta_0}. \tag{6.2.10}$$

We now explicitly assume that $\delta n = n_2 I$, where I is the maximum intensity of the input beam. Equation (6.2.9) then becomes

$$z_f = w_0\left(\frac{n_0}{n_2 I}\right)^{1/2} = w_0^2\left(\frac{\pi n_0}{2n_2 P}\right)^{1/2}, \tag{6.2.11}$$

where we have taken the total power of the input beam to be $P = \frac{1}{2}\pi w_0^2 I$. This

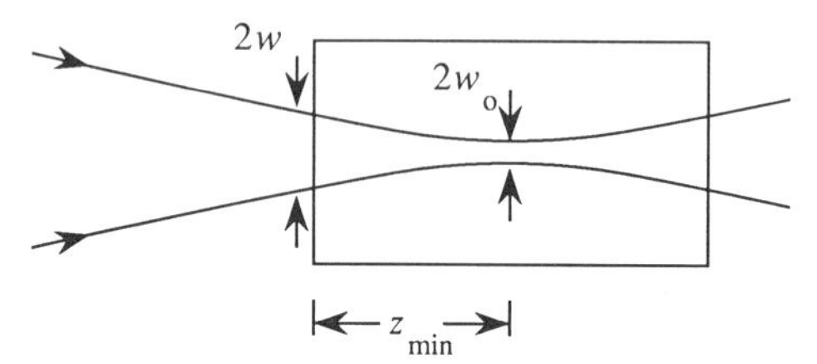

FIGURE 6.2.4 Definitions of the parameters w, w_0, and $z_{\min}$. The rays are shown as unmodified by the nonlinear interaction.

result shows how the self-focusing distance z_f decreases as the power P of the laser beam increases. It is sometimes useful to rewrite this result so that the laser power is given in terms of the critical power for self-focusing, as given by Eq. (6.2.7). We then find that

$$z_f = \frac{2n_0}{0.61} \frac{w_0^2}{\lambda} \frac{1}{(P/P_{cr})^{1/2}}. \tag{6.2.12}$$

As an example of the application of this formula, we consider the case of self-focusing in carbon disulfide, for which $P_{cr} = 33$ kW. A fairly modest Q-switched Nd:YAG laser operating at a wavelength of 1.06 μm might produce an output pulse containing 10 mJ of energy with a pulse duration of 10 ns, and hence with a peak power of the order of 1 MW. If we take w_0 equal to 100 μm, Eq. (6.2.12) predicts that $z_f = 1$ cm.

Yariv (1975) has shown that for the more general case in which the beam has arbitrary power and arbitrary beam waist position, the distance from the entrance face to the position of the self-focus is given by the formula

$$z_f = \frac{\frac{1}{2}kw^2}{(P/P_{cr} - 1)^{1/2} + 2z_{\min}/kw_0^2}, \tag{6.2.13}$$

where $k = n_0\omega/c$. The beam radius parameters w and w_0 (which have their conventional meanings) and $z_{\min}$ are defined in Fig. 6.2.4.

6.3. Optical Bistability

Certain nonlinear optical systems can possess more than one output state for a given input state. The term *optical bistability* refers to the situation in which two different output intensities are possible for a given input intensity, and the more general term *optical multistability* is used to describe the circumstance in which two or more stable output states are possible. Interest in optical bistability stems from its potential usefulness as a switch for use in optical communication and in optical computing.

Optical bistability was first described theoretically by Szöke *et al.* (1969) and was first observed experimentally by Gibbs *et al.* (1976). The bistable optical device described in these works consists of a nonlinear medium placed inside of a Fabry–Perot resonator. Such a device is illustrated schematically in Fig. 6.3.1. Here A_1 denotes the field amplitude of the incident wave, A_1' denotes that of the reflected wave, A_2 and A_2' denote the amplitudes of the forward- and backward-going waves within the interferometer, and A_3 denotes the amplitude of the transmitted wave. The cavity mirrors are assumed to be identical and lossless, with amplitude reflectance ρ and transmittance τ that are related to the intensity reflectance R and transmittance T through

$$R = |\rho|^2 \tag{6.3.1a}$$

and

$$T = |\tau|^2 \tag{6.3.1b}$$

with

$$R + T = 1. \tag{6.3.1c}$$

The incident and internal fields are related to each other through boundary conditions of the form

$$A_2' = \rho A_2 e^{2ikl - \alpha l}, \tag{6.3.2a}$$

$$A_2 = \tau A_1 + \rho A_2'. \tag{6.3.2b}$$

In these equations, we assume that the field amplitudes are measured at the inner surface of the left-hand mirror. The propagation constant $k = n\omega/c$ and intensity absorption coefficient α are taken to be real quantities, which include both their linear and nonlinear contributions. In writing Eq. (6.3.2a) in the form shown, we have implicitly made the *mean-field approximation*, that is, we

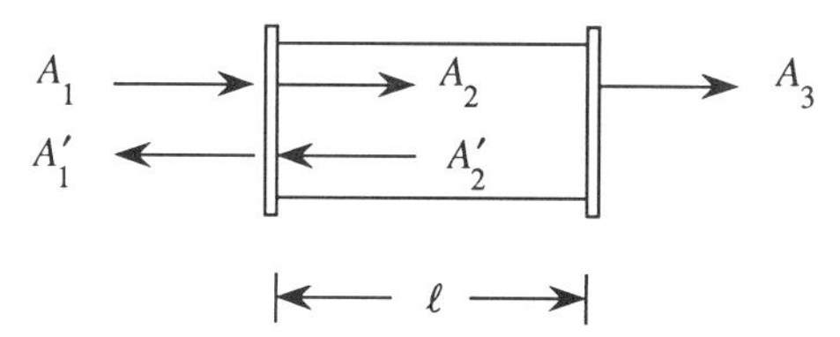

FIGURE 6.3.1 Bistable optical device in the form of a Fabry–Perot interferometer containing a nonlinear medium.

have assumed that the quantities k and α are spatially invariant; if such is not the case, the exponent should be replaced by $\int dz\,[(2ik(z) - \alpha(z)]$. For simplicity we also assume that the nonlinear material and the medium surround the resonator have the same linear refractive indices.

Equations (6.3.2) can be solved algebraically by eliminating A_2' to obtain

$$A_2 = \frac{\tau A_1}{1 - \rho^2 e^{2ikl - \alpha l}}, \tag{6.3.3}$$

which is known as Airy's equation and which describes the properties of a Fabry–Perot interferometer. If k or α (or both) is a sufficiently nonlinear function of the intensity of the light within the interferometer, this equation predicts bistability in the intensity of the transmitted wave. In general, both k and α can display nonlinear behavior; however, we can obtain a better understanding of the nature of optical bistability by considering in turn the limiting cases in which either the absorptive or the dispersive contribution dominates.

Absorptive Bistability

Let us first examine the case in which only the absorption coefficient α depends nonlinearly on the field intensity. The wave vector magnitude k is hence assumed to be constant. To simplify the following analysis, we assume that the mirror separation l is adjusted so that the cavity is tuned to resonance with the applied field; in such a case the factor $\rho^2 e^{2ikl}$ that appears in the denominator of Eq. (6.3.3) is equal to the real quantity R. We also assume that $\alpha l \ll 1$, so that we can ignore the spatial variation of the intensity of the field inside the cavity, which justifies the use of the mean-field approximation. Under these conditions, Airy's equation (6.3.3) reduces to

$$A_2 = \frac{\tau A_1}{1 - R(1 - \alpha l)}. \tag{6.3.4}$$

The analogous equation relating the intensities $I_i = (nc/2\pi)|A_i|^2$ is given by

$$I_2 = \frac{T I_1}{[1 - R(1 - \alpha l)]^2}. \tag{6.3.5}$$

This equation can be simplified by introducing the dimensionless parameter C (known as the cooperation number),

$$C = \frac{R\alpha l}{1 - R}, \tag{6.3.6}$$

which becomes (since $1 + C = (1 - R + R\alpha l)/(1 - R) = [1 - R(1 - \alpha l)]/T$)

$$I_2 = \frac{1}{T}\frac{I_1}{(1 + C)^2}. \tag{6.3.7}$$

We now assume that the absorption coefficient α and hence the value of the parameter C depend upon the intensity of the light within the interferometer. For simplicity, we assume that the absorption coefficient obeys the relation valid for a two-level saturable absorber,

$$\alpha = \frac{\alpha_0}{1 + I/I_s}, \tag{6.3.8}$$

where α_0 denotes the unsaturated absorption coefficient, I the local value of the intensity, and I_s the saturation intensity. For simplicity we also ignore the standing-wave nature of the field within the interferometer and take I equal to $I_2 + I_2' \approx 2I_2$. It is only approximately valid to ignore standing wave effects for the interferometer of Fig. 6.3.1, but it is strictly valid for the traveling-wave interferometer shown in Fig. 6.3.2. Under the assumption that the absorption coefficient depends on the intensity of the internal fields according to Eq. (6.3.8) with $I = 2I_2$, the parameter C is given by

$$C = \frac{C_0}{1 + 2I_2/I_s} \tag{6.3.9}$$

with $C_0 = R\alpha_0 l/(1 - R)$. The relation between I_1 and I_2 given by Eq. (6.3.7) can be rewritten using this expression for C as

$$I_1 = TI_2\left(1 + \frac{C_0}{1 + 2I_2/I_s}\right)^2. \tag{6.3.10}$$

Finally, the output intensity I_3 is related to I_2 by

$$I_3 = TI_2. \tag{6.3.11}$$

The input–output relation implied by Eqs. (6.3.10) and (6.3.11) is illustrated graphically in Fig. 6.3.3 for several different values of the weak-field parameter C_0. For C_0 greater than approximately 8, more than one output intensity can occur for certain values of the input intensity, which shows that the system possesses multiple solutions.

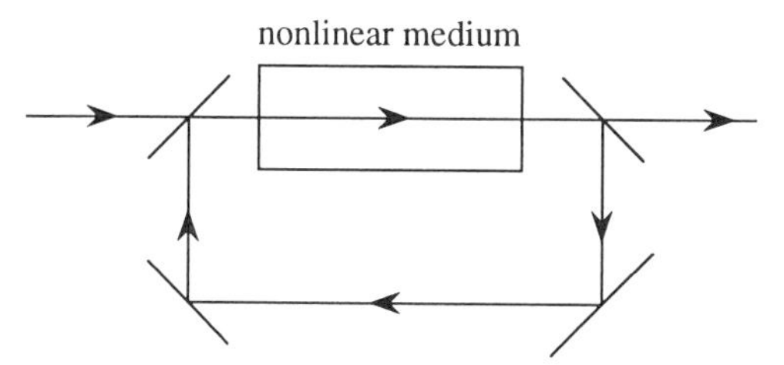

FIGURE 6.3.2 Bistable optical device in the form of a traveling-wave interferometer containing a nonlinear medium.

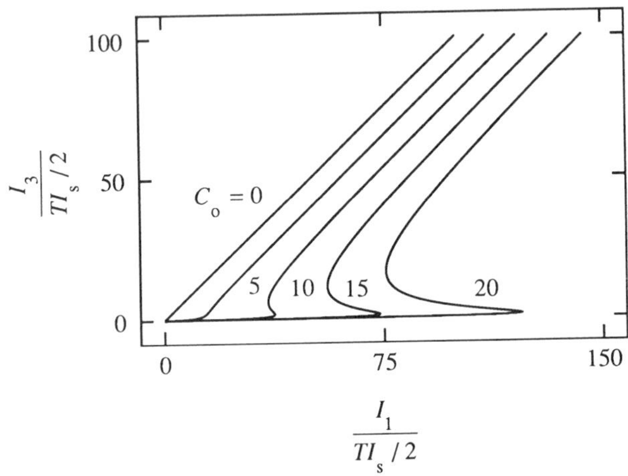

FIGURE 6.3.3 The input–output relation for a bistable optical device described by Eqs. (6.3.10) and (6.3.11).

The input–output characteristics for a system showing optical bistability are shown schematically in Fig. 6.3.4(a). The portion of the curve that has negative slope is shown by a dashed line. This portion corresponds to the branch of the solution to Eq. (6.3.10) for which the output intensity increases as the input intensity decreases. As might be expected on intuitive grounds, and as can be verified by means of a linear stability analysis, this branch of the solution is unstable; if the system is initially in this state, it will rapidly switch to one of the stable solutions due to the growth of small perturbations.

The solution shown in Fig. 6.3.4(a) displays hysteresis in the following sense. We imagine that the input intensity I_1 is initially zero and is slowly increased. As I_1 is increased from zero to I_h (the high jump point), the output intensity is given by the *lower branch* of the solution, that is, by the segment terminated by points *a* and *b*. As the input intensity is increased still further, the output intensity must jump to point *c* and trace out that portion of the curve labeled *c*–*d*. If the intensity is now slowly decreased, the system will remain on the upper branch and the output intensity will be given by the curve segment *e*–*d*. As the input intensity passes through the value I_l (the low jump point), the system makes a transition to point *f* and traces out the curve *f*–*a* as the input intensity is decreased to zero.

The use of such a device as an optical switch is illustrated in part (b) of Fig. 6.3.4. If the input intensity is held fixed at the value I_b (the bias intensity), the two stable output points indicated by the filled dots are possible. The state of the system can be used to store binary information. The system can be forced to make a transition to the upper state by injecting a pulse of light

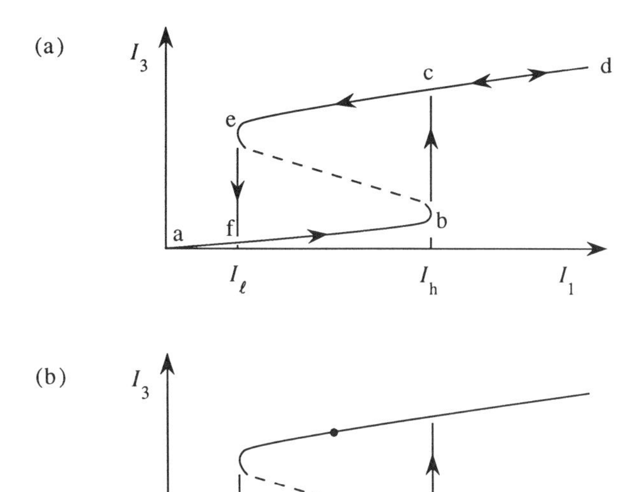

FIGURE 6.3.4 Schematic representation of the input–output characteristics of a system that shows optical bistability.

so that the total input intensity exceeds I_h; the system can be forced to make a transition to the lower state by momentarily blocking the input.

Dispersive Bistability

Let us now consider the case in which the absorption coefficient vanishes but in which the refractive index n depends nonlinearly on the optical intensity. For $\alpha = 0$, Eq. (6.3.3) becomes

$$A_2 = \frac{\tau A_1}{1 - \rho^2 e^{2ikl}} = \frac{\tau A_1}{1 - Re^{i\delta}}. \tag{6.3.12}$$

In obtaining the second form of this equation, we have written ρ^2 in terms of its amplitude and phase as

$$\rho^2 = Re^{i\phi} \tag{6.3.13}$$

and have introduced the total phase shift δ acquired in a round trip through the cavity. This phase shift is the sum

$$\delta = \delta_0 + \delta_2 \tag{6.3.14}$$

of a linear contribution

$$\delta_0 = \phi + 2n_0 \frac{\omega}{c} l \tag{6.3.15}$$

and a nonlinear contribution

$$\delta_2 = 2n_2 I \frac{\omega}{c} l, \tag{6.3.16}$$

where

$$I = I_2 + I_2' \simeq 2I_2. \tag{6.3.17}$$

Equation (6.3.12) can be used to relate the intensities $I_i = (nc/2\pi)|A_i|^2$ of the incident and internal fields as

$$\begin{aligned} I_2 &= \frac{TI_1}{(1 - Re^{i\delta})(1 - Re^{-i\delta})} = \frac{TI_1}{1 + R^2 - 2R\cos\delta} \\ &= \frac{TI_1}{(1-R)^2 + 4R\sin^2\frac{1}{2}\delta} = \frac{TI_1}{T^2 + 4R\sin^2\frac{1}{2}\delta} \\ &= \frac{I_1/T}{1 + (4R/T^2)\sin^2\frac{1}{2}\delta}, \end{aligned} \tag{6.3.18}$$

which shows that

$$\frac{I_2}{I_1} = \frac{1/T}{1 + (4R/T^2)\sin^2\frac{1}{2}\delta}, \tag{6.3.19}$$

where, according to Eqs. (6.3.14) through (6.3.17), the phase shift is given by

$$\delta = \delta_0 + \left(4n_2 \frac{\omega}{c} l\right) I_2, \tag{6.3.20}$$

In order to determine the conditions under which bistability can occur, we solve Eqs. (6.3.19) and (6.3.20) for the internal intensity I_2 as a function of the incident intensity I_1. This procedure is readily performed graphically by plotting each side of Eq. (6.3.19) as a function of I_2. Such a plot is shown in Fig. 6.3.5. We see that the system can possess one, three, five, or more solutions depending on the value of I_1. For the case in which three solutions exist for the range of input intensities I_1 that are available, a plot of I_3 versus I_1 looks very much like the curves shown in Fig. 6.3.4. Hence, the qualitative discussion of optical bistability given above is applicable in this case as well.

More detailed treatments of optical bistability can be found in Lugiato (1984) and Gibbs (1985).

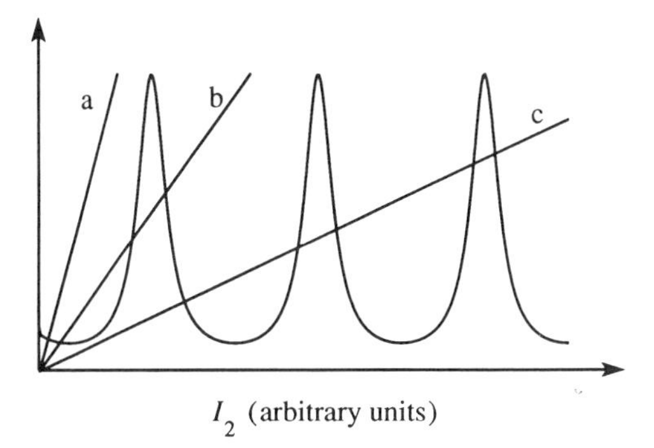

FIGURE 6.3.5 Graphical solution to Eq. (6.3.19). The oscillatory curve represents the right-hand side of this equation, and the straight lines labeled a through c represent the left-hand side for increasing values of the input intensity I_1.

6.4. Two-Beam Coupling

Let us consider the situation shown in Fig. 6.4.1, in which two beams of light (which in general have different frequencies) interact in a nonlinear material. Under certain conditions, the two beams interact in such a manner that energy is transferred from one beam to the other; this phenomenon is known as two-beam coupling. Two-beam coupling is a process that is automatically phase-matched. Consequently the efficiency of the process does not depend critically upon the angle θ between the two beams. The reason why this process is automatically phase-matched will be clarified by the following analysis; for the present it is perhaps helpful to note that the origin of two-beam coupling is that the refractive index experienced by either wave is modified by the *intensity* of the other wave.

Two-beam coupling occurs under several different circumstances in nonlinear optics. We saw in Chapter 5 that the nonlinear response of a two-level atom to pump and probe fields can lead to amplification of the probe wave. Furthermore, we shall see in Chapters 8 and 9 that gain occurs for various

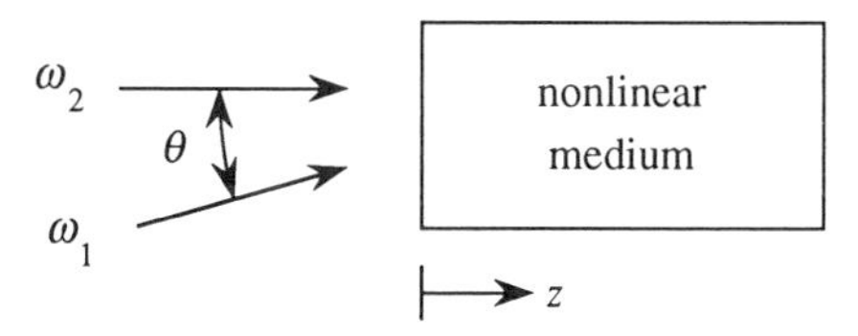

FIGURE 6.4.1 Two-beam coupling.

scattering processes such as stimulated Brillouin scattering and stimulated Raman scattering. Furthermore, in Chapter 10 we shall see that two-beam coupling occurs in many photorefractive materials. In the present section, we examine two-beam coupling from a general point of view that elucidates the conditions under which such energy transfer can occur. Our analysis is similar to that of Silberberg and Bar-Joseph (1982, 1984).

We describe the total optical field within the nonlinear medium as

$$\tilde{E}(\mathbf{r},t) = A_1 e^{i(\mathbf{k}_1\cdot\mathbf{r}-\omega_1 t)} + A_2 e^{i(\mathbf{k}_2\cdot\mathbf{r}-\omega_2 t)} + \text{c.c.}, \tag{6.4.1}$$

where

$$k_i = n_0\omega_i/c \tag{6.4.2}$$

with n_0 denoting the linear part of the refractive index experienced by each wave. We now consider the intensity distribution associated with the interference between the two waves. The intensity is given in general by

$$I = \frac{n_0 c}{4\pi}\overline{\tilde{E}^2}, \tag{6.4.3}$$

where the overbar denotes an average over a time interval of many optical periods. The intensity distribution for $\tilde{E}$ given by Eq. (6.4.1) is hence given by

$$\begin{aligned} I &= \frac{n_0 c}{2\pi}\{A_1A_1^* + A_2A_2^* + [A_1A_2^* e^{i(\mathbf{k}_1-\mathbf{k}_2)\cdot\mathbf{r}-i(\omega_1-\omega_2)t} + \text{c.c.}]\} \\ &= \frac{n_0 c}{2\pi}\{A_1A_1^* + A_2A_2^* + [A_1A_2^* e^{i(\mathbf{q}\cdot\mathbf{r}-\delta t)} + \text{c.c.}]\}, \end{aligned} \tag{6.4.4}$$

where we have introduced the wave vector difference (or "grating" wave vector)

$$\mathbf{q} = \mathbf{k}_1 - \mathbf{k}_2 \tag{6.4.5a}$$

and the frequency difference

$$\delta = \omega_1 - \omega_2. \tag{6.4.5b}$$

For the geometry of Fig. 6.4.1, the interference pattern has the form shown in Fig. 6.4.2, where we have assumed that $|\delta| \ll \omega_1$. Note that the pattern moves upward for $\delta > 0$, moves downward for $\delta < 0$, and is stationary for $\delta = 0$.

A particularly simple example is the special case in which $\theta = 180$ degrees. Then, again assuming that $|\delta| \ll \omega_1$, we find that the wave vector difference is given approximately by

$$\mathbf{q} \simeq -2\mathbf{k}_2 \tag{6.4.6}$$

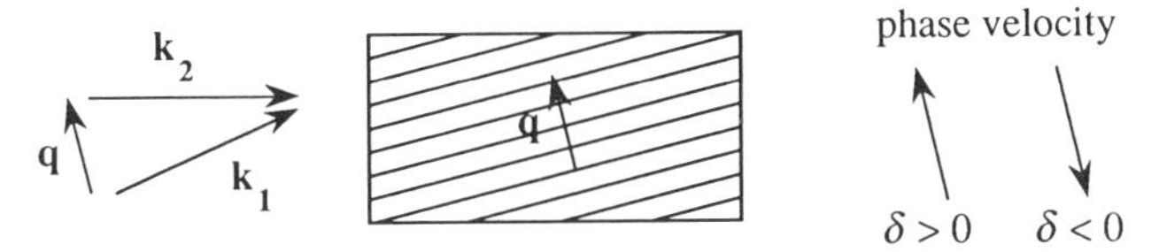

FIGURE 6.4.2 Interference pattern formed by two interacting waves.

and hence that the intensity distribution is given by

$$I = \frac{n_0 c}{2\pi}\{A_1 A_1^* + A_2 A_2^* + [A_1 A_2^* e^{i(-2kz-\delta t)} + \text{c.c.}]\}. \tag{6.4.7}$$

The interference pattern is hence of the form shown in Fig. 6.4.3. If δ is positive, the interference pattern moves to the left, and if δ is negative it moves to the right, in either case with phase velocity $|\delta|/2k$.

Since the material system is nonlinear, a refractive index variation accompanies this intensity variation. Each wave is scattered by this index variation, or grating. We shall show below that no energy transfer accompanies this interaction for the case of a nonlinearity that responds instantaneously to the applied field. In order to allow the possibility of energy transfer, we assume that the nonlinear part of the refractive index (n_{NL}) obeys a Debye relaxation equation of the form

$$\tau \frac{dn_{\text{NL}}}{dt} + n_{\text{NL}} = n_2 I. \tag{6.4.8}$$

Note that this equation predicts that, in steady state, the nonlinear contribution to the refractive index is given simply by $n_{\text{NL}} = n_2 I$, in consistency with Eq. (4.1.15). However, under transient conditions it predicts that the nonlinearity develops in a time interval of the order of τ.

Equation (6.4.8) can be solved (for example, by the method of variation of parameters or by the Green's function method) to give the result

$$n_{\text{NL}} = \frac{n_2}{\tau}\int_{-\infty}^{t} I(t') e^{(t'-t)/\tau}\, dt'. \tag{6.4.9}$$

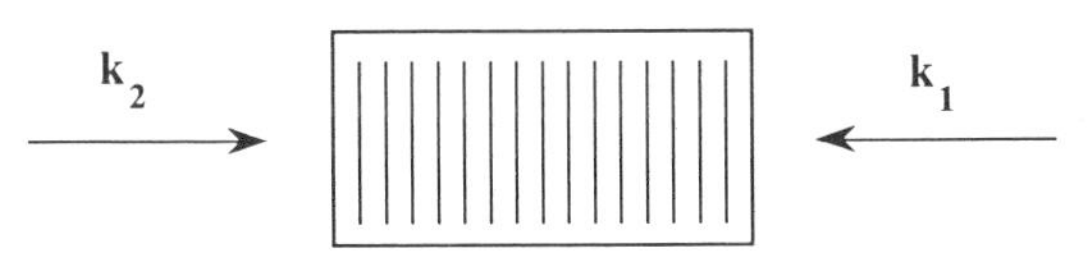

FIGURE 6.4.3 Interference pattern formed by two counterpropagating beams.

The expression (6.4.4) for the intensity $I(t)$ is next introduced into this equation. We find, for example, that the part of $I(t)$ that varies as $e^{-i\delta t}$ leads to an integral of the form

$$\int_{-\infty}^{t} e^{-i\delta t'} e^{(t'-t)/\tau} \, dt' = e^{-t/\tau} \int_{-\infty}^{t} e^{(-i\delta + 1/\tau)t'} \, dt' = \frac{e^{-i\delta t}}{-i\delta + 1/\tau} \tag{6.4.10}$$

Equation (6.4.9) hence shows that the nonlinear contribution to the refractive index is given by

$$n_{\mathrm{NL}} = \frac{n_0 n_2 c}{2\pi} \left[(A_1 A_1^* + A_2 A_2^*) + \frac{A_1 A_2^* e^{i(\mathbf{q}\cdot\mathbf{r} - \delta t)}}{1 - i\delta\tau} + \frac{A_1^* A_2 e^{-i(\mathbf{q}\cdot\mathbf{r} - \delta t)}}{1 + i\delta\tau} \right]. \tag{6.4.11}$$

Due to the imaginary contributions to the denominators, the refractive index variation is not in general in phase with the intensity distribution.

In order to determine the degree of coupling between the two fields, we require that the field given by Eq. (6.4.1) satisfy the wave equation

$$\nabla^2 \tilde{E} - \frac{n^2}{c^2} \frac{\partial^2 \tilde{E}}{\partial t^2} = 0, \tag{6.4.12}$$

where we take the refractive index to have the form

$$n = n_0 + n_{\mathrm{NL}}. \tag{6.4.13}$$

We make the physical assumption that

$$|n_{\mathrm{NL}}| \ll n_0, \tag{6.4.14}$$

in which case it is a good approximation to express n^2 as

$$n^2 = n_0^2 + 2 n_0 n_{\mathrm{NL}}. \tag{6.4.15}$$

Let us consider the part of Eq. (6.4.12) that shows a spatial and temporal dependence given by $\exp[i(\mathbf{k}_2 \cdot \mathbf{r} - \omega_2 t)]$. Using Eqs. (6.4.1), (6.4.11), and (6.4.15), we find that this portion of Eq. (6.4.12) is given by

$$\begin{aligned} &\frac{d^2 A_2}{dz^2} + 2ik_2 \frac{dA_2}{dz} - k_2^2 A_2 + \frac{n_0^2 \omega_2^2}{c^2} A_2 \\ &\qquad = -\frac{n_0^2 n_2 \omega_2^2}{\pi c} (|A_1|^2 + |A_2|^2) A_2 - \frac{n_0^2 n_2 \omega_1^2}{\pi c} \frac{|A_1|^2 A_2}{1 + i\delta\tau}. \end{aligned} \tag{6.4.16}$$

Note that the origin of the last term is the scattering of the field $A_1 \exp[i(\mathbf{k}_1 \cdot \mathbf{r} - \omega_1 t)]$ from the time-varying refractive index distribution (i.e., the moving grating)

$$\frac{n_0 c}{2\pi} n_2 A_1^* A_2 \frac{e^{-i(\mathbf{q}\cdot\mathbf{r} - \delta t)}}{1 + i\delta\tau},$$

whereas the origin of the second-last term is the scattering of the field $A_2 \exp[i(\mathbf{k}_2 \cdot \mathbf{r} - \omega_2 t)]$ from the stationary refractive index variation

$$\frac{n_0 c}{2\pi} n_2 (A_1 A_1^* + A_2 A_2^*).$$

We next drop the first term on the left-hand side of Eq. (6.4.16) by making the slowly-varying-amplitude approximation, and we note that the third and fourth terms exactly cancel. The equation then reduces to

$$\frac{dA_2}{dz} = i \frac{n_0 n_2 \omega}{2\pi} (|A_1|^2 + |A_2|^2) A_2 + i \frac{n_0 n_2 \omega}{2\pi} \frac{|A_1|^2 A_2}{1 + i\delta\tau}, \tag{6.4.17}$$

where, to good approximation, we have replaced ω_1 and ω_2 by ω. We now calculate the rate of change of intensity of the ω_2 field. We introduce the intensities

$$I_1 = \frac{n_0 c}{2\pi} A_1 A_1^* \quad \text{and} \quad I_2 = \frac{n_0 c}{2\pi} A_2 A_2^* \tag{6.4.18}$$

and note that the spatial variation of I_2 is given by

$$\frac{dI_2}{dz} = \frac{n_0 c}{2\pi} \left(A_2^* \frac{dA_2}{dz} + A_2 \frac{dA_2^*}{dz} \right). \tag{6.4.19}$$

We then find from Eqs. (6.4.17) through (6.4.19) that

$$\frac{dI_2}{dz} = \frac{2 n_2 \omega}{c} \frac{\delta\tau}{1 + \delta^2\tau^2} I_1 I_2. \tag{6.4.20}$$

Note that only the last term on the right-hand side of Eq. (6.4.17) contributes to energy transfer.

For the case of a positive value of n_2 (for example, for the molecular orientation Kerr effect, for electrostriction, or for a two-level atom with the optical frequencies above the resonance frequency), Eq. (6.4.20) predicts gain for positive δ, i.e., for $\omega_2 < \omega_1$. The frequency dependence of Eq. (6.4.20) is shown in Fig. 6.4.4.

Note that the ω_2 wave experiences maximum gain for $\delta\tau = 1$, in which case Eq. (6.4.20) becomes

$$\frac{dI_2}{dz} = n_2 \frac{\omega}{c} I_1 I_2. \tag{6.4.21}$$

Note also from Eq. (6.4.20) that in the limit of an infinitely fast nonlinearity, that is, in the limit $\tau \to 0$, the coupling of intensity between the two waves vanishes. The reason for this behavior is that only the imaginary part of the (total) refractive index can lead to a change in intensity of the ω_2 wave. We see from Eq. (6.4.11) that (for a real n_2) the only way in which n_{NL} can become complex is if τ is nonzero. When τ is nonzero, the response can lag in phase

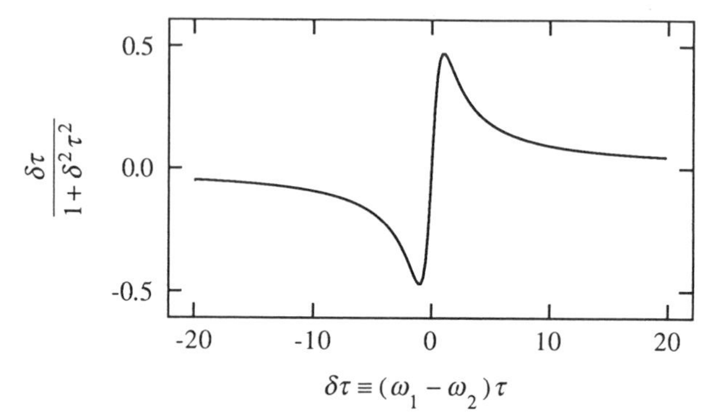

FIGURE 6.4.4 Frequency dependence of the gain for two-beam coupling.

behind the driving term, leading to a complex value of the nonlinear contribution to the refractive index.

The theory just presented predicts that there will be no energy coupling if the product $\delta\tau$ vanishes, either because the nonlinearity has a fast response or because the input waves are at the same frequency. However, two-beam coupling can occur in certain photorefractive crystals even between beams of the same frequency (Feinberg, 1983). In such cases, energy transfer occurs as a result of a *spatial* phase shift between the nonlinear index grating and the optical intensity distribution. The direction of energy flow depends upon the orientation of the wave vectors of the optical beams with respect to some symmetry axis of the photorefractive crystal. The photorefractive effect is described in greater detail in Chapter 10.

6.5. Pulse Propagation and Optical Solitons

In this section we study some of the nonlinear optical effects that can occur when short optical pulses propagate through dispersive nonlinear optical media. We shall see that the spectral content of the pulse can become modified by the nonlinear optical process of self-phase modulation. This process is especially important for pulses of high peak intensity. We shall also see that (even for the case of a medium with a linear response) the shape of the pulse can become modified by means of propagation effects such as disperson of the group velocity within the medium. This process is especially important for very short optical pulses, which necessarily have a broad spectrum. In general, self-phase modulation and group-velocity dispersion occur simultaneously, and both tend to modify the shape of the optical pulse. However, under certain circumstances, which are described below, an exact cancellation of these two effects can occur, allowing a special type of pulse known as an optical soliton to propagate through large distances with no change in shape.

Self-Phase Modulation

Self-phase modulation is the change in the phase of an optical pulse due to the nonlinearity of the refractive index of the material medium. In order to understand the origin of this effect, let us consider the propagation of the optical pulse

$$\tilde{E}(z,t) = \tilde{A}(z,t)e^{i(k_0 z - \omega_0 t)} + \text{c.c.} \tag{6.5.1}$$

through a medium characterized by a nonlinear refractive index of the sort

$$n(t) = n_0 + n_2 I(t), \tag{6.5.2}$$

where $I(t) = (n_0 c/2\pi)|\tilde{A}(z,t)|^2$. Note that for the present we are assuming that the medium can respond essentially instantaneously to the pulse intensity. We also assume that the nonlinear medium is sufficiently short that no reshaping of the optical pulse can occur within the medium; the only effect of the medium is to change the phase of the transmitted pulse by the amount

$$\phi_{\text{NL}}(t) = -n_2 I(t)\omega_0 L/c. \tag{6.5.3}$$

As a result of the time-varying phase of the wave, the spectrum of the transmitted pulse will be modified and typically will be broader than that of the incident pulse. From a formal point of view, we can determine the spectral content of the transmitted pulse by calculating its energy spectrum

$$S(\omega) = \left| \int_{-\infty}^{\infty} \tilde{A}(t)e^{-i\omega_0 t - i\phi_{\text{NL}}(t)}e^{i\omega t}\, dt \right|^2. \tag{6.5.4}$$

However, it is more intuitive to describe the spectral content of the transmitted pulse by introducing the concept of the instantaneous frequency $\omega(t)$ of the pulse, which is described by

$$\omega(t) = \omega_0 + \delta\omega(t) \tag{6.5.5a}$$

where

$$\delta\omega(t) = \frac{d}{dt}\phi_{\text{NL}}(t) \tag{6.5.5b}$$

denotes the variation of the instantaneous frequency. The instantaneous frequency is a well-defined concept and is given by Eqs. (6.5.5) whenever the amplitude $\tilde{A}(t)$ varies slowly compared to an optical period.

As an example of the use of these formulas, we consider the case illustrated in part (a) of Fig. 6.5.1, in which the pulse shape is given by the form

$$I(t) = I_0 \operatorname{sech}^2(t/\tau_0). \tag{6.5.6}$$

We then find from Eq. (6.5.3) that the nonlinear phase shift is given by

$$\phi_{\mathrm{NL}}(t) = -n_2 \frac{\omega_0}{c} L I_0 \operatorname{sech}^2(t/\tau_0), \tag{6.5.7}$$

and from Eq. (6.5.5b) that the change in instantaneous frequency is given by

$$\delta\omega(t) = 2n_2 \frac{\omega_0}{c\tau_0} L I_0 \operatorname{sech}^2(t/\tau_0) \tanh(t/\tau_0). \tag{6.5.8}$$

The variation in the instantaneous frequency is illustrated in part (b) of Fig. 6.5.1, under the assumption that n_2 is positive. We see that the leading edge of the pulse is shifted to lower frequencies and that the trailing edge is

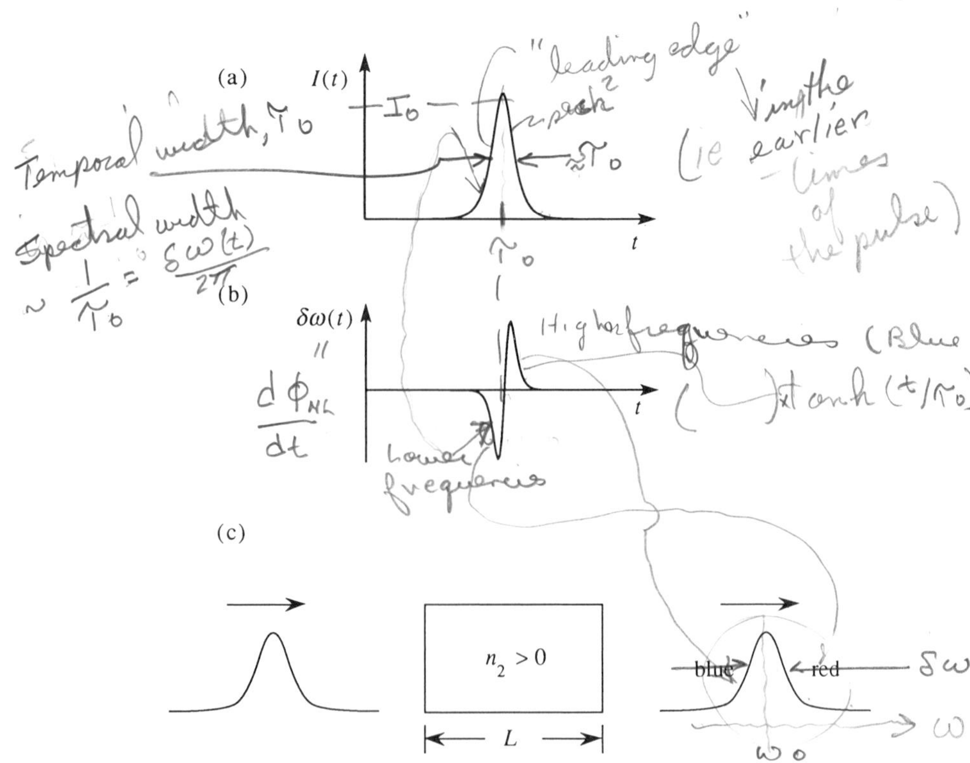

FIGURE 6.5.1 (a) Time dependence of the incident pulse. (b) Change in instantaneous frequency of the transmitted pulse. (c) Experimental arrangement to observe self-phase modulation.

shifted to higher frequencies. This conclusion is summarized schematically in part (c) of the figure. The maximum value of the frequency shift will be of the order of

$$\delta\omega_{\max} \simeq \frac{\Delta\phi_{\mathrm{NL}}^{\max}}{\tau_0}, \qquad \text{where} \quad \Delta\phi_{\mathrm{NL}}^{\max} \simeq n_2 \frac{\omega_0}{c} I_0 L. \tag{6.5.9}$$

We expect that spectral broadening due to self-phase modulation will be important whenever $\delta\omega_{\max}$ exceeds the spectral width of the incident pulse, which for the case of a smooth pulse is of the order of $1/\tau_0$. We thus expect self-phase modulation to be important whenever $\Delta\phi_{\mathrm{NL}}^{\max} \geq 2\pi$.

Self-phase modulation of the sort just described was studied initially by Brewer (1967), by Shimizu (1967), and by Cheung *et al.* (1968).

Pulse Propagation Equation

Let us next consider the equations that govern the propagation of the pulse

$$\tilde{E}(z,t) = \tilde{A}(z,t)e^{i(k_0 z - \omega_0 t)} + \text{c.c.}, \tag{6.5.10}$$

where $k_0 = n_{\text{lin}}(\omega_0)\omega_0/c$, through a dispersive, nonlinear optical medium. We take the wave equation in the form (see also Eq. (2.1.9a))

$$\frac{\partial^2 \tilde{E}}{\partial z^2} - \frac{1}{c^2}\frac{\partial^2 \tilde{D}}{\partial t^2} = 0, \tag{6.5.11}$$

where $\tilde{D}$ represents the total displacement field, including both linear and nonlinear contributions. We now introduce the Fourier transforms of $\tilde{E}(z,t)$ and $\tilde{D}(z,t)$ by the equations

$$\tilde{E}(z,t) = \int_{-\infty}^{\infty} E(z,\omega)e^{-i\omega t}\frac{d\omega}{2\pi}, \qquad \tilde{D}(z,t) = \int_{-\infty}^{\infty} D(z,\omega)e^{-i\omega t}\frac{d\omega}{2\pi}. \tag{6.5.12}$$

The Fourier amplitudes $E(z,\omega)$ and $D(z,\omega)$ are related by

$$D(z,\omega) = \epsilon(\omega)E(z,\omega), \tag{6.5.13}$$

where $\epsilon(\omega)$ is the effective dielectric constant that describes both the linear and nonlinear contributions to the response.

Equations (6.5.12) and (6.5.13) are now introduced into the wave equation (6.5.11), which becomes

$$\frac{\partial^2 E(z,\omega)}{\partial z^2} + \epsilon(\omega)\frac{\omega^2}{c^2}E(z,\omega) = 0. \tag{6.5.14}$$

We now write this equation in terms of the Fourier transform of $\tilde{A}(z,t)$, which is given by

$$A(z,\omega') = \int_{-\infty}^{\infty} \tilde{A}(z,t)e^{i\omega' t}\,dt, \tag{6.5.15}$$

and which is related to $E(z,\omega)$ by

$$\begin{aligned} E(z,\omega) &= A(z,\omega-\omega_0)e^{ik_0 z} + A^*(z,\omega+\omega_0)e^{-ik_0 z} \\ &\simeq A(z,\omega-\omega_0)e^{ik_0 z}. \end{aligned} \tag{6.5.16}$$

The latter expression for $E(z,\omega)$ is now introduced into Eq. (6.5.14), and the slowly-varying-amplitude approximation is made, so that the term containing $\partial^2 A/\partial z^2$ can be dropped. We obtain

$$2ik_0 \frac{\partial A}{\partial z} + (k^2 - k_0^2)A = 0, \tag{6.5.17}$$

where

$$k(\omega) = \sqrt{\epsilon(\omega)}\,\omega/c. \tag{6.5.18}$$

In practice, k typically differs from k_0 by only a small fractional amount, and thus to good approximation $k^2 - k_0^2$ can be replaced by $2k_0(k-k_0)$, so that Eq. (6.5.17) becomes

$$\frac{\partial A(z,\omega-\omega_0)}{\partial z} - i(k-k_0)A(z,\omega-\omega_0) = 0. \tag{6.5.19}$$

Recall that the propagation constant k depends both on the frequency and (through the intensity dependence of ϵ) on the intensity of the optical wave. It is often adequate to describe this dependence in terms of a truncated power series expansion of the form

$$k = k_0 + \Delta k_{\mathrm{NL}} + k_1(\omega-\omega_0) + \tfrac{1}{2}k_2(\omega-\omega_0)^2. \tag{6.5.20}$$

In this expression, we have introduced the nonlinear contribution to the propagation constant, given by

$$\Delta k_{\mathrm{NL}} = \Delta n_{\mathrm{NL}} \frac{\omega_0}{c} = n_2 I \frac{\omega_0}{c}, \tag{6.5.21}$$

with $I = [n_{\mathrm{lin}}(\omega_0)c/2\pi]|\tilde{A}(z,t)|^2$, and have introduced the quantities

$$k_1 = \left(\frac{dk}{d\omega}\right)_{\omega=\omega_0} = \frac{1}{c}\left[n_{\mathrm{lin}}(\omega) + \omega\frac{dn_{\mathrm{lin}}(\omega)}{d\omega}\right]_{\omega=\omega_0} \equiv \frac{1}{v_g(\omega_0)} \tag{6.5.22a}$$

and

$$k_2 = \left(\frac{d^2k}{d\omega^2}\right)_{\omega=\omega_0} = \frac{d}{d\omega}\left[\frac{1}{v_g(\omega)}\right]_{\omega=\omega_0} = \left(-\frac{1}{v_g^2}\frac{dv_g}{d\omega}\right)_{\omega=\omega_0} \tag{6.5.23}$$

Here k_1 is the reciprocal of the group velocity, and k_2 is a measure of the dispersion of the group velocity. As illustrated in Fig. 6.5.2, the long-wavelength components of an optical pulse propagate faster than the short-wavelength components when the group velocity dispersion parameter k_2 is positive, and vice versa.

The expression (6.5.20) for k is next introduced into the reduced wave equation (6.5.19), which becomes

$$\frac{\partial A}{\partial z} - i\,\Delta k_{\mathrm{NL}}\,A - ik_1(\omega-\omega_0)A - \tfrac{1}{2}ik_2(\omega-\omega_0)^2A = 0. \qquad (6.5.24)$$

This equation is now transformed from the frequency domain to the time domain. To do so, we multiply each term by the factor $\exp[-i(\omega-\omega_0)t]$ and integrate the resulting equation over all values of $\omega-\omega_0$. We evaluate the resulting integrals as follows:

$$\int_{-\infty}^{\infty} A(z,\omega-\omega_0)e^{-i(\omega-\omega_0)t}\frac{d(\omega-\omega_0)}{2\pi} = \tilde{A}(z,t), \qquad (6.5.25a)$$

$$\int_{-\infty}^{\infty} (\omega-\omega_0)A(z,\omega-\omega_0)e^{-i(\omega-\omega_0)t}\frac{d(\omega-\omega_0)}{2\pi} \qquad (6.5.25b)$$

$$= \frac{1}{-i}\frac{\partial}{\partial t}\int_{-\infty}^{\infty} A(z,\omega-\omega_0)e^{-i(\omega-\omega_0)t}\frac{d(\omega-\omega_0)}{2\pi} = i\frac{\partial}{\partial t}\tilde{A}(z,t),$$

$$\int_{-\infty}^{\infty} (\omega-\omega_0)^2A(z,\omega-\omega_0)e^{-i(\omega-\omega_0)t}\frac{d(\omega-\omega_0)}{2\pi} = -\frac{\partial^2}{\partial t^2}\tilde{A}(z,t). \qquad (6.5.25c)$$

Equation (6.5.24) then becomes

$$\frac{\partial \tilde{A}}{\partial z} + k_1\frac{\partial \tilde{A}}{\partial t} + \tfrac{1}{2}ik_2\frac{\partial^2\tilde{A}}{\partial t^2} - i\,\Delta k_{\mathrm{NL}}\,\tilde{A} = 0. \qquad (6.5.26)$$

This equation can be simplified by means of a coordinate transformation. In

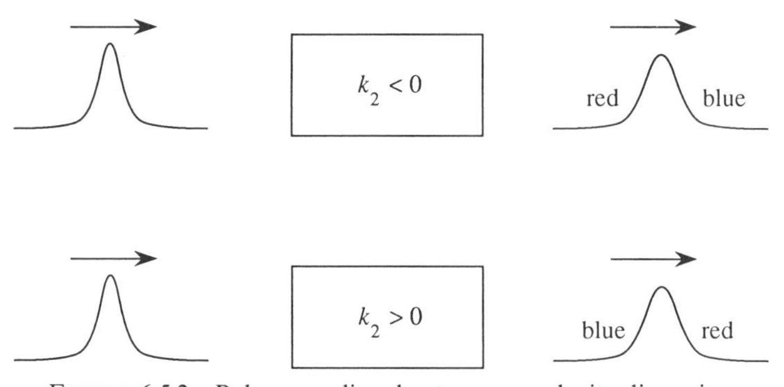

FIGURE 6.5.2 Pulse spreading due to group velocity dispersion.

particular we introduce the retarded time τ by the substitution

$$\tau = t - \frac{z}{v_g} = t - k_1 z, \tag{6.5.27}$$

and we describe the optical pulse by the function $\tilde{A}_s(z, \tau)$, which is related to the function $\tilde{A}(z, t)$ by

$$\tilde{A}_s(z, \tau) = \tilde{A}(z, t). \tag{6.5.28}$$

We next use the chain rule of differentiation to show that

$$\frac{\partial \tilde{A}}{\partial z} = \frac{\partial \tilde{A}_s}{\partial z} + \frac{\partial \tilde{A}_s}{\partial \tau}\frac{\partial \tau}{\partial z} = \frac{\partial \tilde{A}_s}{\partial z} - k_1 \frac{\partial \tilde{A}_s}{\partial \tau} \tag{6.5.29a}$$

$$\frac{\partial \tilde{A}}{\partial t} = \frac{\partial \tilde{A}_s}{\partial z}\frac{\partial z}{\partial t} + \frac{\partial \tilde{A}_s}{\partial \tau}\frac{\partial \tau}{\partial t} = \frac{\partial \tilde{A}_s}{\partial \tau}, \tag{6.5.29b}$$

and analogously that $\partial^2 \tilde{A}/\partial t^2 = \partial^2 \tilde{A}_s/\partial \tau^2$. These expressions are now introduced into Eq. (6.5.26), which becomes

$$\frac{\partial \tilde{A}_s}{\partial z} + \tfrac{1}{2} i k_2 \frac{\partial^2 \tilde{A}_s}{\partial \tau^2} - i\,\Delta k_{\text{NL}}\,\tilde{A}_s = 0. \tag{6.5.30}$$

Finally, we express the nonlinear contribution to the propagation constant as

$$\Delta k_{\text{NL}} = n_2 \frac{\omega_0}{c} I = \frac{n_0 n_2 \omega_0}{2\pi} |\tilde{A}_s|^2 \equiv \gamma |\tilde{A}_s|^2, \tag{6.5.31}$$

so that Eq. (6.5.30) can be expressed as

$$\frac{\partial \tilde{A}_s}{\partial z} + \tfrac{1}{2} i k_2 \frac{\partial^2 \tilde{A}_s}{\partial \tau^2} = i\gamma |\tilde{A}_s|^2 \tilde{A}_s. \tag{6.5.32}$$

This equation describes the propagation of optical pulses through dispersive, nonlinear optical media. Note that the second term on the left-hand side shows how pulses tend to spread due to group velocity dispersion, and that the term on the right-hand side shows how pulses tend to spread due to self-phase modulation. Equation (6.5.32) is sometimes referred to as the nonlinear Schrödinger equation.

Optical Solitons

Note from the form of the pulse propagation equation (6.5.32) that it is possible for the effects of group velocity dispersion to compensate for the effects of self-phase modulation. In fact, under appropriate circumstances the degree of compensation can be complete, and optical pulses can propagate through a dispersive, nonlinear optical medium with an invariant shape. Such pulses are known as optical solitons.

As an example of a soliton, note that Eq. (6.5.32) is solved identically by a pulse whose amplitude is of the form

$$\tilde{A}_s(z, \tau) = A_s^0 \operatorname{sech}(\tau/\tau_0)\, e^{i\kappa z} \tag{6.5.33a}$$

where the pulse amplitude A_s^0 and pulse width τ_0 must be related according to

$$|A_s^0|^2 = \frac{-k_2}{\gamma \tau_0^2} = \frac{-2\pi k_2}{n_0 n_2 \omega_0 \tau_0^2} \tag{6.5.33b}$$

and where

$$\kappa = -k_2/2\tau_0^2 \tag{6.5.33c}$$

represents the phase shift experienced by the pulse upon propagation. One can verify by direct substitution that Eqs. (6.5.33) do in fact satisfy the pulse propagation equation (6.5.32) (see Problem 10 at the end of this chapter).

Note that the condition (6.5.33b) shows that k_2 and n_2 must have opposite signs in order for Eq. (6.5.33a) to represent a physical pulse in which the intensity $|A_s^0|^2$ and the square of the pulse width τ_0^2 are both positive. We can see from Eq. (6.5.32) that in fact k_2 and γ must have opposite signs in order for group velocity dispersion to compensate for self-phase modulation (because $\tilde{A}_s^{-1}(\partial^2 \tilde{A}_s/\partial \tau^2)$ will be negative near the peak of the pulse, where the factor $|\tilde{A}_s|^2 \tilde{A}_s$ is most important).

Expressions (6.5.33) give what is known as the fundamental soliton solution to the pulse propagation equation (6.5.32). Higher-order soliton solutions also are known. These solutions were first obtained through use of inverse scattering methods by Zakharov and Shabat (1972) and are described in more detail by Agrawal (1989).

One circumstance under which k_2 and γ have opposite signs occurs in fused silica optical fibers. In this case, the nonlinearity in the refractive index occurs as the result of electronic polarization, and n_2 is consequently positive. The group velocity dispersion parameter k_2 is positive for visible light, but becomes negative for wavelengths longer than approximately 1.3 μm. This effect is illustrated in Fig. 6.5.3, in which the linear refractive index n_{lin} and the group index $n_g \equiv c/v_g$ are plotted as functions of the vacuum wavelength of the incident radiation. Optical solitons of the sort described by Eq. (6.5.33) have been observed by Mollenauer *et al.* (1980) in the propagation of light pulses at a wavelength of 1.55 μm obtained from a color center laser.

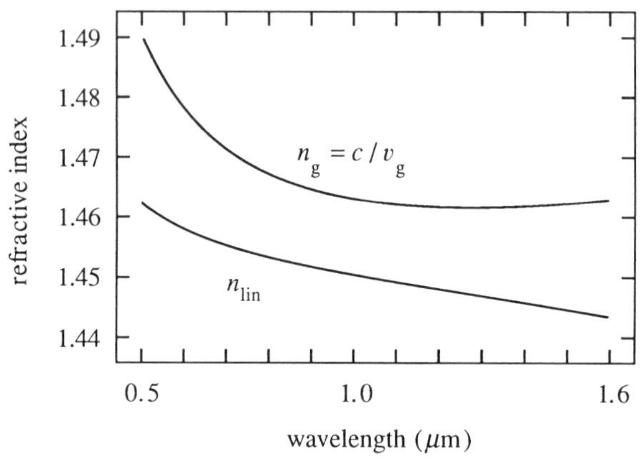

FIGURE 6.5.3 Dependence of the linear refractive index n_{lin} and the group index $n_g = c/v_g$ on the vacuum wavelength for fused silica.

Problems

1. Solve the following coupled equations for the boundary conditions that $A_3(0)$ and $A_4(L)$ are arbitrary:

$$\frac{dA_3}{dz} = -\alpha_3 A_3 - i\kappa_4^* A_4^*,$$

$$\frac{dA_4}{dz} = \alpha_4 A_4 + i\kappa_3^* A_3^*.$$

(These equations generalize Eqs. (6.1.31) and describe four-wave mixing in the usual phase conjugation geometry for the case of a lossy medium.)

2. Same as Problem 1, but with the inclusion of phase-mismatch factors so that the coupled equations are given by

$$\frac{dA_3}{dz} = -\alpha_3 A_3 - i\kappa_4^* A_4^* e^{i\Delta k z},$$

$$\frac{dA_4}{dz} = \alpha_4 A_4 + i\kappa_3^* A_3^* e^{-i\Delta k z}.$$

where $\Delta k = (\mathbf{k}_1 + \mathbf{k}_2 - \mathbf{k}_3 - \mathbf{k}_4) \cdot \hat{\mathbf{z}}$, and $\mathbf{k}_1$ and $\mathbf{k}_2$ are the wave vectors of the two pump waves.

3. Derive an expression for the phase-conjugate reflectivity obtained by degenerate four-wave mixing utilizing the nonlinear response of a collection of "two-level" atoms. You may make the rotating wave and slowly-varying-amplitude approximations and may assume that the amplitudes of the strong pump waves are not modified by the nonlinear interaction.

[Hint: This problem can be solved using the formalism developed in Section 5.3. This problem has been solved in the scientific literature, and the solution is given by R. L. Abrams and R. C. Lind, *Opt. Lett.* **2**, 94 (1978) and **3**, 205 (1978).]

4. Verify Eq. (6.1.41).

5. The discussion of absorptive optical bistability presented in the text assumed that the incident laser frequency was tuned to a cavity resonance. Generalize this treatment by allowing the cavity to be mistuned from resonance, so that the factor $\rho^2 e^{2ikl}$ appearing in the denominator of Eq. (6.3.3) can be set equal to $Re^{i\delta_0}$, where δ_0 is the cavity mistuning in radians.

[Ans: Eq. (6.3.10) must be replaced by

$$TI_1 = I_2\left[1 - 2R\left(1 - \frac{C_0 T/R}{1 + 2I_2/I_s}\right)\cos\delta_0 + R^2\left(1 - \frac{C_0 T/R}{1 + 2I_2/I_s}\right)^2\right].$$

Examination of this expression shows that larger values of C_0 and I_2 are required in order to obtain optical bistability for $\delta \neq 0$.]

6. The treatment of absorptive bistability given in the text assumed that the absorption decreased with increasing laser intensity according to

$$\alpha = \frac{\alpha_0}{1 + I/I_s}.$$

In fact, many saturable absorbers are imperfect in that they do not saturate all the way to zero; the absorption can better be represented by

$$\alpha = \frac{\alpha_0}{1 + I/I_s} + \alpha_1,$$

where α_1 is constant. How large can α_1 be (for given α_0) and still allow the occurrence of bistability? Use the same approximations used in the text, namely that $\alpha l \ll 1$ and that the cavity is tuned to exact resonance. How are the requirements on the intensity of the incident laser beam modified by a nonzero value of α_1?

7. By means of a graphical analysis of the sort illustrated in Fig. 6.3.5, make a plot of the transmitted intensity I_3 as a function of the incident intensity

I_1. Note that more than two stable solutions can occur for a device that displays dispersive bistability.

8. The analysis of self-phase modulation that led to Fig. 6.5.1 assumed that the medium had instantaneous response and that the temporal evolution of the pulse had a symmetric waveform. In this case the spectrum of the pulse is seen to broaden symmetrically. How is the spectrum modified for a medium with a sluggish response (given, for example, by Eq. (6.4.8) with τ much longer than the pulse duration)? How is the spectrum modified if the pulse waveform is not symmetric (a sawtooth waveform, for example)?

9. How is the pulse propagation equation (6.5.32) modified if the quantity $k^2 - k_0^2$ is not approximated by $2k_0(k - k_0)$, as was done in going from Eq. (6.5.18) to Eq. (6.5.19)?

[Ans: k_2 in Eq. (6.5.32) must be replaced by $(k_1^2 + 2k_0k_2)/2k_0$.]

Why is it that this new equation seems to predict that pulses will spread as they propagate, even when both k_2 and γ vanish?

10. Verify that the solution given by Eqs. (6.5.33) does in fact satisfy the pulse propagation equation (6.5.32).

11. *Self-Induced Transparency.* Optical solitons can also be formed as a consequence of the resonant nonlinear optical response of a collection of two-level atoms. Show that, in the absence of damping effects and for the case of exact resonance, the equations describing the propagation of an optical pulse through such a medium are of the form

$$\frac{\partial \tilde{A}}{\partial z} + \frac{1}{c}\frac{\partial \tilde{A}}{\partial t} = \frac{2\pi i \omega N}{c} p,$$

$$\frac{dp}{dt} = -\frac{i}{\hbar}|\mu|^2 \tilde{A} w, \qquad \frac{dw}{dt} = \frac{-4i}{\hbar}\tilde{A}p$$

(where the atomic response is described as in Section 5.4). Show that these equations yield soliton-like solutions of the form

$$\tilde{A}(z,t) = \frac{\hbar}{\mu\tau_0}\operatorname{sech}\left(\frac{t - z/v}{\tau_0}\right),$$

$$w(z,t) = -1 + 2\operatorname{sech}^2\left(\frac{t - z/v}{\tau_0}\right),$$

$$p(z,t) = -i\mu \operatorname{sech}\left(\frac{t - z/v}{\tau_0}\right)\tanh\left(\frac{t - z/v}{\tau_0}\right)$$

as long as the pulse width and pulse velocity are related by

$$\frac{c}{v} = 1 + \frac{2\pi N\mu^2\omega\tau_0^2}{\hbar}.$$

What is the value (and the significance) of the quantity

$$\int_{-\infty}^{\infty} \frac{2\mu}{\hbar} \tilde{A}(z,t)\,dt?$$

(For the case of an inhomogeneously broadened medium, the equations are still satisfied by a sech pulse, but the relation between v and τ_0 is different. See, for example, Allen and Eberly (1975).)

References

Section 6.1. Optical Phase Conjugation

D. M. Bloom and G. C. Bjorklund, *Appl. Phys. Lett.* **31**, 592 (1977).
R. A. Fisher, editor, *Optical Phase Conjugation*, Academic Press, Orlando, 1983.
R. W. Hellwarth, *J. Opt. Soc. Am* **67**, 1 (1977).
A. Yariv and D. M. Pepper, *Opt. Lett.* **1**, 16 (1977).
B. Ya. Zel'dovich, N. F. Pilipetsky, and V. V. Shkunov, *Principles of Phase Conjugation*, Springer-Verlag, Berlin, 1985.

Polarization Properties of Phase Conjugation

M. Ducloy and D. Bloch, *Phys. Rev. A* **30**, 3107 (Dec. 1984).
G. Grynberg, *Opt. Commun.* **48**, 432 (1984).
M. S. Malcuit, D. J. Gauthier, and R. W. Boyd, *Opt. Lett.* **13**, 663 (1988).
G. Martin, L. K. Lam, and R. W. Hellwarth, *Opt. Lett.* **5**, 186 (1980).
S. Saikan, *J. Opt. Soc. Am.* **68**, 1185 (1978).
S. Saikan, *J. Opt. Soc. Am.* **72**, 515 (1982).
S. Saikan and M. Kiguchi, *Opt. Lett.* **7**, 555 (1982).
B. Ya. Zel'dovich and V. V. Shkunov, *Sov. J. Quantum Electron.* **9**, 379 (1979).

Section 6.2. Self-Focusing of Light

S. A. Akhmanov, R. V. Khokhlov, and A. P. Sukhorukov, in *Laser Handbook*, edited by F. T. Arecchi and E. O. Schulz-Dubois, North-Holland, 1972.
V. I. Bespalov and V. I. Talanov, *JETP Lett.* **3**, 471 (1966).
J. E. Bjorkholm and A. Ashkin, *Phys. Rev. Lett.* **32**, 129 (1974).
R. Y. Chiao, E. Garmire, and C. H. Townes, *Phys. Rev. Lett.* **13**, 479 (1964).

P. L. Kelley, *Phys. Rev. Lett.* **15**, 1005 (1965).
M. Hercher, *J. Opt. Soc. Am.* **54**, 563 (1964).
J. H. Marburger, *Prog. Quantum Electron.* **4**, 35 (1975).
Y. R. Shen, *Prog. Quantum Electron.* **4**, 1 (1975).
O. Svelto, in *Progress in Optics XII*, edited by E. Wolf, North-Holland, 1974.
A. Yariv, *Quantum Electronics*, Wiley, New York, 1975, p. 498.

Section 6.3. Optical Bistability

H. M. Gibbs, *Optical Bistability*, Academic Press, Orlando, 1985.
H. M. Gibbs, S. L. McCall, and T. N. C. Venkatesan, *Phys. Rev. Lett.* **36**, 113 (1976).
L. A. Lugiato, "Theory of Optical Bistability," in *Progress in Optics XXI*, edited by E. Wolf. North-Holland, 1984.
A. Szöke, V. Daneu, J. Goldhar, and N. A. Kurnit, *Appl. Phys. Lett.* **15**, 376 (1969).

Section 6.4. Two-Beam Coupling

J. Feinberg, in *Optical Phase Conjugation*, edited by R. A. Fisher, Academic Press, Orlando, 1983.
Y. Silberberg and I. Bar-Joseph, *Phys. Rev. Lett.* **48**, 1541 (1982).
Y. Silberberg and I. Bar-Joseph, *J. Opt. Soc. Am B* **1**, 662 (1984).

Section 6.5. Pulse Propagation and Optical Solitons

G. P. Agrawal, *Nonlinear Fiber Optics*, Academic Press, Boston, 1989.
R. G. Brewer, *Phys. Rev. Lett.* **19**, 8 (1967).
A. C. Cheung, D. M. Rank, R. Y. Chiao, and C. H. Townes, *Phys. Rev. Lett.* **20**, 786 (1968).
L. F. Mollenauer, R. H. Stolen, and J. P. Gordon, *Phys. Rev. Lett.* 45, 1095 (1980).
F. Shimizu, *Phys. Rev. Lett.* **19**, 1097 (1967).
V. E. Zakharov and A. B. Shabat, *Sov. Phys. JETP* **34**, 63 (1972).

Chapter 7

Spontaneous Light Scattering and Acousto-optics

7.1. Features of Spontaneous Light Scattering

In this chapter, we describe spontaneous light scattering; Chapters 8 and 9 present descriptions of various stimulated light scattering processes. By spontaneous light scattering, we mean light scattering under conditions such that the optical properties of the material system are unmodified by the presence of the incident light beam. We shall see in the following two chapters that the character of the light scattering process is profoundly modified whenever the intensity of the incident light is sufficiently large to modify the optical properties of the material system.

Let us first consider the light scattering experiment illustrated in part (a) of Fig. 7.1.1. Under the most general circumstances, the spectrum of the scattered light has the form shown in part (b) of the figure, in which Raman, Brillouin, Rayleigh, and Rayleigh-wing features are present. By definition, those components of the scattered light that are shifted to lower frequencies are known as Stokes components, and those components that are shifted to higher frequencies are known as anti-Stokes components. Table 7.1.1 lists some of the physical processes that can lead to light scattering of the sort shown in the figure and gives some of the physical parameters that describe these processes.

One of these processes is Raman scattering. Raman scattering results from the interaction of light with the vibrational modes of the molecules constituting the scattering medium. Raman scattering can equivalently be described as the scattering of light from optical phonons.

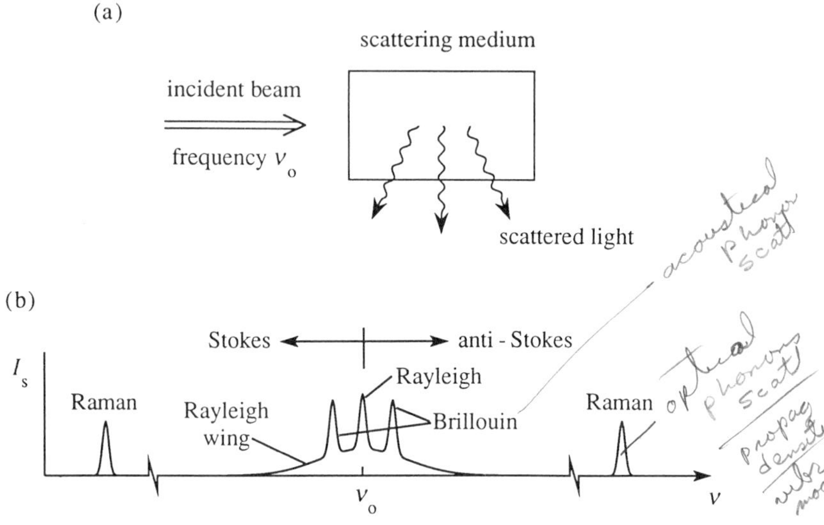

FIGURE 7.1.1 Spontaneous light scattering. (a) Experimental setup. (b) Typical observed spectrum.

Brillouin scattering is the scattering of light from sound waves, that is, from propagating pressure (and hence density) waves. Brillouin scattering can also be considered to be the scattering of light from acoustic phonons.

Rayleigh scattering (or Rayleigh center scattering) is the scattering of light from nonpropagating density fluctuations. Formally, it can be described as scattering from entropy fluctuations. It is known as quasielastic scattering because it induces no frequency shift.

Rayleigh-wing scattering (that is, scattering in the wing of the Rayleigh line) is scattering from fluctuations in the orientation of anisotropic molecules. Since the molecular reorientation process is very rapid, this component is

TABLE 7.1.1 Typical values of the parameters describing several light scattering processes

Process	Shift (cm^{-1})	Linewidth (cm^{-1})	Relaxation time (sec)	Gain* (cm/MW)
Raman	1000	5	10^{-12}	5×10^{-3}
Brillouin	0.1	5×10^{-3}	10^{-9}	10^{-2}
Rayleigh	0	5×10^{-4}	10^{-8}	10^{-4}
Rayleigh-wing	0	5	10^{-12}	10^{-3}

*Gain of the stimulated version of the process.

spectrally very broad. Rayleigh-wing scattering does not occur for molecules with an isotropic polarizability tensor.

Fluctuations as the Origin of Light Scattering

Light scattering occurs as a consequence of fluctuations in the optical properties of a material medium; a completely homogeneous material can scatter light only in the forward direction (see, for example, Fabelinskii, 1968). This conclusion can be demonstrated with the aid of Fig. 7.1.2, which shows a completely homogeneous medium being illuminated by a plane wave. We suppose that the volume element dV_1 scatters light into the θ direction. However, for any direction except the exact forward direction ($\theta = 0$) there must be a nearby volume element (labeled dV_2) whose scattered field interferes destructively with that from dV_1. Since the same argument can be applied to any volume element in the medium, we conclude that there can be no scattering in any direction except $\theta = 0$. Scattering in the direction $\theta = 0$ is known as coherent forward scattering, and is the origin of the index of refraction. (See, for example, the discussion in Section 31 of Feynman *et al.*, 1963.)

Note that the argument that scattering cannot occur (except in the forward direction) requires that the medium be *completely* homogeneous. Scattering can occur as the result of fluctuations in any of the optical properties of the medium. For example, if the density of the medium is nonuniform, then the total number of molecules in the volume element dV_1 may not be equal to the number of molecules in dV_2, and consequently the destructive interference between the fields scattered by these two elements will not be exact.

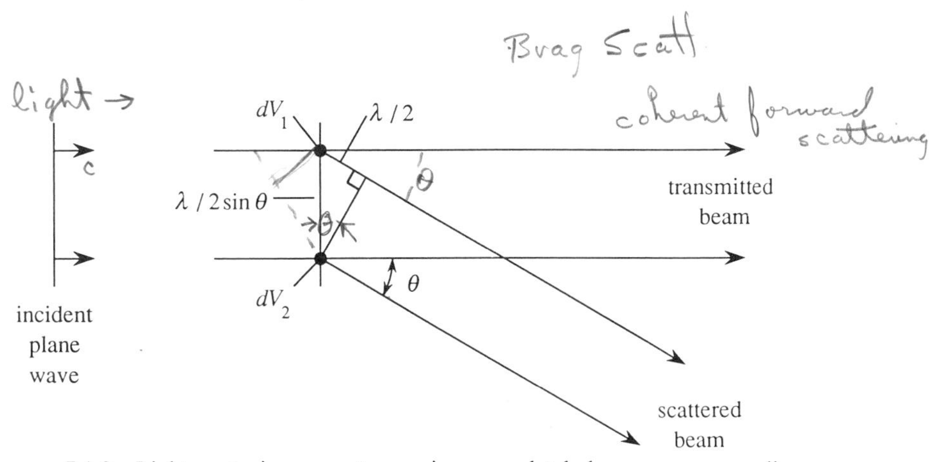

FIGURE 7.1.2 Light scattering cannot occur in a completely homogeneous medium.

Since light scattering results from fluctuations in the optical properties of a material medium, it is useful to represent the dielectric tensor of the medium (which for simplicity we assume to be isotropic in its average properties) as (Landau and Lifshitz, 1960)

$$\epsilon_{ik} = \epsilon_0\,\delta_{ik} + \Delta\epsilon_{ik}, \tag{7.1.1}$$

where ϵ_0 represents the mean dielectric constant of the medium and where $\Delta\epsilon_{ik}$ represents the (temporally and/or spatially varying) fluctuations in the dielectric tensor that lead to light scattering. It is convenient to decompose the fluctuation $\Delta\epsilon_{ik}$ in the dielectric tensor into the sum of a scalar contribution $\Delta\epsilon\,\delta_{ik}$ and a (traceless) tensor contribution $\Delta\epsilon_{ik}^{(t)}$ as

$$\Delta\epsilon_{ik} = \Delta\epsilon\,\delta_{ik} + \Delta\epsilon_{ik}^{(t)}. \tag{7.1.2}$$

The scalar contribution $\Delta\epsilon$ arises from fluctuations in thermodynamic quantities such as the pressure, entropy, density, or temperature. In a chemical solution it also has a contribution from fluctuations in concentration. Scattering that results from $\Delta\epsilon$ is called scalar light scattering; examples of scalar light scattering include Brillouin and Rayleigh scattering.

Scattering that results from $\Delta\epsilon_{ik}^{(t)}$ is called tensor light scattering. The tensor $\Delta\epsilon_{ik}^{(t)}$ can be taken to be traceless (i.e., $\sum_i \Delta\epsilon_{ii}^{(t)} = 0$), since the scalar contribution $\Delta\epsilon$ has been separated out. It is useful to express $\Delta\epsilon_{ik}^{(t)}$ as

$$\Delta\epsilon_{ik}^{(t)} = \Delta\epsilon_{ik}^{(s)} + \Delta\epsilon_{ik}^{(a)}, \tag{7.1.3}$$

where $\Delta\epsilon_{ik}^{(s)}$ is the symmetric part of $\Delta\epsilon_{ik}^{(t)}$ (symmetric in the sense that $\Delta\epsilon_{ik}^{(s)} = \Delta\epsilon_{ki}^{(s)}$) and gives rise to Rayleigh-wing scattering, and where $\Delta\epsilon_{ik}^{(a)}$ is the antisymmetric part of $\Delta\epsilon_{ik}^{(t)}$ (that is, $\Delta\epsilon_{ik}^{(a)} = -\Delta\epsilon_{ki}^{(a)}$) and gives rise to Raman scattering.

It can be shown that the fluctuations $\Delta\epsilon$, $\Delta\epsilon_{ik}^{(s)}$, and $\Delta\epsilon_{ik}^{(a)}$ are statistically independent. Scattering due to $\Delta\epsilon$ is called polarized scattering. Scattering due to $\Delta\epsilon_{ik}^{(t)}$ is called depolarized scattering, because in general the degree of polarization in the scattered light is less than that of the incident light.

Scattering Coefficient

A quantity that is used to describe the efficiency of the scattering process is the scattering coefficient R, which is defined in terms of the quantities shown in Fig. 7.1.3. Here a beam of light of intensity I_0 illuminates a scattering region of volume V, and the intensity I_s of the scattered light is measured at a distance L from the interaction region. It is reasonable to assume that the intensity of the scattered light increases linearly with the intensity I_0 of the incident light and

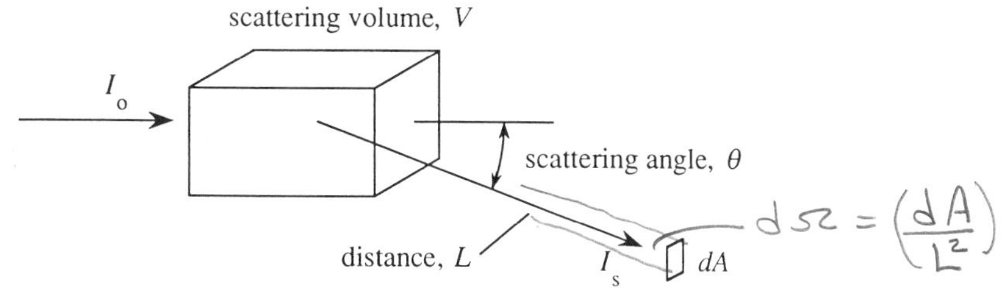

FIGURE 7.1.3 Quantities used to define the scattering coefficient.

with the volume V of the interaction region, and that it obeys the inverse square law with respect to the distance L to the point of observation. We can hence represent I_s as

$$I_s = \frac{I_0 R V}{L^2}, \tag{7.1.4}$$

where the constant of proportionality R is known as the scattering coefficient.

We now assume that the scattered light falls onto a small detector of projected area dA. The power hitting the detector is given by $dP = I_s\, dA$. Since the detector subtends a solid angle at the scattering region given by $d\Omega = dA/L^2$, the scattered power per unit solid angle is given by $dP/d\Omega = I_s L^2$, or by

$$\frac{dP}{d\Omega} = I_0 R V. \tag{7.1.5}$$

Either Eq. (7.1.4) or (7.1.5) can be taken as the definition of the scattering coefficient. For scattering of visible light through an angle of 90 degrees, R has the value $2 \times 10^{-8}\ \text{cm}^{-1}$ for air and $1 \times 10^{-6}\ \text{cm}^{-1}$ for water.

Scattering Cross Section

It is also useful to define the scattering cross section. We consider a beam of intensity I_0 falling onto an individual molecule, as shown in Fig. 7.1.4. We let P denote the total power of the radiation scattered by this molecule. We assume that P increases linearly with I_0 according to

$$P = \sigma I_0, \tag{7.1.6}$$

where the constant of proportionality σ is known as the total scattering cross section. Since I_0 has the dimensions of power per unit area, we see that σ has the dimensions of an area, which is why it is called a cross section. The cross

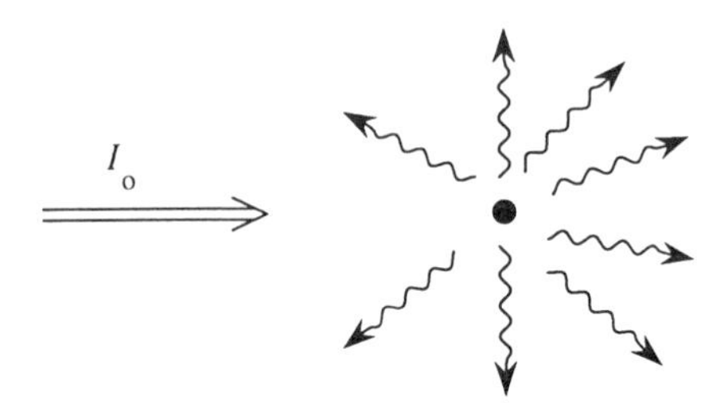

FIGURE 7.1.4 Scattering of light by a molecule.

section can be interpreted as the effective geometrical area of the molecule for removing light from the incident beam.

We also define a differential cross section. Rather than describing the total scattered power, this quantity describes the power dP scattered in some particular direction into the element of solid angle $d\Omega$. We assume that the scattered power per unit solid angle, $dP/d\Omega$, increases linearly with the incident intensity according to

$$\frac{dP}{d\Omega} = I_0 \frac{d\sigma}{d\Omega}, \tag{7.1.7}$$

where $d\sigma/d\Omega$ is known as the differential cross section. Clearly, since P is equal to $\int (dP/d\Omega)\, d\Omega$, it follows from Eqs. (7.1.6) and (7.1.7) that

$$\sigma = \int_{4\pi} \frac{d\sigma}{d\Omega} d\Omega. \tag{7.1.8}$$

Let us next see how to relate the differential scattering cross section to the scattering coefficient. If each of the $\mathcal{N}$ molecules contained in the volume V of Fig. 7.1.3 scatters independently, then the total power per unit solid angle of the scattered light will be $\mathcal{N}$ times larger than the result given in Eq. (7.1.7). Consequently, by comparison with Eq. (7.1.5), we see that the scattering coefficient is given by

$$R = \frac{\mathcal{N}}{V} \frac{d\sigma}{d\Omega}. \tag{7.1.9}$$

One should be wary about taking this equation to be a general result. Recall that a completely homogeneous medium does not scatter light at all, which implies that for such a medium R would be equal to zero and not to $(\mathcal{N}/V)(d\sigma/d\Omega)$. In the next section we examine the conditions under which it is valid to assume that each molecule scatters independently. As a general rule, Eq. (7.1.9) is valid for dilute media and is entirely invalid for condensed matter.

7.2. Microscopic Theory of Light Scattering

Let us now consider light scattering in terms of the field scattered by each molecule contained within the interaction region. Such a treatment is particularly well suited for the case of scattering from a dilute gas, where collective effects due to the interaction of the various molecules are relatively unimportant. (Light scattering from condensed matter is more conveniently treated using the thermodynamic formalism presented in the next section.) As illustrated in Fig. 7.2.1, we assume that the optical field

$$\tilde{\mathbf{E}} = \mathbf{E}_0 e^{-i\omega t} + \text{c.c.} \tag{7.2.1}$$

of intensity $I_0 = (nc/2\pi)|E_0|^2$ is incident on a molecule whose linear dimensions are assumed to be much smaller than the wavelength of light. In response to the applied field, the molecule develops the dipole moment

$$\tilde{\mathbf{p}} = \alpha(\omega)\mathbf{E}_0 e^{-i\omega t} + \text{c.c.}, \tag{7.2.2}$$

where $\alpha(\omega)$ is the polarizability of the particle. Explicit formulas for $\alpha(\omega)$ for certain types of scatterers are given below, but for reasons of generality we leave the form of $\alpha(\omega)$ unspecified for the present.

Due to the time-varying dipole moment given by Eq. (7.2.2), the particle will radiate. The intensity of this radiation at a distance L from the scatterer is given by the magnitude of the Poynting vector (see, for example, Jackson, 1982, Section 9.2) as

$$I_s = \frac{n\langle \ddot{\tilde{p}}^2 \rangle}{4\pi c^3 L^2} \sin^2\phi = \frac{n\omega^4 |\alpha(\omega)|^2 |E_0|^2 \sin^2\phi}{2\pi c^3 L^2}. \tag{7.2.3}$$

The angular brackets in the first form imply that the time average of the enclosed quantity is to be taken. As shown in Fig. 7.2.1, ϕ is the angle between the induced dipole moment of the particle and the direction $\mathbf{r}$ to the point of observation.

We next use Eq. (7.2.3) to derive an expression for the differential scattering cross section. As in the derivation of Eq. (7.1.5), the scattered power per unit

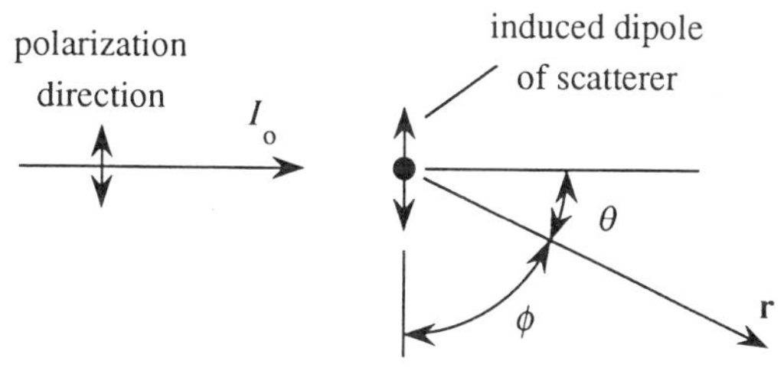

FIGURE 7.2.1 Geometry of light scattering from an individual molecule.

solid angle is given by $dP/d\Omega = I_s L^2$. We introduce the differential cross section of Eq. (7.1.7), $d\sigma/d\Omega = (dP/d\Omega)/I_0 = I_s L^2/I_0$, which through use of Eq. (7.2.3) becomes

$$\frac{d\sigma}{d\Omega} = \frac{\omega^4}{c^4} |\alpha(\omega)|^2 \sin^2 \phi. \tag{7.2.4}$$

We note that this expression for the differential cross section $d\sigma/d\Omega$ predicts a $\sin^2 \phi$ dependence for *any* functional form for $\alpha(\omega)$. This result is a consequence of our assumption that the scattering particle is small compared to an optical wavelength and hence that the scattering is due solely to electric dipole and not to higher-order multipole processes. Since the angular dependence of $d\sigma/d\Omega$ is contained entirely in the $\sin^2 \phi$ term, we can immediately obtain an expression for the total scattering cross section by integrating $d\sigma/d\Omega$ over all solid angles, yielding

$$\sigma = \int_{4\pi} d\Omega \frac{d\sigma}{d\Omega} = \frac{8\pi}{3} \frac{\omega^4}{c^4} |\alpha(\omega)|^2. \tag{7.2.5}$$

In deriving Eq. (7.2.4) for the differential scattering cross section, we assumed that the incident light was linearly polarized, and for convenience we took the direction of polarization to lie in the plane of Fig. 7.2.1. For this direction of polarization, the scattering angle θ and the angle ϕ of Eq. (7.2.3) are related by $\theta + \phi = 90$ degrees, and hence for this direction of polarization Eq. (7.2.4) can be expressed in terms of the scattering angle as

$$\left(\frac{d\sigma}{d\Omega}\right)_{\mathrm{p}} = \frac{\omega^4}{c^4} |\alpha(\omega)|^2 \cos^2 \theta. \tag{7.2.6}$$

Other types of polarization can be treated by allowing the incident field to have a component perpendicular to the plane of Fig. 7.2.1. For this component ϕ is equal to 90 degrees for any value of the scattering angle θ, and hence for this component the differential cross section is given by

$$\left(\frac{d\sigma}{d\Omega}\right)_{\mathrm{s}} = \frac{\omega^4}{c^4} |\alpha(\omega)|^2 \tag{7.2.7}$$

for any value of θ. Since unpolarized light consists of equal intensities in the two orthogonal polarization directions, the differential cross section for unpolarized light is obtained by averaging Eqs. (7.2.6) and (7.2.7), giving

$$\left(\frac{d\sigma}{d\Omega}\right)_{\mathrm{unpolarized}} = \frac{\omega^4}{c^4} |\alpha(\omega)|^2 \tfrac{1}{2}(1 + \cos^2 \theta). \tag{7.2.8}$$

As an example of the use of these equations, we consider scattering from an atom whose optical properties can be described by the Lorentz model of the atom (that is, we model the atom as a simple harmonic oscillator). According to Eqs. (1.4.10) and (1.4.17) and the relation of $\chi(\omega) = N\alpha(\omega)$, the polarizability of such an atom is given by

$$\alpha(\omega) = \frac{e^2/m}{\omega_0^2 - \omega^2 - 2i\omega\gamma}, \tag{7.2.9}$$

where ω_0 is the resonance frequency and γ is the dipole damping rate. Through use of this expression, the total scattering cross section given by Eq. (7.2.5) becomes

$$\sigma = \frac{8\pi}{3}\left(\frac{e^2}{mc^2}\right)^2 \frac{\omega^4}{(\omega_0^2 - \omega^2)^2 + 4\omega^2\gamma^2}. \tag{7.2.10}$$

The frequency dependence of the scattering cross section predicted by this equation is illustrated in Fig. 7.2.2. Equation (7.2.10) can be simplified under several different limiting conditions. In particular, we find that

$$\sigma = \frac{8\pi}{3}\left(\frac{e^2}{mc^2}\right)^2 \frac{\omega^4}{\omega_0^4} \quad \text{for} \quad \omega \ll \omega_0, \tag{7.2.11a}$$

$$\sigma = \frac{2\pi}{3}\left(\frac{e^2}{mc^2}\right)^2 \frac{\omega_0^2}{(\omega_0 - \omega)^2 + \gamma^2} \quad \text{for} \quad \omega \simeq \omega_0, \tag{7.2.11b}$$

$$\sigma = \frac{8\pi}{3}\left(\frac{e^2}{mc^2}\right)^2 \quad \text{for} \quad \omega \gg \omega_0. \tag{7.2.11c}$$

Equation (7.2.11a) shows that the scattering cross section increases as the fourth power of the optical frequency ω in the limit $\omega \ll \omega_0$. This result leads,

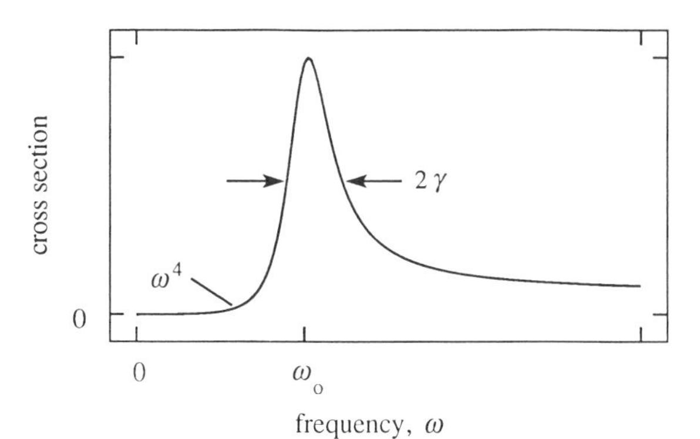

FIGURE 7.2.2 Frequency dependence of the scattering cross section of a Lorentz oscillator.

for example, to the prediction that the sky is blue, since the shorter wavelengths of sunlight are scattered far more efficiently in the earth's atmosphere than are the longer wavelengths. Scattering in this limit is often known as Rayleigh scattering. Equation (7.3.11b) shows that near the atomic resonance frequency the dependence of the scattering cross section on the optical frequency has a Lorentzian lineshape. Equation (7.8.11c) shows that for very large frequencies the scattering cross section approaches a constant value. This value is of the order of the square of the "classical" electron radius, $r_e = e^2/mc^2$. Scattering in this limit is known as Thompson scattering.

As a second example of the application of Eq. (7.2.5), we consider scattering from a collection of small dielectric spheres. We take ϵ_1 to be the dielectric constant of the material within each sphere and ϵ to be that of the surrounding medium. We assume that each sphere is small in the sense that its radius a is much smaller than the wavelength of the incident radiation. We can then calculate the polarizability of each sphere using the laws of electrostatics. It is straightforward to show (see, for example, Stratton (1941) or Jackson (1982)) that the polarizability is given by the expression

$$\alpha = \frac{\epsilon_1 - \epsilon}{\epsilon_1 + 2\epsilon} a^3. \tag{7.2.12}$$

Note that α depends upon frequency only through any possible frequency dependence of ϵ or of ϵ_1. Through use of Eq. (7.2.5), we find that the scattering cross section is given by

$$\sigma = \frac{8\pi}{3} \frac{\omega^4}{c^4} a^6 \left(\frac{\epsilon_1 - \epsilon}{\epsilon_1 + 2\epsilon} \right)^2. \tag{7.2.13}$$

Note that, as in the low-frequency limit of the Lorentz atom, the cross section scales as the fourth power of the frequency.

Let us now consider the rather subtle problem of calculating the total intensity of the light scattered from a collection of molecules. We recall from the discussion of Fig. 7.1.2 that only the fluctuations in the optical properties of the medium can lead to light scattering. As shown in Fig. 7.2.3, we divide the total scattering volume V into a large number of identical small regions of volume V'. We assume that V' is sufficiently small that all of the molecules within V' radiate essentially in phase. The intensity of the light emitted by the atoms in V' in some particular direction can thus be represented as

$$I_{V'} = \nu^2 I_{\text{mol}}, \tag{7.2.14}$$

where ν represents the number of molecules in V' and I_{mol} denotes the intensity of the light scattered by a single molecule.

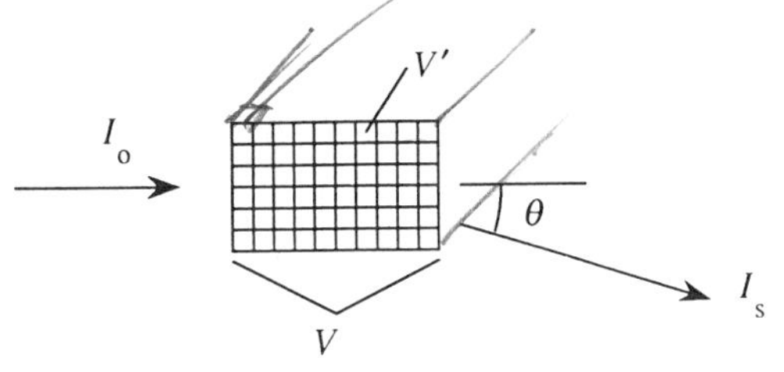

FIGURE 7.2.3 Light scattering from a collection of molecules.

We next calculate the total intensity of the scattered radiation from the entire volume V. We recall from the discussion of Section 7.1 that, for each volume element V', there will be another element whose radiated field tends to interfere destructively with that from V'. Insofar as each volume element contains exactly the same number of molecules, the cancellation will be complete. However, any deviation of ν from its mean value $\bar{\nu}$ can lead to a net intensity of the scattered radiation. The net intensity of the light scattered by V' is thus given by $\overline{\Delta\nu^2} I_{\text{mol}}$, where $\overline{\Delta\nu^2} = \overline{\nu^2} - \bar{\nu}^2$. The intensity of the radiation scattered from the total volume V is then given by

$$I_V = I_{\text{mol}} \overline{\Delta\nu^2} \frac{V}{V'}, \tag{7.2.15}$$

where the last factor V/V' gives the total number of regions of volume V' contained within the interaction volume V. This result shows how the total scattered intensity I_V depends upon the fluctuations in the number density of molecules. We see that the scattered intensity I_V vanishes if the fluctuation $\overline{\Delta\nu^2}$ vanishes.

For the case of a medium sufficiently dilute that the locations of the individual molecules are uncorrelated (that is, for an ideal gas), we can readily calculate the mean fluctuation $\overline{\Delta\nu^2}$ in the number of particles. If N denotes the mean number density of particles, then the mean number of particles in V' is given by

$$\bar{\nu} = NV', \tag{7.2.16}$$

and the mean fluctuation is given by

$$\overline{\Delta\nu^2} = \overline{\nu^2} - \bar{\nu}^2 = \bar{\nu}, \tag{7.2.17}$$

where the last equality follows from the properties of the Poisson probability distribution, which are obeyed by uncorrelated particles. We hence find from Eqs. (7.2.15) through (7.2.17) that

$$I_V = NVI_{\text{mol}} = \mathcal{N} I_{\text{mol}}. \tag{7.2.18}$$

Hence, for an ideal gas the total intensity is simply the intensity of the light scattered by a single molecule multiplied by the total number of molecules, $\mathcal{N} = NV$. Consequently the scattering coefficient R and differential cross section $d\sigma/d\Omega$ introduced in Section 7.1 are related by Eq. (7.1.9), that is, by

$$R = N\frac{d\sigma}{d\Omega}. \tag{7.2.19}$$

By introducing Eq. (7.2.4) into this expression, we find that the scattering coefficient is given by

$$R = N\frac{\omega^4}{c^4}|\alpha(\omega)|^2 \sin^2\phi. \tag{7.2.20}$$

If the scattering medium is sufficiently dilute that its refractive index can be represented as

$$n = 1 + 2\pi N\alpha(\omega), \tag{7.2.21}$$

Eq. (7.2.20) can be rewritten as

$$R = \frac{\omega^4}{c^4}\frac{|n-1|^2}{4\pi^2 N}\sin^2\phi. \tag{7.2.22}$$

This result can be used to determine the number density N of molecules in a gaseous sample in terms of two optical constants: the refractive index n and scattering coefficient R at a fixed angle ϕ. In fact, the first accurate measurement of Loschmidt's number (the number density of molecules at standard temperature and pressure, $N_0 = 2.7 \times 10^{19}\ \text{cm}^{-3}$) was performed through application of Eq. (7.2.22).

7.3. Thermodynamic Theory of Scalar Light Scattering

We next develop a macroscopic description of the light scattering process. We consider the case in which light scattering occurs as the result of fluctuations in the (scalar) dielectric constant and in which these fluctuations are themselves the result of fluctuations in thermodynamic variables, such as the material density and temperature. We assume, as in Fig. 7.2.3 in the preceding section, that the scattering volume V can be divided into a number of smaller volumes V' having the property that all atoms in V' radiate essentially in phase in the θ direction. We let $\Delta\epsilon$ denote the fluctuation of the dielectric constant averaged over the volume V'. Since $\epsilon = 1 + 4\pi\chi$, the fluctuation in the susceptibility is then given by $\Delta\chi = \Delta\epsilon/4\pi$. Due to this change in the susceptibility, the volume V' develops the additional polarization

$$\tilde{\mathbf{P}} = \Delta\chi\,\tilde{\mathbf{E}}_0 = \frac{\Delta\epsilon}{4\pi}\tilde{\mathbf{E}}_0 \tag{7.3.1}$$

and hence the additional dipole moment

$$\tilde{\mathbf{p}} = V'\tilde{\mathbf{P}} = V'\frac{\Delta\epsilon}{4\pi}\tilde{\mathbf{E}}_0. \tag{7.3.2}$$

The intensity $I_s = (nc/4\pi)\langle\tilde{\mathbf{E}}_s^2\rangle$ of the radiation emitted by this oscillating dipole moment is obtained by introducing Eq. (7.3.2) into Eq. (7.2.3), to obtain

$$I_s = I_0\frac{\omega^4 V'^2\langle\Delta\epsilon^2\rangle \sin^2\phi}{16\pi^2 L^2 c^4}, \tag{7.3.3}$$

where, as before, ϕ is the angle between $\tilde{\mathbf{p}}$ and the direction to the point of observation, and where we have introduced the intensity $I_0 = (nc/4\pi)\langle\tilde{\mathbf{E}}_0^2\rangle$ of the incident light. Equation (7.3.3) gives the intensity of the light scattered from one cell. The total intensity from all the cells is V/V' times as large, since the fluctuations in the dielectric constant for different cells are uncorrelated.

We next need to calculate the mean square fluctuation in the dielectric constant, $\langle\Delta\epsilon^2\rangle$, for any one cell. We take the density ρ and temperature T as the independent thermodynamic variables. We then express the change in the dielectric constant as

$$\Delta\epsilon = \left(\frac{\partial\epsilon}{\partial\rho}\right)_T \Delta\rho + \left(\frac{\partial\epsilon}{\partial T}\right)_\rho \Delta T. \tag{7.3.4}$$

To good accuracy (the error is estimated to be of the order of 2%; see Fabelinskii, 1968), we can usually ignore the second term, since the dielectric constant typically depends much more strongly on density than on temperature.* We thus find

$$\langle\Delta\epsilon^2\rangle = \left(\frac{\partial\epsilon}{\partial\rho}\right)^2 \langle\Delta\rho^2\rangle,$$

which can be expressed as

$$\langle\Delta\epsilon^2\rangle = \gamma_e^2\frac{\langle\Delta\rho^2\rangle}{\rho_0^2}, \tag{7.3.5}$$

* For this reason, it is not crucial that we retain the subscript T on $\partial\epsilon/\partial\rho$.

where ρ_0 denotes the mean density of the material and where we have introduced the electrostrictive constant γ_e,* which is defined by

$$\gamma_e = \left(\rho \frac{\partial \epsilon}{\partial \rho} \right) \bigg|_{\rho = \rho_0} . \tag{7.3.6}$$

The quantity $\langle \Delta \rho^2 \rangle / \rho_0^2$ appearing in Eq. (7.3.5) can be calculated using the laws of statistical mechanics. The result (see, for example, Fabelinskii, 1968, Appendix I, Eq. I.13; or Landau and Lifshitz, 1969) is

$$\frac{\langle \Delta \rho^2 \rangle}{\rho_0^2} = \frac{kTC_T}{V'} \tag{7.3.7}$$

where

$$C_T = -\frac{1}{V} \left(\frac{\partial V}{\partial p} \right)_T \tag{7.3.8}$$

is the isothermal compressibility. Note that this result (whose proof is outside the subject area of this book) makes sense: fluctuations are driven by thermal excitation; the larger the compressibility, the larger will be the resulting excursion; and the smaller the volume under consideration, the easier it is to change its mean density.

By introducing Eqs. (7.3.5) and (7.3.7) into Eq. (7.3.3) and multiplying the result by the total number of cells, V/V', we find that the total intensity of the scattered radiation is given by

$$I_s = I_0 \frac{\omega^4 V}{16\pi^2 L^2 c^4} \gamma_e^2 C_T kT \sin^2 \phi. \tag{7.3.9a}$$

We can use this result to find that the scattering coefficient R defined by Eq. (7.1.4) is given by

$$R = \frac{\omega^4}{16\pi^2 c^4} \gamma_e^2 C_T kT \sin^2 \phi. \tag{7.3.9b}$$

Ideal Gas

As an example, let us apply the result given by Eq. (7.3.9) to the case of light scattering from an ideal gas, for which the equation of state is of the form

$$pV = \mathcal{N} kT, \tag{7.3.10}$$

* The reason why γ_e is called the electrostrictive constant will be described in Section 8.1.

where $\mathcal{N}$ denotes the total number of molecules in the gas. We then find that $(\partial V/\partial p)_T = -\mathcal{N}kT/p^2$ and hence that the isothermal compressibility is given by

$$C_T = \frac{\mathcal{N}kT}{Vp^2} = \frac{1}{p} = \frac{V}{\mathcal{N}kT}. \tag{7.3.11}$$

We next assume that $\epsilon - 1$ is linearly proportional to ρ, so that we can represent ϵ as $\epsilon = 1 + A\rho$ for some constant A. We hence find that $\partial\epsilon/\partial\rho = A$, or that $\partial\epsilon/\partial p = (\epsilon - 1)/\rho$, and the electrostrictive constant is given by

$$\gamma_e = \epsilon - 1. \tag{7.3.12}$$

If we now introduce Eqs. (7.3.11) and (7.3.12) into Eq. (7.3.9a), we find that the scattered intensity can be expressed as

$$I_s = I_0 \frac{\omega^4 V}{16\pi^2 L^2 c^4} \frac{(\epsilon - 1)^2}{N} \sin^2\phi, \tag{7.3.13}$$

where we have introduced the mean density of particles $N = \mathcal{N}/V$. Through use of Eq. (7.1.4), we can write this result in terms of the scattering coefficient as

$$R = \frac{(\epsilon - 1)^2 \omega^4 \sin^2\phi}{16\pi^2 c^4 N}. \tag{7.3.14}$$

Note that, since $\epsilon - 1$ is equal to $2(n - 1)$ for a dilute gas (i.e., for $\epsilon - 1 \ll 1$), this result is in agreement with the prediction of the microscopic model of light scattering for an ideal gas, given by Eq. (7.2.22).

Spectrum of the Scattered Light

The analysis just presented has led to an explicit prediction (7.3.9) for the *total* intensity of the light scattered as the result of the fluctuations in the density (and hence the dielectric constant) of a material system in thermal equilibrium. In order to determine the *spectrum* of the scattered light, we have to examine the dynamical behavior of the density fluctuations that give rise to light scattering. As before, we represent the fluctuation in the dielectric constant as

$$\Delta\tilde{\epsilon} = \frac{\partial\epsilon}{\partial\rho}\Delta\tilde{\rho}. \tag{7.3.15}$$

We now choose the entropy s and pressure p (instead of ρ and T) to be our independent thermodynamic variables. We can then represent the variation in density, $\Delta\tilde{\rho}$, as

$$\Delta\tilde{\rho} = \left(\frac{\partial \rho}{\partial p}\right)_s \Delta\tilde{p} + \left(\frac{\partial \rho}{\partial s}\right)_p \Delta\tilde{s}. \tag{7.3.16}$$

Here the first term describes adiabatic density fluctuations (that is, accoustic waves) and leads to Brillouin scattering. The second term describes isobaric density fluctuations (that is, entropy or temperature fluctuations) and leads to Rayleigh-center scattering. The two contributions to $\Delta\rho$ are quite different in character and lead to very different spectral distributions of the scattered light, because (as we shall see) the equations of motion for Δp and Δs are very different.

Brillouin Scattering

The equation of motion for a pressure wave is well known from the field of acoustics and is given by (see, e.g., Fabelinskii, 1968, Section 34.9)

$$\frac{\partial^2 \Delta\tilde{p}}{\partial t^2} - \Gamma' \nabla^2 \frac{\partial \Delta\tilde{p}}{\partial t} - v^2 \nabla^2 \Delta\tilde{p} = 0. \tag{7.3.17}$$

Here v denotes the velocity of sound, which is given in terms of thermodynamic variables by

$$v^2 = \left(\frac{\partial p}{\partial \rho}\right)_s. \tag{7.3.18}$$

The equation for the velocity of sound is conveniently expressed in terms of the compressibility C or in terms of its reciprocal, the bulk modulus K, which are defined by

$$C \equiv \frac{1}{K} = -\frac{1}{V}\frac{\partial V}{\partial p} = \frac{1}{\rho}\frac{\partial \rho}{\partial p}. \tag{7.3.19}$$

The compressibility can be measured either at constant temperature or at constant entropy. The two values of the compressibility, denoted respectively as C_T and C_s, are related by

$$\frac{C_T}{C_s} = \frac{c_p}{c_V} \equiv \gamma, \tag{7.3.20}$$

where c_p is the specific heat (i.e., the heat capacity per unit mass, whose units are erg/g K) at constant pressure, c_V is the specific heat at constant volume, and where their ratio γ is known as the adiabatic index. The velocity of sound as defined by Eq. (7.3.18) can thus be written as

$$v^2 = \frac{K_s}{\rho} = \frac{1}{C_s \rho}. \tag{7.3.21}$$

An important special case of the use of this formula is that of an ideal gas, for which the equation of state is given by Eq. (7.3.10) and the isothermal compressibility is given by Eq. (7.3.11). The adiabatic compressibility is thus given by $C_s = C_T/\gamma = 1/\gamma p$. We hence find from Eq. (7.3.21) that the velocity of sound is given by

$$v = \left(\frac{\gamma p}{\rho}\right)^{1/2} = \left(\frac{\gamma \mathcal{N} kT}{\rho V}\right)^{1/2} = \left(\frac{\gamma kT}{\mu}\right)^{1/2}, \tag{7.3.22}$$

where μ denotes the molecular mass. We thus see that the velocity of sound is of the order of the mean thermal velocity of the molecules of the gas.

The velocity of sound for some common optical materials is listed in Table 7.3.1.

The parameter Γ' appearing in the wave equation (7.3.17) is a damping parameter that can be shown to be expressible as

$$\Gamma' = \frac{1}{\rho}\left[\tfrac{4}{3}\eta_s + \eta_b + \frac{\kappa}{c_p}(\gamma - 1)\right] \tag{7.3.23}$$

where η_s is the shear viscosity coefficient, η_b is the bulk viscosity coefficient,

TABLE 7.3.1 Typical sound velocities

Material	v (cm/s)
Gases	
Dry air	3.31×10^4
He	9.65×10^4
H_2	12.84×10^4
Water vapor	4.94×10^4
Liquids	
CS_2	1.15×10^5
CCl_4	0.93×10^5
Ethanol	1.21×10^5
Water	1.50×10^5
Solids	
Fused silica	5.97×10^5
Lucite	2.68×10^5

and κ is the thermal conductivity. For most materials of interest in optics, the last contribution to Γ' is much smaller than the first two. Conventions involving the naming of the viscosity coefficients are discussed briefly in the Appendix to Section 8.6.

As an illustration of the nature of the acoustic wave equation (7.3.17), we consider the propagation of the wave

$$\Delta\tilde{p} = \Delta p\, e^{i(qz-\Omega t)} + \text{c.c.} \tag{7.3.24}$$

through an acoustic medium. By substituting this form into the acoustic wave equation, we find that q and Ω must be related by a dispersion relation of the form

$$\Omega^2 = q^2(v^2 - i\Omega\Gamma'). \tag{7.3.25}$$

We can rewrite this relation as

$$q^2 = \frac{\Omega^2}{v^2 - i\Omega\Gamma'} = \frac{\Omega^2/v^2}{1 - i\Omega\Gamma'/v^2} \simeq \frac{\Omega^2}{v^2}\left(1 + \frac{i\Omega\Gamma'}{v^2}\right), \tag{7.3.26}$$

which shows that

$$q \simeq \frac{\Omega}{v} + \frac{i\Gamma}{2v}, \tag{7.3.27}$$

where we have introduced the phonon decay rate

$$\Gamma = \Gamma' q^2. \tag{7.3.28}$$

We find by introducing the form for q given by Eq. (7.3.27) into Eq. (7.3.24) that the intensity of the acoustic wave varies spatially as

$$|\Delta p(z)|^2 = |\Delta p(0)|^2 e^{-\alpha_s z}, \tag{7.3.29}$$

where we have introduced the sound absorption coefficient

$$\alpha_s = \frac{q^2\Gamma'}{v} = \frac{\Gamma}{v}. \tag{7.3.30}$$

It is also useful to define the phonon lifetime as

$$\tau_{\text{p}} = \frac{1}{\Gamma} = \frac{1}{q^2\Gamma'}. \tag{7.3.31}$$

Next, we calculate the rate at which light is scattered out of a beam of light by these acoustic waves. We assume that the incident optical field is described by

$$\tilde{E}_0(z,t) = E_0 e^{i(\mathbf{k}\cdot\mathbf{r}-\omega t)} + \text{c.c.}, \tag{7.3.32}$$

and that the scattered field obeys the driven wave equation

$$\nabla^2 \tilde{E} - \frac{n^2}{c^2}\frac{\partial^2 \tilde{E}}{\partial t^2} = \frac{4\pi}{c^2}\frac{\partial^2 \tilde{P}}{\partial t^2}. \tag{7.3.33}$$

We take the polarization $\tilde{P}$ of the medium to be given by Eq. (7.3.1) with the variation $\Delta\tilde{\epsilon}$ in dielectric constant given by Eq. (7.3.15), that is, we take $\tilde{P} = (\partial\epsilon/\partial\rho)\,\Delta\tilde{\rho}\,\tilde{E}_0/4\pi$. We take the variation in density to be given by the first contribution to Eq. (7.3.16), that is, by $\Delta\tilde{\rho} = (\partial\rho/\partial p)\,\Delta\tilde{p}$, where $\Delta\tilde{p}$ denotes the incremental pressure. We thus find that

$$\begin{aligned}\tilde{P}(\mathbf{r}, t) &= \frac{1}{4\pi}\frac{\partial\epsilon}{\partial\rho}\left(\frac{\partial\rho}{\partial p}\right)_s \Delta\tilde{p}(\mathbf{r}, t)\,\tilde{E}_0(z, t)\\ &= \frac{1}{4\pi}\gamma_e C_s\,\Delta\tilde{p}(\mathbf{r}, t)\,\tilde{E}_0(z, t),\end{aligned} \tag{7.3.34}$$

where we have introduced the adiabatic compressibility C_s of Eq. (7.3.19) and the electrostrictive constant of Eq. (7.3.6). We take a typical component of the thermally excited pressure disturbance within the interaction region to be given by

$$\Delta\tilde{p}(\mathbf{r}, t) = \Delta p\, e^{i(\mathbf{q}\cdot\mathbf{r} - \Omega t)} + \text{c.c.} \tag{7.3.35}$$

By combining Eqs. (7.3.33) through (7.3.35), we find that the scattered field must obey the wave equation

$$\begin{aligned}\nabla^2 \tilde{\mathbf{E}} - \frac{n^2}{c^2}\frac{\partial^2 \tilde{\mathbf{E}}}{\partial t^2} = -\frac{\gamma_e C_s}{c^2}\big[&(\omega - \Omega)^2 \mathbf{E}_0\,\Delta p^*\, e^{i(\mathbf{k}-\mathbf{q})\cdot\mathbf{r} - i(\omega - \Omega)t}\\ &+ (\omega + \Omega)^2 \mathbf{E}_0\,\Delta p\, e^{i(\mathbf{k}+\mathbf{q})\cdot\mathbf{r} - i(\omega + \Omega)t} + \text{c.c.}\big]\end{aligned} \tag{7.3.36}$$

The first term in this expression leads to Stokes scattering; the second to anti-Stokes scattering. We study these two contributions in turn.

Stokes Scattering (First Term in Eq. (7.3.36))

The polarization is seen to have a component with wave vector

$$\mathbf{k}' \equiv \mathbf{k} - \mathbf{q} \tag{7.3.37}$$

and frequency

$$\omega' \equiv \omega - \Omega, \tag{7.3.38}$$

where the frequency ω and wave vector $\mathbf{k}$ of the incident optical field are related according to

$$\omega = |\mathbf{k}|c/n, \tag{7.3.39}$$

and where the frequency Ω and wave vector $\mathbf{q}$ of the acoustic wave are related according to

$$\Omega = |\mathbf{q}|v. \tag{7.3.40}$$

This component of the polarization can couple efficiently to the scattered optical wave only if its frequency ω' and wave vector $\mathbf{k}'$ are related by the dispersion relation for optical waves, namely

$$\omega' = |\mathbf{k}'|c/n. \tag{7.3.41}$$

In order for Eqs. (7.3.37) through (7.3.41) to be satisfied simultaneously, the sound wave frequency and wave vector must each have a particular value for any scattering direction. For the case of scattering at angle θ, we must have the situation illustrated in Fig. 7.3.1. Part (a) of this figure shows the relative orientations of the wave vectors of the incident and scattered fields. Part (b) illustrates Eq. (7.3.37) and shows how the wave vector of the acoustic disturbance is related to those of the incident and scattered optical radiation.

Since $|\mathbf{k}|$ is very nearly equal to $|\mathbf{k}'|$ (because Ω is much smaller than ω), diagram (b) shows that

$$|\mathbf{q}| = 2|\mathbf{k}|\sin(\theta/2). \tag{7.3.42}$$

The dispersion relation (7.3.40) then shows that the acoustic frequency is given by

$$\Omega = 2|\mathbf{k}|v\sin(\theta/2) = 2n\omega\frac{v}{c}\sin(\theta/2). \tag{7.3.43}$$

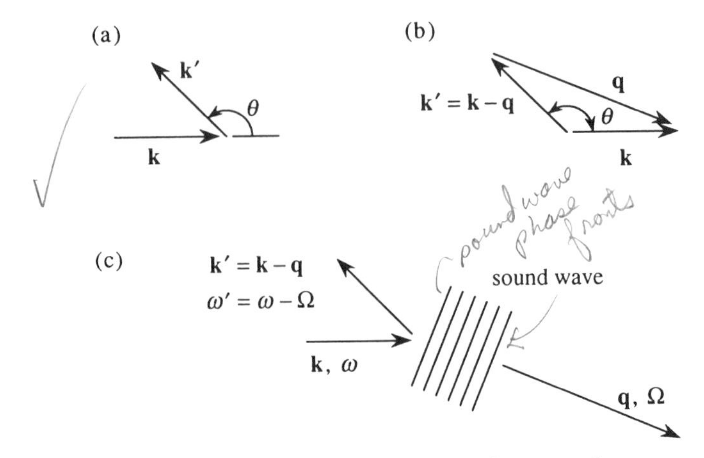

FIGURE 7.3.1 Illustration of Stokes scattering.

We note that the Stokes shift Ω is equal to zero for forward scattering and is maximum for backscattering (i.e., for $\theta = 180$ degrees). The maximum frequency shift is thus given by

$$\Omega_{\text{max}} = 2n\frac{v}{c}\omega. \tag{7.3.44}$$

For $\omega/2\pi = 3 \times 10^{14}$ Hz (i.e., at $\lambda = 1\ \mu\text{m}$), $v = 1 \times 10^5$ cm/s (a typical value), and $n = 1.5$, we obtain $\Omega_{\text{max}}/2\pi = 3 \times 10^9$ Hz.

Stokes scattering can be visualized as the scattering of light from a retreating acoustic wave, as illustrated in part (c) of Fig. 7.3.1.

Anti-Stokes Scattering (Second Term in Eq. (7.3.36))

The analysis here is analogous to that for Stokes scattering. The polarization is seen to have a component with wave vector

$$\mathbf{k}' \equiv \mathbf{k} + \mathbf{q} \tag{7.3.45}$$

and frequency

$$\omega' = \omega + \Omega, \tag{7.3.46}$$

where, as before, $\omega = |\mathbf{k}|c/n$ and $\Omega = |\mathbf{q}|v$. This component of the polarization can couple efficiently to an electromagnetic wave only if ω' and $|\mathbf{k}'|$ are related by $\omega' = |\mathbf{k}'|c/n$. We again assume that θ denotes the scattering angle, as illustrated in Fig. 7.3.2. The condition (7.3.45) is illustrated as part (b) of the

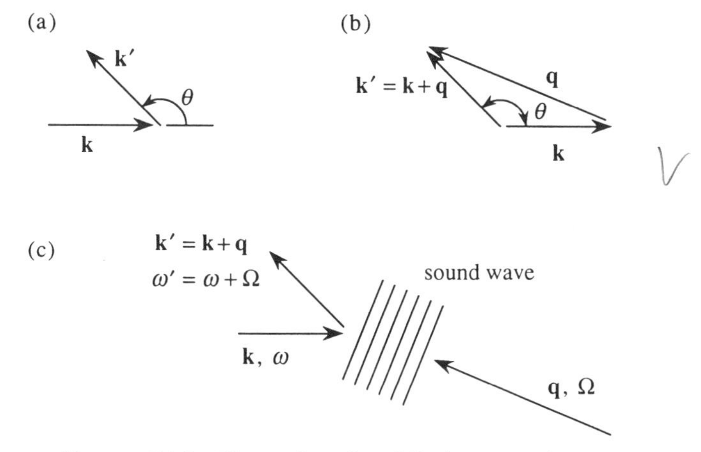

FIGURE 7.3.2 Illustration of anti-Stokes scattering.

figure. Since (as before) $|\mathbf{k}|$ is very nearly equal to $|\mathbf{k}'|$, the length of the acoustic wave vector is given by

$$|\mathbf{q}| = 2|\mathbf{k}|\sin(\theta/2). \tag{7.3.47}$$

Hence, by Eq. (7.3.40), the acoustic frequency is given by

$$\Omega = 2n\omega\frac{v}{c}\sin(\theta/2). \tag{7.3.48}$$

Anti-Stokes scattering can be visualized as scattering from an oncoming sound wave, as shown in part (c) of Fig. 7.3.2.

We have thus far ignored attenuation of the acoustic wave in our analysis. If we include this effect, we find that the light scattered into direction θ is not monochromatic but has a spread in angular frequency whose width (FWHM) is given by

$$\delta\omega = 1/\tau_p = \Gamma' q^2 \tag{7.3.49}$$

which becomes, through use of Eq. (7.3.42),

$$\delta\omega = 4\Gamma'|\mathbf{k}|^2\sin^2(\theta/2) = 4n^2\Gamma'\frac{\omega^2}{c^2}\sin^2(\theta/2). \tag{7.3.50}$$

For the case of backscattering ($\theta = 180$ degrees), $\delta\omega/2\pi$ is typically of the order of 100 MHz for organic liquids. Since the acoustic frequency is given by Eq. (7.3.43), we see that the ratio of the linewidth to the Brillouin frequency shift is given by

$$\frac{\delta\omega}{\Omega} = \frac{2n\Gamma'\omega}{vc}\sin(\theta/2). \tag{7.3.51}$$

The spectrum of the scattered light has the form shown in Fig. 7.3.3.

Rayleigh Center Scattering

We now consider the contribution to $\Delta\tilde{\rho}$ (and hence to $\Delta\tilde{\epsilon}$) resulting from isobaric density fluctuations, which are described by the second term in

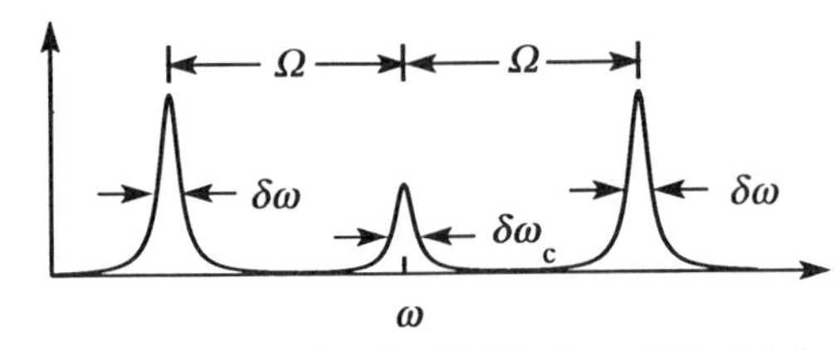

FIGURE 7.3.3 Spectrum showing Brillouin and Rayleigh scattering.

Eq. (7.3.16) and are proportional to the entropy fluctuation $\Delta\tilde{s}$. Entropy fluctuations are described by the same equation as that describing temperature variations:

$$\rho c_p \frac{\partial \Delta\tilde{s}}{\partial t} - \kappa\nabla^2 \Delta\tilde{s} = 0, \tag{7.3.52}$$

where, as before, c_p denotes the specific heat at constant pressure, and where κ denotes the thermal conductivity. Note that these fluctuations obey a diffusion equation and not a wave equation. A solution to the diffusion equation (7.3.52) is

$$\Delta\tilde{s} = \Delta s_0 e^{-\delta t} e^{-i\mathbf{q}\cdot\mathbf{r}}, \tag{7.3.53}$$

where the damping rate of the entropy disturbance is given by

$$\delta = \frac{\kappa}{\rho c_p} q^2. \tag{7.3.54}$$

We see that, unlike pressure waves, entropy waves do not propagate. As a result, the nonlinear polarization proportional to Δs can give rise only to an unshifted component of the scattered light. The width (FWHM) of this component is given by $\delta\omega_c = \delta$, that is, by

$$\delta\omega_c = \frac{4\kappa}{\rho c_p} |\mathbf{k}|^2 \sin^2(\theta/2). \tag{7.3.55}$$

As a representative case, for liquid water $\kappa = 6$ mW/cm K, $\rho = 1$ g/cm^3, $c_p = 4.2$ J/g K, and the predicted width of the central component for backscatttering ($\theta = 180$ degrees) of radiation at 500 nm is $\delta\omega_c/2\pi = 1.4 \times 10^7$ Hz.

It can be shown (Fabelinskii, 1968, Eq. 5.39) that the relative intensities of the Brillouin and Rayleigh center components are given by

$$\frac{I_c}{2I_B} = \frac{c_p - c_v}{c_v} = \gamma - 1. \tag{7.3.56}$$

Here I_c denotes the integrated intensity of the central component, and I_B that of either of the Brillouin components. This result is known as the Landau–Placzek relation.

7.4. Acousto-optics

The analysis given above of the scattering of light from sound waves can be applied to the situation in which the sound wave is applied to the interaction region externally by means of a transducer. Such acousto-optic devices are useful as intensity or frequency modulators for laser beams or as beam deflectors.

Acousto-optic devices are commonly classified as falling into one of two regimes, each of which will be discussed in greater detail below. These regimes are as follows:

Bragg scattering. This type of scattering occurs for the case of interaction lengths that are sufficiently long that phase-matching considerations become important. Bragg scattering leads to a single diffracted beam. The name is given by analogy to the scattering of x-rays from the atomic planes in a crystal. Bragg scattering can lead to an appreciable scattering efficiency (>50%).

Raman–Nath scattering. This type of scattering occurs in cells with a short interaction length. Phase-matching considerations are not important, and several scattered orders are usually present.

We shall first consider the case of Bragg scattering of light waves; a more precise statement of the conditions under which the Bragg or Raman–Nath scattering occurs is given below in connection with the discussion of Raman–Nath scattering.

Bragg Scattering of Light by Sound Waves

The operation of a typical Bragg scattering cell is shown schematically in Figure 7.4.1. A traveling acoustic wave of frequency Ω and wavelength $\Lambda = 2\pi v/\Omega$ (where v denotes the velocity of sound) is established in the scattering medium. The density variation associated with this acoustic wave produces a variation in the dielectric constant of the medium, and the incident

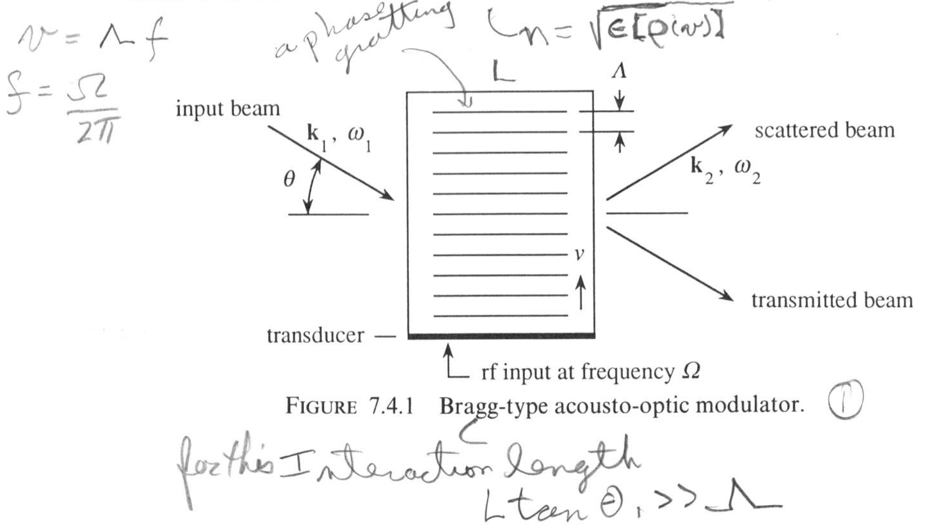

FIGURE 7.4.1 Bragg-type acousto-optic modulator.

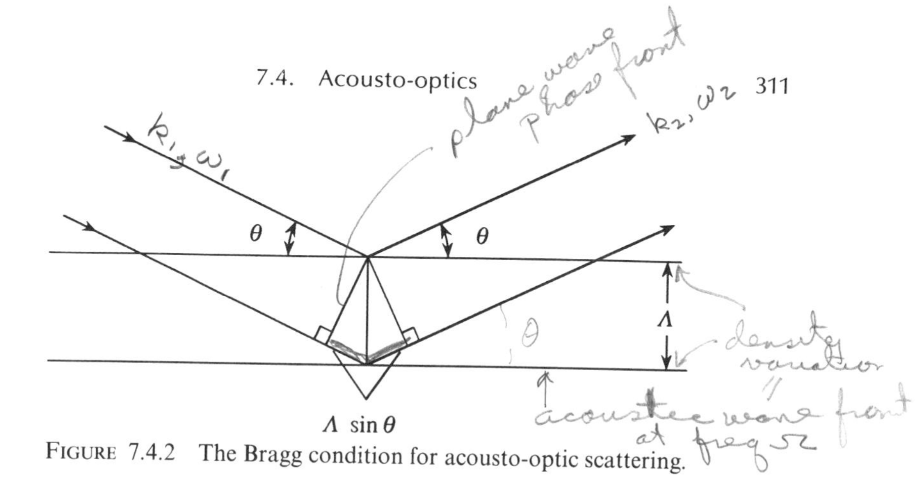

FIGURE 7.4.2 The Bragg condition for acousto-optic scattering.

optical wave scatters from this variation. Although the amplitude of the wave scattered from each acoustic wavefront is typically rather small, the total scattered field can become quite intense if the various contributions add in phase to produce constructive interference. The condition for this to occur is obtained with the help of the construction shown in Fig. 7.4.2, and is given by the relation

$$\lambda = 2\Lambda \sin\theta, \tag{7.4.1}$$

where λ is the wavelength of light in the medium. This condition is known as the Bragg condition. It ensures that the path length difference between rays that reflect from successive acoustic maxima is equal to an optical wavelength. In a typical acousto-optic device, relevant parameters might be $v = 1.5 \times 10^5$ cm/s and $\Omega/2\pi = 200$ MHz, which imply that the acoustic wavelength is equal to $\Lambda = 2\pi v/\Omega = 7.5\ \mu$m. If the optical wavelength is 0.5 μm, we see from Eq. (7.4.1) that $\sin\theta = 1/30$ and hence that the deflection angle is given by $2\theta = 4$ degrees.

The Bragg condition given by Eq. (7.4.1) can alternatively be understood as a phase-matching condition. If $\mathbf{k}_1$ denotes the wave vector of the incident optical wave, $\mathbf{k}_2$ that of the diffracted optical wave, and $\mathbf{q}$ that of the acoustic wave, the Bragg condition can be seen with the help of Fig. 7.4.3(a) to be

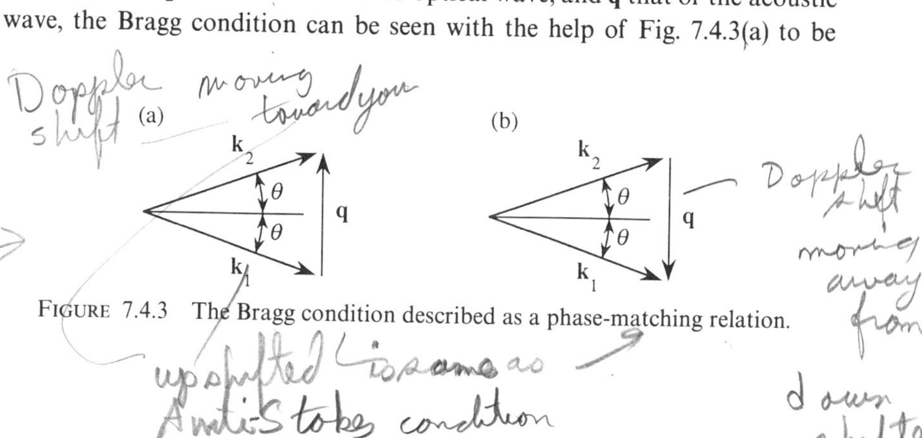

FIGURE 7.4.3 The Bragg condition described as a phase-matching relation.

a statement that

$$\mathbf{k}_2 = \mathbf{k}_1 + \mathbf{q}. \tag{7.4.2}$$

By comparison with the analysis of Section 7.3 for spontaneous Brillouin scattering (and as shown explicitly below), we can see that the frequency of the scattered beam is shifted upward to

$$\omega_2 = \omega_1 + \Omega. \tag{7.4.3}$$

Since Ω is much less than ω_1, we see that ω_2 is approximately equal to ω_1, and hence that $|\mathbf{k}_2| \simeq |\mathbf{k}_1|$. The configuration shown in Fig. 7.4.1 shows the case in which the acoustic wave is advancing toward the incident optical wave. For the case of a sound wave propagating in the opposite direction, Eqs. (7.4.2) and (7.4.3) must be replaced by

$$\mathbf{k}_2 = \mathbf{k}_1 - \mathbf{q}, \tag{7.4.4a}$$

$$\omega_2 = \omega_1 - \Omega. \tag{7.4.4b}$$

Figures 7.4.1 and 7.4.2 are unchanged except for the reversal of the direction of the sound velocity vector, although Fig. 7.4.3(a) must be replaced by Fig. 7.4.3(b).

Bragg scattering of light by sound waves can be treated theoretically by considering the time-varying change $\Delta\tilde{\epsilon}$ in the dielectric constant induced by the acoustic density variation $\Delta\tilde{\rho}$. It is usually adequate to assume that $\Delta\tilde{\epsilon}$ scales linearly with $\Delta\tilde{\rho}$, so that

$$\Delta\tilde{\epsilon} = \frac{\partial\epsilon}{\partial\rho}\Delta\tilde{\rho} = \gamma_e \frac{\Delta\tilde{\rho}}{\rho_0}. \tag{7.4.5}$$

Here ρ_0 denotes the mean density of the material, and γ_e denotes the electrostrictive constant defined by Eq. (7.3.6). Equation (7.4.5) applies rigorously to the case of liquids, and it predicts the correct qualitative behavior for all materials. For the case of anisotropic materials, the change in the optical properties is described more precisely by means of a tensor relation, which conventionally is given by

$$[\Delta(\epsilon^{-1})]_{ij} = \sum_{kl} p_{ijkl} S_{kl}, \tag{7.4.6}$$

where the quantity p_{ijkl} is known as the strain–optic tensor and where

$$S_{kl} = \frac{1}{2}\left(\frac{\partial d_k}{\partial x_l} + \frac{\partial d_l}{\partial x_k}\right) \tag{7.4.7}$$

is the strain tensor, in which d_k is the k component of the displacement of a particle from its equilibrium position. Whenever the change in the inverse of the dielectric tensor $(\epsilon^{-1})_{ij}$ given by the right-hand side of Eq. (7.4.6) is small, the change in the dielectric tensor ϵ_{ij} is given by

$$(\Delta\epsilon)_{il} = -\sum_{jk} \epsilon_{ij}[\Delta(\epsilon^{-1})]_{jk}\epsilon_{kl}. \tag{7.4.8}$$

Our theoretical treatment of Bragg scattering assumes the geometry shown in Figure 7.4.4. The interaction of the incident field

$$\tilde{E}_1 = A_1 e^{i(\mathbf{k}_1 \cdot \mathbf{r} - \omega_1 t)} + \text{c.c.} \tag{7.4.9}$$

with the acoustic wave of wave vector $\mathbf{q}$ produces the diffracted wave

$$\tilde{E}_2 = A_2 e^{i(\mathbf{k}_2 \cdot \mathbf{r} - \omega_2 t)} + \text{c.c.} \tag{7.4.10}$$

with $\omega_2 = \omega_1 + \Omega$. The interaction is assumed to be nearly Bragg-matched (i.e., phase-matched) in the sense that

$$\mathbf{k}_2 \simeq \mathbf{k}_1 + \mathbf{q}. \tag{7.4.11}$$

The variation of the dielectric constant induced by the acoustic wave is represented as

$$\Delta\tilde{\epsilon} = \Delta\epsilon\, e^{i(\mathbf{q}\cdot\mathbf{r} - \Omega t)} + \text{c.c.}, \tag{7.4.12}$$

where the complex amplitude $\Delta\epsilon$ is given by $\Delta\epsilon = \gamma_e \Delta\rho/\rho_0$ under those conditions where the change in dielectric constant is accurately predicted by Eq. (7.4.5). More generally, for anisotropic interactions, $\Delta\epsilon$ is the amplitude of the appropriate tensor component of $\Delta\tilde{\epsilon}_{ij}$ given by Eq. (7.4.6). The total optical field $\tilde{E} = \tilde{E}_1 + \tilde{E}_2$ is required to satisfy the wave equation

$$\nabla^2 \tilde{E} - \frac{n^2 + \Delta\tilde{\epsilon}}{c^2} \frac{\partial^2 \tilde{E}}{\partial t^2} = 0, \tag{7.4.13}$$

where n denotes the refractive index of the material in the absence of the acoustic wave. Since according to Eq. (7.4.12) $\Delta\tilde{\epsilon}$ oscillates at frequency Ω, it couples the optical waves of frequencies ω_1 and $\omega_2 = \omega_1 + \Omega$.

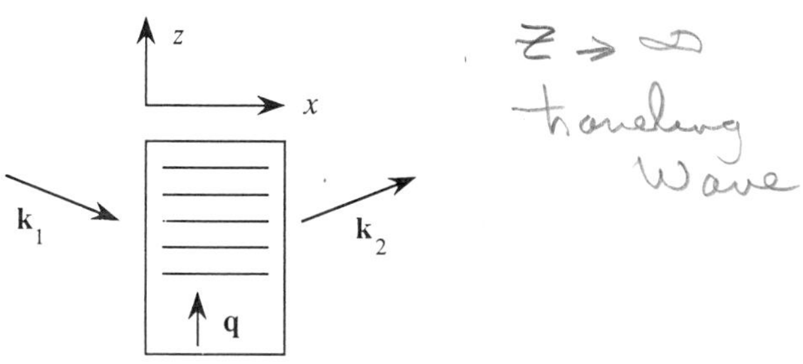

FIGURE 7.4.4 Geometry of a Bragg-type acousto-optic modulator.

We first consider the portion of Eq. (7.4.13) that oscillates at frequency ω_1. This part is given by

$$\frac{\partial^2 A_1}{\partial x^2} + \frac{\partial^2 A_1}{\partial z^2} + 2ik_{1x}\frac{\partial A_1}{\partial x} + 2ik_{1z}\frac{\partial A_1}{\partial z} - (k_{1x}^2 + k_{1z}^2)A_1$$

$$+ \frac{n^2\omega_1^2}{c^2}A_1 + \frac{\omega_2^2}{c^2}A_2\Delta\epsilon^* e^{i(\mathbf{k}_2 - \mathbf{k}_1 - \mathbf{q})\cdot\mathbf{r}} = 0. \quad (7.4.14)$$

This equation can be simplified in the following manner: (1) we introduce the slowly-varying-amplitude approximation, which entails ignoring the second-order derivatives, (2) we note that A_1 depends only on x and not on z, since the interaction is invariant to a translation in the z direction, and so we set $\partial A_1/\partial z$ equal to 0, and (3) we note that $k_{1x}^2 + k_{1z}^2 = n^2\omega_1^2/c^2$. Equation (7.4.14) thus becomes

$$2ik_{1x}\frac{dA_1}{dx} = -\frac{\omega_2^2}{c^2}A_2\Delta\epsilon^* e^{i(\mathbf{k}_2 - \mathbf{k}_1 - \mathbf{q})\cdot\mathbf{r}}. \quad (7.4.15)$$

Next, we note that the propagation vector mismatch $\mathbf{k}_2 - \mathbf{k}_1 - \mathbf{q} \equiv -\Delta\mathbf{k}$ can have a nonzero component only in the x direction, because the geometry we are considering has infinite extent in the z direction, and the z component of the $\mathbf{k}$ wave vector mismatch must therefore vanish. We thus see that

$$(\mathbf{k}_2 - \mathbf{k}_1 - \mathbf{q})\cdot\mathbf{r} \equiv -\Delta k\, x, \quad (7.4.16)$$

and hence that Eq. (7.4.15) can be written as

$$\frac{dA_1}{dx} = \frac{i\omega_2^2\Delta\epsilon^*}{2k_{1x}c^2}A_2 e^{-i\Delta k x}. \quad (7.4.17)$$

By a completely analogous derivation, we find that the portion of the wave equation (7.4.13) that describes a wave at frequency ω_2 is given by

$$\frac{dA_2}{dx} = \frac{i\omega_1^2\Delta\epsilon}{2k_{2x}c^2}A_1 e^{i\Delta k x}. \quad (7.4.18)$$

Finally, we note that since $\omega_1 \simeq \omega_2 \equiv \omega$ and $k_{1x} \simeq k_{2x} \equiv k_x$, the coupled equations (7.4.17) and (7.4.18) can be written as

$$\frac{dA_1}{dx} = i\kappa A_2 e^{-i\Delta k x}, \quad (7.4.19a)$$

$$\frac{dA_2}{dx} = i\kappa^* A_1 e^{i\Delta k x}, \quad (7.4.19b)$$

where we have introduced the coupling constant

$$\kappa = \frac{\omega^2 \Delta\epsilon^*}{2k_x c^2}. \tag{7.4.20}$$

The solution to these coupled-amplitude equations is particularly simple for the case in which $\tilde{E}_1$ is incident at the Bragg angle. In this case, the interaction is perfectly phase-matched, so that $\Delta k = 0$, and hence Eqs. (7.4.19) reduce to the set

$$\frac{dA_1}{dx} = i\kappa A_2, \qquad \frac{dA_2}{dx} = i\kappa^* A_1. \tag{7.4.21}$$

These equations are easily solved using methods similar to those introduced in Chapter 2. The solution appropriate to the boundary conditions illustrated in Fig. 7.4.4 is

$$A_1(x) = A_1(0)\cos(|\kappa|x), \tag{7.4.22a}$$

$$A_2(x) = \frac{i\kappa^*}{|\kappa|} A_1(0)\sin(|\kappa|x). \tag{7.4.22b}$$

Note that these solutions obey the relation

$$|A_1(x)|^2 + |A_2(x)|^2 = |A_1(0)|^2, \tag{7.4.23}$$

which shows that the energy of the optical field is conserved in the Bragg scattering process (since we have assumed that $\Omega \ll \omega$). We define the diffraction efficiency of the Bragg scattering process to be the ratio of the output intensity of the ω_2 wave to the input intensity of the ω_1 wave, and we find that the diffraction efficiency is given by

$$\eta \equiv \frac{|A_2(L)|^2}{|A_1(0)|^2} = \sin^2(|\kappa|L). \tag{7.4.24}$$

For practical purposes, it is useful to express the coupling constant κ defined by Eq. (7.4.20) in terms of the intensity (i.e., power per unit area) of the acoustic wave. The intensity of a sound wave is given by the relation

$$I = Kv\frac{\langle \Delta\tilde{\rho}^2 \rangle}{\rho_0^2} = 2Kv\left|\frac{\Delta\rho}{\rho_0}\right|^2, \tag{7.4.25}$$

where, as before, $K = 1/C$ is the bulk modulus, v is the sound velocity, and $\Delta\rho$ is the complex amplitude of the density disturbance associated with the acoustic wave. It follows from Eq. (7.4.5) that $\Delta\epsilon$ is equal to $\gamma_e \Delta\rho/\rho$, and hence the acoustic intensity can be written as $I = 2Kv|\Delta\epsilon|^2/\gamma_e^2$. The coupling constant $|\kappa|$ (see Eq. (7.4.20)) can thus be expressed as

$$|\kappa| = \frac{\omega\gamma_e}{2nc\cos\theta}\left(\frac{I}{2Kv}\right)^{1/2}, \tag{7.4.26}$$

where we have replaced k_x by $n(\omega/c)\cos\theta$.

As an example, we evaluate Eq. (7.4.26) for the case of Bragg scattering in water, which is characterized by the following physical constants: $n = 1.33$, $\gamma_e = 0.82$, $v = 1.5 \times 10^5$ cm/s, and $K = 2.19 \times 10^{10}$ cm^2/dyne. We assume that $\cos\theta \simeq 1$, as is usually the case; that the vacuum optical wavelength is 0.5 μm, so that $\omega = 3.8 \times 10^{15}$ rad/s; and that the acoustic intensity is 1.0 W/cm^2 (as might be obtained using 1 W of acoustic power and an acoustic beam diameter of approximately 1 cm), or $I = 10^7$ erg/cm^2 s. Under these conditions, Eq. (7.4.26) gives the value $|\kappa| = 1.5$ cm^{-1}. According to Eq. (7.4.24), 100% conversion of the incident beam into the diffracted beam is predicted for $|\kappa| L = \pi/2$, or under the present conditions for a path length through the acoustic beam of $L = 1.0$ cm.

For the case in which the incident beam does not intercept the acoustic wavefronts at the Bragg angle, the theoretical analysis is more complicated because the wave vector mismatch Δk does not vanish. The phase-matching diagrams for the cases of Bragg-angle and non-Bragg-angle incidence are contrasted in Fig. 7.4.5. As discussed above in connection with Eq. (7.4.16), the wave vector mismatch can have a component only in the x direction, since the medium is assumed to have infinite extent in the z direction.

We first determine the relationship between the wave vector mismatch Δk and the angle of incidence θ_1. We note that the x and z components of the vectors of diagram (b) obey the relations

$$k\cos\theta_1 - k\cos\theta_2 = \Delta k, \tag{7.4.27a}$$

$$k\sin\theta_1 + k\sin\theta_2 = q, \tag{7.4.27b}$$

where we have let $k_1 \simeq k_2 = k$. We note that if the angle of incidence θ_1 is

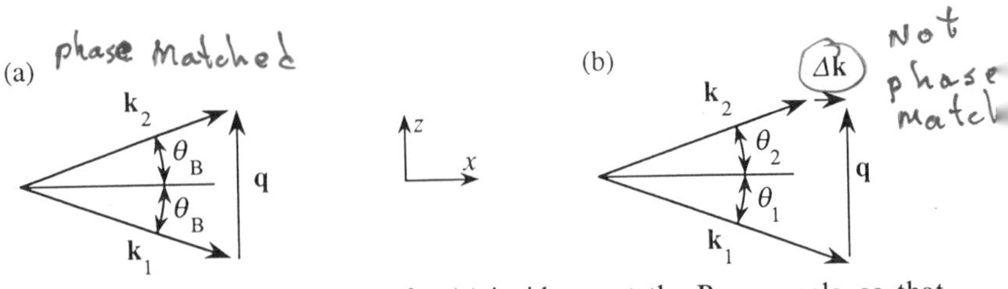

FIGURE 7.4.5 Wave vector diagrams for (a) incidence at the Bragg angle, so that $\Delta k = 0$, and (b) non-Bragg-angle incidence, so that $\Delta k \neq 0$.

equal to the Bragg angle

$$\theta_B = \sin^{-1}\frac{q}{2k} = \sin^{-1}\frac{\lambda}{2\Lambda}, \tag{7.4.28}$$

then Eqs. (7.4.27) imply that the diffraction angle θ_2 is also equal to θ_B and that $\Delta k = 0$. For the case in which the light is not incident at the Bragg angle, we set

$$\theta_1 = \theta_B + \Delta\theta, \tag{7.4.29a}$$

where we assume that $\Delta\theta \ll 1$. We note that Eq. (7.4.27b) will be satisfied so long as

$$\theta_2 = \theta_B - \Delta\theta. \tag{7.4.29b}$$

These values of θ_1 and θ_2 are now introduced into Eq. (7.4.27a). The cosine functions are expanded to lowest order in $\Delta\theta$ as

$$\cos(\theta_B \pm \Delta\theta) = \cos\theta_B \mp (\sin\theta_B)\Delta\theta,$$

and we obtain $(2k\sin\theta_B)\Delta\theta = \Delta k$, which through use of Eq. (7.4.28) shows that the wave vector mismatch Δk that occurs as the result of an angular misalignment $\Delta\theta$ is given by

$$\Delta k = -\Delta\theta\, q. \tag{7.4.30}$$

We next solve Eqs. (7.4.19) for arbitrary values of Δk. The solution for the case in which no field at frequency ω_2 is applied externally is

$$A_1(x) = e^{-i(1/2)\Delta k x} A_1(0)\left(\cos sx + i\frac{\Delta k}{2s}\sin sx\right), \tag{7.4.31a}$$

$$A_2(x) = ie^{i(1/2)\Delta k x} A_1(0)\frac{\kappa^*}{s}\sin sx, \tag{7.4.31b}$$

where

$$s^2 = |\kappa|^2 + (\tfrac{1}{2}\Delta k)^2. \tag{7.4.32}$$

The diffraction efficiency for arbitrary Δk is now given by

$$\eta(\Delta k) \equiv \frac{|A_2(L)|^2}{|A_1(0)|^2} = \frac{|\kappa|^2}{|\kappa|^2 + (\frac{1}{2}\Delta k)^2}\sin^2\{[|\kappa|^2 + (\tfrac{1}{2}\Delta k)^2]^{1/2}L\}. \tag{7.4.33}$$

We see that for $\Delta k \neq 0$ the maximum efficiency is always less than 100%. Let us examine the rate at which the efficiency decreases as the phase

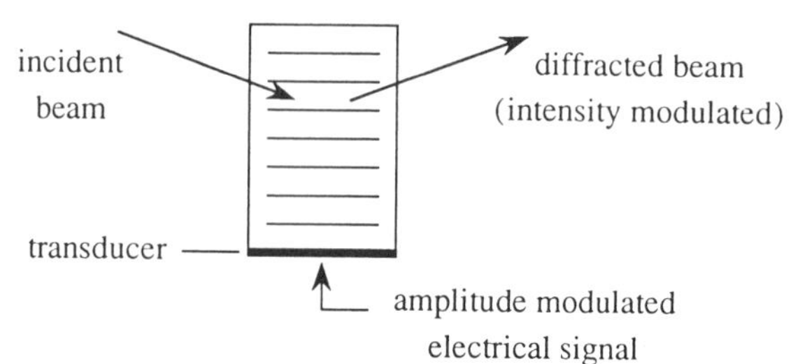

FIGURE 7.4.6 Acousto-optic amplitude modulator.

mismatch Δk is increased. We expand $\eta(\Delta k)$ as a power series in Δk as

$$\eta(\Delta k) = \eta(0) + \Delta k \left.\frac{d\eta}{d(\Delta k)}\right|_{\Delta k = 0} + \tfrac{1}{2}(\Delta k)^2 \left.\frac{d^2\eta}{d(\Delta k)^2}\right|_{\Delta k = 0} + \cdots. \tag{7.4.34}$$

By performing these differentiations, we find that, correct to second order in Δk, the efficiency is given by

$$\eta(\Delta k) = \eta(0)\left[1 - \frac{(\Delta k)^2}{4|\kappa|^2}\left(1 - \frac{|\kappa|L\cos(|\kappa|L)}{\sin(|\kappa|L)}\right)\right], \tag{7.4.35a}$$

where

$$\eta(0) = \sin^2(|\kappa|L). \tag{7.4.35b}$$

One common use of the Bragg acousto-optic effect is to produce an amplitude-modulated laser beam, as illustrated in Fig. 7.4.6. In such a device, the frequency of the electrical signal that is fed to the acoustic transducer is held fixed, but the amplitude of this wave is modulated. As a result, the depth of modulation of the acoustic grating is varied, leading to a modulation of the intensity of the scattered wave.

Another application of Bragg acousto-optic scattering is to produce a beam deflector (Fig. 7.4.7). In such a device, the frequency Ω of the electrical signal that is fed to the acoustic transducer is allowed to vary. As a result, the acoustic wavelength Λ is varied, and hence the diffraction angle θ_2 given by Eq. (7.4.29b) can be controlled. It should be noted the diffraction efficiency given by Eq. (7.4.33) decreases for diffraction at angles different from the Bragg angle, and this effect places limitations on the range of deflection angles that are achievable by means of this technique.

Raman–Nath Effect

The description of Bragg scattering given in the preceeding subsection implicitly assumed that the width L of the interaction region was sufficiently

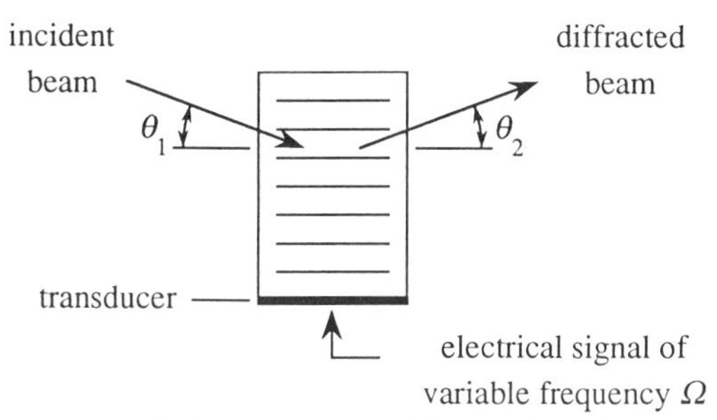

FIGURE 7.4.7 Acousto-optic beam deflector. The angle θ_2 depends on the frequency of Ω of the electrical signal.

large that an incident ray of light would interact with a large number of acoustic wavefronts. As illustrated in Fig. 7.4.8, this condition requires that

$$L \tan \theta_1 \gg \Lambda, \tag{7.4.36}$$

where Λ is the acoustic wavelength. However, the angle of incidence θ_1 must satisfy the Bragg condition

$$\sin \theta_1 = \frac{\lambda}{2\Lambda} \tag{7.4.37}$$

if efficient scattering is to occur. In most cases of interest, θ_1 is much smaller than unity, and hence $\tan \theta_1 \simeq \sin \theta_1$. Equation (7.4.37) can then be used to eliminate θ_1 from Eq. (7.4.36), which becomes

$$\frac{\lambda L}{\Lambda^2} \gg 1. \tag{7.4.38}$$

If this condition is satisfied, Bragg scattering can occur. Scattering in the opposite limit is known as Raman–Nath scattering.

Raman–Nath scattering can be understood in terms of the diagram shown in Fig. 7.4.9. A beam of light falls onto the scattering cell, typically at

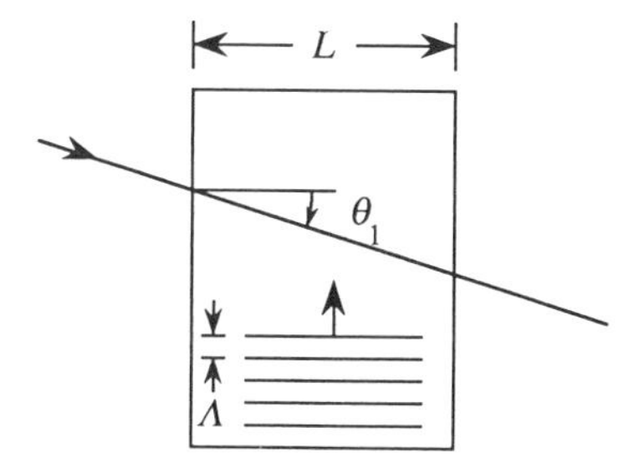

FIGURE 7.4.8 Illustration of the condition under which Bragg scattering occurs.

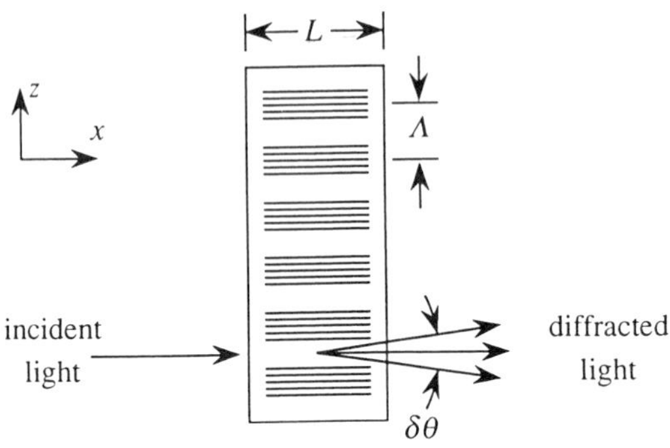

FIGURE 7.4.9 Raman–Nath diffraction.

near-normal incidence. Due to the presence of the acoustic wave, whose wavelength is denoted Λ, the refractive index of the medium varies spatially with period Λ. The incident light diffracts off this index grating; the characteristic angular spread of the diffracted light is

$$\delta\theta = \frac{\lambda}{\Lambda}. \tag{7.4.39}$$

We now assume that the cell is sufficiently thin that multiple scattering *cannot* occur. This condition can be stated as

$$\delta\theta \, L < \Lambda. \tag{7.4.40}$$

If $\delta\theta$ is eliminated from this inequality through use of Eq. (7.4.39), we find that

$$\frac{\lambda L}{\Lambda^2} < 1. \tag{7.4.41}$$

We note that this condition is the opposite of the inequality (7.4.38) for the occurrence of Bragg scattering.

We now present a mathematical analysis of Raman–Nath scattering. We assume that the acoustic wave within the scattering cell can be represented as the density variation

$$\Delta\tilde{\rho} = \Delta\rho \, e^{i(qz-\Omega t)} + \text{c.c.} \tag{7.4.42}$$

A refractive index variation

$$\Delta\tilde{n} = \Delta n \, e^{i(qz-\Omega t)} + \text{c.c.} \tag{7.4.43}$$

is associated with this acoustic wave. We relate the complex amplitude Δn of the refractive index disturbance to the amplitude $\Delta\rho$ of the acoustic wave as follows: We let $\tilde{n} = n_0 + \Delta\tilde{n}$, where $\tilde{n} = \tilde{\epsilon}^{1/2}$ with $\tilde{\epsilon} = \epsilon_0 + \Delta\tilde{\epsilon}$. We thus find that $n_0 = \epsilon_0^{1/2}$ and that $\Delta\tilde{n} = \Delta\tilde{\epsilon}/2n_0$. We now represent $\Delta\tilde{\epsilon}$ as $\Delta\tilde{\epsilon} = (\partial\epsilon/\partial\rho)\Delta\tilde{\rho} = \gamma_e\Delta\tilde{\rho}/\rho_0$ and find that $\Delta\tilde{n} = \gamma_e\Delta\tilde{\rho}/2n_0\rho_0$, and hence that

$$\Delta n = \frac{\gamma_e \Delta\rho}{2n_0\rho_0}. \tag{7.4.44}$$

The ensuing analysis is simplified by representing $\Delta\tilde{n}$ using real quantities; we assume that the phase conventions are chosen such that

$$\Delta\tilde{n}(z, t) = 2\,\Delta n \sin(qz - \Omega t). \tag{7.4.45}$$

The electric field of the incident optical wave is represented as

$$\tilde{E}(\mathbf{r}, t) = Ae^{i(kx - \omega t)} + \text{c.c.} \tag{7.4.46}$$

After passing through the acoustic wave, the optical field will have experienced a phase shift

$$\phi = \Delta\tilde{n}\frac{\omega}{c}L = 2\,\Delta n\frac{\omega}{c}L\sin(qz - \Omega t) \equiv \delta\sin(qz - \Omega t), \tag{7.4.47}$$

where the quantity

$$\delta = 2\,\Delta n\,\omega L/c \tag{7.4.48}$$

is known as the modulation index. The transmitted field can hence be represented as $\tilde{E}(\mathbf{r}, t) = A\exp[i(kx - \omega t + \phi)] + \text{c.c.}$, or as

$$\tilde{E}(\mathbf{r}, t) = Ae^{i[kx - \omega t + \delta\sin(qz - \Omega t)]} + \text{c.c.} \tag{7.4.49}$$

We see that the transmitted field is phase-modulated in time. To determine the consequences of this form of modulation, we note that Eq. (7.4.49) can be transformed through use of the Bessel function identity

$$e^{i\delta\sin y} = \sum_{l=-\infty}^{\infty} J_l(\delta)e^{ily} \tag{7.4.50}$$

so that the transmitted field can be expressed as

$$\tilde{E}(\mathbf{r}, t) = A\sum_{l=-\infty}^{\infty} J_l(\delta)e^{i[(kx + lqz) - (\omega + l\Omega)t]} + \text{c.c.} \tag{7.4.51}$$

We see that the transmitted field is a linear superposition of plane wave components with frequencies $\omega + l\Omega$ and wave vectors $\mathbf{k} + l\mathbf{q}$. As shown in Fig. 7.4.10 (for the case $l = 2$), the lth-order diffracted wave is emitted at angle

$$\theta_l = \tan^{-1}\left(\frac{lq}{k}\right) \simeq \frac{lq}{k} = \frac{l\lambda}{\Lambda}. \tag{7.4.52}$$

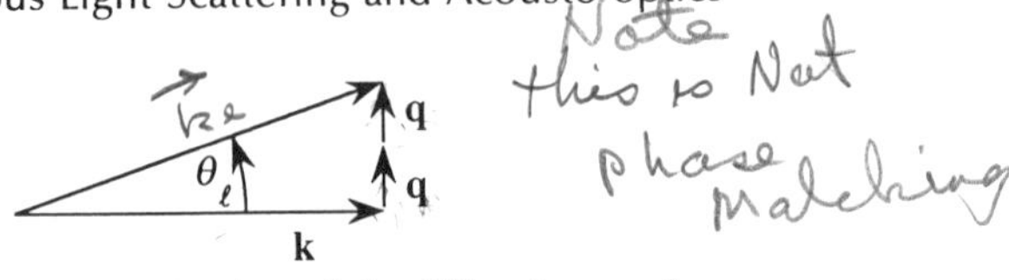

FIGURE 7.4.10 Determination of the diffraction angle.

The intensity of the light in this diffraction order is

$$I_l = |A|^2 J_l(\delta)^2, \tag{7.4.53}$$

where, as before, $\delta \equiv 2\,\Delta n\,(\omega/c)L$. Equations (7.4.48) and (7.4.53) constitute the Raman–Nath equations.

It is instructive to repeat this analysis for the case of a standing sound wave. For convenience, we take the resulting modulation of the refractive index to be of the form

$$\tilde{n}(z,t) = 2\,\Delta n \cos \Omega t \sin qz. \tag{7.4.54}$$

The phase shift induced in the optical wave is then given by

$$\begin{aligned} \phi &= 2\,\Delta n \frac{\omega}{c} L \cos \Omega t \sin qz \\ &\equiv \delta \cos \Omega t \sin qz, \end{aligned} \tag{7.4.55}$$

and the transmitted optical field is given by

$$\tilde{E}(\mathbf{r},t) = A e^{i(kx - \omega t + \delta \cos \Omega t \sin qz)} + \text{c.c.} \tag{7.4.56}$$

We now use the Bessel function identity (7.4.50) to transform the factor $\sin qz$ that appears in the exponent of this expression. We find that

$$\tilde{E}(\mathbf{r},t) = A \sum_{l=-\infty}^{\infty} J_l(\delta \cos \Omega t) \exp[i(kx + lqz) - i\omega t]. \tag{7.4.57}$$

We see that once again the transmitted field is composed of plane wave components; the lth diffracted order makes an angle

$$\theta_l \simeq \frac{lq}{k} = \frac{l\lambda}{\Lambda} \tag{7.4.58}$$

with the forward direction. The intensity of the lth order is now given by

$$I_l = |A|^2 J_l(\delta \cos \Omega t)^2. \tag{7.4.49}$$

We see that in this case each component is amplitude-modulated.

Problems

1. Estimate numerically, using the Lorentz model of the atom, the value of the scattering cross section for molecular nitrogen (N_2) for visible light at a wavelength of 500 nm. Use this result to estimate the value of the scattering coefficient R for air at STP. Compare this value with that obtained using Eq. (7.2.22) and the known refractive index of air. (The measured value for 90-degree scattering of unpolarized light, R_{90}^{u}, is approximately 2×10^{-8} cm^{-1}.) Also, estimate numerically the attenuation distance of light in air, that is, the propagation distance through which the intensity falls to $1/e$ of its initial value due to scattering losses.

   [Ans: $\alpha = (16\pi/3)R_{90}^{u} = 3 \times 10^{-7}$ cm^{-1} = (33 km)$^{-1}$.]

2. Through use of Eq. (7.3.9b) and handbook values of γ_e and C_T, estimate numerically the value of the scattering coefficient R for liquid water at room temperature for 90-degree scattering of visible light at a wavelength of 500 nm. Use this result to estimate the attenuation distance of light in water.

   [Ans: Using the values $C_T = 4.5 \times 10^{-11}$ cm^2/dyne, $n = 1.33$, $\gamma_e = (n^2 - 1)(n^2 + 2)/3 = 0.98$, and $T = 300$ K, we find that $R = 2.8 \times 10^{-6} \sin^2\phi$ cm^{-1}, and hence that for 90-degree scattering of unpolarized light $R_{90}^{u} = 1.4 \times 10^{-6}$ cm^{-1}. Thus the attenuation constant is given by $\alpha = (16\pi/3)R_{90}^{u} = 2.34 \times 10^{-5}$ cm^{-1} = (426 m)$^{-1}$.]

3. Verify Eq. (7.2.12).

4. Estimate numerically the value of the acoustic absorption coefficient α_s for propagation through water at frequencies of 10^3, 10^6, and 10^9 Hz.

   [Ans: The low-frequency shear viscosity coefficient of water is $\eta_s = 0.01$ dyne s/cm^2, and the Stokes relation tells us that $\eta_d = -\frac{2}{3}\eta_s$. We find that $\Gamma' = 0.66 \times 10^{-2}$ cm^2/s. Since $\alpha_s = q^2\Gamma'/v$ and $q = \Omega/v$, where $v = 1.5 \times 10^6$ cm/s, we find that $\alpha_s = 7.7 \times 10^{-14}$ cm^{-1} at 1 kHz and $\alpha_s = 7.7 \times 10^{-2}$ cm^{-1} at 1 GHz.]

5. Verify Eq. (7.4.8).

6. Verify Eqs. (7.4.31) through (7.4.35).

7. Consider an acousto-optic beam deflector. The incidence angle θ_1 remains fixed while the acoustic frequency Ω is varied to control the deflection angle θ_2. Derive a formula that predicts the maximum useful deflection angle, defined arbitrarily to be that deflection angle for which the diffraction efficiency drops to 50% of its maximum value. Evaluate this formula

numerically for the case treated following Eq. (7.4.26), where $|\kappa|L = \pi/2$, $L = 1.1$ cm, and $\Lambda = 30\ \mu$m.

[Ans: Starting with Eq. (7.4.25), and the readily derived relation $\Delta k = -\frac{1}{2}q\,\delta\theta$, we find that the efficiency drops by 50% when the incidence angle is increased by an amount

$$\delta\theta = \frac{2\sqrt{2}|\kappa|}{q}\left[1 - \frac{|\kappa|L\cos|\kappa|L}{\sin|\kappa|L}\right]^{-1/2}.$$

For the case $|\kappa|L = \pi/2$, where the efficiency for $\Delta k = 0$ is 100%, this result simplifies to

$$\delta\theta = \frac{2\sqrt{2}|\kappa|}{q}.$$

For the numerical example, $2\,\delta\theta = 0.22$ degree.]

References

I. L. Fabelinskii, *Molecular Scattering of Light*, Plenum Press, New York, 1968.

R. P. Feynman, R. B. Leighton, and M. Sands, *The Feynman Lectures on Physics*, Vol. I, Addison-Wesley, Reading Mass., 1963.

J. D. Jackson, *Classical Electrodynamics*, Wiley, New York, 1982.

L. D. Landau and E. M. Lifshitz, *Electrodynamics of Continuous Media*, Addison-Wesley, Reading, Mass., 1960; see especially Chapter 14.

L. D. Landau and E. M. Lifshitz, *Statistical Physics*, Addison-Wesley, Reading, Mass., 1969.

J. A. Stratton, *Electromagnetic Theory*, McGraw-Hill, New York, 1941.

A. Yariv, *Quantum Electronics*, Wiley, New York, 1975.

A. Yariv and P. Yeh, *Optical Waves in Crystals*, Wiley, New York, 1984.

Chapter 8

Stimulated Brillouin and Stimulated Rayleigh Scattering

8.1. Stimulated Scattering Processes

We saw in Section 7.1 of the previous chapter that light scattering can occur only as the result of fluctuations in the optical properties of a material system. A light scattering process is said to be *spontaneous* if the fluctuations (typically in the dielectric constant) that cause the light scattering are excited by thermal or by quantum-mechanical zero-point effects. In contrast, a light scattering process is said to be *stimulated* if the fluctuations are induced by the presence of the light field. Stimulated light scattering is typically very much more efficient than spontaneous light scattering. For example, approximately one part in 10^5 of the power contained in a beam of visible light would be scattered out of the beam by spontaneous scattering in passing through 1 cm of liquid water.* In this chapter, we shall see that when the intensity of the incident light is sufficiently large, essentially 100% of a beam of light can be scattered in a 1-cm path due to stimulated scattering processes.

In the present chapter we study stimulated light scattering due to induced density variations of a material system. The most important example of such a process is stimulated Brillouin scattering (SBS), which is illustrated schematically in Fig. 8.1.1. This figure shows an incident laser beam of frequency ω_L scattering from the refractive index variation associated with a sound wave of frequency Ω. Since the acoustic wavefronts are moving away from the incident laser wave, the scattered light is shifted downward in

* Recall that the scattering coefficient R is of the order of 10^{-6} cm^{-1} for water.

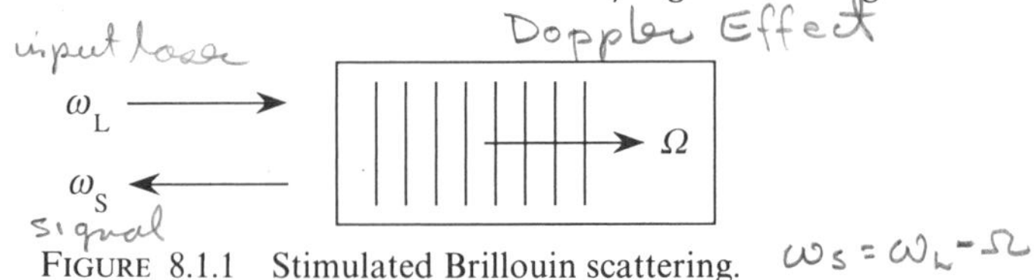

FIGURE 8.1.1 Stimulated Brillouin scattering.

frequency to the Stokes frequency $\omega_S = \omega_L - \Omega$. The reason why this interaction can lead to stimulated light scattering is that the interference of the laser and Stokes fields contains a frequency component at the difference frequency $\omega_L - \omega_S$, which of course is equal to the frequency Ω of the sound wave. The response of the material system to this interference term can act as a source that tends to increase the amplitude of the sound wave. Thus the beating of the laser wave with the sound wave tends to reinforce the Stokes wave, whereas the beating of the laser wave and Stokes waves tends to reinforce the sound wave. Under proper circumstances, the positive feedback described by these two interactions can lead to exponential growth of the amplitude of the Stokes wave.

There are two different physical mechanisms by which the interference of the laser and Stokes waves can drive the acoustic wave. One mechanism is electrostriction, that is, the tendency of materials to become more dense in regions of high optical intensity; this process is described in detail in the next section. The other mechanism is optical absorption. The heat evolved by absorption in regions of high optical intensity tends to cause the material to expand in those regions. The density variation induced by this effect can excite an acoustic disturbance. Absorptive SBS is less commonly used than electrostrictive SBS, since it can occur only in lossy optical media. For this reason we shall treat the electrostrictive case first and return to the case of absorptive coupling in Section 8.6 of this chapter.

There are two conceptually different configurations in which SBS can be studied. One is the SBS generator shown in part (a) of Fig. 8.1.2. In this configuration only the laser beam is applied externally, and both the Stokes and acoustic fields grow from noise within the interaction region. The noise process that initiates SBS is typically the scattering of laser light from thermally generated phonons. For the generator configuration, the Stokes radiation is created at frequencies near that for which the gain of the SBS process is largest. We shall see in Section 8.3 how to calculate this frequency.

Part (b) of Fig. 8.1.2 shows an SBS amplifier. In this configuration both the laser and Stokes fields are applied externally. Strong coupling occurs in this case only if the frequency of the injected Stokes wave is approximately equal to the frequency that would be created by an SBS generator.

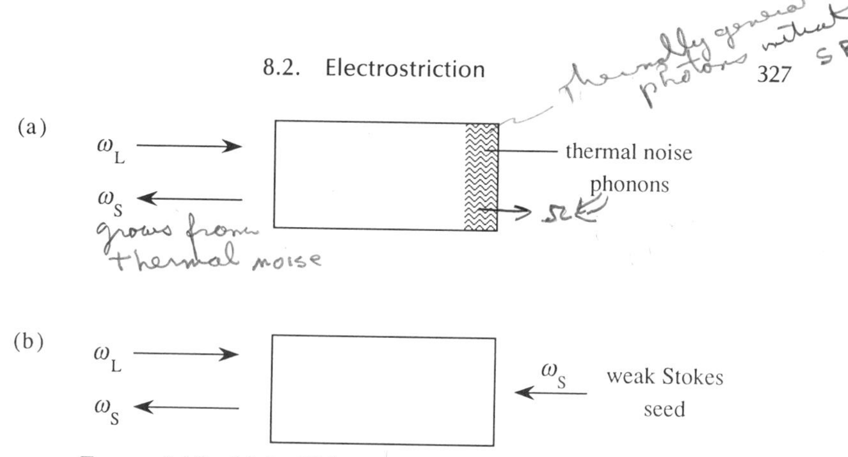

FIGURE 8.1.2 (a) An SBS generator; (b) an SBS amplifier.

In drawing Figs. 8.1.1 and 8.1.2, we have assumed that the laser and Stokes waves are counterpropagating. In fact, the SBS process leads to amplification of a Stokes wave propagating in any direction except for the propagation direction of the laser wave.* However, SBS is usually observed only in the backwards direction, because the spatial overlap of the laser and Stokes beams is largest under these conditions.

8.2. Electrostriction

Electrostriction is the tendency of materials to become compressed in the presence of an electric field.

The origin of the effect can be explained in terms of the behavior of a dielectric slab placed in the fringing field of a plane-parallel capacitor. As illustrated in part (a) of Fig. 8.2.1, the slab will experience a force tending to pull it into the region of maximum field strength. The nature of this force can be understood either globally or locally.

We can understand the origin of the electrostrictive force from a global point of view as being a consequence of the maximization of potential energy. The potential energy per unit volume of a material located in an electric field of field strength E is changed with respect to its value in the absence of the field by the amount

$$u = \frac{\epsilon E^2}{8\pi}, \tag{8.2.1}$$

* We shall see in Section 8.3 that copropagating laser and Stokes waves could interact only by means of acoustic waves of infinite wavelength, which cannot occur in a medium of finite spatial extent.

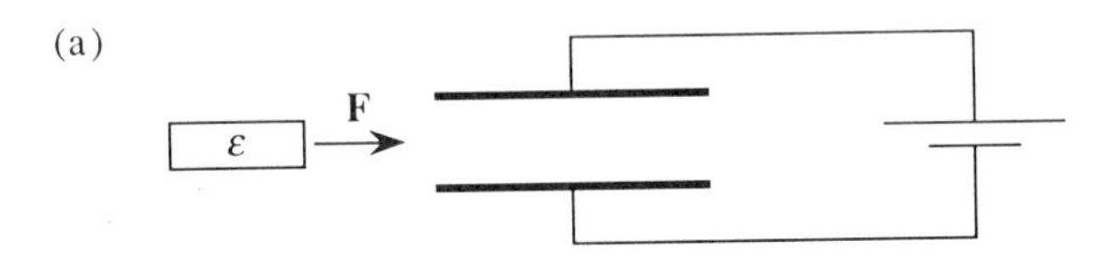

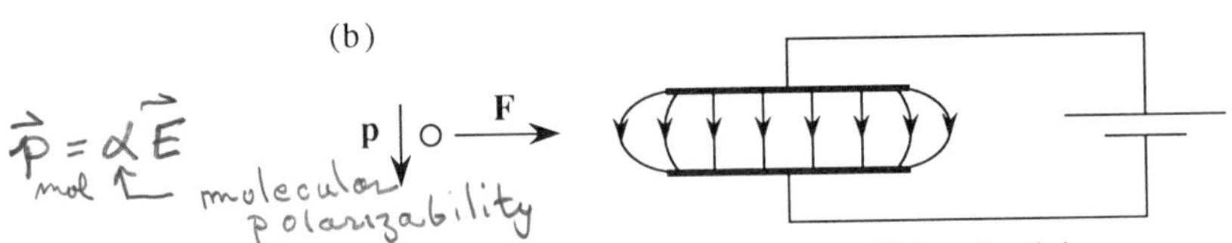

FIGURE 8.2.1 Origin of electrostriction: (a) a dielectric slab near a parallel plate capacitor; (b) a molecule near a parallel plate capacitor.

where ϵ is the dielectric constant of the material. Consequently the total potential energy of the system $\int u\, dV$, is maximized by allowing the slab to move into the region between the capacitor plates where the field strength is largest.

From a microscopic point of view, we can consider the force acting on an individual molecule placed in the fringing field of the capacitor, as shown in part (b) of Fig. 8.2.1. In the presence of the field **E**, the molecule develops the dipole moment $\mathbf{p} = \alpha\mathbf{E}$, where α is the molecular polarizability. The energy stored in the polarization of the molecule is given by

$$U = -\int_0^{\mathbf{E}} \mathbf{p} \cdot d\mathbf{E}' = -\int_0^{\mathbf{E}} \alpha\mathbf{E}' \cdot d\mathbf{E}' = -\tfrac{1}{2}\alpha\mathbf{E}\cdot\mathbf{E} \equiv -\tfrac{1}{2}\alpha E^2. \tag{8.2.2}$$

The force acting on the molecule is then given by

$$\mathbf{F} = -\nabla \mathrm{U} = \tfrac{1}{2}\alpha\nabla(E^2). \tag{8.2.3}$$

We see that each molecule is pulled into the region of increasing field strength.

Next we consider the situation illustrated in Fig. 8.2.2, in which the capacitor is immersed in the dielectric liquid. Molecules are pulled from the surrounding medium into the region between the capacitor plates, thus increasing the density in this region by an amount that we call $\Delta\rho$. We calculate the value of $\Delta\rho$ by means of the following argument: As a result of the increase in density of the material, its dielectric constant changes from its

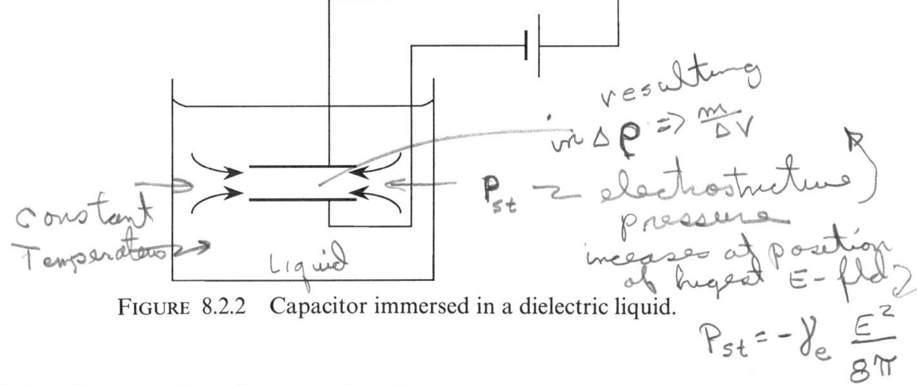

FIGURE 8.2.2 Capacitor immersed in a dielectric liquid.

origin value ϵ_0 to the value $\epsilon_0 + \Delta\epsilon$, where

$$\Delta\epsilon = \frac{\partial\epsilon}{\partial\rho}\Delta\rho. \tag{8.2.4}$$

Consequently the field energy density changes by the amount

$$\Delta u = \frac{E^2}{8\pi}\Delta\epsilon = \frac{E^2}{8\pi}\left(\frac{\partial\epsilon}{\partial\rho}\right)\Delta\rho. \tag{8.2.5}$$

However, according to the first law of thermodynamics, this change in energy Δu must be equal to the work Δw performed in compressing the material; the work done per unit volume is given by

$$\Delta w = p_{\text{st}}\frac{\Delta V}{V} = -p_{\text{st}}\frac{\Delta\rho}{\rho}. \tag{8.2.6}$$

Here p_{st} is the contribution to the pressure of the material that is due to the presence of the electric field. Since $\Delta u = \Delta w$, by equating Eqs. (8.2.5) and (8.2.6) we find that the electrostrictive pressure is given by

$$p_{\text{st}} = -\left(\rho\frac{\partial\epsilon}{\partial\rho}\right)\frac{E^2}{8\pi} \equiv -\gamma_e\frac{E^2}{8\pi}, \tag{8.2.7}$$

where $\gamma_e = \rho\,\partial\epsilon/\partial\rho$ is known as the electrostrictive constant (see also Eq. (7.3.6)). Since p_{st} is negative, the total pressure is reduced in regions of high field strength. The fluid tends to be drawn into these regions, and the density is increased. We calculate the change in density as $\Delta\rho = -(\partial\rho/\partial p)\,\Delta p$, where we equate Δp with the electrostrictive pressure of Eq. (8.2.7). We write this result as

$$\Delta\rho = -\rho\left(\frac{1}{\rho}\frac{\partial\rho}{\partial p}\right)p_{\text{st}} \equiv -\rho C p_{\text{st}}, \tag{8.2.8}$$

where $C = (1/\rho)\,\partial\rho/\partial p$ is the compressibility. Combining this result with Eq. (8.2.7), we find that

$$\Delta\rho = \rho C \gamma_e \frac{E^2}{8\pi}. \tag{8.2.9}$$

The derivation of this expression for $\Delta\rho$ has implicitly assumed that the electric field E is a static field. In such a case, the derivatives that appear in the expressions for C and γ_e are to be performed with the temperature T held constant. However, our primary interest is in the case where E represents an optical frequency field; in such a case Eq. (8.2.9) should be replaced by

$$\Delta\rho = \rho C \gamma_e \frac{\langle \tilde{\mathbf{E}}^2 \rangle}{8\pi}, \tag{8.2.10}$$

where the angular brackets denote a time average over an optical period. If $\tilde{\mathbf{E}}(t)$ contains more than one frequency component, so that $\langle \tilde{\mathbf{E}}^2 \rangle$ contains both static components and hypersonic components (as in the case of SBS), C and γ_e should be evaluated at constant entropy to determine the response for the hypersonic components and at constant temperature to determine the response for the static components.

Let us consider the modification of the optical properties of a material system that occurs as a result of electrostriction. We represent the change in the susceptibility in the presence of an optical field as $\Delta\chi = \Delta\epsilon/4\pi$, where $\Delta\epsilon$ is calculated as $(\partial\epsilon/\partial\rho)\Delta\rho$, with $\Delta\rho$ given by Eq. (8.2.10). We thus find that

$$\Delta\chi = \frac{1}{32\pi^2} C \gamma_e^2 \langle \tilde{\mathbf{E}}^2 \rangle. \tag{8.2.11}$$

For simplicity, we consider the case of the monochromatic field

$$\tilde{\mathbf{E}}(t) = \mathbf{E} e^{-i\omega t} + \text{c.c.}; \tag{8.2.12}$$

the case in which $\tilde{\mathbf{E}}(t)$ contains two frequency components that differ by approximately the Brillouin frequency is treated in the next section, on SBS. Then, since $\langle \tilde{\mathbf{E}}^2 \rangle = 2\mathbf{E}\cdot\mathbf{E}^*$, we see that

$$\Delta\chi = \frac{1}{16\pi^2} C_T \gamma_e^2 \mathbf{E}\cdot\mathbf{E}^*. \tag{8.2.13}$$

The complex amplitude of the nonlinear polarization that results from this change in the susceptibility can be represented as $\mathbf{P} = \Delta\chi\,\mathbf{E}$, that is, as

$$\mathbf{P} = \frac{1}{16\pi^2} C_T \gamma_e^2 |\mathbf{E}|^2 \mathbf{E}. \tag{8.2.14}$$

If we write this result in terms of a conventional third-order susceptibility, defined through

$$\mathbf{P} = 3\chi^{(3)}(\omega = \omega + \omega - \omega)|\mathbf{E}|^2\mathbf{E}, \tag{8.2.15}$$

we find that

$$\chi^{(3)}(\omega = \omega + \omega - \omega) = \frac{1}{48\pi^2} C_T \gamma_e^2. \tag{8.2.16}$$

For simplicity, we have suppressed the tensor nature of the nonlinear susceptibility in the foregoing discussion. However, we can see from the form of Eq. (8.2.14) that, for an isotropic material, the nonlinear coefficients of Maker and Terhune (see Eq. (4.2.10)) have the form $A = C_T\gamma_e^2/16\pi^2$ and $B = 0$. Let us estimate numerically the value of $\chi^{(3)}$. We saw in Eq. (7.3.12) that for a dilute gas the electrostrictive constant $\gamma_e \equiv \rho\, \partial\epsilon/\partial\rho$ is given by $\gamma_e = n^2 - 1$. More generally, we can estimate γ_e through use of the Lorentz–Lorenz law (Eq. 3.8.8a),which leads to the prediction

$$\gamma_e = (n^2 - 1)(n^2 + 2)/3. \tag{8.2.17}$$

This result shows that γ_e will be of the order of unity for condensed matter. The compressibility $C_T = (\partial\rho/\partial p)/\rho$ is approximately equal to 10^{-10} cm^2/dyne for CS_2 and is of the same order of magnitude for all condensed matter. We thus find that $\chi^{(3)}(\omega = \omega + \omega - \omega)$ is of the order of 2×10^{-13} esu for condensed matter. For ideal gases, the compressibility C_T is equal to $1/p$, where at one atmosphere $p = 10^6$ dyne/cm^2. The electrostrictive constant $\gamma_e = n^2 - 1$ for air at one atmosphere is approximately equal to 6×10^{-4}. We thus find that $\chi^{(3)}(\omega = \omega + \omega - \omega)$ is of the order of 1×10^{-15} for gases at one atmosphere of pressure.

In comparison with other types of optical nonlinearities, the value of $\chi^{(3)}$ resulting from electrostriction is not unusually large. However, we shall see in the next section that electrostriction provides the nonlinear coupling that leads to stimulated Brillouin scattering, which is often an extremely strong process.

8.3. Stimulated Brillouin Scattering (Induced by Electrostriction)*

Our discussion of spontaneous Brillouin scattering in Chapter 7 presupposes that the applied optical fields are sufficiently weak that they do not alter the

* Stimulated Brillouin scattering can also be induced by absorptive effects. This less commonly studied case is examined in Section 8.6.

acoustic properties of the material system. Spontaneous Brillouin scattering then results from the scattering of the incident radiation off the sound waves that are present in thermal equilibrium.

For an incident laser field of sufficient intensity, even the spontaneously scattered light can become quite intense. The incident and scattered light fields can then beat together, giving rise to density and pressure variations by means of electrostriction. The incident laser field can then scatter off the refractive index variation that accompanies these density variations. The scattered light will be at the Stokes frequency and will add constructively with the Stokes radiation that produced the acoustic disturbance. In this manner, the acoustic and Stokes waves mutually reinforce each other's growth, and each can grow to a large amplitude. This circumstance is depicted in Fig. 8.3.1. Here an incident wave of amplitude E_1, angular frequency ω_1, and wave vector $\mathbf{k}_1$ scatters off a retreating sound wave of amplitude ρ, frequency Ω, and wave vector $\mathbf{q}$ to form a scattered wave of amplitude E_2, frequency ω_2, and wave vector $\mathbf{k}_2$.*

Let us next deduce the frequency ω_2 of the Stokes field that is created by the SBS process for the case of an SBS generator (see also part (a) of Fig. 8.1.2). Since the laser field at frequency ω_1 is scattered from a retreating sound wave, the scattered radiation will be shifted downward in frequency to

$$\omega_2 = \omega_1 - \Omega_B. \tag{8.3.1}$$

Here Ω_B is called the Brillouin frequency, and its value will be determined in order to find ω_2. The Brillouin frequency is related to the acoustic wave vector

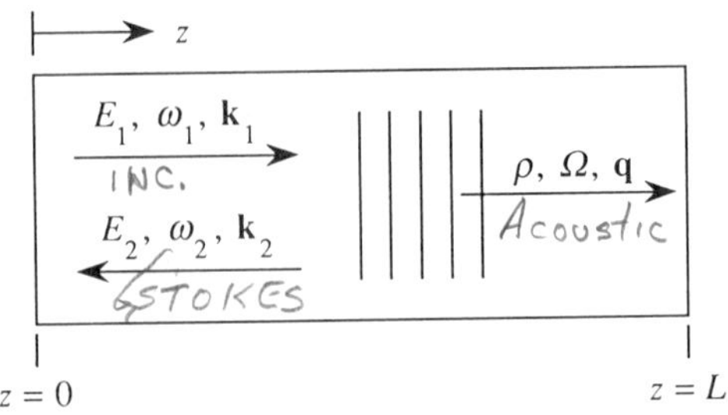

FIGURE 8.3.1 Schematic representation of the stimulated Brillouin scattering process.

* We denote the field frequencies as ω_1 and ω_2 rather than ω_L and ω_S so that we can later apply the results of the present treatment to the case of anti-Stokes scattering by identifying ω_1 with ω_{AS} and ω_2 with ω_L. The treatment of the present section assumes only that $\omega_2 < \omega_1$.

$\mathbf{q}_B$ by the phonon dispersion relation

$$\Omega_B = |\mathbf{q}_B| v, \tag{8.3.2}$$

where v is the velocity of sound. By assumption, this sound wave is driven by the beating of the laser and Stokes fields, and its wave vector is therefore given by

$$\mathbf{q}_B = \mathbf{k}_1 - \mathbf{k}_2. \tag{8.3.3}$$

Since the wave vectors and frequencies of the optical waves are related in the usual manner, that is, by $|\mathbf{k}_i| = n\omega_i/c$, we can use Eq. (8.3.3) and the fact that the laser and Stokes waves are counterpropagating to express the Brillouin frequency of Eq. (8.3.2) as

$$\Omega_B = \frac{nv}{c}(\omega_1 + \omega_2). \tag{8.3.4}$$

Equation (8.3.1) and (8.3.4) are now solved simultaneously to obtain an expression for the Brillouin frequency in terms of the frequency ω_1 of the applied field only, that is, we eliminate ω_2 from these equations to obtain

$$\Omega_B = \frac{\dfrac{2nv}{c}\omega_1}{\left(1 + \dfrac{nv}{c}\right)}. \tag{8.3.5}$$

However, since v is very much smaller than c/n for all known materials, it is an excellent approximation to take the Brillouin frequency to be

$$\Omega_B = \frac{2nv}{c}\omega_1. \tag{8.3.6}$$

At this same level of approximation, the acoustic wave vector is given by

$$\mathbf{q}_B = 2\mathbf{k}_1. \tag{8.3.7}$$

For the case of the SBS amplifier configuration (see part (b) of Fig. 8.1.2), the Stokes wave is imposed externally and its frequency ω_2 is known *a priori*. The frequency of the driven acoustic wave will then be given by

$$\Omega = \omega_1 - \omega_2, \tag{8.3.8}$$

which in general will be different from the Brillouin frequency of Eq. (8.3.6). As we shall see below, the acoustic wave will be excited efficiently under these circumstances only when ω_2 is chosen such that the frequency difference $|\Omega - \Omega_B|$ is less than or of the order of the Brillouin linewidth Γ_B.

Let us next see how to treat the nonlinear coupling among the three interacting waves. We represent the optical field within the Brillouin medium as $\tilde{E}(z,t) = \tilde{E}_1(z,t) + \tilde{E}_2(z,t)$, where

$$\tilde{E}_1(z,t) = A_1(z,t)e^{i(k_1 z - \omega_1 t)} + \text{c.c.} \tag{8.3.9a}$$

and

$$\tilde{E}_2(z,t) = A_2(z,t)e^{i(-k_2 z - \omega_2 t)} + \text{c.c.} \tag{8.3.9b}$$

Similarly, we describe the acoustic field in terms of the material density distribution

$$\tilde{\rho}(z,t) = \rho_0 + [\rho(z,t)e^{i(qz - \Omega t)} + \text{c.c.}], \tag{8.3.10}$$

where $\Omega = \omega_1 - \omega_2$, $q = 2k_1$, and ρ_0 denotes the mean density of the medium.

We assume that the material density obeys the acoustic wave equation (see also Eq. (7.3.17))

$$\frac{\partial^2 \tilde{\rho}}{\partial t^2} - \Gamma' \nabla^2 \frac{\partial \tilde{\rho}}{\partial t} - v^2 \nabla^2 \tilde{\rho} = \nabla \cdot \mathbf{f}, \tag{8.3.11}$$

where v is the velocity of sound and Γ' is a damping parameter given by Eq. (7.3.23). The source term on the right-hand side of this equation consists of the divergence of the force per unit volume **f**, which is given explicitly by

$$\mathbf{f} = \nabla p_{\text{st}}, \qquad p_{\text{st}} = -\gamma_e \frac{\langle \tilde{E}^2 \rangle}{8\pi}. \tag{8.3.12}$$

For the fields given by Eq. (8.3.9), this source term is given by

$$\nabla \cdot \mathbf{f} = \frac{\gamma_e q^2}{4\pi} [A_1 A_2^* e^{i(qz - \Omega t)} + \text{c.c.}]. \tag{8.3.13}$$

If we now introduce Eqs. (8.3.10) and (8.3.13) into the acoustic wave equation (8.3.11) and assume that the acoustic amplitude varies slowly (if at all) in space and time, we obtain the result

$$-2i\Omega \frac{\partial \rho}{\partial t} + (\Omega_B^2 - \Omega^2 - i\Omega\Gamma_B)\rho - 2iqv^2 \frac{\partial \rho}{\partial z} = \frac{\gamma_e q^2}{4\pi} A_1 A_2^*, \tag{8.3.14a}$$

where we have introduced the Brillouin linewidth

$$\Gamma_B = q^2 \Gamma'; \tag{8.3.14b}$$

its reciprocal $\tau_p = \Gamma_B^{-1}$ gives the phonon lifetime.

Eq. (8.3.14a) can often be simplified substantially by omitting the last term on its the left-hand side. This term describes the propagation of phonons. However, hypersonic phonons are strongly damped and thus propagate only over very short distances before being absorbed.* Since the phonon propagation distance is typically small compared to the distance over which the source term on the right-hand side of Eq. (8.3.14) varies significantly, it is conventional to drop the term containing $\partial\rho/\partial z$ in describing SBS. This approximation can break down, however, as discussed by Chiao (1965) and by Kroll and Kelley (1971). If we drop the spatial derivative term in Eq. (8.3.14) and assume steady-state conditions so that $\partial\rho/\partial t$ also vanishes, we find that the acoustic amplitude is given by

$$\rho(z, t) = \frac{\gamma_e q^2}{4\pi} \frac{A_1 A_2^*}{\Omega_B^2 - \Omega^2 - i\Omega\Gamma_B}. \tag{8.3.15}$$

The spatial evolution of the optical fields is described by the wave equation

$$\frac{\partial^2 \tilde{E}_i}{\partial z^2} - \frac{1}{(c/n)^2} \frac{\partial^2 \tilde{E}_i}{\partial t^2} = \frac{4\pi}{c^2} \frac{\partial^2 \tilde{P}_i}{\partial t^2}, \qquad i = 1, 2. \tag{8.3.16}$$

The nonlinear polarization, which acts as a source term in this equation, is given by

$$\tilde{P} = \Delta\chi\, \tilde{E} = \frac{\Delta\epsilon}{4\pi} \tilde{E} = \frac{1}{4\pi\rho_0} \gamma_e \tilde{\rho} \tilde{E}. \tag{8.3.17}$$

We next determine the components of $\tilde{P}$ that can act as phase-matched source terms for the laser and Stokes fields. These components are given by

$$\tilde{P}_1 = p_1 e^{i(k_1 z - \omega_1 t)} + \text{c.c.}, \qquad \tilde{P}_2 = p_2 e^{i(-k_2 z - \omega_2 t)} + \text{c.c.}, \tag{8.3.18}$$

where

$$p_1 = \frac{\gamma_e}{4\pi\rho_0} \rho A_2, \qquad p_2 = \frac{\gamma_e}{4\pi\rho_0} \rho^* A_1. \tag{8.3.19}$$

We introduce Eqs. (8.3.9) into the wave equation (8.3.16) along with Eqs. (8.3.18) and (8.3.19), make the slowly-varying-amplitude approximation, and obtain the equations

$$\frac{\partial A_1}{\partial z} + \frac{1}{c/n} \frac{\partial A_1}{\partial t} = \frac{i\omega\gamma_e}{2nc\rho_0} \rho A_2, \tag{8.3.20a}$$

$$-\frac{\partial A_2}{\partial z} + \frac{1}{c/n} \frac{\partial A_2}{\partial t} = \frac{i\omega\gamma_e}{2nc\rho_0} \rho^* A_1. \tag{8.3.20b}$$

* We can estimate this distance as follows: According to Eq. (7.3.30), the sound absorption coefficient is given by $\alpha_s = \Gamma_B/v$, where by Eqs. (7.3.23) and (7.3.28) Γ_B is of the order of $\eta_s q^2/\rho_0$. For the typical values $v = 1 \times 10^5$ cm/s, $\eta_s = 10^{-2}$ dyne cm/s^2, $q = 4\pi \times 10^4$ cm^{-1}, and $\rho = 1$ cm^{-3}, we find that $\Gamma_B = 1.6 \times 10^8$ s^{-1} and $\alpha_s^{-1} = 6.3$ μm.

In these equations ρ is given by the solution to Eq. (8.3.14). Furthermore, we have dropped the distinction between ω_1 and ω_2 by setting $\omega = \omega_1 \simeq \omega_2$.

Let us now consider steady-state conditions. In this case the time derivatives appearing in Eqs. (8.3.20) can be dropped, and ρ is given by Eq. (8.3.15). The coupled-amplitude equations then become

$$\frac{dA_1}{dz} = \frac{i\omega q^2\gamma_e^2}{8\pi nc\rho_0} \frac{|A_2|^2 A_1}{\Omega_B^2 - \Omega^2 - i\Omega\Gamma_B}, \tag{8.3.21a}$$

$$\frac{dA_2}{dz} = \frac{-i\omega q^2\gamma_e^2}{8\pi nc\rho_0} \frac{|A_1|^2 A_2}{\Omega_B^2 - \Omega^2 + i\Omega\Gamma_B}. \tag{8.3.21b}$$

We see from the form of these equations that SBS is a pure gain process, that is, that the SBS process is automatically phase-matched. For this reason, it is possible to introduce coupled equations for the intensities of the two interacting optical waves. Defining the intensities as $I_i = (nc/2\pi)A_i A_i^*$, we find from Eqs. (8.3.21) that

$$\frac{dI_1}{dz} = -gI_1I_2 \tag{8.3.22a}$$

and

$$\frac{dI_2}{dz} = -gI_1I_2. \tag{8.3.22b}$$

In these equations g is the SBS gain factor, which to good approximation is given by

$$g = g_0 \frac{(\Gamma_B/2)^2}{(\Omega_B - \Omega)^2 + (\Gamma_B/2)^2}, \tag{8.3.23}$$

where the line-center gain factor is given by

$$g_0 = \frac{\gamma_e^2\omega^2}{nvc^3\rho_0\Gamma_B}. \tag{8.3.24}$$

The solution to Eqs. (8.3.22) under general conditions will be described below. Note, however, that in the constant-pump limit I_1 = constant, the solution to Eq. (8.3.22b) is

$$I_2(z) = I_2(L)e^{gI_1(L-z)}. \tag{8.3.25}$$

In this limit a Stokes wave injected into the medium at $z = L$ experiences exponential growth as it propagates through the medium. It should be noted that the line-center gain factor of Eq. (8.3.24) is actually independent of the laser frequency ω, since the Brillouin linewidth Γ_B is proportional to ω^2 (recall that, according to Eq. (7.3.28), Γ_B is proportional to q^2 and that q is proportional to ω). An estimate of the size of g_0 for the case of CS_2 at a wavelength of 1 μm can be made as follows: $\omega = 2\pi \times 3 \times 10^{14}$ rad/s, $n = 1.67$, $v = 1.1 \times 10^5$ cm/s, $\rho_0 = 1.26$ g/cm^3, $\gamma_e = 2.4$, and $\tau_p = \Gamma_B^{-1} = 4 \times 10^{-9}$ s, giving $g_0 = 1.5 \times 10^{-14}$ cm s/erg, which in conventional laboratory units becomes $g_0 = 0.15$ cm/MW. The Brillouin gain factors and spontaneous linewidths $\Delta\nu = \Gamma_B/2\pi$ are listed in Table 8.3.1 for a variety of materials.

The theoretical treatment just presented can also be used to describe the propagation of a wave at the anti-Stokes frequency, $\omega_{as} \simeq \omega_L + \Omega_B$. Equations (8.3.22) were derived for the geometry of Fig. 8.3.1 under the assumption that $\omega_1 > \omega_2$. We can treat anti-Stokes scattering by identifying ω_1 with ω_{as} and ω_2 with ω_L. We then find that the constant-pump approximation corresponds to the case $I_2(z) =$ constant, and that the solution to Eq. (8.3.22a) is $I_1(z) = I_1(0)e^{-gI_2z}$. Since the anti-Stokes wave at frequency ω_1 propagates in the positive z direction, we see that it experiences attenuation due to the SBS process.

Pump Depletion Effects in SBS

We have seen (Eq. 8.3.25) that, in the approximation in which the pump intensity is taken to be spatially invariant, the Stokes wave experiences

TABLE 8.3.1 Properties of stimulated Brillouin scattering for a variety of materials*

Substance	$\Omega_B/2\pi$ (MHz)	$\Gamma/2\pi$ (MHz)	g_0 (cm/MW)	$g_B^a(\max)/\alpha$ (cm^2/MW)
CS_2	5850	52.3	0.15	0.14
Acetone	4600	224	0.02	0.022
Toluene	5910	579	0.013	
CCl_4	4390	520	0.006	0.013
Methanol	4250	250	0.013	0.013
Ethanol	4550	353	0.012	0.010
Benzene	6470	289	0.018	0.024
H_2O	5690	317	0.0048	0.0008
Cyclohexane	5550	774	0.0068	
CH_4 (140 atm)	150	10	0.1	
Optical glasses	11,000–16,000	10–106	0.004–0.025	
SiO_2	17,000	78	0.0045	

* Values are quoted for a wavelength of 0.694 μm. The last column gives a parameter used to describe the process of absorptive SBS, which is discussed in Section 8.6. To convert to other laser frequencies ω, recall that Ω_B is proportional to ω, Γ is proportional to ω^2, g_0 is independent of ω, and $g_B^a(\max)$ is proportional to ω^{-3}.

exponential growth as it propagates through the Brillouin medium. Once the Stokes wave has grown to an intensity comparable to that of the pump wave, significant depletion of the pump wave must occur, and under these conditions we must solve the coupled-intensity equations (8.3.22) simultaneously in order to describe the SBS process. We see from these equations that $dI_1/dz = dI_2/dz$ and hence that

$$I_1(z) = I_2(z) + C, \tag{8.3.26}$$

where the value of the integration constant C depends on the boundary conditions. Using this result, Eq. (8.3.22b) can be expressed as

$$\frac{dI_2}{I_2(I_2 + C)} = -g\,dz. \tag{8.3.27}$$

This equation can be integrated formally as

$$\int_{I_2(0)}^{I_2(z)} \frac{dI_2}{I_2(I_2 + C)} = -\int_0^z g\,dz', \tag{8.3.28}$$

which implies that

$$\ln\left\{\frac{I_2(z)[I_2(0) + C]}{I_2(0)[I_2(z) + C]}\right\} = -gCz. \tag{8.3.29}$$

Since we have specified the value of I_1 at $z = 0$, it is convenient to express the constant C defined by Eq. (8.3.26) as $C = I_1(0) - I_2(0)$. Equation (8.3.29) is now solved algebraically for $I_2(z)$, yielding

$$I_2(z) = \frac{I_2(0)[I_1(0) - I_2(0)]}{I_1(0)\exp\{gz[I_1(0) - I_2(0)]\} - I_2(0)}. \tag{8.3.30a}$$

According to Eq. (8.3.26), $I_1(z)$ can be found in terms of this expression as

$$I_1(z) = I_2(z) + I_1(0) - I_2(0). \tag{8.3.30b}$$

Equation (8.3.30) give the spatial distribution of the field intensities in terms of the boundary values $I_1(0)$ and $I_2(0)$. However, the boundary values that are known physically are $I_1(0)$ and $I_2(L)$; see Fig. 8.3.2. In order to find the unknown quantity $I_2(0)$ in terms of the known quantities $I_1(0)$ and $I_2(L)$, we set z equal to L in Eq. (8.3.30a) and write the resulting expression as follows:

$$I_2(L) = \frac{I_1(0)[I_2(0)/I_1(0)][1 - I_2(0)/I_1(0)]}{\exp\{gI_1(0)L[1 - I_2(0)/I_1(0)]\} - I_2(0)/I_1(0)}. \tag{8.3.31}$$

This expression is a transcendental equation giving the unknown quantity $I_2(0)/I_1(0)$ in terms of the known quantities $I_1(0)$ and $I_2(L)$.

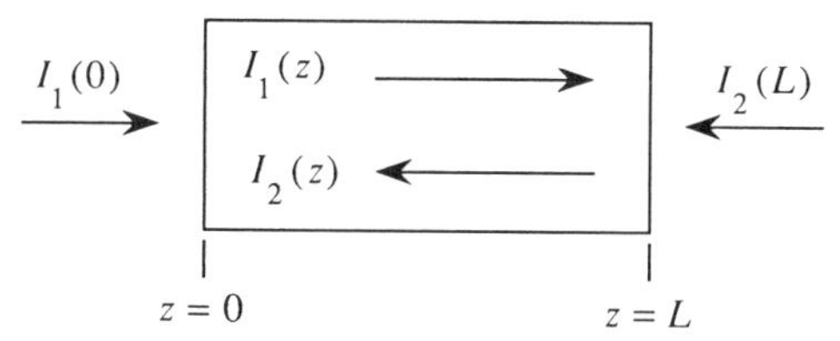

FIGURE 8.3.2 Geometry of an SBS amplifier. The boundary value $I_1(0)$ and $I_2(L)$ are known.

The results given by Eqs. (8.3.30) and (8.3.31) can be used to analyze the SBS amplifier shown in Fig. 8.3.2. The transfer characteristics of such an amplifier are illustrated in Fig. 8.3.3. Here the vertical axis gives the fraction of the laser intensity that is transferred to the Stokes wave, and the horizontal axis gives the exponential gain $G = gI_1(0)L$ experienced by a *weak* Stokes input. The various curves are labeled according to the ratio of input intensities, $I_2(L)/I_1(0)$. For sufficiently large values of the exponential gain, essentially complete transfer of the pump energy to the Stokes beam is possible.

SBS Generator

For the case of an SBS generator, no Stokes field is injected externally into the interaction region, and hence the value of the Stokes intensity near the Stokes input face $z = L$ is not known *a priori*. In this case, the SBS process is initiated by Stokes photons that are created by spontaneous Brillouin scattering involving the laser beam near its exit plane $z = L$. We therefore expect that the effective Stokes input intensity $I_2(L)$ will be proportional to the local value of

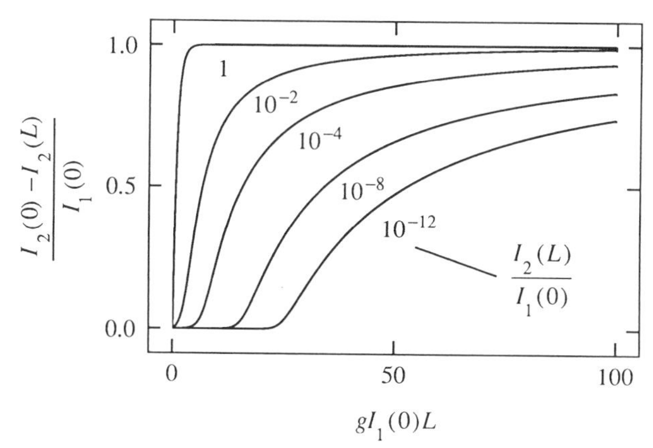

FIGURE 8.3.3 Intensity transfer characteristics of an SBS amplifier.

the laser intensity $I_1(L)$; we designate the constant of proportionality as f, so that

$$I_2(L) = fI_1(L). \tag{8.3.32}$$

We estimate the value of f as follows: We consider the conditions that apply near but below the threshold for SBS, such that the SBS reflectivity $R = I_2(0)/I_1(0)$ is much smaller than unity. Under these conditions the laser intensity is essentially constant throughout the medium, and the Stokes output intensity is related to the Stokes input intensity by $I_2(0) = I_2(L)e^G$, where $G = gI_1(0)L$. However, since $I_2(L) = fI_1(0)$ (since $I_1(z)$ is constant), the SBS reflectivity can be expressed as

$$R \equiv \frac{I_2(0)}{I_1(0)} = fe^G. \tag{8.3.33}$$

Laboratory experience has shown that the SBS threshold condition (i.e., R of the order of unity) requires that G attain a specific value G_{th}, which for most material systems is approximately 25–30. We thus see from Eq. (8.3.33) that f is of the order of $\exp(-G_{th})$, or approximately 10^{-12} to 10^{-11}. An order-of-magnitude estimate based on the properties of spontaneous scattering performed by Zel'dovich *et al.* (1985) reaches the same conclusion.

We next calculate the SBS reflectivity R for the general case $G > G_{th}$ (i.e., above threshold) through use of Eq. (8.3.31), which we write as

$$\frac{I_2(L)}{I_1(0)} = \frac{R(1-R)}{\exp[G(1-R)] - R}. \tag{8.3.34}$$

To good approximation, the term $-R$ can be dropped from the denominator of the right-hand side of this equation. In order to determine the ratio $I_2(L)/I_1(0)$ that appears on the left-hand side of Eq. (8.3.34), we express Eq. (8.3.30b) as

$$I_1(L) - I_2(L) = I_1(0) - I_2(0).$$

Through use of Eq. (8.3.32) and the smallness of f, we can replace the left-hand side of this equation by $f^{-1}I_2(L)$. We now multiply both sides of the resulting equation by $f/I_1(0)$ to obtain the result $I_2(L)/I_1(0) = f(1-R)$. This expression is substituted for the left-hand side of Eq. (8.3.34), which is then solved for G, yielding the result

$$\frac{G}{G_{th}} = \frac{G_{th}^{-1}\ln R + 1}{1-R}, \tag{8.3.35}$$

where we have substituted G_{th} for $-\ln f$.

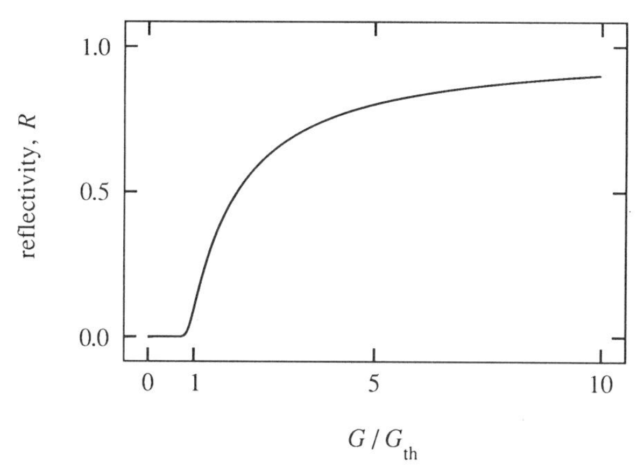

FIGURE 8.3.4 Dependence of the SBS reflectivity on the weak-signal gain $G = gI_1(0)L$.

The nature of this solution is illustrated in Fig. 8.3.4, where the SBS reflectivity $R = I_2(0)/I_1(0)$ is shown plotted as a function of $G = gI_f(0)L$ for the value $G_{\text{th}} = 25$. We see that essentially no Stokes light is created for G less than G_{th} and that the reflectivity rises rapidly for laser intensitites slightly above this threshold value. In addition, for $G \gg G_{\text{th}}$ the reflectivity asympototically approaches 100%. Well above the threshold for SBS (i.e., for $G \gtrsim 3G_{\text{th}}$), Eq. (8.3.35) can be approximated as $G/G_{\text{th}} \simeq 1/(1-R)$, which shows that the SBS reflectivity in this limit can be expressed as

$$R = 1 - \frac{1}{G/G_{\text{th}}} \quad (\text{for } G \gg G_{\text{th}}). \tag{8.3.36}$$

Since the intensity $I_1(L)$ of the transmitted laser beam is given by $I_1(L) = I_1(0)(1-R)$, in the limit of validity of Eq. (8.3.36) the intensity of the transmitted beam is given by

$$I_1(L) = \frac{G_{\text{th}}}{gL}; \tag{8.3.37}$$

here G_{th}/gL can be interpreted as the input laser intensity at the threshold for SBS. Hence the transmitted intensity is "clamped" at the threshold value for the occurrence of SBS.

Once the value of the Stokes intensity at the plane $z = 0$ is known from Eq. (8.3.35), the distributions of the intensities within the interaction region

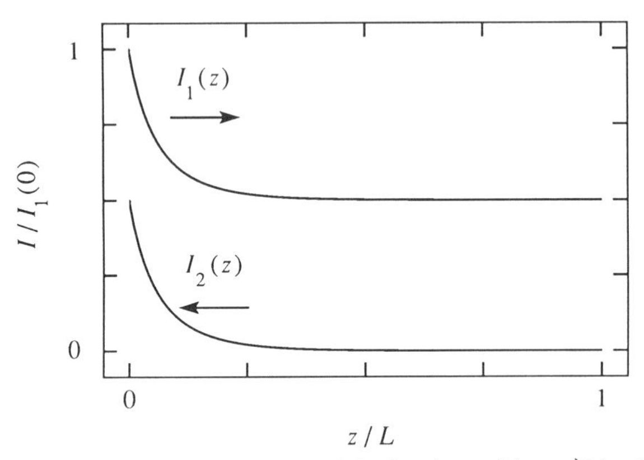

FIGURE 8.3.5 Distribution of the laser and Stokes intensities within the interaction region of an SBS generator.

can be obtained from Eqs. (8.3.30). Figure 8.3.5 shows the distribution of intensities within an SBS generator.*

Let us next estimate the minimum laser power P_{th} required to excite SBS under optimum conditions. We assume that a laser beam having a gaussian transverse profile is focused tightly into a cell containing a Brillouin-active medium. The characteristic intensity of such a beam at the beam waist is given by $I = P/\pi w_0^2$, where w_0 is the beam waist radius. The interaction length L is limited to the characteristic diffraction length $b = 2\pi w_0^2/\lambda$ of the beam. The product $G = gIL$ is thus given by $G = 2gP/\lambda$, and by equating this expression with the threshold value G_{th} we find that the minimum laser power required to excite SBS is of the order of

$$P_{th} = \frac{G_{th}\lambda}{2g}. \tag{8.3.38}$$

For $\lambda = 1.06\ \mu\text{m}$, $G_{th} = 25$, and $g = 0.15$ cm/MW (the value for CS_2) we find that P_{th} is equal to 9 kW. For other organic liquids the minimum power is approximately ten times larger.

The phonon lifetime for stimulated Brillouin scattering in liquids is of the order of several nanoseconds. Since Q-switched laser pulses have a duration of the order of several nanoseconds, and mode-locked laser pulses can be much shorter, it is normal for experiments on SBS to be performed in the transient

* Figure 8.3.5 is plotted for the case $G_{th} = 10$. The physically realistic case of $G_{th} = 25$ produces a much less interesting graph because the perceptible variation in intensities occurs in a small region near $z = 0$.

regime. The problem of transient SBS has been treated by Kroll (1965), by Pohl, Maier, and Kaiser (1968), and by Pohl and Kaiser (1970).

8.4. Phase Conjugation by Stimulated Brillouin Scattering

It was noted even in the earliest experiments on stimulated Brillouin scattering (SBS) that the Stokes radiation was emitted in a highly collimated beam in the backward direction. In fact, the Stokes radiation was found to be so well collimated that it was efficiently fed back into the exciting laser, often leading to the generation of new spectral components in the output of the laser (Goldblatt and Hercher, (1968)). These effects were initially explained as a purely geometrical effect resulting from the long but thin shape of the interaction region.

The first indication that the backscattered light was in fact the phase conjugate of the input was due to an experiment by Zel'dovich *et al.* (1972). The setup used in this experiment is shown in Fig. 8.4.1. The output of a single-mode ruby laser was focused into a cell containing methane gas at a pressure of 125 atmospheres. This cell was constructed in the shape of a cylindrical, multimode waveguide, and served to confine the radiation in the transverse dimension. A strong SBS signal was generated from within this cell. A glass plate that had been etched in hydrofluoric acid was placed in the incident beam to serve as an aberrator. Two cameras were used to monitor the transverse intensity distributions of the incident laser beams and of the Stokes return.

The results of this experiment are summarized in the photographs taken by V. V. Ragulsky that are reproduced in Fig. 8.4.2. Part (a) of this figure shows the laser beam shape as recorded by camera 1, and part (b) shows the Stokes beam shape as recorded by camera 2. The similarity of the spot sizes and shapes indicates that the return beam is the phase conjugate of the incident beam. These highly elongated beam shapes are a consequence of the

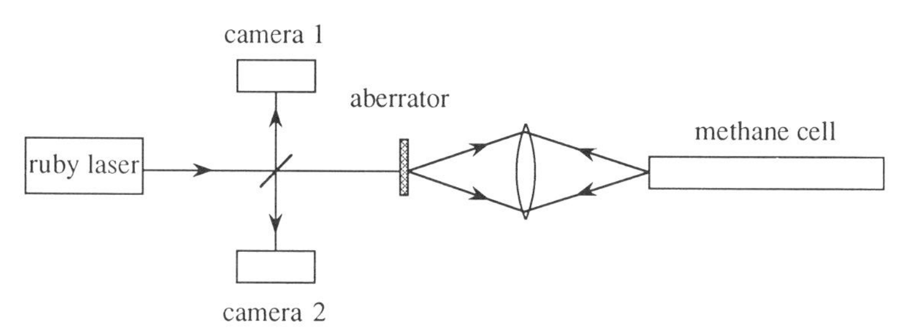

FIGURE 8.4.1 Setup of first experiment on phase conjugation by stimulated Brillouin scattering.

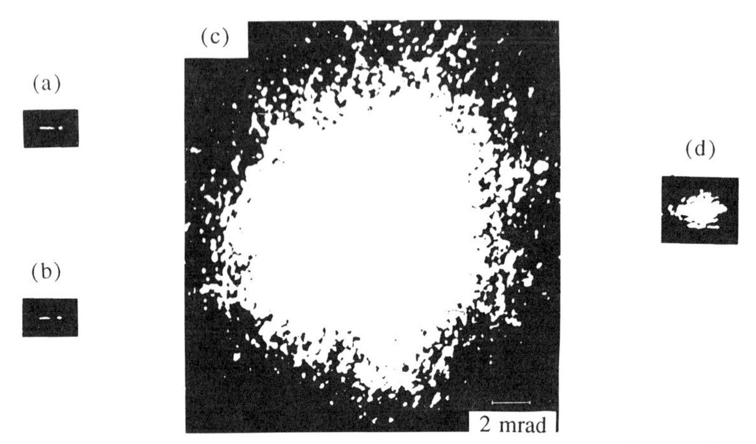

FIGURE 8.4.2 Results of the first experiment demonstrating SBS phase conjugation.

unusual mode pattern of the laser used in these experiments. Part (c) of the figure shows the spot size recorded by camera 2 when the SBS cell was replaced by a conventional mirror. The spot size in this case is very much larger than that of the incident beam; this result shows the severity of the distortions impressed on the beam by the aberrator. Part (d) of the figure shows the spot size of the return beam when the aberrator was removed from the beam path. This spot size is larger than that shown in part (b). This result shows that SBS forms a more accurate conjugate of the incident light when the beam is highly distorted than when the beam is undistorted.*

The results of the experiment of Zel'dovich *et al.* are somewhat surprising, because it is not clear from inspection of the coupled-amplitude equations that describe SBS why the SBS process should lead to phase conjugation. We recall that the reason why degenerate four-wave mixing leads to phase conjugation is that the source term driving the output wave A_4 in the coupled-amplitude equations describing four-wave mixing (see, for example, Eq. (6.1.31b)) is proportional to the complex conjugate of the input wave amplitude, that is, to A_3^*. However, for the case of SBS, Eq. (8.3.21b) shows that the output wave amplitude A_2 is driven by a term proportional to $|A_1|^2 A_2$, which contains no information regarding the phase of the input wave.

The reason why SBS leads to the generation of a phase-conjugate wave is in fact rather subtle (Zel'dovich *et al.*, 1972; Sidorovich, 1976). As illustrated in Fig. 8.4.3, we consider a badly aberrated optical wave that is focused into

* The conclusion that SBS forms a better phase conjugate of an aberrated beam than of an unaberrated beam is not true in all cases, and appears to be a consequence of the details of the geometry of the experiment of Zel'dovich *et al.*

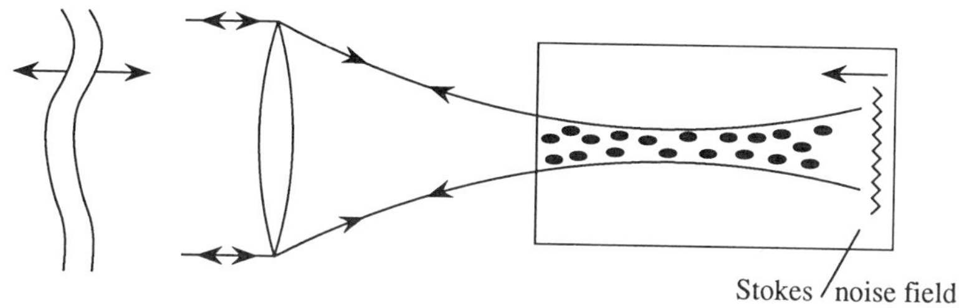

FIGURE 8.4.3 Origin of phase conjugation by SBS. The highly aberrated incident wavefront produces a highly nonuniform intensity distribution (and hence a nonuniform gain distribution) in the focal region of the lens.

the SBS interaction region. Since the wave is highly aberrated, a highly nonuniform intensity distribution (i.e., a volume speckle pattern) is created in the focal region of the wave. Since the gain experienced by the Stokes wave depends upon the local value of the laser intensity (see, for example, Eq. (8.3.22b)), a highly nonuniform gain distribution for the Stokes wave is therefore present in the focal volume. We recall that SBS is initiated by noise, that is, by spontaneously generated Stokes photons. The noise field that leads to SBS initially contains all possible spatial Fourier components. However, the portion of the noise field that experiences the maximum amplification will be the portion whose intensity distribution best matches the nonuniform gain distribution. This portion of the noise field must have wavefronts that match those of the incident laser beam, and hence corresponds to the phase conjugate of the incident laser field.

In order to make this argument more precise, we consider the intensity equation satisfied by the Stokes field (see also Eq. (8.3.22b)),

$$\frac{dI_S}{dz} = -gI_L I_S. \tag{8.4.1}$$

Since we are now considering the case where I_L and I_S possess nonuniform transverse distributions, it is useful to consider the total power in each wave (at fixed z), defined by

$$P_L = \int I_L \, dA, \qquad P_S = \int I_S \, dA, \tag{8.4.2}$$

where the integrals are to be carried out over an area large enough to include essentially all of the power contained in each beam. Equation (8.4.1) can then be rewritten in the form

$$\frac{dP_S}{dz} = -g\frac{P_L P_S}{A}C, \tag{8.4.3}$$

where $A = \int dA$ and where

$$C = \frac{\langle I_L I_S \rangle}{\langle I_L \rangle \langle I_S \rangle} \tag{8.4.4}$$

represents the normalized cross correlation function of the laser and Stokes field intensity distributions. Here the angular brackets are defined so that $\langle x \rangle = \int x \, dA/A$, where x denotes I_L, I_S, or the product $I_L I_S$.

We see that the power gain experienced by the Stokes wave depends not only on the total power in the laser wave, but also on the degree of correlation between the laser and Stokes wave intensity distributions. If I_L and I_S are completely uncorrelated, so that $\langle I_L I_S \rangle = \langle I_L \rangle \langle I_S \rangle$, the correlation function C takes on the value one. C is also equal to one for the case in which both I_L and I_S are spatially uniform. However, if I_L and I_S are correlated, for example because the laser and Stokes fields are phase conjugates of one another, the correlation function can be greater than one.

A limiting case is that in which the laser field is so badly aberrated that the transverse variations in the complex field amplitude obey gaussian statistics. In such a case, the probability density function for the laser intensity fluctuations is given by (see, for example, Goodman, 1985)

$$P(I) = \frac{1}{I_0} e^{-I/I_0}. \tag{8.4.5}$$

The moments of this distribution are given in general by $\langle I^n \rangle = n! \langle I \rangle^n$, and in particular the second moment is given by

$$\langle I^2 \rangle = 2 \langle I \rangle^2. \tag{8.4.6}$$

For that portion of the Stokes field that is the phase conjugate of the laser field, the intensity I_S will be proportional to I_L, and we see from Eqs. (8.4.4) and (8.4.6) that C will be equal to 2. Hence the exponential gain $G \equiv g P_L C L/A$ experienced by the phase-conjugate portion of the noise field will be two times larger than that experienced by any other mode of the noise field. Since the threshold for SBS corresponds to G of the order of 30, the phase-conjugate portion of the SBS signal at threshold will be approximately e^{15} times larger than that of any other component.

On the basis of the argument just presented, we expect that high-quality phase conjugation will occur only if a large number of speckles of the laser intensity distribution are present within the interaction volume. We now determine the conditions under which the number of speckles will be large. We assume that in the focal region the incident laser field has transverse wavefront irregularities on a distance scale as small as a. Each such region will diffract the incident beam into a cone with a characteristic angular spread of $\theta = \lambda/a$. Hence the speckle pattern will look appreciably different after the beam has propagated through the longitudinal distance Δz such that $\theta\,\Delta z = a$. These considerations show that $\Delta z = a^2/\lambda$. We hence expect that SBS will lead to a high-quality phase-conjugate signal only if the transverse extent of the interaction region is much larger than a and if the longitudinal extent of the interaction region is much longer than Δz. In addition, the quality of the phase-conjugate signal can be degraded if there is poor spatial overlap of the various spatial Fourier components of the laser beam. For example, a highly aberrated beam will spread with a large angular divergence $\theta = \lambda/a$. If those components of the beam with large divergence angle θ fail to overlap the strong central portion of the beam, they will be reflected with low efficiency, leading to a degradation of the quality of the phase conjugation process. To avoid the possibility of such effects, SBS phase conjugation is often performed using the waveguide geometry shown in Fig. 8.4.1.

8.5. Stimulated Brillouin Scattering in Gases

We next consider stimulated Brillouin scattering (SBS) in gases. As before (Eq. (8.3.24)), the steady-state line-center gain factor for SBS is given by

$$g_0 = \frac{\gamma_e^2 \omega^2}{\rho_0 n v c^3 \Gamma_B} \tag{8.5.1}$$

with the electrostrictive constant γ_e given by Eq. (7.3.12) and with the Brillouin linewidth given to good approximation by (see also Eqs. (7.3.23) and (8.3.14b))

$$\Gamma_B = (2\eta_s + \eta_d)\frac{q^2}{\rho_0}. \tag{8.5.2}$$

For the case of an ideal gas, we can readily predict the values of the material parameters appearing in these equations (Loeb, 1961). First, we can assume the validity of the Stokes relation (see also the discussion in the Appendix to Section 8.6), which states that the shear and dilation viscosity coefficients are related by $\eta_d = -\frac{2}{3}\eta_s$, and hence we find that

$$\Gamma_B = \tfrac{4}{3}\eta_s \frac{q^2}{\rho_0}. \tag{8.5.3}$$

The shear viscosity coefficient η_s can be shown from kinetic theory to be given by

$$\eta_s = \tfrac{1}{3} N m \bar{v} L, \tag{8.5.4}$$

where N is the atomic number density, m is the molecular mass, $\bar{v}$ is the mean molecular velocity given by $\bar{v} = (8kT/\pi m)^{1/2}$, and L is the mean free path given by $L = (\sqrt{2}\pi d^2 N)^{-1}$ with d denoting the molecular diameter. We hence find that the shear viscosity coefficient is given by

$$\eta_s = \frac{2}{3\pi^{3/2}} \frac{\sqrt{kTm}}{d^2}. \tag{8.5.5}$$

Note that the shear viscosity coefficient is independent of the molecular number density N. The measured (and theoretical) value of the shear viscosity coefficient for N_2 at standard temperature and pressure is $\eta_s = 1.8 \times 10^{-4}$ dyne s/cm^2.

By introducing the expression (8.5.5) for the viscosity into Eq. (8.5.2) and replacing q by $2n\omega/c$, we find that the Brillouin linewidth is given by

$$\Gamma_B = \frac{32}{9\pi^{3/2}} \frac{n^2\omega^2}{c^2} \frac{\sqrt{kT/m}}{d^2 N}. \tag{8.5.6}$$

If we assume that the incident optical radiation has a wavelength λ of 1.06 μm, we find that the Brillouin linewidth for N_2 at standard temperature and pressure is equal to $\Gamma_B = 2.77 \times 10^9$ rad/s and hence that the Brillouin linewidth in ordinary frequency units is given by $\delta\nu(\text{FWHM}) = \Gamma_B/2\pi = 440$ MHz.

The velocity of sound v, which appears in Eq. (8.5.1), is given for an ideal gas by $v = (\gamma kT/m)^{1/2}$, where the ratio of specific heats, γ, is equal to $\frac{5}{3}$ for a monatomic gas and $\frac{7}{5}$ for a diatomic gas. In addition, the electrostrictive constant γ_e can be estimated as $\gamma_e = \rho\, \partial\epsilon/\partial\rho$ with $\partial\epsilon/\partial\rho$ taken as the essentially constant quantity $(\epsilon - 1)/\rho$.

The dependence of g_0 on material parameters can be determined by combining these results with Eq. (8.5.1) to obtain

$$g_0 = \frac{9\pi^{3/2} N^2 m^2 d^2 (\partial\epsilon/\partial\rho)^2}{32\gamma^{1/2} n^3 ckT}. \tag{8.5.7}$$

However, in order to obtain a numerical estimate of g_0, it is often more convenient to evaluate the expression (8.5.1) for g_0 directly with the numerical value of Γ_B obtained from Eq. (8.5.6). For N_2 gas at standard temperature and pressure and for a wavelength of 1.06 μm, we take the values $\omega = 1.8 \times 10^{15}$ rad/s, $n = 1.0003$, $v = 3.3 \times 10^4$ cm/s, $\gamma_e = n^2 - 1 = 6 \times 10^{-4}$, and we thereby obtain

$$g_0 = 3.8 \times 10^{-19} \frac{\text{cm s}}{\text{erg}} = 3.8 \frac{\text{cm}}{\text{TW}}. \tag{8.5.8}$$

Note from Eq. (8.5.7) that g_0 scales quadratically with molecular density.

Hence, at a pressure of 100 atmospheres the gain factor of N_2 is equal to $g_0 = 0.042$ cm/MW, which is comparable to that of typical organic liquids.

One advantage of the use of gases as the active medium for Brillouin scattering is that the gain for SBS scales with molecular number density as N^2, whereas the gain for stimulated Raman scattering, which is often a competing process, scales as N (see, for example, Eqs. 9.3.19 and 9.3.20). At pressures greater than 10 atmospheres, the gain for SBS typically exceeds that of stimulated Raman scattering. Moreover, through the use of rare gases (which have no vibrational modes), it is possible to suppress the occurrence of stimulated Raman scattering altogether.

Some parameters relevant to SBS at the 249 nm wavelength of the KrF laser have been compiled by Damzen and Hutchinson (1983) and are presented in Table 8.5.1.

8.6. General Theory of Stimulated Brillouin and Stimulated Rayleigh Scattering

In this section we develop a theoretical model that can treat both stimulated Brillouin and stimulated Rayleigh scattering. These two effects can conveniently be treated together because they both entail the scattering of light from inhomogeneities in thermodynamic quantities. For convenience, we choose the temperature T and density ρ to be the independent thermodynamic variables. The theory that we present incorporates both electrostrictive

TABLE 8.5.1 Gain Factors, Phonon Lifetimes, and Frequency Shifts for Some Compressed Brillouin-Active Gases at a Wavelength of 249 nm*

Gas	p (atm)	g_0 (cm/MW)	τ (ns)	$\Omega_B/2\pi$ (GHz)	g_R (cm/MW)
SF_6	15.5	2.5×10^{-2}	1	0.9	3×10^{-4}
	10	0.9×10^{-2}	0.6		2×10^{-4}
Xe	39	4.4×10^{-2}	2	0.12	0
	10	1.8×10^{-2}	0.4		
Ar	10	1.5×10^{-4}	0.1	3	0
N_2	10	1.7×10^{-4}	0.2	3	3×10^{-5}
CH_4	10	8×10^{-4}	0.1	3	1×10^{-3}

* For comparison, the gain factor g_R for forward stimulated Raman scattering is also listed. (After Damzen and Hutchinson, 1983.) (© 1983 IEEE)

and absorptive coupling of the radiation to the material system. Our analysis therefore describes the following four scattering processes:

1. Electrostrictive stimulated Brillouin scattering: the scattering of light from sound waves that are driven by the interference of the laser and Stokes fields through the process of electrostriction.
2. Thermal stimulated Brillouin scattering: the scattering of light from sound waves that are driven by the absorption and subsequent thermalization of the optical energy, leading to temperature and hence to density variations within the medium.
3. Electrostrictive stimulated Rayleigh scattering: the scattering of light from isobaric density fluctuations that are driven by the process of electrostriction.
4. Thermal stimulated Rayleigh scattering: the scattering of light from isobaric density fluctuations that are driven by the process of optical absorption.

Our analysis is based on the three equations of hydrodynamics (Hunt, 1955; Kaiser and Maier, 1972). The first of these equations is the equation of continuity

$$\frac{\partial \tilde{\rho}_t}{\partial t} + \tilde{\mathbf{u}}_t \cdot \nabla \tilde{\rho}_t + \tilde{\rho}_t \nabla \cdot \tilde{\mathbf{u}}_t = 0, \tag{8.6.1}$$

where $\tilde{\rho}_t$ is the mass density of the fluid and $\tilde{\mathbf{u}}_t$ is the velocity of some small volume element of the fluid.*

The second equation is the equation of momentum transfer. It is a generalization of the Navier–Stokes equation, and is given by

$$\tilde{\rho}_t \frac{\partial \tilde{\mathbf{u}}_t}{\partial t} + \tilde{\rho}_t(\tilde{\mathbf{u}}_t \cdot \nabla)\tilde{\mathbf{u}}_t = \tilde{\mathbf{f}} - \nabla \tilde{p}_t + (2\eta_s + \eta_d)\nabla(\nabla \cdot \tilde{\mathbf{u}}_t) - \eta \nabla \times (\nabla \times \tilde{\mathbf{u}}_t). \tag{8.6.2}$$

Here $\tilde{\mathbf{f}}$ represents the force per unit volume of any externally imposed forces; for the case of electrostriction, $\tilde{\mathbf{f}}$ is given by (see also Eq. (8.3.12))

$$\tilde{\mathbf{f}} = -\frac{\gamma_e}{8\pi} \nabla \langle \tilde{\mathbf{E}}^2 \rangle, \tag{8.6.3}$$

where $\tilde{\mathbf{E}}$ denotes the instantaneous value of the time-varying applied total electric field and γ_e represents the electrostrictive coupling constant

$$\gamma_e = \rho \frac{\partial \epsilon}{\partial \rho}. \tag{8.6.4}$$

* The subscript t stands for *total*; we shall later linearize these equations to find the equations satisfied by the linearized quantities, which we shall designate by nonsubscripted symbols.

The second term on the right-hand side of Eq. (8.6.2) denotes the force due to the gradient of the pressure $\tilde{p}_t$. In the third term, η_s denotes the shear viscosity coefficient and η_d denotes the dilational viscosity coefficient. When the Stokes relation is satisfied, as it is for example for an ideal gas, these coefficients are related by

$$\eta_d = -\tfrac{2}{3}\eta_s. \tag{8.6.5}$$

The coefficients are defined in detail in the Appendix at the end of this section.

The last of three principal equations of hydrodynamics is the equation of heat transport, given by

$$\tilde{\rho}_t c_v \frac{\partial \tilde{T}_t}{\partial t} + \tilde{\rho}_t c_v (\tilde{\mathbf{u}} \cdot \nabla \tilde{T}_t) + \tilde{\rho}_t c_v \left(\frac{\gamma - 1}{\beta_p} \right) (\nabla \cdot \tilde{\mathbf{u}}_t) = -\nabla \cdot \tilde{\mathbf{Q}} + \tilde{\phi}_\eta + \tilde{\phi}_{\text{ext}}. \tag{8.6.6}$$

Here $\tilde{T}_t$ denotes the local value of the temperature, c_v the specific heat at constant volume, $\gamma = c_p/c_v$ the adiabatic index, $\beta_p = -\tilde{\rho}^{-1}(\partial\tilde{\rho}/\partial\tilde{T})_p$ the thermal expansion coefficient, and $\tilde{\mathbf{Q}}$ the heat flux vector. For heat flow due to thermal conduction, $\tilde{\mathbf{Q}}$ satisfies the equation

$$\nabla \cdot \tilde{\mathbf{Q}} = -\kappa \nabla^2 \tilde{T}_t, \tag{8.6.7}$$

where κ denotes the thermal conductivity. $\tilde{\phi}_\eta$ denotes the viscous energy deposited within the medium per unit volume per unit time, and is given by

$$\tilde{\phi}_\eta = \sum_{ij} (2\eta_s d_{ij} d_{ji} + \eta_d d_{ii} d_{jj}), \tag{8.6.8a}$$

where

$$d_{ij} = \frac{1}{2}\left(\frac{\partial \tilde{u}_i}{\partial x_j} + \frac{\partial \tilde{u}_j}{\partial x_i} \right) \tag{8.6.8b}$$

is the rate-of-dilation tensor. Finally, $\tilde{\phi}_{\text{ext}}$ gives the energy per unit time per unit volume delivered to the medium from external sources. Absorption of the optical wave provides the contribution

$$\tilde{\phi}_{\text{ext}} = \alpha \frac{nc}{4\pi} \langle \tilde{E}^2 \rangle, \tag{8.6.9}$$

to this quantity, where α is the optical absorption coefficient.

The acoustic equations are now derived by linearizing the hydrodynamic equations about the nominal conditions of the medium. In particular, we take

$$\tilde{\rho}_t = \rho_0 + \tilde{\rho} \qquad \text{with} \quad |\tilde{\rho}| \ll \rho_0, \tag{8.6.10a}$$

$$\tilde{T}_t = T_0 + \tilde{T} \qquad \text{with} \quad |\tilde{T}| \ll T_0, \tag{8.6.10b}$$

$$\tilde{\mathbf{u}}_t = \tilde{\mathbf{u}} \qquad \text{with} \quad |\tilde{\mathbf{u}}| \ll v, \tag{8.6.10c}$$

where v denotes the velocity of sound. Note that we have assumed that the medium is everywhere motionless in the absence of the acoustic disturbance. We can reliably use the linearized form of the resulting equations so long as the indicated inequalities are satisfied.

We substitute the expansions (8.6.10) into the hydrodynamic equations (8.6.1), (8.6.2), and (8.6.6), drop any term that contains more than one small quantity, and subtract the unperturbed, undriven solution containing only $\tilde{\rho}_0$ and $\tilde{T}_0$. The continuity equation (8.6.1) then becomes

$$\frac{\partial \tilde{\rho}}{\partial t} + \rho_0 \nabla \cdot \tilde{\mathbf{u}} = 0. \tag{8.6.11}$$

In order to linearize the momentum transport equation (8.6.2), we first express $\tilde{p}_t$ as

$$\tilde{p}_t = p_0 + \tilde{p} \qquad \text{with} \quad |\tilde{p}| \ll p_0. \tag{8.6.12}$$

Since we have taken T and ρ as the independent thermodynamic variables, we can express $\tilde{p}$ as

$$\tilde{p} = \left(\frac{\partial p}{\partial \rho}\right)_T \tilde{\rho} + \left(\frac{\partial p}{\partial T}\right)_\rho \tilde{T}, \tag{8.6.13}$$

or as

$$\tilde{p} = \frac{v^2}{\gamma}(\tilde{\rho} + \beta_p \rho_0 \tilde{T}), \tag{8.6.14}$$

where we have expressed $(\partial p/\partial \rho)_T$ as $\gamma^{-1}(\partial p/\partial \rho)_s = v^2/\gamma$ with $v^2 = (\partial p/\partial \rho)_s$ representing the square of the velocity of sound, and where we have expressed $(\partial p/\partial T)_\rho$ as $\gamma^{-1}(\partial p/\partial \rho)_s(\partial \rho/\partial T)_p = v^2 \beta_p \rho_0/\gamma$ with β_p representing the thermal expansion coefficient at constant pressure. Through use of Eq. (8.6.14), the linearized form of Eq. (8.6.2) becomes

$$\begin{aligned}\rho_0 \frac{\partial \tilde{\mathbf{u}}}{\partial t} + \frac{v^2}{\gamma} \nabla \tilde{\rho} + \frac{v^2 \beta_p \rho_0}{\gamma} \nabla \tilde{T} - (2\eta_s + \eta_d) \nabla (\nabla \cdot \tilde{\mathbf{u}}) \\ + \eta_s \nabla \times (\nabla \times \tilde{\mathbf{u}}) = \tilde{\mathbf{f}}.\end{aligned} \tag{8.6.15}$$

Finally, the linearized form of the energy transport equation, Eq. (8.6.6), becomes

$$\rho_0 c_v \frac{\partial \tilde{T}}{\partial t} + \frac{\rho_0 c_v (\gamma - 1)}{\beta_p} (\nabla \cdot \tilde{\mathbf{u}}) - \kappa \nabla^2 \tilde{T} = \tilde{\phi}_{\text{ext}}. \tag{8.6.16}$$

Note that the viscous contribution to the heat input, $\tilde{\phi}_\eta$, does not contribute in the linear approximation.

Equations (8.6.11), (8.6.15), and (8.6.16) constitute the three linearlized equations of hydrodynamics for the quantities $\tilde{\mathbf{u}}$, $\tilde{\rho}$, and $\tilde{T}$. The continuity equation in its linearized form (Eq. (8.6.11)) can be used to eliminate the variable $\tilde{\mathbf{u}}$ from the remaining two equations. To do so, we take the divergence of the equation of momentum transfer (8.6.15) and use Eq. (8.6.11) to eliminate the terms containing $\nabla \cdot \tilde{\mathbf{u}}$. We obtain

$$-\frac{\partial^2 \tilde{\rho}}{\partial t^2} + \frac{v^2}{\gamma}\nabla^2 \tilde{\rho} + \frac{v^2 \beta_p \rho_0}{\gamma}\nabla^2 \tilde{T} + \frac{2\eta_s + \eta_d}{\rho_0}\frac{\partial}{\partial t}(\nabla^2 \tilde{\rho}) = \frac{\gamma_e}{8\pi}\nabla^2 \langle \tilde{\mathbf{E}}^2 \rangle, \quad (8.6.17)$$

where we have explicitly introduced the form of $\tilde{\mathbf{f}}$ from Eq. (8.6.3). Also, the energy transport equation (8.6.16) can then be expressed through use of Eqs. (8.6.9) and (8.6.11) as

$$\rho_0 c_v \frac{\partial \tilde{T}}{\partial t} - \frac{c_v(\gamma - 1)}{\beta_p}\frac{\partial \tilde{\rho}}{\partial t} - \kappa \nabla^2 \tilde{T} = \frac{nc\alpha}{4\pi}\langle \tilde{\mathbf{E}}^2 \rangle. \quad (8.6.18)$$

Equations (8.6.17) and (8.6.18) constitute two coupled equations for the thermodynamic variables $\tilde{\rho}$ and $\tilde{T}$, and they show how these quantities are coupled to one another and are driven by the applied optical field.

In the absence of the driving terms appearing on their right-hand sides, Eqs. (8.6.17) and (8.6.18) allow solutions of the form of freely propagating acoustic waves

$$\tilde{F}(z,t) = Fe^{-i\Omega(t - z/v)}e^{-\alpha_s z} + \text{c.c.} \quad (8.6.19)$$

where F denotes either ρ or T, and where the sound absorption coefficient α_s is given for low frequencies ($\Omega \ll \rho_0 v^2/(2\eta_s + \eta_d)$) by

$$\alpha_s = \frac{\Omega^2}{2\rho_0 v^3}\left[(2\eta_s + \eta_d) + (\gamma - 1)\frac{\kappa}{c_p}\right]. \quad (8.6.20)$$

For details, see the article by Sette (1961).

We next study the nature of the solution to Eqs. (8.6.17) and (8.6.18) in the presence of their driving terms. We assume that the total optical field can be represented as

$$\tilde{E}(z,t) = A_1 e^{i(k_1 z - \omega_1 t)} + A_2 e^{i(-k_2 z - \omega_2 t)} + \text{c.c.} \quad (8.6.21)$$

We first determine the response of the medium at the beat frequency between these two applied field frequencies. This disturbance will have frequency

$$\Omega = \omega_1 - \omega_2 \quad (8.6.22)$$

and wave number

$$q = k_1 + k_2, \quad (8.6.23)$$

and can be taken to be of the form

$$\tilde{\rho}(z,t) = \rho e^{i(qz-\Omega t)} + \text{c.c.}, \tag{8.6.24}$$

$$\tilde{T}(z,t) = T e^{i(qz-\Omega t)} + \text{c.c.} \tag{8.6.25}$$

For the present, we are interested only in the steady-state response of the medium, and hence we assume that the amplitudes A_1, A_2, ρ, and T are time-independent. We introduce the fields $\tilde{E}$, $\tilde{\rho}$, and $\tilde{T}$ given by Eqs. (8.6.21) through (8.6.25) into the coupled acoustic equations (8.6.17) and (8.6.18). The parts of these equations that oscillate at frequency Ω are given respectively by

$$-\left(\Omega^2 + i\Omega\Gamma_B - \frac{v^2 q^2}{\gamma}\right)\rho + \frac{v^2 \beta_p \rho_0 q^2}{\gamma} T = \frac{\gamma_e q^2}{4\pi} A_1 A_2^* \tag{8.6.26}$$

and

$$-(i\Omega - \tfrac{1}{2}\gamma\Gamma_R)T + \frac{i(\gamma - 1)\Omega}{\beta_p \rho_0}\rho = \frac{nc\alpha}{2\pi c_v \rho_0} A_1 A_2^*. \tag{8.6.27}$$

Here we have introduced the Brillouin linewidth

$$\Gamma_B = (2\eta_s + \eta_d) q^2/\rho_0, \tag{8.6.28}$$

whose reciprocal $\tau_p = \Gamma_B^{-1}$ is the phonon lifetime, and the Rayleigh linewidth

$$\Gamma_R = \frac{2\kappa q^2}{\rho_0 c_p}, \tag{8.6.29}$$

whose reciprocal $\tau_R = \Gamma_R^{-1}$ is the characteristic decay time of the isobaric density disturbances that give rise to Rayleigh scattering.

In deriving Eqs. (8.6.26) and (8.6.27) we have ignored those terms that contain the spatial derivatives of ρ and T. This approximation is equivalent to assuming that the material excitations are strongly damped and hence do not propagate over any appreciable distances. This approximation is valid so long as

$$q \gg \left|\frac{1}{\rho}\frac{\partial \rho}{\partial z}\right|, \left|\frac{1}{T}\frac{\partial T}{\partial z}\right| \quad \text{and} \quad q^2 \gg \left|\frac{1}{\rho}\frac{\partial^2 \rho}{\partial z^2}\right|, \left|\frac{1}{T}\frac{\partial^2 T}{\partial z^2}\right|.$$

These inequalities are usually satisfied. Recall that a similar approximation was introduced in Section 8.3 in the derivation of Eq. (8.3.15).

We next solve Eq. (8.6.27) algebraically for T and introduce the resulting expression into Eq. (8.6.26). We obtain the equation

$$\left[-\left(\Omega^2 + i\Omega\Gamma_B - \frac{v^2 q^2}{\gamma}\right) + \frac{v^2 q^2 \Omega(\gamma - 1)}{(\Omega + \frac{1}{2}i\gamma\Gamma_R)\gamma}\right]\rho = \left[\gamma_e - \frac{i\gamma_a q v}{\Omega + \frac{1}{2}i\gamma\Gamma_R}\right]\frac{q^2}{4\pi} A_1 A_2^*, \tag{8.6.30}$$

where we have introduced the absorptive coupling constant

$$\gamma_a = \frac{2\alpha n v^2 c\beta_p}{c_p \Omega_B} \tag{8.6.31}$$

with $\Omega_B = qv$. Equation (8.6.30) shows how the amplitude ρ of the acoustic disturbance depends on the amplitudes A_1 and A_2 of the two optical fields. Both Brillouin and Rayleigh contributions to ρ are contained in Eq. (8.6.30).

It is an empirical fact (see, for example, Fig. 7.1.1) that the spectrum for Brillouin scattering does not appreciably overlap that for Rayleigh scattering. Equation (8.6.30) can thus be simplified by considering the resonant contributions to the two processes separately. First, we consider the case of stimulated Brillouin scattering (SBS). In this case Ω^2 is approximately equal to $\Omega_B^2 = v^2q^2$, and hence the denominator $\Omega + \frac{1}{2}i\gamma\Gamma_R$ is nonresonant. We can thus drop the contribution $\frac{1}{2}i\gamma\Gamma_R$ in comparison with Ω in these denominators. Equation (8.6.30) then shows that the Brillouin contribution to ρ is given by

$$\rho_B = \frac{-(\gamma_e - i\gamma_a qv/\Omega)q^2}{4\pi(\Omega^2 + i\Omega\Gamma_B - v^2q^2)} A_1A_2^*. \tag{8.6.32}$$

The other resonance in Eq. (8.6.30) occurs at $\Omega = 0$, and leads to stimulated Rayleigh scattering (SRLS). For $|\Omega| \lesssim \Gamma_R$, the Brillouin denominator $\Omega^2 + i\Omega\Gamma_B - v^2q^2/\gamma$ is nonresonant and can be approximated by $-v^2q^2/\gamma$. Equation (8.6.30) thus becomes

$$\rho_R = \left[\frac{\gamma_e(\Omega + \frac{1}{2}i\gamma\Gamma_R) - i\gamma_a\Omega_B}{\Omega + \frac{1}{2}i\Gamma_R}\right]\frac{1}{4\pi v^2} A_1A_2^*. \tag{8.6.33}$$

We next calculate the nonlinear polarization as

$$\tilde{p}^{NL} = \Delta\chi\,\tilde{E} = \frac{\Delta\epsilon}{4\pi}\tilde{E} = \frac{1}{4\pi}\left(\frac{\partial\epsilon}{\partial\rho}\right)_T \tilde{\rho}\tilde{E} = \frac{\gamma_e}{4\pi\rho_0}\tilde{\rho}\tilde{E}, \tag{8.6.34}$$

where $\tilde{\rho}$ and $\tilde{E}$ are given by Eqs. (8.6.24) and (8.6.21) respectively. We represent the nonlinear polarization in terms of its complex amplitudes as

$$\tilde{P}^{NL} = p_1 e^{i(k_1z - \omega_1 t)} + p_2 e^{i(-k_2z - \omega_2 t)} + \text{c.c.} \tag{8.6.35}$$

with

$$p_1 = \frac{\gamma_e}{4\pi\rho_0}\rho A_2, \qquad p_2 = \frac{\gamma_e}{4\pi\rho_0}\rho^* A_1. \tag{8.6.36}$$

This form of the nonlinear polarization is now introduced into the wave equation, which we write in the form (see also Eq. (2.1.21))

$$-\nabla^2(A_n(\mathbf{r})e^{i\mathbf{k}_n\cdot\mathbf{r}}) - \frac{\epsilon\omega_n^2}{c^2}A_n(\mathbf{r})e^{i\mathbf{k}_n\cdot\mathbf{r}} = \frac{4\pi\omega_n^2}{c^2}p_n e^{i\mathbf{k}_n\cdot\mathbf{r}}. \tag{8.6.37}$$

We next make the slowly-varying-amplitude approximation and find that the field amplitudes obey the equations

$$\left(\frac{d}{dz} + \tfrac{1}{2}\alpha\right) A_1 = \frac{2\pi i\omega}{nc} p_1, \tag{8.6.38a}$$

$$\left(\frac{d}{dz} - \tfrac{1}{2}\alpha\right) A_2 = \frac{-2\pi i\omega}{nc} p_2, \tag{8.6.38b}$$

where we have introduced the real part of the refractive index $n = \operatorname{Re}\sqrt{\epsilon}$ and the optical absorption coefficient $\alpha = (2\omega/c)\operatorname{Im}\sqrt{\epsilon}$. Equations (8.6.38) can be used to describe either SBS or SRLS, depending on whether form (8.6.32) or (8.6.33) is used to determine the factor ρ that appears in the expression (8.3.36) for the nonlinear polarization. Since in either case ρ is proportional to the produce $A_1 A_2^*$, Eqs. (8.6.38) can be written as

$$\frac{dA_1}{dz} = \kappa |A_2|^2 A_1 - \tfrac{1}{2}\alpha A_1, \tag{8.6.39a}$$

$$\frac{dA_2}{dz} = \kappa^* |A_1|^2 A_2 + \tfrac{1}{2}\alpha A_2, \tag{8.6.39b}$$

where κ for SBS is given by

$$\kappa_B = -\frac{q^2\omega}{8\pi\rho_0 nc} \frac{i\gamma_e(\gamma_e - i\gamma_a)}{(\Omega^2 + i\Omega\Gamma_B - v^2q^2)}, \tag{8.6.40a}$$

and for SRWS is given by

$$\kappa_R = \frac{i\gamma_e\omega}{8\pi\rho_0 ncv^2}\left[\frac{\gamma_e(\Omega + \frac{1}{2}i\gamma\Gamma_R) - i\gamma_a\Omega_B}{\Omega + \frac{1}{2}i\Gamma_R}\right]. \tag{8.6.40b}$$

We now introduce the intensities

$$I_i = \frac{nc}{2\pi}|A_i|^2 \tag{8.6.41}$$

of the two interacting optical waves and use Eqs. (8.6.39) to calculate the spatial rate of change of the intensities as

$$\frac{dI_1}{dz} = -gI_1I_2 - \alpha I_1, \tag{8.6.42a}$$

$$\frac{dI_2}{dz} = -gI_1I_2 + \alpha I_2, \tag{8.6.42b}$$

where we have introduced the gain factor

$$g = \frac{-4\pi}{nc}\operatorname{Re}\kappa. \tag{8.6.43}$$

For the case of SBS, we find that the gain factor can be expressed as

$$g_B = g_B^e + g_B^a, \tag{8.6.44a}$$

where

$$g_B^e = \frac{\omega^2 \gamma_e^2}{\rho_0 n v c^2 \Gamma_B} \frac{1}{1 + (2\,\Delta\Omega/\Gamma_B)^2} \tag{8.6.44b}$$

and

$$g_B^a = \frac{-\omega^2 \gamma_e \gamma_a}{2\rho_0 n v c^2 \Gamma_B} \frac{4\,\Delta\Omega/\Gamma_B}{1 + (2\,\Delta\Omega/\Gamma_B)^2} \tag{8.6.44c}$$

denote the electrostrictive and absorptive contributions to the SBS gain factor, respectively. Here we have introduced the detuning from the Brillouin resonance given by $\Delta\Omega = \Omega_B - \Omega$, where $\Omega_B = qv = (k_1 + k_2)v$ and where $\Omega = \omega_1 - \omega_2$. The electrostrictive contribution is maximum for $\Delta\Omega = 0$, where it attains the value

$$g_B^e(\max) = \frac{\omega^2 \gamma_e^2}{\rho_0 n v c^3 \Gamma_B}. \tag{8.6.45}$$

Since (according to Eq. (8.6.28)) Γ_B is proportional to q^2 and hence to ω^2, the gain for electrostrictive SBS is independent of the laser frequency. The absorptive contribution is maximum for $\Delta\Omega = -\Gamma_B/2$, that is, when the Stokes wave (at frequency ω_2) is detuned by one-half the spontaneous Brillouin linewidth Γ_B to the low-frequency side of resonance. The maximum value of the gain for this process is

$$g_B^a(\max) = \frac{\omega^2 \gamma_e \gamma_a}{2\rho_0 n v c^3 \Gamma_B}. \tag{8.6.46}$$

Note that since Γ_B is proportional to q^2 and (according to Eq. (8.6.31) γ_a is proportional to q^{-1}, the absorptive SBS gain factor is proportional to q^{-3} and hence depends on the laser frequency as ω^{-3}. Since the gain factor for thermal SBS is linearly proportional to the optical absorption coefficient α (by Eqs. (8.6.31) and (8.6.46), the gain for thermal SBS can be made to exceed that for electrostrictive SBS by adding an absorber such as a dye to the Brillouin-active medium. As shown in Table 8.3.1, this effect occurs roughly for absorption coefficients greater than 1 cm^{-1}.*

* The quantity $g_B^e(\max)$ is designated g_0 in Table 8.3.1

The spectral dependence of the two contributions to the SBS gain is shown schematically in Fig. 8.6.1.

For the case of stimulated Rayleigh scattering, we can express the gain factor appearing in Eq. (8.6.42) through use of Eqs. (8.6.40b) and (8.6.43) as

$$g_R = g_R^e + g_R^a, \tag{8.6.47}$$

where

$$g_R^e = \frac{-\omega\gamma_e^2(\gamma - 1)}{4\rho_0 n^2 c^2 v^2}\left[\frac{4\Omega/\Gamma_R}{1 + (2\Omega/\Gamma_R)^2}\right] \tag{8.6.48}$$

and

$$g_R^a = \frac{\omega\gamma_e\gamma_a\Omega_B}{2\rho_0 n^2 c^2 v^2 \Gamma_R}\left[\frac{4\Omega/\Gamma_R}{1 + (2\Omega/\Gamma_R)^2}\right] \tag{8.6.49}$$

denote the electrostrictive and absorptive contributions to the gain factor

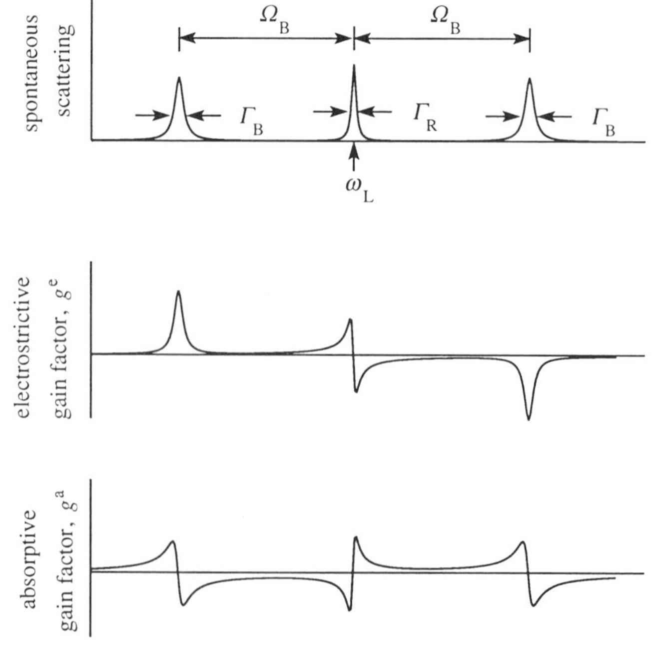

FIGURE 8.6.1 Gain spectra for stimulated Brillouin scattering and stimulated Rayleigh scattering, showing their electrostrictive and absorptive contributions. For comparison, the spectrum of spontaneous Brillouin and Rayleigh scattering is also shown.

respectively. The contribution g_R^e gives rise to electrostrictive stimulated Rayleigh scattering (SRLS). The gain factor for this process is maximum for $\Omega = -\Gamma_R/2$ and has the value

$$g_R^e(\max) = \frac{\omega\gamma_e^2(\gamma - 1)}{4\rho_0 n^2 c^2 v^2}. \tag{8.6.50}$$

Note that this quantity scales linearly with laser frequency. The absorptive contribution g_R^a gives rise to thermal SRLS. The gain for this process is maximum for $\Omega = \Gamma_R/2$ and has the value

$$g_R^a(\max) = \frac{\omega\gamma_e\gamma_a\Omega_B}{2\rho_0 n^2 c^2 v^2 \Gamma_R}. \tag{8.6.51}$$

Since Γ_R scales with the laser frequency as ω^2, γ_a scales as $1/\omega$, and Ω_B scales as ω, we see that the gain factor for thermal SRLS scales with the laser frequency as $1/\omega$.

As can be seen from Table 8.6.1, Γ_R is often of the order of 10 MHz, which is much narrower than the linewidths of pulsed lasers. In such cases, laser linewidth effects can often be treated in an approximate fashion by convolving the gain predicted by Eqs. (8.6.48) and (8.6.49) with the laser lineshape. If the laser linewidth Γ_L is much broader than Γ_R, the maximum gain for absorptive SRLS is then given by Eq. (8.6.51) with Γ_R replaced by Γ_L. Under these conditions $g_R^a(\max)$ is independent of the laser frequency.

We note by inspection of Table 8.6.1 that $g_R^a(\max)$ is very much larger than $g_R^e(\max)$ except for extremely small values of the absorption coefficient. The two gains become comparable for $\alpha \simeq 10^{-3}\ \text{cm}^{-1}$, which occurs only for unusually pure materials.

TABLE 8.6.1 Properties of stimulated Rayleigh scattering for a variety of materials at a wavelength of 694 nm*

	Gain factor		Linewidth
Substance	$g_e(\max)$ (cm/MW)	$g_a(\max)/\alpha$ (cm^2/MW)	$\delta\nu_R$ (MHz)
CCl_4	2.6×10^{-4}	0.82	17
Methanol	8.4×10^{-4}	0.32	20
CS_2	6.0×10^{-4}	0.62	36
Benzene	2.2×10^{-4}	0.57	24
Acetone	2.0×10^{-4}	0.47	21
H_2O	0.02×10^{-4}	0.019	27.5
Ethanol		0.38	18

* After Kaiser and Maier (1972).

We also see by comparison of Eqs. (8.6.51) and (8.6.46) that the ratio of the two thermal gain factors is given by

$$\frac{g_{\mathrm{R}}^{\mathrm{a}}(\max)}{g_{\mathrm{B}}^{\mathrm{a}}(\max)} = \frac{2\Gamma_{\mathrm{B}}}{\Gamma_{\mathrm{R}}}. \tag{8.6.52}$$

Comparison of Tables 8.3.1 and 8.6.1 shows that for a given material the ratio $\Gamma_{\mathrm{B}}/\Gamma_{\mathrm{R}}$ is typically of the order of 100. Hence, when thermal stimulated scattering occurs, the gain for thermal SRLS is much larger than that for thermal SBS, and most of the energy is emitted by this process.

The frequency dependence of the gain for stimulated Rayleigh scattering is shown in Fig. 8.6.1. Note that electrostrictive SRLS gives rise to gain for Stokes shifted light but that thermal SRLS gives rise to gain for anti-Stokes scattering (Herman and Gray, 1967). This result can be understood from the point of view that n_2 is positive for electrostriction but is negative for the process of heating and subsequent thermal expansion. We saw in the discussion of two-beam coupling presented in Section 6.4 that the lower-frequency wave experiences gain for n_2 positive and loss for n_2 negative.

Appendix: Definition of the Viscosity Coefficients

The viscosity coefficients are defined as follows: The component t_{ij} of the stress tensor gives the i component of the force per unit area on an area element whose normal is in the j direction. We represent the stress tensor as

$$t_{ij} = -p\,\delta_{ij} + \sigma_{ij},$$

where p is the pressure and σ_{ij} is the contribution to the stress tensor due to viscosity. If we assume that σ_{ij} is linearly proportional to the rate of deformation

$$d_{ij} = \frac{1}{2}\left[\frac{\partial \tilde{u}_i}{\partial x_j} + \frac{\partial \tilde{u}_j}{\partial x_i}\right],$$

we can represent σ_{ij} as

$$\sigma_{ij} = 2\eta_{\mathrm{s}} d_{ij} + \eta_{\mathrm{d}}\,\delta_{ij} \sum_k d_{kk},$$

where η_{s} is the shear viscosity coefficient and η_{d} is the dilational viscosity coefficient. The quantity $\sum_k d_{kk}$ can be interpreted as follows:

$$\sum_k d_{kk} = \sum_k \frac{\partial u_k}{\partial x_k} = \nabla \cdot \mathbf{u}.$$

In general, η_s and η_d are independent parameters. However, for certain physical systems they are related to one another through a relationship first formulated by Stokes. One assumes that the viscous stress tensor σ_{ij} is traceless. In this case the trace of t_{ij} is unaffected by viscous effects; in other words, the mean pressure $-\frac{1}{3}\sum_i t_{ii}$ is unaffected by the effects of viscosity. The condition that σ_{ij} is traceless implies that the combination

$$\sum_i \sigma_{ii} = 2\eta_s \sum_i d_{ii} + 3\eta_d \sum_k d_{kk} = (2\eta_s + 3\eta_d) \sum_k d_{kk}$$

vanishes, or that

$$\eta_d = -\tfrac{2}{3}\eta_s.$$

This result is known as the Stokes relation.

The viscosity coefficients η_s and η_d often appear in the combination $2\eta_s + \eta_d$, as they do in Eq. (8.6.2). When the Stokes relation is satisfied, this combination takes the value

$$2\eta_s + \eta_d = \tfrac{4}{3}\eta_s \qquad \text{(Stokes relation valid).}$$

Under general conditions, such that the Stokes relation is not satisfied, one often defines the bulk viscosity coefficient η_b by

$$\eta_b = \tfrac{2}{3}\eta_s + \eta_d,$$

in terms of which the quantity $2\eta_s + \eta_d$ can be represented as

$$2\eta_s + \eta_d = \tfrac{4}{3}\eta_s + \eta_b \qquad \text{(in general).}$$

Note that η_b vanishes identically when the Stokes relation is valid, for example, for the case of an ideal gas.

As an example of the use of these relations, we note that the Brillouin linewidth Γ_B introduced in Eqs. (7.3.23), (8.5.2), and (8.6.28) can be represented (ignoring the contribution due to thermal conduction) either as

$$\Gamma_B = (2\eta_s + \eta_d)q^2/\rho_0$$

or as

$$\Gamma_B = (\tfrac{4}{3}\eta_s + \eta_b)q^2/\rho_0.$$

Problems

1. Verify Eq. (8.2.17).

2. Generalize the discussion of Section 8.3 to allow for an arbitrary angle θ between the laser and Stokes propagation directions. In particular,

determine how the Brillouin frequency Ω_B, the steady-state line-center gain factor g_0, and the phonon lifetime τ_p depend on the angle θ.

[Ans:

$$\Omega_B(\theta) = \Omega_B(\theta = 180^\circ)\sin(\tfrac{1}{2}\theta)$$

$$g_0(\theta) = g_0(\theta = 180^\circ)/\sin(\tfrac{1}{2}\theta)$$

$$\tau_p(\theta) = \tau_p(\theta = 180^\circ)/\sin^2(\tfrac{1}{2}\theta).]$$

3. Consider the possibility of exciting SBS in the transverse direction by a laser beam passing through a fused silica window at near-normal incidence. Assume conditions appropriate to a large fusion laser. In particular, assume that the window is 70 cm in diameter and is uniformly filled with a laser pulse of 10-ns duration at a wavelength of 350 nm. What is the minimum value of the laser pulse energy for which SBS can be excited? (In fact, transverse SBS has been observed under such conditions similar to those assumed in this problem; see, for example, J. R. Murray, J. R. Smith, R. B. Ehrlich, D. T. Kyrazis, C. E. Thompson, T. L. Weiland, and R. B. Wilcox, *J. Opt. Soc. Am.* **6**, 2402 (1989).)

[Ans: ~2 kJ.]

4. The threshold intensity for optical damage to fused silica is approximately 3 GW/cm^2, and is of the same order of magnitude for most optical materials. (See, for example, W. H. Lowdermilk and D. Milam, *IEEE J. Quantum. Electron.* **17**, 1888 (1981).) Use this fact and the value of the SBS gain factor at line center quoted in Table 8.3.1 to determine the minimum length of a cell utilizing fused silica windows that can be used to excite SBS in acetone with a collimated laser beam. Assume that the laser intensity is restricted to 50% of the threshold intensity as a safety factor to avoid damage to the windows. If the laser pulse length is 20 ns, what is the minimum value of the laser pulse energy per unit area that can be used to excite SBS? (SBS is often excited by tightly focused laser beams rather than by collimated beams to prevent optical damage to the windows of the cell.)

5. *Pulse compression by SBS.* Explain qualitatively why the Stokes radiation excited by SBS in the backward direction can be considerably shorter in duration than the exciting radiation. How must the physical length of the interaction region be related to the duration of the laser pulse in order to observe this effect? Write down the coupled-amplitude equations that are needed to describe this effect, and, if you wish, solve these equations numerically by computer. What determines the minimum value of the duration of the output pulse?

[Hint: Pulse compression by SBS is described in the scientific literature by D. T. Hon, *Opt. Lett.* **5**, 516 (1980) and by S. S. Gulidov, A. A. Mak, and S. B. Papernyi, JETP Lett. *47* 394 (1988).]

6. *Brillouin-Enhanced Four-Wave Mixing.* In addition to SBS, light beams can interact in a Brillouin medium by means of the process known as Brillouin-enhanced four-wave mixing (BEFWM), which is illustrated in the figure shown below.

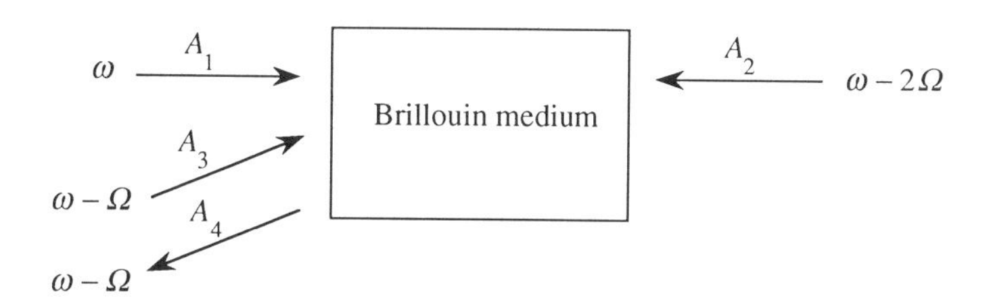

In this process, the incoming signal wave A_3 interferes with the backward-going pump wave A_2 to generate an acoustic wave propagating in the forward direction. The forward-going pump wave scatters from the acoustic wave to generate the phase-conjugate wave A_4. Since A_4 is at the Stokes sideband of A_1, it also undergoes amplification by the usual SBS process. Phase-conjugate reflectivities much larger than 100% have been observed in the BEFWM process. Using the general formalism outlined in Section 8.3, derive the form of the four coupled-amplitude equations that describe BEFWM under steady-state conditions. Solve these equations analytically in the constant-pump approximation.

[Hint: BEFWM has been discussed in the scientific literature. See, for example, M. D. Skeldon, P. Narum, and R. W. Boyd, *Opt. Lett.* **12**, 243 (1987), or P. Narum and R. W. Boyd, *IEEE J. Quantum Electron.* **23**, 1211 (1987).]

References

I. P. Batra, R. H. Enns, and D. Pohl, *Phys. Stat. Sol. (6)* **48**, 11 (1971).
R. Y. Chiao, Ph.D. dissertation, Massachusetts Institute of Technology, 1965.
E. U. Condon, in *Handbook of Physics*, edited by E. U. Condon and H. Odishaw, McGraw-Hill, New York, 1967, Chapter 1.
M. J. Damzen and H. Hutchinson, *IEEE J. Quantum Electron.* **QE-19**, 7 (1983).
N. Goldblatt and M. Hercher, *Phys. Rev. Lett.* **20**, 310 (1968).
J. W. Goodman, *Statistical Optics*, Wiley, New York, 1985.
E. E. Hagenlocker, R. W. Minck, and W. G. Rado, *Phys. Rev.* **154**, 226 (1967).

R. M. Herman and M. A. Gray, *Phys. Rev. Lett.* **19**, 824 (1967).
F. V. Hunt, *J. Acoust. Soc. Am.* **27**, 1019 (1955); see also *American Institute of Physics Handbook*, McGraw-Hill, New York, 1972, p. 3-37 ff.
N. M. Kroll, *J. Appl. Phys.* **36**, 34 (1965).
N. M. Kroll and P. L. Kelly, *Phys. Rev. A* **4**, 763 (1971).
H. Lamb, *Hydrodynamics*, Dover, New York, 1945.
L. D. Landau and E. M. Lifshitz, *Fluid Mechanics*, Pergamon, London, 1959.
L. B. Loeb, *The Kinetic Theory of Gases*, Dover, New York, 1961.
D. Pohl and W. Kaiser, *Phys. Rev. B* **1**, 31 (1970).
D. Pohl, M. Maier, and W. Kaiser, *Phys. Rev. Lett.* **20**, 366 (1968).
D. Sette, in *Handbuch der Physik, XI/1, Acoustics I*, edited by S. Flügge, Springer-Verlag, Berlin, 1961.
V. G. Sidorovich, *Sov. Phys. Tech. Phys.* **21**, 1270 (1976).
B. Ya. Zel'dovich, V. I. Popovichev, V. V. Ragulsky, and F. S. Faizullov, *JETP Lett.* **15**, 109 (1972).

Reviews of Stimulated Scattering

I. L. Fabelinskii, "Stimulated Mandlestam–Brillouin Process," in *Quantum Electronics: A Treatise*, edited by H. Rabin and C. L. Tang, Academic Press, New York, 1975, Vol. I, Part A.
I. L. Fabelinskii, *Molecular Scattering of Light*, Plenum Press, New York, 1968.
R. A. Fisher, editor, *Optical Phase Conjugation*, Academic Press, Orlando, 1983, especially Chapters 6 and 7.
W. Kaiser and M. Maier, in *Laser Handbook*, edited by F. T. Arecchi and E. O. Schulz-DuBois, North-Holland, 1972.
Y. R. Shen, *Principles of Nonlinear Optics*, Wiley, New York, 1984.
B. Ya. Zel'dovich, N. F. Pilipetsky, and V. V. Shkunov, *Principles of Phase Conjugation*, Springer-Verlag, Berlin, 1985.

Chapter 9

Stimulated Raman Scattering and Stimulated Rayleigh-Wing Scattering

9.1. The Spontaneous Raman Effect

The spontaneous Raman effect was discovered by C. V. Raman in 1928. To observe this effect, a beam of light illuminates a material sample (which can be a solid, liquid, or gas), and the scattered light is observed spectroscopically, as illustrated in Fig. 9.1.1. In general, the scattered light contains frequencies different from those of the excitation source. Those new components shifted to lower frequencies are called Stokes lines, and those shifted to higher frequencies are called anti-Stokes lines. The Stokes lines are typically orders of magnitude more intense than the anti-Stokes lines.

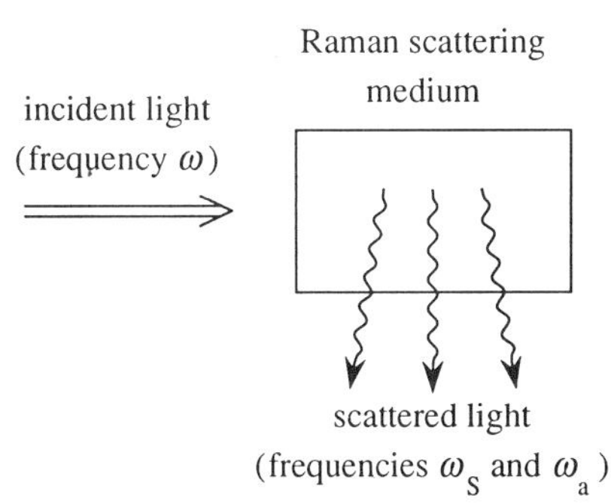

FIGURE 9.1.1 Spontaneous Raman scattering.

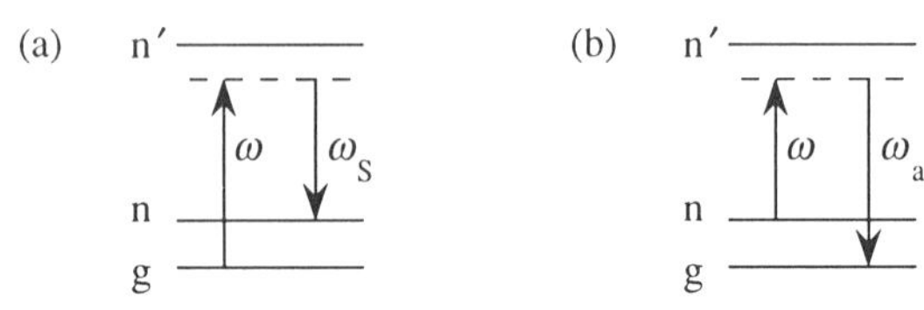

FIGURE 9.1.2 Energy level diagrams describing (a) Raman Stokes scattering and (b) Raman anti-Stokes scattering.

These properties of Raman scattering can be understood through use of the energy level diagrams shown in Fig. 9.1.2. Raman Stokes scattering consists of a transition from the ground state g to a virtual level associated with the excited state n' followed by a transition from the virtual level to the final state n. Raman anti-Stokes scattering entails a transition from level n to level g with n' serving as the intermediate level. The anti-Stokes lines are typically much weaker than the Stokes lines because, in thermal equilibrium, the population of level n is smaller than the population in level g by the Boltzmann factor $\exp(-\hbar\omega_{ng}/kT)$.

The Raman effect has important spectroscopic applications because transitions that are one-photon forbidden can often be studied using Raman scattering. For example, the Raman transitions illustrated in Fig. 9.1.2 can occur only if the matrix elements $\langle g|\hat{\mathbf{r}}|n'\rangle$ and $\langle n'|\hat{\mathbf{r}}|n\rangle$ are both nonzero, and this fact implies (for a material system that possesses inversion symmetry, so that the energy eigenstates possess definite parity) that the states g and n must possess the same parity. But under these conditions the $g \to n$ transition is forbidden for single-photon electric dipole transitions because the matrix element $\langle g|\hat{\mathbf{r}}|n\rangle$ must necessarily vanish.

9.2. Spontaneous versus Stimulated Raman Scattering

The spontaneous Raman scattering process described in the previous section is typically a rather weak process. Even for condensed matter, the scattering cross section per unit volume for Raman Stokes scattering is only about 10^{-6} cm^{-1}. Hence, in propagating through 1 cm of the scattering medium, only approximately 1 part in 10^6 of the incident radiation will be scattered into the Stokes frequency.

However, under excitation by an intense laser beam, highly efficient scattering can occur as a result of the stimulated version of the Raman scattering process. Stimulated Raman scattering is typically a very strong scattering process: ten percent or more of the energy of the incident laser beam is often converted into the Stokes frequency. Another difference between spontaneous and stimulated Raman scattering is that the spontaneous process leads to nearly isotropic emission, whereas the stimulated process leads to emission in a narrow cone in the forward and backward directions. Stimulated Raman scattering was discovered by Woodbury and Ng (1962) and was described more fully by Eckardt *et al.* (1962). The properties of stimulated Raman scattering have been reviewed by Bloembergen (1967), Kaiser and Maier (1972), Penzkofer *et al.* (1979), and Raymer and Walmsley (1990).

The relation between spontaneous and stimulated Raman scattering can be understood in terms of an argument (Hellwarth, 1963) that considers the process from the point of view of the photon occupation numbers of the various field modes. One postulates that the probability per unit time that a photon will be emitted into Stokes mode S is given by

$$P_S = D m_L (m_S + 1). \tag{9.2.1}$$

Here m_L is the mean number of photons per mode in the laser radiation, m_S is the mean number of photons in Stokes mode S, and D is a proportionality constant whose value depends on the physical properties of the material medium. This functional form is assumed because the factor m_L leads to the expected linear dependence of the transition rate on the laser intensity, and the factor $m_S + 1$ leads to stimulated scattering through the contribution m_S and to spontaneous scattering through the contribution of unity. This dependence on the factor $m_S + 1$ is reminiscent of the stimulated and spontaneous contributions to the total emission rate for a single-photon transition of an atomic system as treated by the Einstein A and B coefficients. Equation (9.2.1) can be justified by more rigorous treatments; note, for example, that the results of our analysis are consistent with those of the fully quantum-mechanical treatment of Raymer and Mostowski (1981).

By the definition of P_S as a probability per unit time for emitting a photon into mode S, the time rate of change of the mean photon occupation number for the Stokes mode is given by $dm_S/dt = P_S$ or, through use of Eq. (9.2.1), by

$$\frac{dm_S}{dt} = D m_L (m_S + 1). \tag{9.2.2}$$

If we now assume that the Stokes mode corresponds to a wave traveling in the positive z direction at the velocity c/n, as illustrated in Fig. 9.2.1, we see that the time rate of change given by Eq. (9.2.2) corresponds to a spatial growth rate given by

$$\frac{dm_S}{dz} = \frac{1}{c/n}\frac{dm_S}{dt} = \frac{1}{c/n} D m_L (m_S + 1). \tag{9.2.3}$$

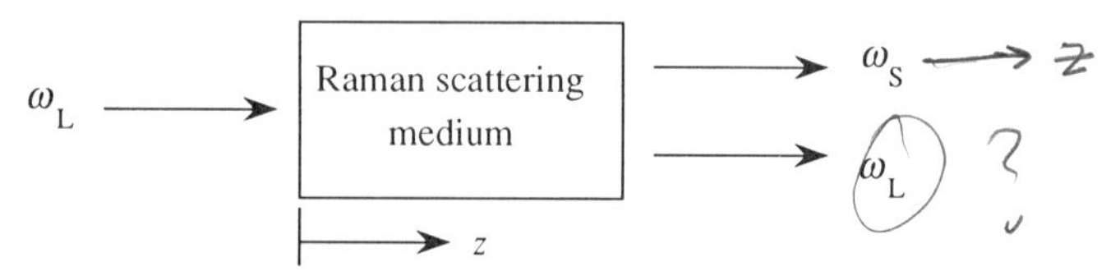

FIGURE 9.2.1 Geometry describing Raman Stokes generation.

For definiteness, Fig. 9.2.1 shows the laser and Stokes beams propagating in the same direction; in fact, Eq. (9.2.3) applies even if the angle between the propagation directions of the laser and Stokes waves is arbitrary, as long as z is measured along the propagation direction of the Stokes wave.

It is instructive to consider Eq. (9.2.3) in the two opposite limits of $m_S \ll 1$ and $m_S \gg 1$. In the first limit, where the occupation number of the Stokes mode is much less than unity, Eq. (9.2.3) becomes simply

$$\frac{dm_S}{dz} = \frac{1}{c/n} Dm_L \qquad (\text{for } m_S \ll 1). \tag{9.2.4}$$

The solution to this equation for the geometry of Fig. 9.2.1 and under the assumption that the laser field is unaffected by the interaction (and hence that m_L is independent of z) is

Spont. Raman

$$m_S(z) = \frac{1}{c/n} Dm_L z \qquad (\text{for } m_S \ll 1). \tag{9.2.5}$$

This limit corresponds to spontaneous Raman scattering; the Stokes intensity here is proportional to the length of the Raman medium and hence to the total number of molecules contained in the interaction region.

The opposite limiting case is that in which there are many photons in the Stokes mode. In this case Eq. (9.2.3) becomes

Stim Raman

$$\frac{dm_S}{dz} = \frac{1}{c/n} Dm_L m_S \qquad (\text{for } m_S \gg 1), \tag{9.2.6}$$

whose solution (again under the assumption of an undepleted input field) is

$$m_S(z) = m_S(0) e^{Gz} \qquad (\text{for } m_S \gg 1), \tag{9.2.7}$$

where we have introduced the Raman gain coefficient

$$G = \frac{Dnm_L}{c}. \tag{9.2.8}$$

Here $m_S(0)$ denotes the photon occupation number associated with the Stokes field at the input to the Raman medium. If no field is injected into the Raman medium, $m_S(0)$ represents the quantum noise associated with the vacuum state, which is equivalent to one photon per mode. Emission of the sort described by Eq. (9.2.7) is called stimulated Raman scattering. The Stokes intensity is seen to grow exponentially with propagation distance through the medium, and large values of the Stokes intensity are routinely observed at the output of the interaction region.

We see from Eq. (9.2.8) that the Raman gain coefficient can be related simply to the phenomenological constant D introduced earlier. However, we see from Eq. (9.2.5) that the strength of spontaneous Raman scattering is also proportional to D. Since the strength of spontaneous Raman scattering is often described in terms of a scattering cross section, it is thus possible to determine a relation between the gain coefficient G for stimulated Raman scattering and the cross section for spontaneous Raman scattering. This relationship is derived as follows:

Since one laser photon is lost for each Stokes photon that is created, the occupation number of the laser field changes as the result of spontaneous scattering into one particular Stokes mode in accordance with the relation $dm_L/dz = -dm_S/dz$, with dm_S/dz given by Eq. (9.2.4). However, since the system can radiate into a large number of Stokes modes, the total rate of loss of laser photons is given by

$$\frac{dm_L}{dz} = -Mb\frac{dm_S}{dz} = \frac{-Dnm_L Mb}{c}, \tag{9.2.9}$$

where M is the number of modes into which the system can radiate and where b is a geometrical factor that accounts for the fact that the angular distribution of scattered radiation may be nonuniform and hence that the scattering rate into different Stokes modes may be different. Explicitly, b is the ratio of the angularly averaged Stokes emission rate to the rate in the direction of the particular Stokes mode S for which D (and hence the Raman gain coefficient) is to be determined. If $|f(\theta, \phi)|^2$ denotes the angular distribution of the Stokes radiation, b is then given by

$$b = \frac{\int |f(\theta, \phi)|^2 \, d\Omega/4\pi}{|f(\theta_S, \phi_S)|^2}, \tag{9.2.10}$$

where (θ_S, ϕ_S) gives the direction of the particular Stokes mode for which D is to be determined.

The total number of Stokes modes into which the system can radiate is given by the expression (see, for example, Boyd, 1983, Eq. (3.4.4))

$$M = \frac{V\omega_S^2 \, \Delta\omega}{\pi^2 (c/n)^3}, \tag{9.2.11}$$

where V denotes the volume of the region in which the modes are defined and where $\Delta\omega$ denotes the linewidth of the scattered Stokes radiation. The rate of loss of laser photons is conventionally described by the cross section σ for Raman scattering, which is defined by the relation

$$\frac{dm_{\mathrm{L}}}{dz} = -N\sigma m_{\mathrm{L}}, \tag{9.2.12}$$

where N is the number density of molecules. By comparison of Eqs. (9.2.9) and (9.2.12), we see that we can express the parameter D in terms of the cross section σ by

$$D = \frac{N\sigma(c/n)}{Mb}. \tag{9.2.13}$$

This expression for D, with M given by Eq. (9.2.11), is now substituted into expression (9.2.8) for the Raman gain coefficient to give the result

$$G = \frac{N\sigma\pi^2 c^3 m_{\mathrm{L}}}{V\omega_{\mathrm{S}}^2\,\Delta\omega\, bn^3} \equiv \frac{N\pi^2 c^3 m_{\mathrm{L}}}{V\omega_{\mathrm{S}}^2 bn^3}\left(\frac{\partial\sigma}{\partial\omega}\right)_0, \tag{9.2.14}$$

where in obtaining the second form we have used the definition of the spectral density of the scattering cross section to express σ in terms of its line-center value $(\partial\sigma/\partial\omega)_0$ as

$$\sigma = \left(\frac{\partial\sigma}{\partial\omega}\right)_0 \Delta\omega. \tag{9.2.15}$$

Equation (9.2.14) gives the Raman gain coefficient in terms of the number of laser photons per mode, m_{L}. In order to express the gain coefficient in terms of the laser intensity, which can be measured directly, we assume the geometry shown in Fig. 9.2.2. The laser intensity I_{L} is equal to the number of photons contained in this region multiplied by the energy per photon and divided by the cross-sectional area of the region and by the transit time through the region, that is,

$$I_{\mathrm{L}} = \frac{m_{\mathrm{L}}\hbar\omega_{\mathrm{L}}}{A(nL/c)} = \frac{m_{\mathrm{L}}\hbar\omega_{\mathrm{L}} c}{Vn}, \tag{9.2.16}$$

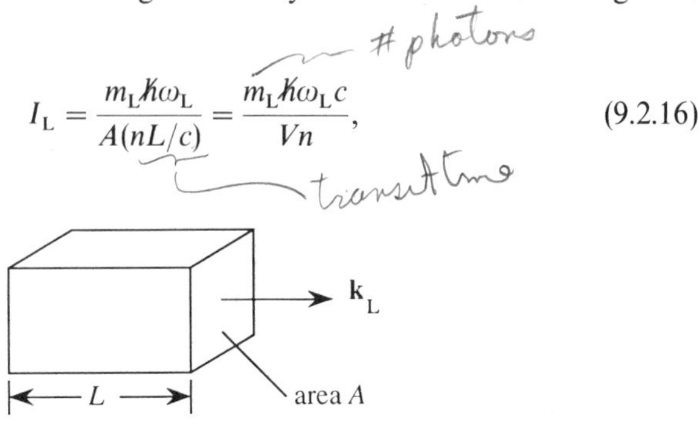

FIGURE 9.2.2 Geometry of the region within which the laser and Stokes modes are defined.

where $V = AL$. Through use of this result, the Raman gain coefficient of Eq. (9.2.14) can be expressed as

$$G = \frac{N\pi^2 c^2}{\omega_S^2 b n^2 \hbar \omega_L}\left(\frac{\partial\sigma}{\partial\omega}\right)_0 I_L. \tag{9.2.17}$$

It is sometimes convenient to express the Raman gain coefficient not in terms of the spectral cross section $(\partial\sigma/\partial\omega)_0$ but in terms of the differential spectral cross section $(\partial^2\sigma/\partial\omega\,\partial\Omega)_0$, where $d\Omega$ is an element of solid angle. These quantities are related by

$$\left(\frac{\partial\sigma}{\partial\omega}\right)_0 = 4\pi b\left(\frac{\partial^2\sigma}{\partial\omega\,\partial\Omega}\right)_0, \tag{9.2.18}$$

where b is the factor defined in Eq. (9.2.10) that accounts for the possible nonuniform angular distribution of the scattered Stokes radiation. Through use of this relation, Eq. (9.2.17) becomes

$$G = \frac{4\pi^3 N c^2}{\omega_S^2 \hbar \omega_L n_S^2}\left(\frac{\partial^2\sigma}{\partial\omega\,\partial\Omega}\right)_0 I_L. \tag{9.2.19}$$

Some of the parameters describing stimulated Raman scattering are listed in Table 9.2.1 for a number of materials.

TABLE 9.2.1 Properties of stimulated Raman scattering for several materials*

Substance	Frequency shift ν_0 (cm^{-1})	Linewidth $\Delta\nu$ (cm^{-1})	Cross section $N(d\sigma/d\Omega)_0$ ($10^{-8}\ cm^{-1}\ sr^{-1}$)	Gain factor† G/I_L (10^{-3} cm/MW)
Liquid O_2	1552	0.117	0.48 ± 0.14	14.5 ± 4
Liquid N_2	2326.5	0.067	0.29 ± 0.09	16 ± 5
Benzene	992	2.15	3.06	2.8
CS_2	655.6	0.50	7.55	24
Nitrobenzene	1345	6.6	6.4	2.1
Bromobenzene	1000	1.9	1.5	1.5
Chlorobenzene	1002	1.6	1.5	1.9
Toulene	1003	1.94	1.1	1.2
$LiNbO_3$	256	23	381	8.9
	637	20	231	9.4
$Ba_2NaNb_5O_{15}$	650			6.7
$LiTaO_3$	201	22	238	4.4
SiO_2	467			0.8
H_2 gas (P > 10 atm)	4155			1.5

† To obtain the gain constant G in units of cm^{-1} at frequency ν_S, multiply the entry by the ratio ν_S/ν_L and by the intensity I in units of MW/cm^2.

* After Kaiser and Maier, 1972.

9.3. Stimulated Raman Scattering Described through Use of the Nonlinear Polarization

Next, we develop a classical model that describes stimulated Raman scattering (Garmire *et al.*, 1963). We assume that the optical radiation interacts with a vibrational mode of a molecule, as illustrated in Fig. 9.3.1. We assume that the vibrational mode can be described as a simple harmonic oscillator of resonance frequency ω_v and damping constant γ, and we denote by $\tilde{q}$ the deviation of the internuclear distance from its equilibrium value q_0. The equation of motion describing the molecular vibration is thus

$$\frac{d^2\tilde{q}}{dt^2} + 2\gamma\frac{d\tilde{q}}{dt} + \omega_v^2\tilde{q} = \frac{\tilde{F}(t)}{m}, \tag{9.3.1}$$

where $\tilde{F}$ denotes any force that acts on the vibrational degree of freedom and where m represents the reduced nuclear mass.

The key assumption of the theory is that the optical polarizability of the molecule is not constant, but depends on the internuclear distance according to the equation

$$\tilde{\alpha}(t) = \alpha_0 + \left(\frac{\partial\alpha}{\partial q}\right)_0 \tilde{q}(t). \tag{9.3.2}$$

Here α_0 is the polarizability of a molecule in which the internuclear distance is held fixed at its equilibrium value. According to Eq. (9.3.2), when the molecule is set into oscillation its polarizability will be modulated in time, and hence the refractive index of a collection of coherently oscillating molecules will be modulated in accordance with the relations

$$\tilde{n}(t) = \sqrt{\tilde{\epsilon}(t)} = [1 + 4\pi N\tilde{\alpha}(t)]^{1/2}. \tag{9.3.3}$$

The temporal modulation of the refractive index will modify a beam of light in passing through the medium. In particular, frequency sidebands separated from the laser frequency by $\pm\omega_v$ will be impressed upon the transmitted laser beam.

Next, we examine how the molecular vibrations can be driven coherently by an applied optical field. In the presence of the optical field $\tilde{\mathbf{E}}(z, t)$, each

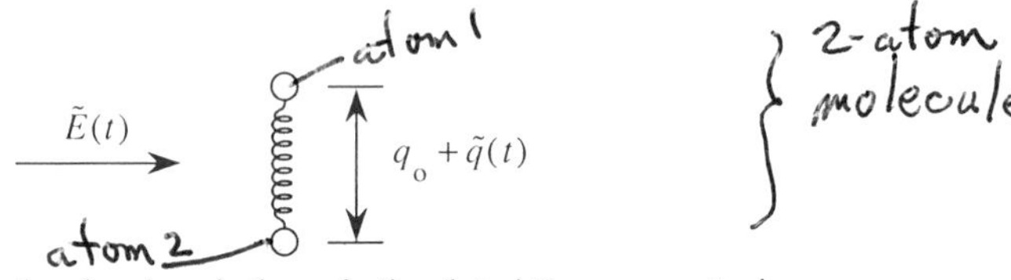

FIGURE 9.3.1 Molecular description of stimulated Raman scattering.

molecule will become polarized, and the induced dipole moment of a molecule located at coordinate z will be given by

$$\tilde{\mathbf{p}}(z,t) = \alpha\tilde{\mathbf{E}}(z,t). \tag{9.3.4}$$

The energy required to establish this oscillating dipole moment is given by

$$W = \tfrac{1}{2}\langle\tilde{\mathbf{p}}(z,t)\cdot\tilde{\mathbf{E}}(z,t)\rangle = \tfrac{1}{2}\alpha\langle\tilde{E}^2(z,t)\rangle, \tag{9.3.5}$$

where the angular brackets denote a time average over an optical period. The applied optical field hence exerts a force given by

$$\tilde{F} = \frac{dW}{dq} = \frac{1}{2}\left(\frac{d\alpha}{dq}\right)_0 \langle\tilde{E}^2(z,t)\rangle \tag{9.3.6}$$

on the vibrational degree of freedom. In particular, if the applied field contains two frequency components, Eq. (9.3.6) shows that the molecular coordinate will experience a time-varying force at the beat frequency between the two field components.

The origin of stimulated Raman scattering can be understood schematically in terms of the interactions shown in Fig. 9.3.2. Part (a) of the figure shows how molecular vibrations modulate the refractive index of the medium at frequency ω_v and thereby impress frequency sidebands onto the laser field. Part (b) shows how the Stokes field at frequency $\omega_S = \omega_L - \omega_v$ can beat with the laser field to produce a modulation of the total intensity of the form

$$\tilde{I}(t) = I_0 + I_1 \cos(\omega_L - \omega_S)t. \tag{9.3.7}$$

(a)

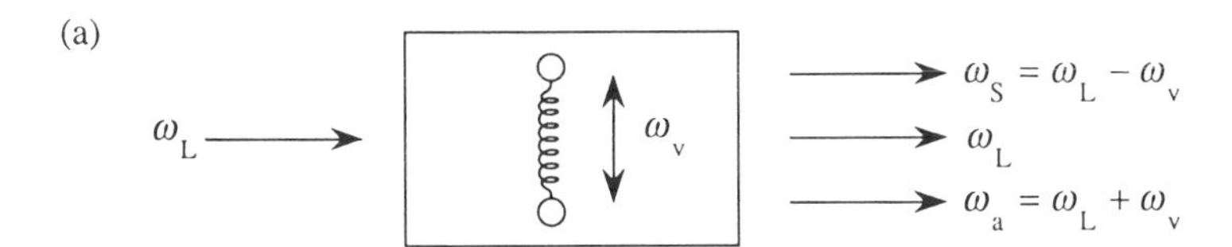

(b)

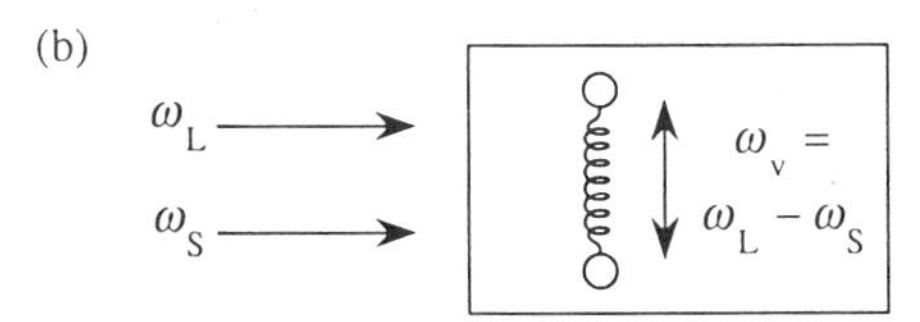

FIGURE 9.3.2 Stimulated Raman scattering.

This modulated intensity coherently excites the molecular oscillation at frequency $\omega_v = \omega_L - \omega_S$. The two processes shown in parts (a) and (b) of the figure reinforce one another in the sense that the interaction shown in part (b) leads to a stronger molecular vibration, which by the interaction shown in part (a) leads to a stronger Stokes field, which in turn leads to a stronger molecular vibration.

To make these ideas quantitative, let us assume that the total optical field can be represented as

$$\tilde{E}(z,t) = A_L e^{i(k_L z - \omega_L t)} + A_S e^{i(k_S z - \omega_S t)} + \text{c.c.} \tag{9.3.8}$$

According to Eq. (9.3.6) the time-varying part of the applied force is then given by

$$\tilde{F}(z,t) = \left(\frac{\partial \alpha}{\partial q}\right)_0 [A_L A_S^* e^{i(Kz - \Omega t)} + \text{c.c.}], \tag{9.3.9}$$

where we have introduced the notation

$$K = k_L - k_S \quad \text{and} \quad \Omega = \omega_L - \omega_S. \tag{9.3.10}$$

We next find the solution to Eq. (9.3.1) with a force term of the form of Eq. (9.3.9). We adopt a trial solution of the form

$$\tilde{q} = q(\Omega) e^{i(Kz - \Omega t)} + \text{c.c.} \tag{9.3.11}$$

We insert Eqs. (9.3.9) and (9.3.11) into Eq. (9.3.1), which becomes

$$-\omega^2 q(\Omega) - 2i\omega\gamma q(\Omega) + \omega_v^2 q(\Omega) = \frac{1}{m}\left(\frac{\partial \alpha}{\partial q}\right)_0 A_L A_S^*,$$

and hence we find that the amplitude of the molecular vibration is given by

$$q(\Omega) = \frac{(1/m)[\partial\alpha/\partial q]_0 A_L A_S^*}{\omega_v^2 - \Omega^2 - 2i\Omega\gamma}. \tag{9.3.12}$$

Since the polarization of the medium is given according to Eqs. (9.3.2) and (9.3.4) by

$$\begin{aligned}\tilde{P}(z,t) &= N\tilde{p}(z,t) = N\tilde{\alpha}(z,t)\tilde{E}(z,t) \\ &= N\left\{\alpha_0 + \left(\frac{\partial \alpha}{\partial q}\right)_0 \tilde{q}(z,t)\right\}\tilde{E}(z,t),\end{aligned} \tag{9.3.13}$$

the nonlinear part of the polarization is given by

$$\begin{aligned}\tilde{P}^{\text{NL}}(z,t) = N\left(\frac{\partial \alpha}{\partial q}\right)_0 &[q(\omega)e^{i(Kz - \Omega t)} + \text{c.c.}] \\ &\times [A_L e^{i(k_L z - \omega_L t)} + A_S e^{i(K_S z - \omega_S t)} + \text{c.c.}].\end{aligned} \tag{9.3.14}$$

The nonlinear polarization is seen to contain several different frequency components. The part of this expression that oscillates at frequency ω_S is known as the Stokes polarization and is given by

$$\tilde{P}_S^{NL}(z,t) = P(\omega_S)e^{-i\omega_S t} + \text{c.c.}, \tag{9.3.15}$$

with the complex amplitude of the Stokes polarization given by

$$P(\omega_S) = N\left(\frac{\partial\alpha}{\partial q}\right)_0 q^*(\Omega)A_L e^{ik_S z}. \tag{9.3.16}$$

By introducing the expression (9.3.12) for $q(\Omega)$ into this equation, we find that the complex amplitude of the Stokes polarization is given by

$$P(\omega_S) = \frac{(N/m)(\partial\alpha/\partial q)_0^2 |A_L|^2 A_S}{\omega_v^2 - \Omega^2 + 2i\Omega\gamma} e^{ik_S z}. \tag{9.3.17}$$

We now define the Raman susceptibility through the expression

$$P(\omega_S) \equiv 6\chi_R(\omega_S)|A_L|^2 A_S e^{ik_S z}, \tag{9.3.18}$$

where for notational convenience we have introduced $\chi_R(\omega_S)$ as a shortened form of $\chi^{(3)}(\omega_S = \omega_S + \omega_L - \omega_L)$. By comparison of Eqs. (9.3.17) and (9.3.18), we find that the Raman susceptibility is given by

$$\chi_R(\omega_S) = \frac{(N/6m)(\partial\alpha/\partial q)_0^2}{\omega_v^2 - (\omega_L - \omega_S)^2 + 2i(\omega_L - \omega_S)\gamma}. \tag{9.3.19a}$$

The real and imaginary parts of $\chi_R(\omega_S) \equiv \chi_R'(\omega_S) + i\chi_R''(\omega_S)$ are illustrated in Fig. 9.3.3.

Near the Raman resonance, the Raman susceptibility can be approximated as

$$\chi_R(\omega_S) = \frac{(N/12m\omega_v)(\partial\alpha/\partial q)_0^2}{[\omega_S - (\omega_L - \omega_v)] + i\gamma}. \tag{9.3.19b}$$

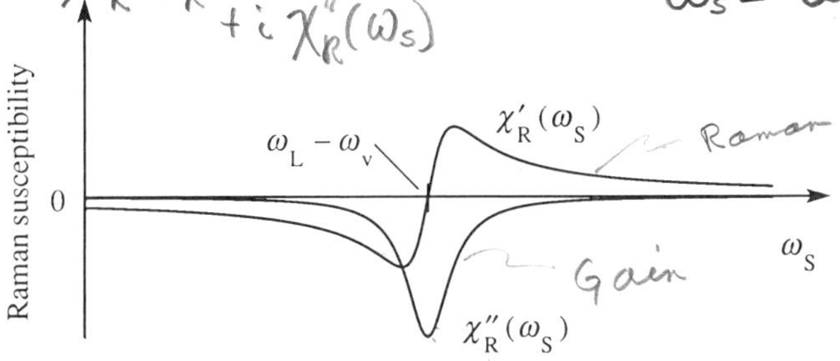

FIGURE 9.3.3 Resonance structure of the Raman susceptibility.

Note that, at the exact Raman resonance (that is, for $\omega_S = \omega_L - \omega_v$), the Raman susceptibility is negative imaginary. (We shall see below that consequently the Stokes wave experiences amplification.)

In order to describe explicitly the spatial evolution of the Stokes wave, we use Eqs. (9.3.8), (9.3.15), (9.3.18), and (9.3.19) for the nonlinear polarization for the driven wave equation (2.1.15). We then find that the evolution of the field amplitude A_S is given in the slowly-varying-amplitude approximation by

$$\frac{dA_S}{dz} = -\alpha_S A_S, \tag{9.3.20}$$

where

$$\alpha_S = -12\pi i \frac{\omega_S}{n_S c} \chi_R(\omega_S)|A_L|^2 \tag{9.3.21}$$

is the Stokes wave "absorption" coefficient. Since the imaginary part of $\chi_R(\omega_S)$ is negative, the real part of the absorption coefficient is negative, implying that the Stokes wave actually experiences exponential growth. Note that α_S depends only on the modulus of the complex amplitude of the laser field. Raman Stokes amplification is thus a process for which the phase-matching condition is automatically satisfied. Alternatively, Raman Stokes amplification is said to be a pure gain process.

We can also predict the spatial evolution of a wave at the anti-Stokes frequency through use of the results of the calculation just completed. In the derivation of Eq. (9.3.19), no assumptions were made regarding the sign of $\omega_L - \omega_S$. We can thus deduce the form of the anti-Stokes susceptibility by formally replacing ω_S by ω_a in Eq. (9.3.19) to obtain the result

$$\chi_R(\omega_a) = \frac{(N/6m)(\partial\alpha/\partial q)_0^2}{\omega_v^2 - (\omega_L - \omega_a)^2 + 2i(\omega_L - \omega_a)\gamma}. \tag{9.3.22}$$

Since ω_S and ω_a are related through

$$(\omega_L - \omega_S) = -(\omega_L - \omega_a), \tag{9.3.23}$$

we see that

$$\chi_R(\omega_a) = \chi_R(\omega_S)^*. \tag{9.3.24}$$

The relation between the Stokes and anti-Stokes Raman susceptibilities is illustrated in Fig. 9.3.4. Near the Raman resonance, Eq. (9.3.22) can be approximated by

$$\chi_R(\omega_a) = -\frac{(N/12m\omega_v)(\partial\alpha/\partial q)_0^2}{[\omega_a - (\omega_L + \omega_v)] + i\gamma}, \tag{9.3.25}$$

and at the exact resonance the Raman susceptibility is positive imaginary. The amplitude of the anti-Stokes wave hence obeys the propagation equation

$$\frac{dA_a}{dz} = -\alpha_a A_a, \tag{9.3.26}$$

where

$$\alpha_a = -12\pi i \frac{\omega_a}{n_a c} \chi_R(\omega_a) |A_L|^2. \tag{9.3.27}$$

For a positive imaginary value $\chi_R(\omega_a)$, α_a is positive real, implying that the anti-Stokes wave experiences attenuation.

However, it is found experimentally (Terhune, 1963) that the anti-Stokes wave is generated with appreciable efficiency, at least in certain directions. The origin of anti-Stokes generation is an additional contribution to the nonlinear polarization beyond that described by the Raman susceptibility of Eq. (9.3.25). Inspection of Eq. (9.3.14) shows that there is a contribution to the anti-Stokes polarization

$$\tilde{P}_a^{NL}(z,t) = P(\omega_a) e^{-i\omega_a t} + \text{c.c.} \tag{9.3.28}$$

given by

$$P(\omega_a) = N\left(\frac{\partial \alpha}{\partial q}\right)_0 q(\Omega) A_L = \frac{(N/m)(\partial\alpha/\partial q)_0^2 A_L^2 A_S^*}{\omega_v^2 - \Omega^2 - 2i\Omega\gamma} e^{i(2k_L - k_S)z}. \tag{9.3.29}$$

(Recall that $\Omega \equiv \omega_L - \omega_S = \omega_a - \omega_L$.) This contribution to the nonlinear polarization can be described in terms of a four-wave mixing susceptibility $\chi_F(\omega_a) \equiv \chi^{(3)}(\omega_a = \omega_L + \omega_L - \omega_S)$, which is defined by the relation

$$P(\omega_a) = 3\chi_F(\omega_a) A_L^2 A_S^* e^{i(2k_L - k_S)z} \tag{9.3.30}$$

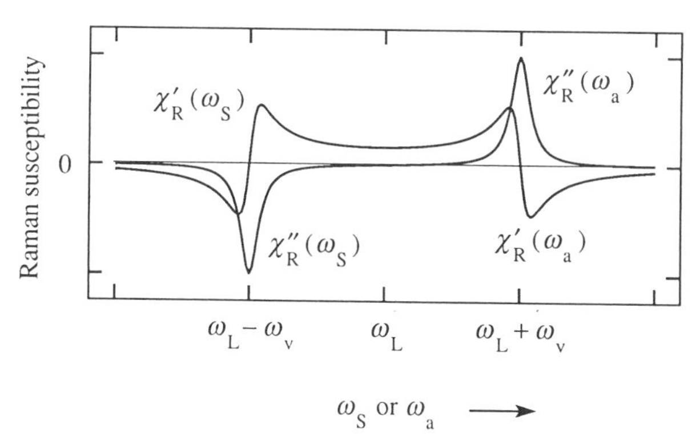

FIGURE 9.3.4 Relation between Stokes and anti-Stokes Raman susceptibilities.

and which is hence equal to

$$\chi_{\mathrm{F}}(\omega_{\mathrm{a}}) = \frac{(N/3m)(\partial\alpha/\partial q)_0^2}{\omega_{\mathrm{v}}^2 - (\omega_{\mathrm{L}} - \omega_{\mathrm{a}})^2 + 2i(\omega_{\mathrm{L}} - \omega_{\mathrm{a}})\gamma}. \tag{9.3.31}$$

We can see by comparison with Eq. (9.3.22) that

$$\chi_{\mathrm{F}}(\omega_{\mathrm{a}}) = 2\chi_{\mathrm{R}}(\omega_{\mathrm{a}}). \tag{9.3.32}$$

The total polarization at the anti-Stokes frequency is the sum of the contributions described by Eqs. (9.3.22) and (9.3.31), and is hence given by

$$P(\omega_{\mathrm{a}}) = 6\chi_{\mathrm{R}}(\omega_{\mathrm{a}})|A_{\mathrm{L}}|^2 A_{\mathrm{a}} e^{ik_{\mathrm{a}}z} + 3\chi_{\mathrm{F}}(\omega_{\mathrm{a}}) A_{\mathrm{L}}^2 A_{\mathrm{S}}^* e^{i(2k_{\mathrm{L}} - k_{\mathrm{S}})z}. \tag{9.3.33}$$

Similarly, there is a four-wave mixing contribution to the Stokes polarization described by

$$\chi_{\mathrm{F}}(\omega_{\mathrm{S}}) = \frac{(N/3m)(\partial\alpha/\partial q)_0^2}{\omega_{\mathrm{v}}^2 - (\omega_{\mathrm{L}} - \omega_{\mathrm{a}})^2 + 2i(\omega_{\mathrm{L}} - \omega_{\mathrm{S}})\gamma}, \tag{9.3.34}$$

so that the total polarization at the Stokes frequency is given by

$$P(\omega_{\mathrm{S}}) = 6\chi_{\mathrm{R}}(\omega_{\mathrm{S}})|A_{\mathrm{L}}|^2 A_{\mathrm{S}} e^{ik_{\mathrm{S}}z} + 3\chi_{\mathrm{F}}(\omega_{\mathrm{S}}) A_{\mathrm{L}}^2 A_{\mathrm{a}}^* e^{i(2k_{\mathrm{L}} - k_{\mathrm{a}})z}. \tag{9.3.35}$$

The Stokes four-wave mixing susceptibility is related to the Raman Stokes susceptibility by

$$\chi_{\mathrm{F}}(\omega_{\mathrm{S}}) = 2\chi_{\mathrm{R}}(\omega_{\mathrm{S}}) \tag{9.3.36}$$

and to the anti-Stokes susceptibility through

$$\chi_{\mathrm{F}}(\omega_{\mathrm{S}}) = \chi_{\mathrm{F}}(\omega_{\mathrm{a}})^*. \tag{9.3.37}$$

The spatial evolution of the Stokes and anti-Stokes fields is now obtained by introducing Eqs. (9.3.33) and (9.3.35) into the driven wave equation (2.1.15). We assume that the medium is optically isotropic and that the slowly-varying-amplitude and constant-pump approximations are valid. We find that the field amplitudes obey the set of coupled equations

$$\frac{dA_{\mathrm{S}}}{dz} = -\alpha_{\mathrm{S}} A_{\mathrm{S}} + \kappa_{\mathrm{S}} A_{\mathrm{a}}^* e^{i\Delta k z}, \tag{9.3.38a}$$

$$\frac{dA_{\mathrm{a}}}{dz} = -\alpha_{\mathrm{a}} A_{\mathrm{a}} + \kappa_{\mathrm{a}} A_{\mathrm{S}}^* e^{i\Delta k z}, \tag{9.3.38b}$$

where we have introduced nonlinear absorption and coupling coefficients

$$\alpha_j = \frac{-12\pi i\omega_j}{n_j c}\chi_{\mathrm{R}}(\omega_j)|A_L|^2, \qquad j = \mathrm{S, a}, \tag{9.3.39a}$$

$$\kappa_j = \frac{6\pi i \omega_j}{n_j c} \chi_F(\omega_j) A_L^2, \qquad j = S, a, \tag{9.3.39b}$$

and have defined the wave vector mismatch

$$\Delta k = \Delta \mathbf{k} \cdot \hat{\mathbf{z}} = (2\mathbf{k}_L - \mathbf{k}_S - \mathbf{k}_a) \cdot \hat{\mathbf{z}}. \tag{9.3.40}$$

The form of Eqs. (9.3.38) shows that each of the Stokes and anti-Stokes amplitudes is driven by a Raman gain or loss term (the first term on the right-hand side) and by a phase-matched four-wave mixing term (the second). The four-wave mixing term is an effective driving term only when the wave vector mismatch Δk is small. For a material with normal dispersion, the refractive index experienced by the laser wave is always less than the mean of those experienced by the Stokes and anti-Stokes waves, as illustrated in part (a) of Fig. 9.3.5. For this reason, perfect phase matching ($\Delta k = 0$) can always be achieved if the Stokes wave propagates at some nonzero angle with respect to the laser wave, as illustrated in part (b) of the figure. For angles appreciably different from this phase-matching angle, Δk is large, and only the first term on the right-hand side of each of Eqs. (9.3.38) is important. For these directions, the two equations decouple, and the Stokes sideband experiences gain and the anti-Stokes sideband experiences loss. However, for directions such that Δk is small, both driving terms on the right-hand sides of Eqs. (9.3.38) are important, and the two equations must be solved simultaneously. In the next section, we shall see how to solve these equations, and shall see that both Stokes and anti-Stokes radiation can be generated in directions for which Δk is small.

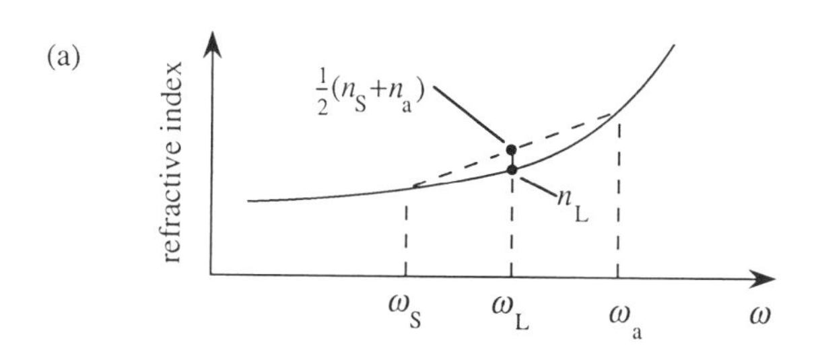

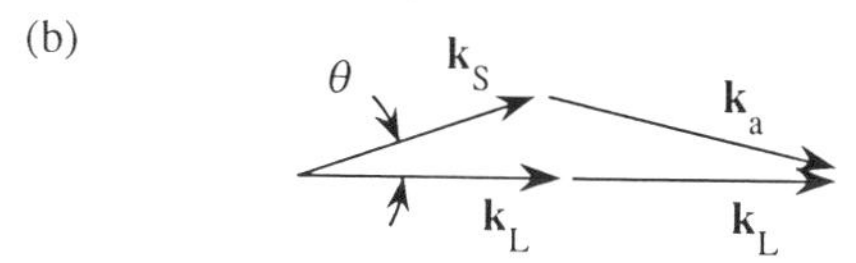

FIGURE 9.3.5 Phase-matching relations for Stokes and anti-Stokes coupling in stimulated Raman scattering.

9.4. Stokes–Anti-Stokes Coupling in Stimulated Raman Scattering

In this section, we study the nature of the solution to the equations describing the propagation of the stokes and anti-Stokes waves. We have just seen that these equations are of the form

$$\frac{dA_1}{dz} = -\alpha_1 A_1 + \kappa_1 A_2^* e^{i\Delta k z}, \tag{9.4.1a}$$

$$\frac{dA_2^*}{dz} = -\alpha_2^* A_2^* + \kappa_2^* A_1 e^{-i\Delta k z}. \tag{9.4.1b}$$

In fact, equations of this form are commonly encountered in nonlinear optics, and also describe, for example, any forward four-wave mixing process in the constant-pump approximation. The ensuring discussion of the solution to these equations is simplified by first rewriting Eqs. (9.4.1) as

$$e^{-i\Delta k z/2}\left(\frac{dA_1}{dz} + \alpha_1 A_1\right) = \kappa_1 A_2^* e^{i\Delta k z/2}, \tag{9.4.2a}$$

$$e^{i\Delta k z/2}\left(\frac{dA_2^*}{dz} + \alpha_2^* A_2^*\right) = \kappa_2^* A_1 e^{-i\Delta k z/2}, \tag{9.4.2b}$$

from which it follows that the equations can be expressed as

$$\left(\frac{d}{dz} + \alpha_1 + \frac{i\Delta k}{2}\right) A_1 e^{-i\Delta k z/2} = \kappa_1 A_2^* e^{i\Delta k z/2}, \tag{9.4.3a}$$

$$\left(\frac{d}{dz} + \alpha_2^* - \frac{i\Delta k}{2}\right) A_2^* e^{i\Delta k z/2} = \kappa_2^* A_1 e^{-i\Delta k z/2}. \tag{9.4.3b}$$

The form of these equations suggests that we introduce the new variables F_1 and F_2 defined by

$$F_1 = A_1 e^{-i\Delta k z/2} \quad \text{and} \quad F_2^* = A_2^* e^{i\Delta k z/2}, \tag{9.4.4}$$

so that Eqs. (9.4.3) become

$$\left(\frac{d}{dz} + \alpha_1 + i\frac{\Delta k}{2}\right) F_1 = \kappa_1 F_2^*, \tag{9.4.5a}$$

$$\left(\frac{d}{dz} + \alpha_2^* - i\frac{\Delta k}{2}\right) F_2^* = \kappa_2^* F_1. \tag{9.4.5b}$$

We now eliminate F_2^* algebraically from this set of equations to obtain the single equation

$$\left(\frac{d}{dz} + \alpha_2^* - i\frac{\Delta k}{2}\right)\left(\frac{d}{dz} + \alpha_1 + i\frac{\Delta k}{2}\right)F_1 = \kappa_1\kappa_2^* F_1. \tag{9.4.6}$$

We solve this equation by adopting a trial solution of the form

$$F_1(z) = F_1(0)e^{gz}, \tag{9.4.7}$$

where g represents an unknown spatial growth rate. We substitute this form into Eq. (9.4.6) and find that this equation is satisfied by the trial solution if g satisfies the algebraic equation

$$\left(g + \alpha_2^* - \frac{i\,\Delta k}{2}\right)\left(g + \alpha_1 + \frac{i\,\Delta k}{2}\right) = \kappa_1\kappa_2^*. \tag{9.4.8}$$

In general, this equation possesses two solutions, which are given by

$$g_\pm = -\tfrac{1}{2}(\alpha_1 + \alpha_2^*) \pm \tfrac{1}{2}[(\alpha_1 - \alpha_2^* + i\,\Delta k)^2 + 4\kappa_1\kappa_2^*]^{1/2}. \tag{9.4.9}$$

Except for special values of $\alpha_1, \alpha_2, \kappa_1, \kappa_2$, and Δk, the two values of g given by Eq. (9.4.9) are distinct. Whenever the two values of g are distinct, the general solution for F is given by

$$F_1 = F_1^+(0)e^{g_+ z} + F_1^-(0)e^{g_- z}, \tag{9.4.10}$$

and hence through use of Eq. (9.4.4) we see that the general solution for A_1 is of the form

$$A_1(z) = (A_1^+ e^{g_+ z} + A_1^- e^{g_- z})e^{i\Delta k z/2}. \tag{9.4.11}$$

Here A_1^+ and A_1^- are constants of integration whose values must be determined from the relevant boundary conditions. The general form of the solution for $A_2^*(z)$ is readily found by substituting Eq. (9.4.11) into Eq. (9.4.3a), which becomes

$$\left(g_+ + \alpha_1 + i\frac{\Delta k}{2}\right)A_1^+ e^{g_+ z} + \left(g_- + \alpha_1 + i\frac{\Delta k}{2}\right)A_1^- e^{g_- z} = \kappa_1 A_2^* e^{i\Delta k z/2}.$$

This equation is now solved for $A_2^*(z)$ to obtain

$$A_2^*(z) = \left[\left(\frac{g_+ + \alpha_1 + i\,\Delta k/2}{\kappa_1}\right)A_1^+ e^{g_+ z} + \left(\frac{g_- + \alpha_1 + i\,\Delta k/2}{\kappa_1}\right)A_1^- e^{g_- z}\right] e^{-i\Delta k z/2}. \tag{9.4.12}$$

If we define constants A_2^+ and A_2^- by means of the equation

$$A_2^*(z) = (A_2^{+*}e^{g_+ z} + A_2^{-*}e^{g_- z})e^{-i\Delta k z/2}, \tag{9.4.13}$$

we see that the amplitudes $A_1^\pm$ and $A_2^\pm$ are related by

$$\frac{A_2^{\pm *}}{A_1^\pm} = \frac{g_\pm + \alpha_1 + i\,\Delta k/2}{\kappa_1}. \tag{9.4.14}$$

This equation shows how the amplitudes A_2^+ and A_1^+ are related in the part of the solution that grows as $\exp(g_+ z)$, and similarly how the amplitudes A_2^- and A_1^- are related in the part of the solution that grows as $\exp(g_- z)$. We can think of Eq. (9.4.14) as specifying the eigenmodes of propagation of the Stokes and anti-Stokes waves.

As written, Eq. (9.4.14) appears to be asymmetric with respect to the roles of the ω_1 and ω_2 fields. However, this asymmetry occurs in appearance only. Since $g_\pm$ depends upon $\alpha_1, \alpha_2, \kappa_1, \kappa_2$, and Δk, the right-hand side of Eq. (9.4.14) can be written in a variety of equivalent ways, some of which display the symmetry of the interaction more explicitly. We next rewrite Eq. (9.4.14) in such a way.

One can show by explicit calculation using Eq. (9.4.9) that the quantities g_+ and g_- are related by

$$\left(g_+ + \alpha_1 + \frac{i\,\Delta k}{2}\right)\left(g_- + \alpha_1 + \frac{i\,\Delta k}{2}\right) = -\kappa_1 \kappa_2^*. \tag{9.4.15}$$

In addition, one can see by inspection of Eq. (9.4.9) that the difference $g_+ - g_-$ is given by

$$g_+ - g_- = [(\alpha_1 - \alpha_2^* + i\,\Delta k)^2 + 4\kappa_1\kappa_2^*]^{1/2}. \tag{9.4.16a}$$

By substitution of Eq. (9.4.9) into this last equation, it follows that

$$g_+ - g_- = \pm[2g_\pm + (\alpha_1 + \alpha_2^*)], \tag{9.4.16b}$$

where on the right-hand side either both pluses or both minuses must be used. Furthermore, one can see from Eq. (9.4.9) that

$$g_+ + g_- = -(\alpha_1 + \alpha_2^*). \tag{9.4.17a}$$

By rearranging this equation and adding $i\,\Delta k/2$ to each side, it follows that

$$\left(g_\pm + \alpha_1 + \frac{i\,\Delta k}{2}\right) = -\left(g_\mp + \alpha_2^* - \frac{i\,\Delta k}{2}\right), \tag{9.4.17b}$$

$$\left(g_\pm + \alpha_2^* + \frac{i\,\Delta k}{2}\right) = -\left(g_\mp + \alpha_1 - \frac{i\,\Delta k}{2}\right). \tag{9.4.17c}$$

Through use of Eq. (9.4.15) and (9.4.17b), Eq. (9.4.14) can be expressed as

$$\frac{A_2^{\pm *}}{A_1^\pm} = \frac{g_\pm + \alpha_1 + i\,\Delta k/2}{\kappa_1} = \frac{-\kappa_2^*}{g_\mp + \alpha_1 + i\,\Delta k/2} = \frac{\kappa_2^*}{g_\pm + \alpha_2^* - i\,\Delta k/2}. \tag{9.4.18}$$

By taking the geometric mean of the last and third-last forms of this expression, we find that the ratio $A_2^{\pm *}/A_1^{\pm}$ can be written as

$$\frac{A_2^{\pm *}}{A_1^{\pm}} = \left[\frac{\kappa_2^*(g_\pm + \alpha_1 + i\,\Delta k/2)}{\kappa_1(g_\pm + \alpha_2^* - i\,\Delta k/2)}\right]^{1/2}; \tag{9.4.19}$$

this form shows explicitly the symmetry between the roles of the ω_1 and ω_2 fields.

Next, we find the form of the solution when the boundary conditions are such that the input fields are known at the plane $z = 0$, that is, when $A_1(0)$ and $A_2^*(0)$ are known. We proceed by finding the values of the constants of integration A_1^+ and A_1^-. Equation (9.4.11) is evaluated at $z = 0$ to give the result

$$A_1(0) = A_1^+ + A_1^-, \tag{9.4.20a}$$

and Eq. (9.4.12) is evaluated at $z = 0$ to give the result

$$A_2^*(0) = \left(\frac{g_+ + \alpha_1 + i\,\Delta k/2}{\kappa_1}\right)A_1^+ + \left(\frac{g_- + \alpha_1 + i\,\Delta k/2}{\kappa_1}\right)A_1^-. \tag{9.4.20b}$$

We rearrange Eq. (9.4.20a) to find that $A_1^- = A_1(0) - A_1^+$, and we substitute this result into Eq. (9.4.20b) to obtain the equation

$$A_2^*(0) = \left(\frac{g_+ - g_-}{\kappa_1}\right)A_1^+ + \left(\frac{g_- + \alpha_1 + i\,\Delta k/2}{\kappa_1}\right)A_1(0).$$

We solve this equation for A_1^+ to obtain

$$A_1^+ = \left(\frac{\kappa_1}{g_+ - g_-}\right)A_2^*(0) - \left(\frac{g_- + \alpha_1 + i\,\Delta k/2}{g_+ - g_-}\right)A_1(0). \tag{9.4.21a}$$

If instead we solve Eq. (9.4.20a) for A_1^+ and substitute the result $A_1^+ = A_1(0) - A_1^-$ into Eq. (9.4.20b), we find that

$$A_2^*(0) = \left(\frac{g_- - g_+}{\kappa_1}\right)A_1^- - \left(\frac{g_+ + \alpha_1 + i\,\Delta k/2}{\kappa_1}\right)A_1(0),$$

which can be solved for A_1^- to obtain

$$A_1^- = -\left(\frac{\kappa_1}{g_+ - g_-}\right)A_2^*(0) + \left(\frac{g_+ + \alpha_1 + i\,\Delta k/2}{g_+ - g_-}\right)A_1(0). \tag{9.4.21b}$$

The expressions (9.4.21a) and (9.4.21b) for the constants A_1^+ and A_1^- are now substituted into Eqs. (9.4.11) and (9.4.12) to give the solution for the spatial evolution of the two interacting fields in terms of their boundary values as

$$A_1(z) = \frac{1}{g_+ - g_-}\left\{\left[\kappa_1 A_2^*(0) - \left(g_- + \alpha_1 + \frac{i\,\Delta k}{2}\right)A_1(0)\right]e^{g_+ z} - \left[\kappa_1 A_2^*(0) - \left(g_+ + \alpha_1 + \frac{i\,\Delta k}{2}\right)A_1(0)\right]e^{g_- z}\right\}e^{i\,\Delta k\, z/2}. \tag{9.4.22}$$

and

$$A_2^*(z) = \frac{1}{g_+ - g_-}\left\{\left[\left(g_+ + \alpha_1 + \frac{i\,\Delta k}{2}\right)A_2^*(0) + \kappa_2^* A_1(0)\right]e^{g_+ z} - \left[\left(g_- + \alpha_1 + \frac{i\,\Delta k}{2}\right)A_2^*(0) + \kappa_2^* A_1(0)\right]e^{g_- z}\right\}e^{-i\,\Delta k\, z/2}. \tag{9.4.23}$$

Through use of Eqs. (9.4.17b) and (9.4.17c), the second form can be written in terms of α_2 instead of α_1 as

$$A_2^*(z) = \frac{1}{g_+ - g_-}\left\{\left[-\left(g_- + \alpha_2^* - \frac{i\,\Delta k}{2}\right)A_2^*(0) + \kappa_2^* A_1(0)\right]e^{g_+ z} - \left[-\left(g_+ + \alpha_2^* - \frac{i\,\Delta k}{2}\right)A_2^*(0) + \kappa_2^* A_1(0)\right]e^{g_- z}\right\}e^{-i\,\Delta k\, z/2}. \tag{9.4.24}$$

Before applying the results of the derivation just performed to the case of stimulated Raman scattering, let us make sure that the solution makes sense by applying it to several specific limiting cases.

Dispersionless Medium without Gain or Loss

For a medium without gain (or loss), we set $\alpha_1 = \alpha_2 = 0$. Also, since the medium is lossless and dispersionless, $\chi^{(3)}(\omega_1 = 2\omega_0 - \omega_2)$ must equal $\chi^{(3)}(\omega_2 = 2\omega_0 - \omega_1)$, and hence the product $\kappa_1\kappa_2^*$ that appears in the solution is equal to

$$\kappa_1\kappa_2^* = \frac{36\pi^2\omega_1\omega_2}{c^2}|\chi^{(3)}(\omega_1 = 2\omega_0 - \omega_2)|^2|A_0|^4, \tag{9.4.25}$$

which is a real, positive quantity. We allow Δk to be arbitrary, to allow the possibility that $\mathbf{k}_1$ and $\mathbf{k}_2$ are not parallel to $\mathbf{k}_0$. Under these conditions, the coupled gain coefficient of Eq. (9.4.9) reduces to

$$g_{\pm} = \pm[\kappa_1\kappa_2^* - (\Delta k/2)^2]^{1/2}. \tag{9.4.26}$$

We see that, so long as Δk is not too large, the root g_+ will be a positive real number corresponding to amplification, whereas the root g_- will be a negative real number corresponding to attenuation. However, if the wave vector mismatch becomes so large that Δk^2 exceeds $4\kappa_1\kappa_2^*$, both roots will become pure imaginary, indicating that each eigensolution shows oscillatory spatial behavior. According to Eq. (9.4.14), the ratio of amplitudes corresponding to each eigensolution is given by

$$\frac{A_2^{\pm *}}{A_1^{\pm}} = \frac{g_{\pm} + i\,\Delta k/2}{\kappa_1}. \tag{9.4.27}$$

The right-hand side of this expression simplifies considerably for the case of perfect phase matching ($\Delta k = 0$) and becomes $\pm(\kappa_2^*/\kappa_1)^{1/2}$. If we also choose our phase conventions so that A_0 is purely real, we find that expression reduces to

$$\frac{A_2^{\pm *}}{A_1^{\pm}} = \pm i\left(\frac{\omega_2}{\omega_1}\right)^{1/2} \simeq \pm i, \tag{9.4.28}$$

which shows that the two frequency sidebands are phased by $\pm\pi/2$ radians in each of the eigensolutions.

Medium without a Nonlinearity

One would expect on physical grounds that, for a medium in which $\chi^{(3)}$ vanishes, the solution would reduce to the usual case of the free propagation of the ω_1 and ω_2 waves. By setting $\kappa_1 = \kappa_2 = 0$ in Eq. (9.4.9), and assuming for simplicity that Δk vanishes, we find that

$$g_+ = -\alpha_2^* \quad \text{and} \quad g_- = -\alpha_1^*. \tag{9.4.29}$$

The eigenamplitudes are found most readily from Eq. (9.4.19). If we assume that κ_1 and κ_2 approach zero in such a manner that κ_2^*/κ_1 remains finite, we find from Eq. (9.4.19) that

$$\frac{A_2^{+*}}{A_1^{+}} = \infty, \qquad \frac{A_2^{-*}}{A_1^{-}} = 0. \tag{9.4.30}$$

Thus the positive root corresponds to a wave at frequency ω_2, which propagates according to

$$A_2^*(z) = A_2^*(0)e^{g_+ z} = A_2^*(0)e^{-\alpha_2^*}(z), \tag{9.4.31a}$$

whereas the negative root corresponds to a wave at frequency ω_1, which propagates according to

$$A_1(z) = A_1(0)e^{g_- z} = A_1(0)e^{-\alpha_1 z}. \tag{9.4.31b}$$

Stokes–Anti-Stokes Coupling in Stimulated Raman Scattering

Let us now apply the analysis to the case of stimulated Raman scattering (see also Bloembergen and Shen, 1964). For definiteness, we associate ω_1 with the Stokes frequency ω_S and ω_2 with the anti-Stokes frequency ω_a. The nonlinear absorption coefficients α_S and α_a and coupling coefficients κ_S and κ_a are given by Eqs. (9.3.39) with the nonlinear susceptibilities given by Eqs. (9.3.19b), (9.3.25), (9.3.31), and (9.3.34). In light of the relations

$$\chi_F(\omega_S) = \chi_F(\omega_a)^* = 2\chi_R(\omega_S) = 2\chi_R(\omega_a)^* \tag{9.4.32}$$

among the various elements of the susceptibility, we see that the absorption and coupling coefficients can be related to each other as follows:

$$\alpha_a = -\alpha_S^*\left(\frac{n_S\omega_a}{n_a\omega_S}\right), \tag{9.4.33a}$$

$$\kappa_S = -\alpha_S e^{2i\phi_L}, \tag{9.4.33b}$$

$$\kappa_a = \alpha_S^*\left(\frac{n_S\omega_a}{n_a\omega_S}\right)e^{2i\phi_L}, \tag{9.4.33c}$$

where ϕ_L is the phase of the pump laser defined through

$$A_L = |A_L|e^{i\phi_L}, \tag{9.4.34}$$

and where the Stokes amplitude absorption coefficient is given explicitly by

$$\alpha_S = \frac{-i\pi\omega_S N(\partial\alpha/\partial q)_0^2|A_L|^2}{2n_S c\omega_v[\omega_S - (\omega_L - \omega_v) + i\gamma]}. \tag{9.4.35}$$

If we now introduce the relations (9.4.33) into the expression (9.4.9) for the coupled gain coefficient, we find the gain eigenvalues are given by

$$g_\pm = -\tfrac{1}{2}\alpha_S\left(1 - \frac{n_S\omega_a}{n_a\omega_S}\right) \pm \frac{1}{2}\left\{\left[\alpha_S\left(1 + \frac{n_S\omega_a}{n_a\omega_S}\right) + i\,\Delta k\right]^2 - 4\alpha_S^2\frac{n_S\omega_a}{n_a\omega_S}\right\}^{1/2}. \tag{9.4.36}$$

It is usually an extremely good approximation to set the factor $n_S\omega_L/n_a\omega_S$ equal to unity. In this case Eq. (9.4.36) simplifies to

$$g_\pm = \pm[i\alpha_S\,\Delta k - (\Delta k/2)^2]^{1/2}. \tag{9.4.37}$$

The dependence of $g_\pm$ on the phase mismatch is shown graphically in Fig. 9.4.1.* Equation (9.4.37) leads to the perhaps surprising result that the coupled gain $g_\pm$ vanishes in the limit of perfect phase matching. The reason for this behavior is that, for sufficiently small Δk, the anti-Stokes wave (which normally experiences loss) is so strongly coupled to the Stokes wave (which normally experiences gain) that is prevents the Stokes wave from growing exponentially.

It is also instructive to study the expression (9.4.37) for the coupled gain in the limit in which $|\Delta k|$ is very large. For $|\Delta k| \gg |\alpha_S|$, Eq. (9.4.37) becomes

$$g_\pm = \pm i \frac{\Delta k}{2}\left(1 - \frac{4i\alpha_S}{\Delta k}\right) \simeq \pm(\alpha_S + \tfrac{1}{2} i\,\Delta k). \qquad (9.4.38)$$

Through use of Eq. (9.4.14), we find that the ratio of sidemode amplitudes associated with each of these gain eigenvalues is given by

$$\frac{A_a^{+*}}{A_S^+} = -2 - i\frac{\Delta k}{\alpha_S} \simeq i\frac{\Delta k}{\alpha_S}, \qquad (9.4.39a)$$

$$\frac{A_a^{-*}}{A_S^-} = 0. \qquad (9.4.39b)$$

Since we have assumed that $|\Delta k|$ is much larger than $|\alpha_S|$, we see that the + mode is primarily anti-Stokes, whereas the − mode is primarily Stokes.†

Let us now examine more carefully the nature of the decreased gain that occurs near $\Delta k = 0$. By setting $\Delta k = 0$ in the exact expression (9.4.36) for the

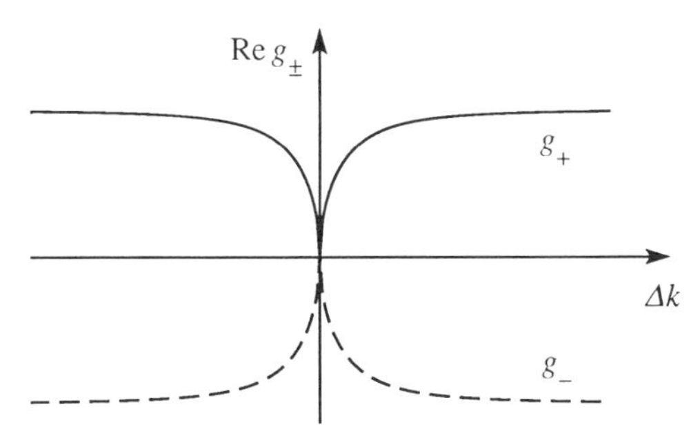

FIGURE 9.4.1 Dependence of the coupled gain on the wave vector mismatch.

* The graph has the same visual appearance whether the approximate form (9.4.37) or the exact form (9.4.36) is plotted.

† Recall that at resonance α_S is real and *negative*; hence $g_- = -\alpha_S - \frac{1}{2}i\,\Delta k$ has a positive real part and leads to amplification.

coupled gain, we find that the gain eigenvalues become

$$g_+ = 0, \qquad g_- = -\alpha_S\left(1 - \frac{n_S\omega_a}{n_a\omega_S}\right). \tag{9.4.40}$$

Note that $|g_-|$ is much smaller than $|\alpha_S|$ but does not vanish identically. We find from Eq. (9.4.14) that to good approximation

$$\frac{A_a^{\pm *}}{A_S^{\pm}} = -1; \tag{9.4.41}$$

thus each eigensolution is seen to be an equal combination of Stokes and anti-Stokes components, as mentioned above in our discussion of Fig. 9.4.1.

Next, let us consider the spatial evolution of the field amplitudes under the assumptions that $\Delta k = 0$ and that their values are known at $z = 0$. We find from Eqs. (9.4.22) and (9.4.23) that

$$\begin{aligned} A_S(z) = \frac{-1}{1 - n_S\omega_a/n_a\omega_S} \Bigg\{ &\left[A_a^*(0)e^{2i\phi_L} + \frac{n_S\omega_a}{n_a\omega_S} A_S(0) \right] \\ &- [A_a^*(0)e^{2i\phi_L} + A_S(0)]e^{g_- z} \Bigg\}, \end{aligned} \tag{9.4.42a}$$

$$\begin{aligned} A_a^*(z) = \frac{1}{1 - n_S\omega_a/n_a\omega_S} \Bigg\{ &\left[A_a^*(0) + \frac{n_S\omega_a}{n_a\omega_S} A_S(0)e^{-2i\phi_L} \right] \\ &- \frac{n_S\omega_a}{n_a\omega_S} [A_a^*(0) + A_S(0)e^{-2i\phi_L}]e^{g_- z} \Bigg\}. \end{aligned} \tag{9.4.42b}$$

Note that, since g_- is negative, the second term in each expression experiences exponential decay, and as $z \to \infty$ the field amplitudes approach the asymptotic values

$$A_S(z \to \infty) = \frac{-1}{1 - n_S\omega_a/n_a\omega_S} \left[A_a^*(0)e^{2i\phi_L} + \frac{n_S\omega_a}{n_a\omega_S} A_S(0) \right], \tag{9.4.43a}$$

$$A_a^*(z \to \infty) = \frac{1}{1 - n_S\omega_a/n_a\omega_S} \left[A_a^*(0) + \frac{n_S\omega_a}{n_a\omega_S} A_S(0)e^{-2i\phi_L} \right]. \tag{9.4.43b}$$

Note that each field is amplified by the factor $(1 - n_S\omega_a/n_a\omega_S)^{-1}$. The nature of this amplification is illustrated in Fig. 9.4.2. We see that after propagating through a distance of several times $1/g_-$, the field amplitudes attain constant values and no longer change with propagation distance.

To see why the field amplitudes remain constant, it is instructive to consider the nature of the molecular vibration in the simultaneous presence of the

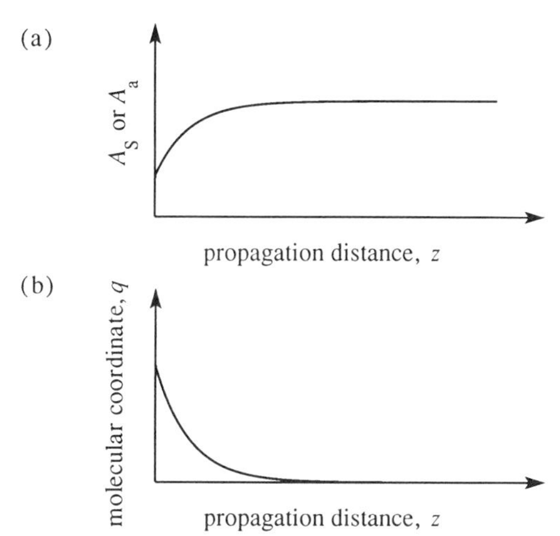

FIGURE 9.4.2 Nature of Raman amplification for the case of perfect phase matching ($\Delta k = 0$).

laser, Stokes, and anti-Stokes fields, that is, in the field

$$\tilde{E}(z,t) = A_L e^{i(k_L z - \omega_L t)} + A_S e^{i(k_S z - \omega_S t)} + A_a e^{i(k_a z - \omega_a t)} + \text{c.c.}, \quad (9.4.44)$$

where $k_L - k_S = k_a - k_L \equiv K$ and $\omega_L - \omega_S = \omega_a - \omega_L \equiv \Omega$. The solution to the equation of motion (9.3.1) for the molecular vibration with the force term given by Eqs. (9.3.6) and (9.4.44) is given by

$$\tilde{q}(z,t) = q(\Omega)e^{i(Kz - \Omega t)} + \text{c.c.},$$

where

$$q(\Omega) = \frac{(1/m)(\partial\alpha/\partial q)_0 (A_L A_S^* + A_a A_L^*)}{\omega_v^2 - \Omega^2 - 2i\Omega\gamma}. \quad (9.4.45)$$

We can see from Eq. (9.4.43) that, once the field amplitudes have attained their asymptotic values, the combination $A_L A_S^* + A_a A_L^*$ vanishes, implying that the amplitude $q(\omega)$ of the molecular vibration also vanishes asymptotically, as illustrated in part (b) of Fig. 9.4.2.

9.5. Stimulated Rayleigh-Wing Scattering

Stimulated Rayleigh-wing scattering is the light scattering process that results from the tendency of anisotropic molecules to become aligned along the electric field vector of an optical wave. Stimulated Rayleigh-wing scattering was described theoretically by Bloembergen and Lallemand (1966) and by Chiao *et al.* (1966), and was observed experimentally by Mash *et al.* (1965) and Cho *et al.* (1967). Other early studies were conducted by Denariez and Bret (1968) and by Foltz *et al.* (1968). The molecular orientation effect was described in Section 4.4 for the case in which the applied optical field $\tilde{E}(t)$ contains a single frequency component, and it was found that the average molecular polarizability is modified by the presence of the applied field. The molecular polarizability can be expressed as

$$\langle \alpha \rangle = \alpha_0 + \alpha_{\mathrm{NL}}, \tag{9.5.1}$$

where the usual, weak-field polarizability is given by

$$\alpha_0 = \tfrac{1}{3}\alpha_{\parallel} + \tfrac{2}{3}\alpha_{\perp}, \tag{9.5.2}$$

where $\alpha_{\parallel}$ and $\alpha_{\perp}$ denote the polarizabilities measured parallel to and perpendicular to the symmetry axis of the molecule, respectively (see Fig. 9.5.1). In addition, the lowest-order nonlinear contribution to the polarizability is given by

$$\alpha_{\mathrm{NL}} = \bar{\alpha}_2 \overline{\tilde{E}^2}, \tag{9.5.3}$$

where

$$\bar{\alpha}_2 = \frac{8\pi}{45} \frac{(\alpha_{\parallel} - \alpha_{\perp})^2}{kT}. \tag{9.5.4}$$

In order to describe stimulated Rayleigh-wing scattering, we need to determine the response of the molecular system to an optical field that contains both laser and Stokes components, which we describe by the equation

$$\tilde{E}(\mathbf{r}, t) = A_{\mathrm{L}} e^{i(k_{\mathrm{L}} z - \omega_{\mathrm{L}} t)} + A_{\mathrm{S}} e^{i(-k_{\mathrm{S}} z - \omega_{\mathrm{S}} t)} + \mathrm{c.c.} \tag{9.5.5}$$

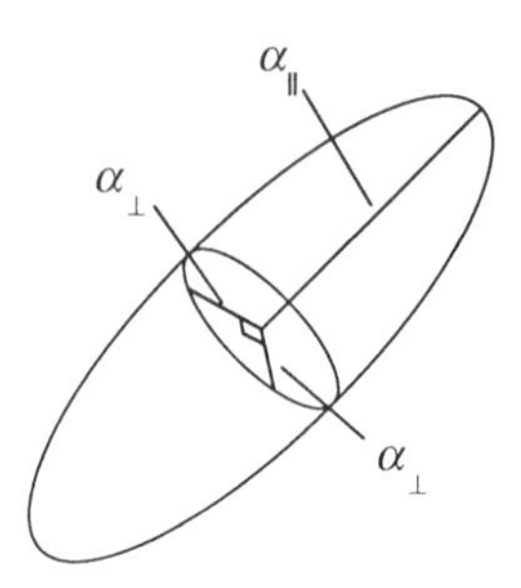

FIGURE 9.5.1 Illustration of the polarizabilities of an anisotropic molecule for the case $\alpha_{\parallel} > \alpha_{\perp}$.

For the present, we assume that the laser and Stokes waves are linearly polarized in the same direction and are counterpropagating. The analysis for the case in which the waves have arbitrary polarization and/or are copropagating is somewhat more involved and is discussed briefly below.

Since the intensity, which is proportional to $\overline{\tilde{E}^2}$, now contains a component at the beat frequency $\omega_L - \omega_S$, the nonlinear contribution to the mean polarizibility $\langle\alpha\rangle$ is no longer given by Eq. (9.5.3), which was derived for the case of a monochromatic field. We assume that, in general, α_{NL} is described by the equation

$$\tau\dot{\alpha}_{NL} + \alpha_{NL} = \overline{\alpha_2}\overline{\tilde{E}^2}. \tag{9.5.6}$$

In this equation τ represents the molecular orientation relaxation time and is the characteristic response time of the SRWS process; see Table 9.5.1 for typical values of τ. Equation (9.5.6) has the form of a Debye relaxation equation; recall that we have studied equations of this sort in our general discussion of two-beam coupling in Section 6.4.

If Eq. (9.5.6) is solved in steady state with $\tilde{E}(t)$ given by Eq. (9.5.5), we find that the nonlinear contribution to the polarizability of a molecule located at position z is given by

$$\alpha_{NL}(z,t) = 2\overline{\alpha_2}(A_L A_L^* + A_S A_S^*) + \left(\frac{2\overline{\alpha_2} A_L A_S^* e^{iqz - \Omega t}}{1 - i\Omega\tau} + \text{c.c.}\right), \tag{9.5.7}$$

where we have introduced the wave vector magnitude q and frequency Ω associated with the material excitation, which are given by

$$q = k_L + k_S, \qquad \Omega = \omega_L - \omega_S. \tag{9.5.8}$$

Note that, due to the complex nature of the denominator of the second term in the expression for $\alpha_{NL}(z,t)$, the nonlinear response will in general be shifted

TABLE 9.5.1 Properties of SRWS for several materials

Substance	G (cm/GW)	τ (ps)	$\Delta\nu = \frac{1}{2}\pi\tau$ (GHz)
CS_2	3	2	80
Nitrobenzene	3	48	3.3
Bromobenzene	1.4	15	10
Clorobenzene	1.0	8	20
Toluene	1.0	2	80
Benzene	0.6	3	53

in phase with respect to the intensity distribution associated with the interference of the laser and Stokes fields. We shall see below that this phase shift is the origin of the gain of the stimulated Rayleigh-wing scattering process.

We next derive the equation describing the propagation of the Stokes field. This derivation is formally identical to that presented in Section 6.4 in our general discussion of two-beam coupling. To apply that treatment to the present case, we need to determine the values of the refractive indices n_0 and n_2 that are relevant to the problem at hand. We find that n_0 is obtained from the usual Lorentz–Lorenz law as (see also Eq. (3.8.8))

$$\frac{n_0^2 - 1}{n_0^2 + 2} = \frac{4\pi}{3} N\alpha_0 \tag{9.5.9a}$$

and that the nonlinear refractive index is given by (see also Eqs. (4.1.18) and (4.4.26))

$$n_2 = \left(\frac{n_0^2 + 2}{3}\right)^4 \frac{2\pi}{n_0^2 c} \bar{\alpha}_2 N. \tag{9.5.9b}$$

Then, as in Eq. (6.4.17), we find that the spatial evolution of the Stokes wave is described by

$$\frac{dA_S}{dz} = \frac{in_0 n_2 \omega_S}{2\pi}(A_L A_L^* + A_S A_S^*)A_S + \frac{in_0 n_2 \omega_S}{2\pi} \frac{A_L A_L^* A_S}{1 + i\Omega\tau}. \tag{9.5.10}$$

Here the first term on the right-hand side describes a nonlinear contribution to the phase of the Stokes wave, whereas the second term leads to the gain of the stimulated Rayleigh-wing scattering process. The nature of the gain can be seen more clearly in terms of the equation relating the intensities of the two waves, which are defined by

$$I_j = \frac{n_0 c}{2\pi} |A_j|^2, \qquad j = \text{L, S}. \tag{9.5.11}$$

The spatial variation of the intensity of the Stokes wave is therefore described by

$$\frac{dI_S}{dz} = \frac{n_0 c}{2\pi}\left[A_S \frac{dA_S^*}{dz} + A_S^* \frac{dA_S}{dz}\right]. \tag{9.5.12}$$

Through use of Eq. (9.5.10), we can write this result as

$$\frac{dI_S}{dz} = g_{RW} I_L I_S, \tag{9.5.13}$$

where we have introduced the gain factor g_{RW} for stimulated Rayleigh-wing scattering, which is given by

$$g_{\mathrm{RW}} = g_{\mathrm{RW}}^{\max}\left(\frac{2\Omega\tau}{1+\Omega^2\tau^2}\right), \tag{9.5.14a}$$

where $g_{\mathrm{RW}}^{\max}$ denotes the maximum value of the gain factor, which is given by

$$g_{\mathrm{RW}}^{\max} = \frac{n_2\omega_S}{c} = \left(\frac{n_0^2+2}{3}\right)^4 \frac{16\pi^2\omega_S N(\alpha_\parallel - \alpha_\perp)^2}{45kTn_0^2c^2}. \tag{9.5.14b}$$

We have made use of Eqs. (9.5.4) and (9.5.9b) in obtaining the second form of the expression for $g_{\mathrm{RW}}^{\max}$.

The frequency dependence of the gain factor for stimulated Rayleigh-wing scattering as predicted by Eq. (9.5.14a) is illustrated in Fig. 9.5.2. We see that amplification of the ω_S wave occurs for $\omega_S < \omega_L$ and that attenuation occurs for $\omega_S > \omega_L$. The maximum gain occurs when $\Omega \equiv \omega_L - \omega_S$ is equal to $1/\tau$.

The nature of the stimulated Rayleigh-wing scattering process is illustrated schematically in Fig. 9.5.3. The interference of the forward-going wave of frequency ω_L and wave vector magnitude k_L and the backward-going wave of frequency ω_S and wave vector magnitude k_S produces a fringe pattern that moves slowly through the medium in the forward direction with phase velocity $v = \Omega/q$. The tendency of the molecules to become aligned along the electric field vector of the total optical wave leads to planes of maximum molecular alignment alternating with planes of minimum molecular alignment. As mentioned above, these planes are shifted in phase with respect to the maxima and minima of the intensity distributions. The scattering of the laser field from this periodic array of aligned molecules leads to the generation of the Stokes wave. The scattered radiation is shifted to lower frequencies because the material disturbance causing the scattering is moving in the forward direction. The scattering process shows gain because the generation of Stokes radiation tends to reinforce the modulated portion of the interference

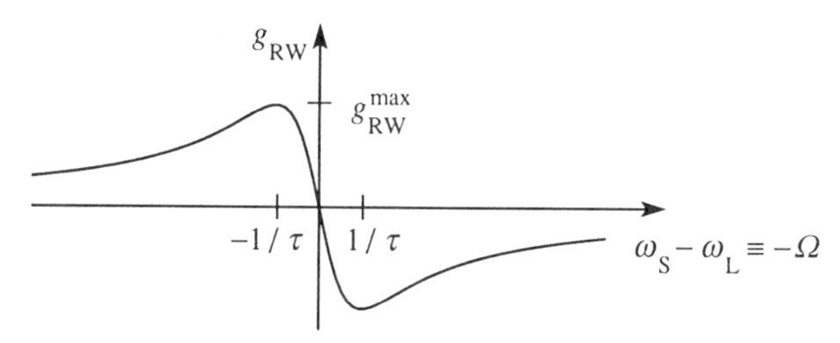

FIGURE 9.5.2 Frequency dependence of the gain factor for stimulated Rayleigh-wing scattering.

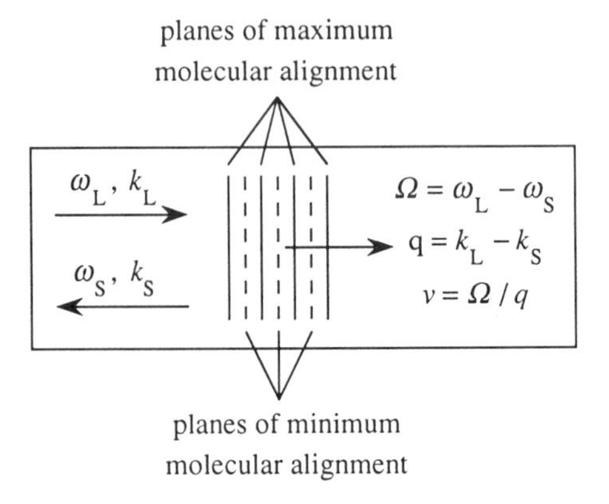

FIGURE 9.5.3 Nature of stimulated Rayleigh-wing scattering.

pattern, which leads to increased molecular alignment and thus to increased scattering of Stokes radiation.

Polarization Properties of Stimulated Rayleigh-Wing Scattering

A theoretical analysis of the polarization properties of stimulated Rayleigh-wing scattering has been conducted by Chiao and Godine (1969). The details of their analysis are quite complicated; here we shall simply quote some of their principal results.

In order to treat the polarization properties of stimulated Rayleigh-wing scattering, one must consider the tensor properties of the material response. The analysis of Chiao and Godine presupposes that the nonlinear contribution to the susceptibility obeys the equation of motion

$$\tau \frac{d}{dt} \Delta\chi_{ik} + \Delta\chi_{ik} = C(\overline{\tilde{E}_i \tilde{E}_k} - \tfrac{1}{3}\delta_{ik} \overline{\tilde{\mathbf{E}} \cdot \tilde{\mathbf{E}}}), \tag{9.5.15}$$

where, ignoring local-field corrections, the proportionality constant C is given by

$$C = \frac{N(\alpha_{\parallel} - \alpha_{\perp})^2}{15kT}. \tag{9.5.16}$$

Note that the trace of the right-hand side of Eq. (9.5.15) vanishes, as required by the fact that Rayleigh-wing scattering is described by a traceless, symmetric permittivity tensor.

By requiring that the Stokes wave obey the wave equation with a susceptibility given by the solution to Eq. (9.5.15), and taking account of ellipse rotation of the pump laser polarization (see, for example, the discussion in Section 4.2), Chiao and Godine calculate the gain factor for stimulated Rayleigh-wing scattering for arbitrary polarization of the laser and Stokes fields. Some of their results for special polarization cases are summarized in Table 9.5.2.

For any state of polarization of the pump wave, some particular polarization of the Stokes wave will experience maximum gain. Due to the large value of the gain required to observe stimulated light scattering ($g_{\mathrm{RW}} I_{\mathrm{L}} L \simeq 25$), the light generated by stimulated Rayleigh-wing scattering will have a polarization that is nearly equal to that for which the gain is maximum. The relation between the laser polarization and the Stokes polarization for which the gain is maximum is illustrated in Table 9.5.3. Note that the generated wave will be nearly, but not exactly, the polarization conjugate (in the sense of vector phase conjugation, as discussed in Section 6.1) of the incident laser wave. In particular, the polarization ellipse of the generated wave will be rounder and tilted with respect to that of the laser wave.

Zel'dovich and Yakovleva (1980) have studied theoretically the polarization properties of stimulated Rayleigh-wing scattering for the case in which the pump radiation is partially polarized. They predict that essentially perfect vector phase conjugation can be obtained by stimulated Rayleigh-wing scattering for the case in which the pump radiation is completely depolarized in the sense that the state of polarization varies randomly over the transverse dimensions of the laser beam. The wavefront-reconstructing properties of stimulated Rayleigh-wing scattering have been studied experimentally by

TABLE 9.5.2 Dependence of the gain factor for stimulated Rayleigh-wing scattering in the backward direction on the polarization of the laser and Stokes waves for the cases of linear and circular polarization*

laser polarization	↕	↕	↻	↻
Stokes polarization	↕	↔	↺	↻
gain factor	1	3/4	3/2	1/6

* The arrows on the circles denote the direction in which the electric field vector rotates in time at a fixed position in space. The gain factors are given relative to that given by Eq. (9.5.14) for the case of linear and parallel polarization.

TABLE 9.5.3 Relation between laser polarization and the Stokes polarization experiencing maximum gain in backward stimulated Rayleigh-wing scattering

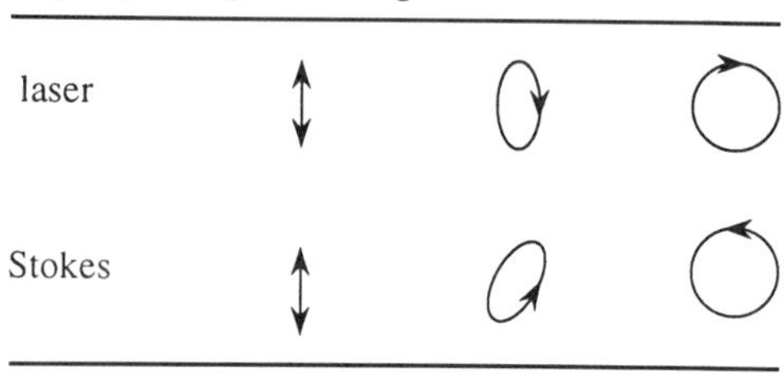

Kudriavtseva *et al.* (1978), and the vector phase conjugation properties have been studied experimentally by Miller *et al.* (1990).

The analysis of stimulated Rayleigh-wing scattering in the forward and near-forward direction is much more complicated than that of backward stimulated Rayleigh-wing scattering because the possibility of Stokes–anti-Stokes coupling (as described in Section 9.4 for stimulated Raman scattering) must be included in the analysis. This situation has been described by Chiao *et al.* (1966) and by Chiao and Godine (1969).

Problem

1. By carrying out the prescription described in the first full paragraph following Eq. (9.5.16), verify that the entries in Table 9.5.2 are correct.

References

Stimulated Raman Scattering

N. Bloembergen and Y. R. Shen, *Phys. Rev. Lett.* **12**, 504 (1964).

N. Bloembergen, *Am. J. Phys.* **35**, 989 (1967).

R. W. Boyd, *Radiometry and the Detection of Optical Radiation*, Wiley, New York, 1983.

G. Eckhardt, R. W. Hellwarth, F. J. McClung, S. E. Schwarz, D. Weiner, and E. J. Woodbury, *Phys. Rev. Lett.* **9**, (1962).

E. Garmire, F. Pandarese, and C. H. Townes, *Phys. Rev. Lett.* **11**, 160 (1963).

R. W. Hellwarth, *Phys. Rev.* **130**, 1850 (1963).

W. Kaiser and M. Maier, in *Laser Handbook*, edited by F. T. Arecchi and E. O. Schulz-DuBois, North-Holland, 1972.

M. G. Raymer and J. Mostowski, *Phys. Rev. A* **24**, 1980 (1981).
M. G. Raymer and I. A. Walmsley, in *Progress in Optics*, Vol. 28, edited by E. Wolf, North-Holland, Amsterdam, 1990.
A. Penzkofer, A. Laubereau, and W. Kaiser, *Prog. Quantum Electron.* **6**, 55 (1979).
Y. R. Shen and N. Bloembergen, *Phys. Rev.* **137**, 1787 (1965).
R. W. Terhune, *Bull. Am. Phys. Soc.* **8**, 359 (1963).
E. J. Woodbury and W. K. Ng, *Proc. I.R.E.* **50**, 2367 (1962).

Stimulated Rayleigh-Wing Scattering

N. Bloembergen and P. Lallemand, *Phys. Rev. Lett.* **16**, 81 (1966).
R. Y. Chiao and J. Godine, *Phys. Rev.* **185**, 430 (1969).
R. Y. Chiao, P. L. Kelley, and E. Garmire, *Phys. Rev. Lett.* **17**, 1158 (1966).
C. W. Cho, N. D. Foltz, D. H. Rank, and T. A. Wiggins, *Phys. Rev. Lett.* **18**, 107 (1967).
M. Denariez and G. Bret, *Phys. Rev.* **177**, 171 (1968).
N. D. Foltz, C. W. Cho, D. H. Rank, and T. A. Wiggins, *Phys. Rev.* **165**, 369 (1968).
A. D. Kudriavtseva, A. I. Sokolovskaia, J. Gazengel, N. Phu Xuan, and G. Rivore, *Opt. Commun.* **26**, 446 (1978).
D. I. Mash, V. V. Morozov, V. S. Starunov, and I. L. Fabelinskii, *JETP Lett.* **2**, 25 (1965).
E. J. Miller, M. S. Malcuit, and R. W. Boyd, *Opt. Lett.* **15**, 1189 (1990).
B. Ya. Zel'dovich and T. V. Yakovleva, *Sov. J. Quantum Electron.* **10**, 501 (1980).

Chapter 10

The Electrooptic and Photorefractive Effects

10.1. Introduction to the Electrooptic Effect

The electrooptic effect is the change in refractive index of a material induced by the presence of a dc (or low-frequency) electric field.

In some materials, the change in refractive index depends linearly on the strength of the applied electric field. This change is known as the linear electrooptic, or Pockels, effect. The linear electrooptic effect can be described in terms of a nonlinear polarization given by

$$P_i(\omega) = 2\sum_{jk} \chi_{ijk}^{(2)}(\omega = \omega + 0)E_j(\omega)E_k(0). \tag{10.1.1}$$

Since the linear electrooptic effect can be described by a second-order nonlinear susceptibility, it follows from the general discussion of Section 1.5 that a linear electrooptic effect can occur only for materials that are noncentrosymmetric. Although the linear electrooptic effect can be described in terms of a second-order nonlinear susceptibility, a very different mathematical formalism has historically been used to describe the electrooptic effect; this formalism is described in Section 10.2 of this chapter.

In centrosymmetric materials (such as liquids and glasses), the lowest-order change in the refractive index depends quadratically on the strength of the applied dc (or low-frequency) field. This effect is known as the Kerr electrooptic effect* or as the quadratic electrooptic effect. It can be described in terms

* The quadratic electrooptic effect is often referred to simply as the Kerr effect. More precisely, it is called the Kerr electrooptic effect to distinguish it from the Kerr magnetooptic effect.

of a nonlinear polarization given by

$$P_i(\omega) = 3\sum_{jkl} \chi_{ijkl}^{(3)}(\omega = \omega + 0 + 0) E_j(\omega) E_k(0) E_l(0). \tag{10.1.2}$$

10.2. Linear Electrooptic Effect

In this section we develop a mathematical formalism that describes the linear electrooptic effect. In an anisotropic material, the constitutive relation between the field vectors **D** and **E** has the form

$$D_i = \sum_j \epsilon_{ij} E_j \tag{10.2.1a}$$

or, explicitly,

$$\begin{bmatrix} D_x \\ D_y \\ D_z \end{bmatrix} = \begin{bmatrix} \epsilon_{xx} & \epsilon_{xy} & \epsilon_{xz} \\ \epsilon_{yx} & \epsilon_{yy} & \epsilon_{yz} \\ \epsilon_{zx} & \epsilon_{zy} & \epsilon_{zz} \end{bmatrix} \begin{bmatrix} E_x \\ E_y \\ E_z \end{bmatrix}. \tag{10.2.1b}$$

For a lossless, non-optically-active material, the dielectric permeability tensor ϵ_{ij} is represented by a real symmetric matrix, which therefore has six independent elements, i.e., ϵ_{xx}, ϵ_{yy}, ϵ_{zz}, $\epsilon_{xy} = \epsilon_{yx}$, $\epsilon_{xz} = \epsilon_{zx}$, and $\epsilon_{yz} = \epsilon_{zy}$. A general mathematical result states that any real, symmetric matrix can be expressed in diagonal form by means of an orthogonal transformation. Physically, this result implies that there exists some new coordinate system (X, Y, Z), related to the coordinate system (x, y, z) of Eq. (10.2.1b) by rotation of the coordinate axes, in which Eq. (10.2.1b) has the much simpler form

$$\begin{bmatrix} D_X \\ D_Y \\ D_Z \end{bmatrix} = \begin{bmatrix} \epsilon_{XX} & 0 & 0 \\ 0 & \epsilon_{YY} & 0 \\ 0 & 0 & \epsilon_{ZZ} \end{bmatrix} \begin{bmatrix} E_X \\ E_Y \\ E_Z \end{bmatrix}. \tag{10.2.2}$$

This new coordinate system is known as the principal-axis system, because in it the dielectric tensor is represented as a diagonal matrix.

We next consider the energy density per unit volume,

$$U = \frac{1}{8\pi} \mathbf{D} \cdot \mathbf{E} = \frac{1}{8\pi} \sum_{ij} \epsilon_{ij} E_i E_j, \tag{10.2.3}$$

associated with a wave propagating through the anistropic medium. In the principal-axis coordinate system, the energy density can be represented as

$$U = \frac{1}{8\pi} \left[\frac{D_X^2}{\epsilon_{XX}} + \frac{D_Y^2}{\epsilon_{YY}} + \frac{D_Z^2}{\epsilon_{ZZ}} \right]. \tag{10.2.4}$$

This result shows that the surfaces of constant energy density in **D** space are ellipsoids. The shape of these ellipsoids can be described in terms of the coordinates (X, Y, Z) themselves. If we let

$$X = \left(\frac{1}{8\pi U}\right)^{1/2} D_X, \qquad Y = \left(\frac{1}{8\pi U}\right)^{1/2} D_Y, \qquad Z = \left(\frac{1}{8\pi U}\right)^{1/2} D_Z, \quad (10.2.5)$$

Eq. (10.2.4) becomes

$$\frac{X^2}{\epsilon_{XX}} + \frac{Y^2}{\epsilon_{YY}} + \frac{Z^2}{\epsilon_{ZZ}} = 1. \quad (10.2.6)$$

The surface described by this equation is known as the *optical indicatrix* or as the *index ellipsoid*. The equation describing the index ellipsoid takes on its simplest form in the principal-axis system; in other coordinate systems it is given by the general expression for an ellipsoid, which we write in the form

$$\begin{aligned} &\left(\frac{1}{n^2}\right)_1 x^2 + \left(\frac{1}{n^2}\right)_2 y^2 + \left(\frac{1}{n^2}\right)_3 z^2 + 2\left(\frac{1}{n^2}\right)_4 yz \\ &+ 2\left(\frac{1}{n^2}\right)_5 xz + 2\left(\frac{1}{n^2}\right)_6 xy = 1. \end{aligned} \quad (10.2.7)$$

The coefficients $(1/n^2)_i$ are optical constants that describe the optical indicatrix in the new coordinate system; they can be expressed in terms of the coefficients ϵ_{XX}, ϵ_{YY}, and ϵ_{ZZ} by means of the standard transformation laws for coordinate transformations, but the exact nature of the relationship is of no interest for our present purposes.

The index ellipsoid can be used to describe the optical properties of an anistropic material by means of the following procedure (Born and Wolf, 1975). For any given direction of propagation within the crystal, a plane perpendicular to the propagation vector and passing through the center of the ellipsoid is constructed. The curve formed by the intersection of this plane with the index ellipsoid forms an ellipse. The semimajor and semiminor axes of this ellipse give the two allowed values of the refractive index for this particular direction of propagation; the orientations of these axes give the polarization directions of the **D** vector associated with these refractive indices.

We next study how the optical indicatrix is modified when the material system is subjected to a constant or low-frequency electric field. This modification is conveniently described in terms of the *impermeability tensor* η_{ij}, which is defined by the relation

$$E_i = \sum_j \eta_{ij} D_j. \quad (10.2.8)$$

Note that this relation is the inverse of that given by Eq. (10.2.1), and hence that η_{ij} is the matrix inverse of ϵ_{ij}, that is, that $\eta_{ij} = (\epsilon^{-1})_{ij}$. We can express the optical indicatrix in terms of the elements of the impermeability tensor by noting that the energy density is equal to $U = (1/8\pi)\mathbf{D}\cdot\mathbf{E} = (1/8\pi)\sum_{ij}\eta_{ij}D_iD_j$. If we now define coordinates x, y, z by means of relations $x = D_x/(8\pi U)^{\frac{1}{2}}$, etc., we find that the expression for U as a function of $\mathbf{D}$ becomes

$$1 = \eta_{11}x^2 + \eta_{22}y^2 + \eta_{33}z^2 + 2\eta_{12}xy + 2\eta_{23}yz + 2\eta_{13}xz. \qquad (10.2.9)$$

By comparison of this expression for the optical indicatrix with that given by Eq. (10.2.7), we find that

$$\begin{aligned} &\left(\frac{1}{n^2}\right)_1 = \eta_{11}, \qquad \left(\frac{1}{n^2}\right)_2 = \eta_{22}, \qquad \left(\frac{1}{n^2}\right)_3 = \eta_{33} \\ &\left(\frac{1}{n^2}\right)_4 = \eta_{23} = \eta_{32}, \qquad \left(\frac{1}{n^2}\right)_5 = \eta_{13} = \eta_{31}, \qquad \left(\frac{1}{n^2}\right)_6 = \eta_{12} = \eta_{21}. \end{aligned} \qquad (10.2.10)$$

We next assume that η_{ij} can be expressed as a power series in the strength of the components E_k of the applied electric field as

$$\eta_{ij} = \eta_{ij}^{(0)} + \sum_k r_{ijk}E_k + \sum_{kl} s_{ijkl}E_kE_l + \cdots. \qquad (10.2.11)$$

Here r_{ijk} is the tensor that describes the linear electrooptic effect, s_{ijkl} is the tensor that describes the quadratic electrooptic effect, etc. Since the dielectric permeability tensor ϵ_{ij} is real and symmetric, its inverse η_{ij} must also be real and symmetric, and consequently the electrooptic tensor r_{ijk} must be symmetric in its first two indices. For this reason, it is often convenient to represent the third-rank tensor r_{ijk} as a two-dimensional matrix r_{hk} using contracted notation according to the prescription

$$h = \begin{cases} 1 & \text{for} \quad ij = 11, \\ 2 & \text{for} \quad ij = 22, \\ 3 & \text{for} \quad ij = 33, \\ 4 & \text{for} \quad ij = 23 \text{ or } 32, \\ 5 & \text{for} \quad ij = 13 \text{ or } 31, \\ 6 & \text{for} \quad ij = 12 \text{ or } 21. \end{cases} \qquad (10.2.12)$$

In terms of this contracted notation, we can express the lowest-order modification of the optical constants $(1/n^2)_i$ that appears in expression (10.2.7) for the optical indicatrix as

$$\Delta\left(\frac{1}{n^2}\right)_i = \sum_j r_{ij}E_j, \qquad (10.2.13a)$$

where we have made use of Eqs. (10.2.10) and (10.2.11). This relationship can be written explicitly as

$$\begin{bmatrix} \Delta(1/n^2)_1 \\ \Delta(1/n^2)_2 \\ \Delta(1/n^2)_3 \\ \Delta(1/n^2)_4 \\ \Delta(1/n^2)_5 \\ \Delta(1/n^2)_6 \end{bmatrix} = \begin{bmatrix} r_{11} & r_{12} & r_{13} \\ r_{21} & r_{22} & r_{23} \\ r_{31} & r_{32} & r_{33} \\ r_{41} & r_{42} & r_{43} \\ r_{51} & r_{52} & r_{53} \\ r_{61} & r_{62} & r_{63} \end{bmatrix} \begin{bmatrix} E_x \\ E_y \\ E_z \end{bmatrix}. \tag{10.2.13b}$$

The quantities r_{ij} are known as the electrooptic coefficients and give the rate at which the coefficients $(1/n^2)_i$ change with increasing electric field strength.

We remarked earlier that the linear electrooptic effect vanishes for materials possessing inversion symmetry. Even for materials lacking inversion symmetry, where the coefficients do not necessarily vanish, the form of r_{ij} is restricted by any symmetry properties that the material may possess. For example, for any material (such as ADP and KDP) possessing the point group symmetry $\bar{4}2m$, the electrooptic coefficients must be of the form

$$r_{ij} = \begin{bmatrix} 0 & 0 & 0 \\ 0 & 0 & 0 \\ 0 & 0 & 0 \\ r_{41} & 0 & 0 \\ 0 & r_{41} & 0 \\ 0 & 0 & r_{63} \end{bmatrix} \quad (\text{for class } \bar{4}2m), \tag{10.2.14}$$

where we have expressed r_{ij} in the standard crystallographic coordinate system, in which the Z direction represents the optic axis of the crystal. We see from Eq. (10.2.14) that the form of the symmetry properties of the point group $\bar{4}2m$ requires fifteen of the electrooptic coefficients to vanish and two of the remaining coefficients to be equal. Hence r_{ij} possesses only two independent elements in this case.

Similarly, the electrooptic coefficients of crystals of class $3m$ (such as lithium niobate) must be of the form

$$r_{ij} = \begin{bmatrix} 0 & -r_{22} & r_{13} \\ 0 & r_{22} & r_{13} \\ 0 & 0 & r_{33} \\ 0 & r_{42} & 0 \\ r_{42} & 0 & 0 \\ -r_{22} & 0 & 0 \end{bmatrix} \quad (\text{for class } 3m), \tag{10.2.15}$$

TABLE 10.2.1 Properties of several electrooptic materials†

Material	Point group	Electrooptic coefficients (10^{-12} m/V)	Refractive index
Ammonium dihydrogen phosphate, $NH_4H_2PO_4$ (ADP)	$\bar{4}2m$	$r_{41} = 24.5$ $r_{63} = 8.5$	$n_o = 1.530$ $n_e = 1.483$ (at 0.5461 μm)
Potassium dihydrogen phosphate KH_2PO_4 (KDP)	$\bar{4}2m$	$r_{41} = 8.77$ $r_{63} = 10.5$	$n_o = 1.514$ $n_e = 1.472$ (at 0.5461 μm)
Potassium dideuterium phosphate KD_2PO_4 (KD*P)	$\bar{4}2m$	$r_{41} = 8.8$ $r_{63} = 26.4$	$n_o = 1.508$ $n_e = 1.468$ (at 0.5461 μm)
Lithium niobate, $LiNbO_3$	$3m$	$r_{13} = 9.6$ $r_{22} = 6.8$ $r_{33} = 30.9$ $r_{42} = 32.6$	$n_o = 2.3410$ $n_e = 2.2457$ (at 0.5 μm)
Lithium tantalate, $LiTaO_3$	$3m$	$r_{13} = 8.4$ $r_{22} = -0.2$ $r_{33} = 30.5$ $r_{51} = 20$	$n_o = 2.176$ $n_e = 2.180$ (at 0.633 nm)
Barium titanate, $BaTiO_3$‡	$4mm$	$r_{13} = 19.5$ $r_{33} = 97$ $r_{42} = 1640$	$n_o = 2.488$ $n_e = 2.424$ (at 514 nm)
Strontium barium niobate, $Sr_{0.6}Ba_{0.4}NbO_6$ (SBN:60)	$4mm$	$r_{13} = 55$ $r_{33} = 224$ $r_{42} = 80$	$n_o = 2.367$ $n_e = 2.337$ (at 514 nm)

† From a variety of sources. See, for example B. J. Thompson and E. Hartfield in the Handbook of Optics, edited by W. G. Driscoll and W. Vaughan, McGraw-Hill, New York, 1978 and W. R. Cook, Jr., and H. Jaffe, "Electrooptic Coefficients," in *Landolt-Bornstein, New Series*, Vol. II, edited by K.-H. Hellwege, Springer-Verlag, 1979, pp. 552–651. The electrooptic coefficients are given in the MKS units of m/V. To convert to the cgs units of cm/statvolt, each entry should be multiplied by 3×10^4.

‡ $\epsilon_{dc}^{\parallel} = 135$, $\epsilon_{dc}^{\perp} = 3700$.

and the electrooptic coefficients of crystals of the class 4*mm* (such as barium titanate) must be of the form

$$r_{ij} = \begin{bmatrix} 0 & 0 & r_{13} \\ 0 & 0 & r_{13} \\ 0 & 0 & r_{33} \\ 0 & r_{42} & 0 \\ r_{42} & 0 & 0 \\ 0 & 0 & 0 \end{bmatrix} \quad \text{(for class } 4mm\text{).} \tag{10.2.16}$$

The properties of several electrooptic materials are summarized in Table 10.2.1.

10.3. Electrooptic Modulators

As an example of the application of the formalism developed in the last section, we now consider how to construct an electrooptic modulator using the material potassium dihydrogen phosphate (KDP). Of course, the analysis is formally identical for any electrooptic material of point group $\bar{4}2m$.

KDP is a uniaxial crystal, and hence in the absence of an applied electric field the index ellipsoid is given in the standard crystallographic coordinate system by the equation

$$\frac{X^2}{n_o^2} + \frac{Y^2}{n_o^2} + \frac{Z^2}{n_e^2} = 1. \tag{10.3.1}$$

Note that this (X, Y, Z) coordinate system is the principal-axis coordinate system in the absence of an applied electric field. If an electric field is applied to the crystal, the index ellipsoid becomes modified according to Eqs. (10.2.13) and (10.2.14) and takes the form

$$\frac{X^2}{n_o^2} + \frac{Y^2}{n_o^2} + \frac{Z^2}{n_e^2} + 2r_{41}E_X YZ + 2r_{41}E_Y XZ + 2r_{63}E_Z XY = 1. \tag{10.3.2}$$

Note that (since cross terms containing YZ, XZ, and XY appear in this equation) the (X, Y, Z) coordinate system is not the principal-axis coordinate system when an electric field is applied to the crystal. Note also that the crystal will no longer necessarily be uniaxial in the presence of a dc electric field.

Let us now assume that the applied electric field has only a Z component, so that Eq. (10.3.2) reduces to

$$\frac{X^2}{n_o^2} + \frac{Y^2}{n_o^2} + \frac{Z^2}{n_e^2} + 2r_{63}E_Z XY = 1. \tag{10.3.3}$$

This special case is the one most often encountered in device applications. The new principal-axis coordinate system can now be found by inspection. If we let

$$X = \frac{x - y}{\sqrt{2}}, \qquad Y = \frac{x + y}{\sqrt{2}}, \qquad Z = z, \tag{10.3.4}$$

we find that Eq. (10.3.3) becomes

$$\left(\frac{1}{n_o^2} + r_{63}E_z\right)x^2 + \left(\frac{1}{n_o^2} - r_{63}E_z\right)y^2 + \frac{z^2}{n_e^2} = 1, \tag{10.3.5}$$

which describes an ellipsoid in its principal-axis system. This ellipsoid can alternatively be written as

$$\frac{x^2}{n_x^2} + \frac{y^2}{n_y^2} + \frac{z^2}{n_e^2} = 1, \tag{10.3.6}$$

where, in the physically realistic limit $r_{63}E_z \ll 1$, the new principal values of the refractive index are given by

$$n_x = n_o - \tfrac{1}{2}n_o^3 r_{63}E_z, \tag{10.3.7a}$$

$$n_y = n_o + \tfrac{1}{2}n_o^3 r_{63}E_z. \tag{10.3.7b}$$

Figure 10.3.1 shows how to construct a modulator based on the electro-optic effect in KDP. Part (a) shows a crystal that has been cut so that the optic axis (Z axis) is perpendicular to the plane of the entrance face, which contains the X and Y crystalline axes. Part (b) of the figure shows the same crystal in the presence of a longitudinal (z-directed) electric field $E_z = \mathrm{V/L}$, which is established by applying a voltage V between the front and rear faces. The principal axes (x, y, z) of the index ellipsoid in the presence of this field are also indicated. In practice, the potential difference is applied by coating the front and rear faces with a thin film of a conductive coating. Historically, thin layers of gold have often been used, although more recently the transparent conducting material indium tin oxide has successfully been used.

Part (c) of Fig. 10.3.1 shows the curve formed by the intersection of the plane perpendicular to the direction of propagation (i.e., the plane $z = Z = 0$) with the index ellipsoid. For the case in which no static field is applied, the curve has the form of a circle, showing that the refractive index has the value n_o for any direction of polarization. For the case in which a field is applied, this curve has the form of an ellipse. In drawing the figure, we have arbitrarily assumed that the factor $r_{63}E_z$ is negative; consequently the semimajor and

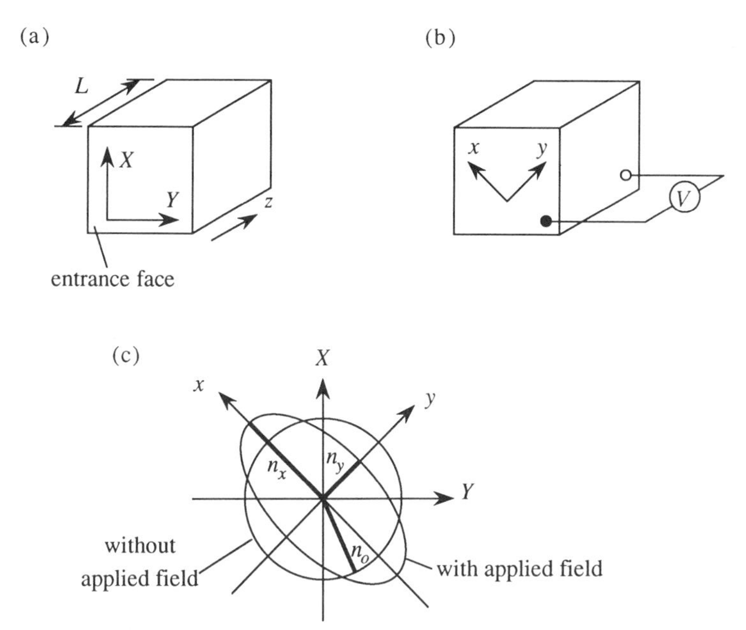

FIGURE 10.3.1 The electrooptic effect in KDP. (a) Principal axes in the absence of an applied field. (b) Principal axes in the presence of an applied field. (c) The intersection of the index ellipsoid with the plane $z = Z = 0$.

semiminor axes of this ellipse are along the x and y directions and have lengths n_x and n_y respectively.

Let us next consider a beam of light propagating in the $z = Z$ direction through the modulator crystal shown in Fig. 10.3.1. A wave polarized in the x direction propagates with a different phase velocity than a wave polarized in the y direction. In propagating through the length L of the modulator crystal, the x and y polarization components will thus acquire the phase difference

$$\Gamma = (n_y - n_x)\frac{\omega L}{c}, \tag{10.3.8}$$

which is known as the retardation. By introducing Eqs. (10.3.7) into this expression we find that

$$\Gamma = \frac{n_o^3 r_{63} E_z \omega L}{c}.$$

Since $E_z = V/L$, this result shows that the retardation introduced by a longitudinal electrooptic modulator depends only on the voltage V applied to the modulator and is independent of the length of the modulator. In particular, the retardation can be represented as

$$\Gamma = \frac{n_o^3 r_{63} \omega V}{c}. \tag{10.3.9}$$

It is convenient to express this result in terms of the quantity

$$V_{\lambda/2} = \frac{\pi c}{\omega n_o^3 r_{63}}, \tag{10.3.10}$$

which is known as the half-wave voltage. Equation (10.3.9) then becomes

$$\Gamma = \pi \frac{V}{V_{\lambda/2}}. \tag{10.3.11}$$

Note that a half-wave (π radians) of retardation is introduced when the applied voltage is equal to the half-wave voltage. Half-wave voltages of typical electrooptic materials are of the order of 10 kV for visible light.

Since the x and y polarization components of a beam of light generally experience different phase shifts in propagating through an electrooptic crystal, the state of polarization of the light leaving the modulator will generally be different from that of the incident light. Figure 10.3.2 shows how the state of

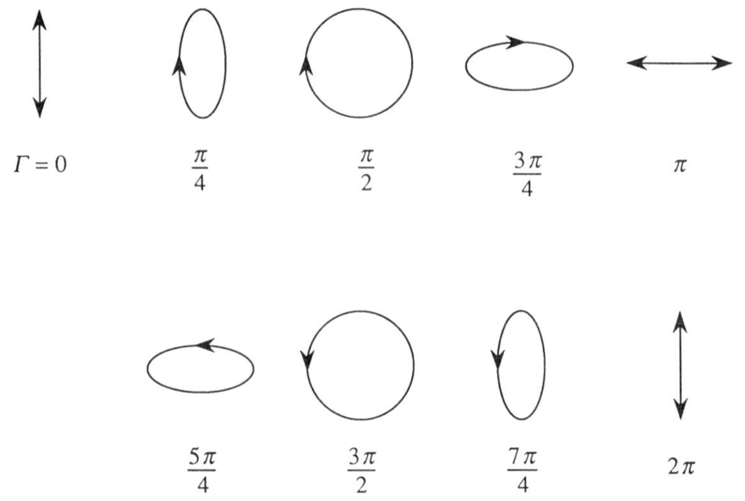

FIGURE 10.3.2 Polarization ellipses describing the light leaving the modulator of Fig. 10.3.1 for various values of the retardation. In all cases, the input light is linearly polarized in the vertical (X) direction.

polarization of the light leaving the modulator depends upon the value of the retardation Γ for the case in which vertically (X) polarized light is incident on the modulator. Note that light of any ellipticity can be produced by controlling the voltage V applied to the modulator.

Figure 10.3.3 shows one way of constructing an intensity modulator based on the configuration shown in Fig. 10.3.1. The incident light is passed through a linear polarizer whose transmission axis is oriented in the X direction. The light then enters the modulator crystal, where its x and y polarization components propagate with different velocities and acquire a phase difference, whose value is given by Eq. (10.3.11). The light leaving the modulator then passes through a quarter-wave plate oriented so that its fast and slow axes coincide with the x and y axes of the modulator crystal, respectively. The beam of light thereby acquires the additional retardation $\Gamma_B = \pi/2$. For reasons that will become apparent below, Γ_B is called the bias retardation. The total retardation is then given by

$$\Gamma = \pi \frac{V}{V_{\lambda/2}} + \frac{\pi}{2}. \tag{10.3.12}$$

In order to analyze the operation of this modulator, let us represent the electric field of the incident radiation after passing through the initial polarizer as

$$\tilde{\mathbf{E}} = \mathbf{E}_{\text{in}} e^{-i\omega t} + \text{c.c.}, \tag{10.3.13a}$$

where

$$\mathbf{E}_{\text{in}} = E_{\text{in}} \hat{\mathbf{X}} = \frac{E_{\text{in}}}{\sqrt{2}} (\hat{\mathbf{x}} + \hat{\mathbf{y}}). \tag{10.3.13b}$$

After the beam passes through the modulator crystal and quarter-wave plate, the phase of the y polarization component will be shifted with respect to that

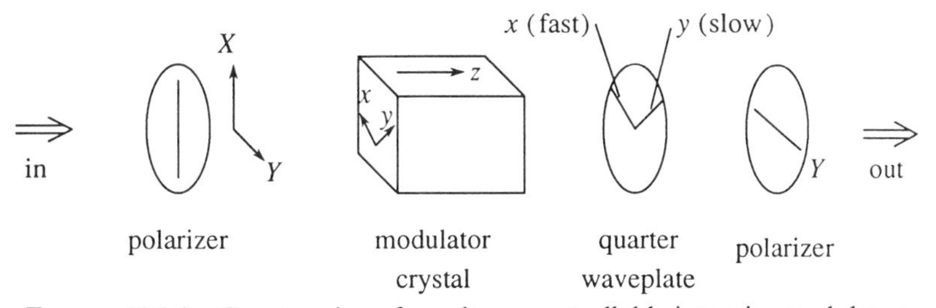

FIGURE 10.3.3 Construction of a voltage-controllable intensity modulator.

of the x polarization component by an amount Γ, so that (to within an unimportant overall phase factor) the complex field amplitude becomes

$$\mathbf{E} = \frac{E_{\text{in}}}{\sqrt{2}}(\hat{\mathbf{x}} + e^{i\Gamma}\hat{\mathbf{y}}). \tag{10.3.14}$$

Only the $\hat{\mathbf{Y}} = (-\hat{\mathbf{x}} + \hat{y})/\sqrt{2}$ component of this field will be transmitted by the final polarizer. The field amplitude measured after this polarizer is hence given by $\mathbf{E}_{\text{out}} = (\mathbf{E} \cdot \hat{\mathbf{Y}})\hat{\mathbf{Y}}$, or as

$$\mathbf{E}_{\text{out}} = \frac{E_{\text{in}}}{2}(-1 + e^{i\Gamma})\hat{\mathbf{Y}}. \tag{10.3.15}$$

If we now define the transmission T of the modulator of Fig. 10.3.3 as

$$T = \frac{|\mathbf{E}_{\text{out}}|^2}{|\mathbf{E}_{\text{in}}|^2}, \tag{10.3.16}$$

we find through use of Eq. (10.3.15) that the transmission is given by

$$T = \sin^2(\Gamma/2). \tag{10.3.17}$$

The functional form of these transfer characteristics is shown in Fig. 10.3.4. We see that the transmission can be made to vary from zero to one by varying the total retardation between zero and π radians. We can also see the motivation for inserting the quarter-wave plate into the setup of Fig. 10.3.3 in order to establish the bias retardation $\Gamma_B = \pi/2$. For the case in which the applied voltage V vanishes, the total retardation will be equal to the bias retardation, and the transmission of the modulator will be 50%. Since the transmission T varies approximately linearly with the retardation Γ for retardations near $\Gamma = \pi/2$, the transmission will vary nearly linearly with the value V of the

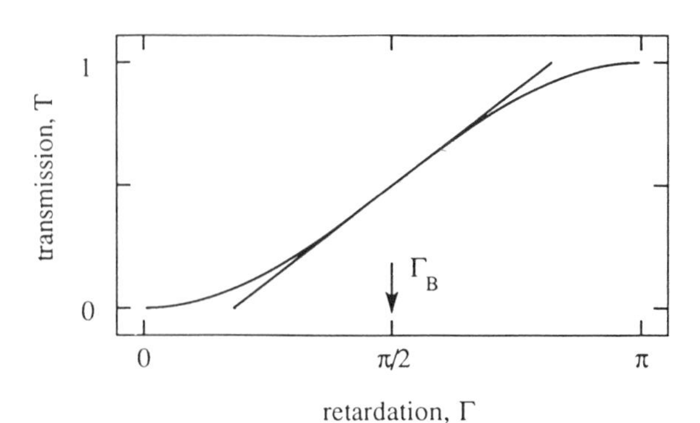

FIGURE 10.3.4 Transmission characteristics of the electrooptic modulator shown in Fig. 10.3.3.

applied voltage. For example, if the applied voltage is given by

$$V(t) = V_{\mathrm{m}} \sin \omega_{\mathrm{m}} t, \tag{10.3.18}$$

the retardation will be given by

$$\Gamma = \frac{\pi}{2} + \frac{\pi V_{\mathrm{m}}}{V_{\lambda/2}} \sin \omega_{\mathrm{m}} t. \tag{10.3.19}$$

The transmission predicted by Eq. (10.3.17) is hence given by

$$\begin{aligned} T &= \sin^2\left(\frac{\pi}{4} + \frac{\pi V_{\mathrm{m}}}{2V_{\lambda/2}} \sin \omega_{\mathrm{m}} t\right) \\ &= \frac{1}{2}\left[1 + \sin\left(\frac{\pi V_{\mathrm{m}}}{V_{\lambda/2}} \sin \omega_{\mathrm{m}} t\right)\right], \end{aligned}$$

which, for $\pi V_{\mathrm{m}}/V_{\lambda/2} \ll 1$, becomes

$$T = \frac{1}{2}\left(1 + \frac{\pi V_{\mathrm{m}}}{V_{\lambda/2}} \sin \omega_{\mathrm{m}} t\right). \tag{10.3.20}$$

The electrooptic effect can also be used to construct a phase modulator for light. For example, if the light incident on the electrooptic crystal of Fig. 10.3.1 is linearly polarized along the x (or the y) axis of the crystal, the light will propagate with its state of polarization unchanged but with its phase shifted by an amount that depends on the value of the applied voltage. The voltage-dependent part of the phase shift is hence given by

$$\phi = (n_x - n_{\mathrm{o}})\frac{\omega L}{c} = -\frac{n_{\mathrm{o}}^3 r_{63} E_z \omega L}{2c} = \frac{n_{\mathrm{o}}^3 r_{63} V \omega}{2c}. \tag{10.3.21}$$

10.4. Introduction to the Photorefractive Effect

The photorefractive effect is the change in refractive index of an optical material that results from the optically induced redistribution of electrons and holes. The photorefractive effect is quite different from most of the other nonlinear-optical effects described in this book in that it cannot be described by a nonlinear susceptibility $\chi^{(n)}$ for any value of n. The reason is that, under a wide range of conditions, the change in refractive index in steady state is independent of the intensity of the light that induces the change. Because the photorefractive effect cannot be described by means of a nonlinear susceptibility, special methods must be employed to describe it; these methods are described in the next several sections. The photorefractive effect tends to give rise to a strong optical nonlinearity; experiments are routinely performed using milliwatts of laser power. However, the effect tends to be rather slow, with response times of 0.1 s being typical.

The origin of the photorefractive effect is illustrated schematically in Fig. 10.4.1. We imagine that a photorefractive crystal is illuminated by two beams of light of the same frequency. These beams interfere to produce the spatially modulated intensity distribution $I(x)$ shown in the upper graph. Free charge carriers, which we assume to be electrons, are generated through photoionization at a rate that is proportional to the local value of the optical intensity. These carriers can diffuse through the crystal or can drift in response to a static electric field. Both processes are observed experimentally. In drawing the figure we have assumed that diffusion is the dominant process, in which case the electron density is smallest in the regions of maximum optical intensity, because electrons have preferentially diffused away from these regions. The spatially varying charge distribution $\rho(x)$ gives rise to a spatially

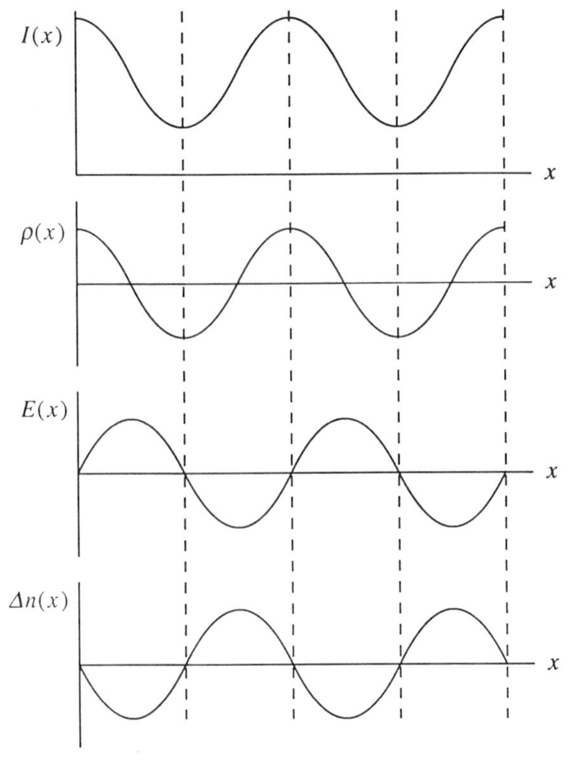

FIGURE 10.4.1 Origin of the photorefractive effect.

TABLE 10.4.1 Properties of some photorefractive crystals*.

Material	Useful wavelength range (μm)	Carrier drift length $\mu\tau E$ at $E = 2$ kV/cm (μm)	τ_d (s)	$n^3 r_{eff}$ (pm/V)	$n^3 r_{eff}/\epsilon_{dc}$ (pm/V)
InP:Fe	0.85–1.3	3	10^{-4}	52	4.1
GaAs:Cr	0.8–1.8	3	10^{-4}	43	3.3
$LiNbO_3:Fe^{+3}$	0.4–0.7	$< 10^{-4}$	300	320	11
$Bi_{12}SiO_{20}$	0.4–0.7	3	10^5	82	1.8
$Sr_{0.4}Ba_{0.6}Nb_2O_6$	0.4–0.6	—	10^2	2,460	4.0
$BaTiO_3$	0.4–0.9	0.1	10^2	11,300	4.9
$KNbBO_3$	0.4–0.7	0.3	10^{-3}	690	14

* τ is the carrier recombination time; τ_d is the dielectric relaxation time in the dark. Adapted from Glass *et al.* (1984).

varying electric field distribution, whose form is shown in the third graph. Note that the maxima of the field $E(x)$ are shifted by 90 degrees with respect to those of the charge density distribution $\rho(x)$. The reason for this behavior is that the Maxwell equation $\nabla \cdot \mathbf{D} = 4\pi\rho$ when applied to the present situation implies that $dE/dx = 4\pi\rho/\epsilon$, and the spatial derivative that appears in this equation leads to a 90-degree phase shift between $E(x)$ and $\rho(x)$.

The last graph in the figure shows the refractive index variation $\Delta n(x)$ that is produced through the linear electrooptic effect (Pockels effect) by the field $E(x)$.* Note that $\Delta n(x)$ is shifted by 90 degrees with respect to the intensity distribution $I(x)$ that produces it. This phase shift has the important consequence that it can lead to the transfer of energy between the two incident beams. This transfer of energy is described in Section 10.6.

The properties of some photorefractive crystals are summarized in Table 10.4.1.

10.5. Photorefractive Equations of Kukhtarev *et al.*

In this section we see how to describe the photorefractive effect by means of a model (see Fig. 10.5.1) due to Kukhtarev and coworkers.† This model presupposes that the photorefractive effect is due solely to one type of charge

* In drawing the figure, we have assumed that the electrooptic coefficient is positive. Note that the relation $\Delta(1/n^2) = r_{eff} E$ implies that $\Delta n = -\frac{1}{2} n^3 r_{eff} E$.

† See Kukhtarev *et al.* (1977, 1979). This model is also described in many of the chapters of the book edited by Günter and Huignard (1988).

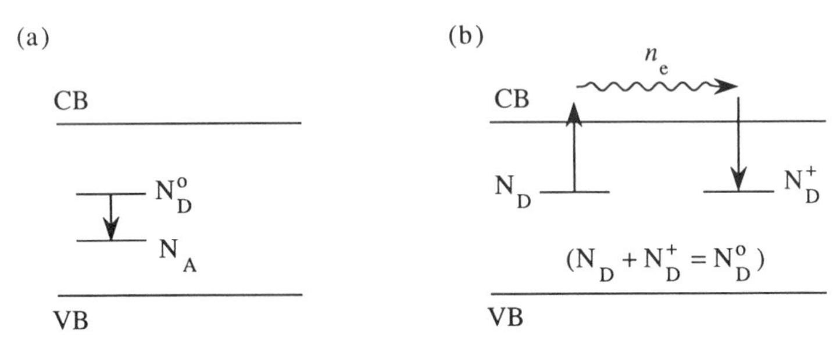

FIGURE 10.5.1 Energy levels and populations of the model of the photorefractive effect due to Kukhtarev *et al.*

carrier, which for definiteness we assume to be the electron. As illustrated in part (a) of the figure, we assume that the crystal contains N_A acceptors and N_D^0 donors per unit volume, with $N_A \ll N_D^0$. We assume that the acceptor levels are completely filled with electrons that have fallen from the donor levels, and that these filled acceptor levels cannot be ionized by thermal or optical effects. Thus, at temperature $T = 0$ and in the absence of an optical field, each unit volume of the crystal contains N_A ionized donors, N_A electrons bound to acceptor impurities, and $N_D^0 - N_A$ neutral donor levels that can participate in the photorefractive effect. We further assume that electrons can be excited thermally or optically from the donor levels into the conduction band, as illustrated in part (b) of the figure. We let n_e, N_D^+, and N_D denote the number densities of conduction band electrons, ionized donors, and un-ionized donors, respectively. Note that $N_D + N_D^+$ must equal N_D^0, but that N_D^+ is not necessarily equal to n_e, because some donors lose their electrons to the acceptors and because electrons can migrate within the crystal, leading to regions that are not electrically neutral.

We next assume that the variation in level populations can be described by the rate equations

$$\frac{\partial N_D^+}{\partial t} = (sI + \beta)(N_D^0 - N_D^+) - \gamma n_e N_D^+, \tag{10.5.1}$$

$$\frac{\partial n_e}{\partial t} = \frac{\partial N_D^+}{\partial t} + \frac{1}{e}(\nabla \cdot \mathbf{j}), \tag{10.5.2}$$

where s is a constant proportional to the photoionization cross section of a donor, β is the thermal generation rate, γ is the recombination coefficient, $-e$ is the charge of the electron, and $\mathbf{j}$ is the electrical current density. Equation (10.5.1) states that the ionized donor concentration can increase by thermal ionization or photoionization of un-ionized donors and can decrease by recombination. Equation (10.5.2) states that the mobile electron concentration can increase in any small region due either to the ionization of donor atoms or to the flow of electrons into the local region. The flow of current is described by the equation

$$\mathbf{j} = n_e e \mu \mathbf{E} + eD\nabla n_e + \mathbf{j}_{\mathrm{ph}}, \tag{10.5.3}$$

where μ is the electron mobility, D is the diffusion constant (which by the Einstein relation is equal to $k_B T\mu/e$), and $\mathbf{j}_{\mathrm{ph}}$ is the photovoltaic (also known as the photogalvanic) contribution to the current. The last contribution results from the tendency of the photoionization process to eject the electron in a preferred direction in anistropic crystals. For some materials (such as barium titanate and bismuth silicon oxide) this contribution to $\mathbf{j}$ is negligible, although for others (such as lithium niobate) it is very important. For lithium niobate, $\mathbf{j}_{\mathrm{ph}}$ has the form $\mathbf{j}_{\mathrm{ph}} = pI\hat{\mathbf{c}}$, where $\hat{\mathbf{c}}$ is a unit vector in the direction of the optic axis of the crystal and p is a constant. The importance of the photovoltaic current has been discussed by Glass (1978).

The field $\mathbf{E}$ appearing in Eq. (10.5.3) is the static (or possibly low-frequency) electric field appearing within the crystal due to any applied voltage or to any charge separation within the crystal. It must satisfy the Maxwell equation

$$\epsilon_{\mathrm{dc}} \nabla \cdot \mathbf{E} = -4\pi e(n_e + N_A - N_D^+), \tag{10.5.4}$$

where ϵ_{dc} is the static dielectric constant of the crystal. The modification of the optical properties is described by assuming that the optical-frequency dielectric constant is changed by an amount

$$\Delta\epsilon = -\epsilon^2 r_{\mathrm{eff}} |\mathbf{E}|. \tag{10.5.5}$$

For simplicity, here we are treating the dielectric properties in the scalar approximation; the tensor properties can be treated explicitly using the formalism developed in Section 10.2* Note that the scalar form of Eq. (10.2.13a) is $\Delta(1/\epsilon) = r_{\mathrm{eff}}|\mathbf{E}|$, from which Eq. (10.5.5) follows directly. The optical field $\tilde{E}_{\mathrm{opt}}$ is assumed to obey the wave equation

$$\nabla^2 \tilde{E}_{\mathrm{opt}} + \frac{1}{c^2}\frac{\partial^2}{\partial t^2}(\epsilon + \Delta\epsilon)\tilde{E}_{\mathrm{opt}} = 0. \tag{10.5.6}$$

Equations (10.5.1) through (10.5.6) constitute the photorefractive equations of Kukhtarev *et al.* They have been solved in a variety of special cases and

* See also the calculation of r_{eff} for one particular case in Eq. (10.6.14) in the next section.

have been found to provide an adequate description of most photorefractive phenomena. We shall consider their solution in special cases in the next two sections.

10.6. Two-Beam Coupling in Photorefractive Materials

Under certain circumstances, two beams of light can interact in a photorefractive crystal in such a manner that energy is transferred from one beam to the other. This process, which is often known as two-beam coupling, can be used, for example, to amplify a weak, image-bearing signal beam by means of an intense pump beam. Exponential gains of 10 per centimeter are routinely observed.

A typical geometry for studying two-beam coupling is show in Fig. 10.6.1. Signal and pump waves, of amplitudes A_s and A_p respectively, interfere to form a nonuniform intensity distribution within the crystal. Due to the nonlinear response of the crystal, this nonuniform intensity distribution produces a refractive index grating within the material. However, this grating is displaced from the intensity distribution in the direction of the positive (or negative, depending on the sign of the dominant charge carrier and the sign of the effective electrooptic coefficient) crystalline c axis. As a result of this phase shift, the light scattered from A_p into A_s interferes constructively with A_s, whereas the light scattered from A_s into A_p interferes destructively with A_p, and consequently the signal wave is amplified whereas the pump wave is attenuated.

In order to describe this process mathematically, we assume that the optical field within the crystal can be represented as

$$\tilde{E}_{\text{opt}}(\mathbf{r}, t) = [A_p(z)e^{i\mathbf{k}_p\cdot\mathbf{r}} + A_s(z)e^{i\mathbf{k}_s\cdot\mathbf{r}}]e^{-i\omega t} + \text{c.c.} \tag{10.6.1}$$

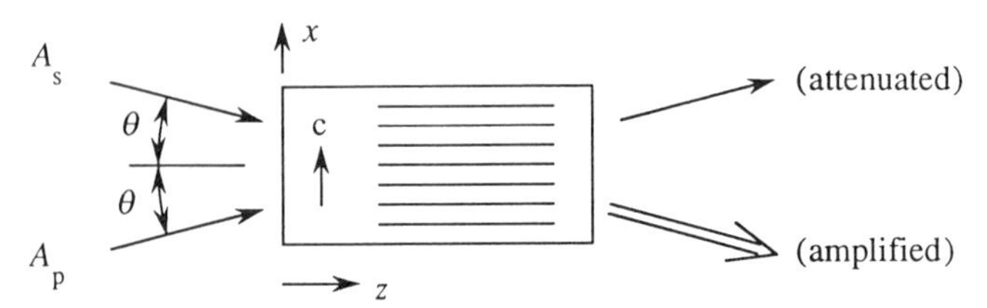

FIGURE 10.6.1 Typical geometry for studying two-beam coupling in a photorefractive crystal.

We assume that $A_p(z)$ and $A_s(z)$ are slowly varying functions of the coordinate z. The intensity distribution of the light within the crystal can be expressed as $I = (n_o c/4\pi)\overline{\tilde{E}^2_{opt}}$ or as

$$I = I_0 + (I_1 e^{iqx} + \text{c.c.}), \tag{10.6.2a}$$

where

$$I_0 = \frac{n_o c}{2\pi}(|A_p|^2 + |A_s|^2),$$

$$I_1 = \frac{n_o c}{2\pi}(A_p A_s^*)(\hat{\mathbf{e}}_p \cdot \hat{\mathbf{e}}_s), \quad \text{and} \quad \mathbf{q} \equiv q\hat{\mathbf{x}} = \mathbf{k}_p - \mathbf{k}_s. \tag{10.6.2b}$$

Here $\hat{\mathbf{e}}_p$ and $\hat{\mathbf{e}}_s$ are the polarization unit vectors of the pump and signal waves, which are assumed to be linearly polarized. The quantity $\mathbf{q}$ is known as the grating wave vector. Note that the intensity distribution can also be described by the expression

$$I = I_0[1 + m\cos(qx + \phi)], \tag{10.6.3}$$

where $m = 2|I_1|/I_0$ is known as the modulation index and where $\phi = \tan^{-1}(\text{Im}\, I_1/\text{Re}\, I_1)$.

In order to determine how the optical properties of the photorefractive material are modified by the presence of the pump and signal fields, we first solve Eqs. (10.5.1) through (10.5.4) of Kukhtarev *et al.* to find the static electric field $\mathbf{E}$ induced by the intensity distribution of Eq. (10.6.2). This static electric field can then be used to calculate the change in the optical-frequency dielectric constant through use of Eq. (10.5.5). Since Eqs. (10.5.1) through (10.5.4) are nonlinear (i.e., they contain products of the unknown quantities n_e, N_D^+, $\mathbf{j}$, and $\mathbf{E}$), they cannot easily be solved exactly. For this reason, we assume that the depth of modulation m is small (i.e., $|I_1| \ll I_0$) and seek an approximate steady-state solution of Eqs. (10.5.1) through (10.5.4) in the form

$$\begin{aligned} E &= E_0 + (E_1 e^{iqx} + \text{c.c.}), \qquad & j &= j_0 + (j_1 e^{iqx} + \text{c.c.}), \\ n_e &= n_{e0} + (n_{e1} e^{iqx} + \text{c.c.}), \qquad & N_D^+ &= N_{D0}^+ + (N_{D1}^+ e^{iqx} + \text{c.c.}), \end{aligned} \tag{10.6.4}$$

where $\mathbf{E} = E\hat{\mathbf{x}}$ and $\mathbf{j} = j\hat{\mathbf{x}}$. We assume that the quantities E_1, j_1, n_{e1}, and N_{D1} are small in the sense that the product of any two of them can be neglected.

We next introduce Eqs. (10.6.4) into Eqs. (10.5.1) through (10.5.4) and equate terms with common x dependences. We thereby find several sets of equations. The set that is independent of the x coordinate depends only on the large quantities (subscript zero) and is given (in the same order as Eqs. (10.5.1) through (10.5.4)) by

$$(sI_0 + \beta)(N_D^0 - N_{D0}^+) = \gamma n_{e0} N_{D0}^+, \tag{10.6.5a}$$

$$j_0 = \text{constant}, \tag{10.6.5b}$$

$$j_0 = n_{e0} e \mu E_0 + j_{\text{ph},0}, \tag{10.6.5c}$$

$$N_{D0}^+ = n_{e0} + N_A. \tag{10.6.5d}$$

Equations (10.6.5a) and (10.6.5d) can be solved directly to determine the mean electron density n_{e0} and mean ionized donor density N_{D0}^+. Since in most realistic cases the inequality $n_{e0} \ll N_A$ is satisfied, the densities are given simply by

$$N_{D0}^+ = N_A, \tag{10.6.6a}$$

$$n_{e0} = \frac{(sI_0 + \beta)(N_D^0 - N_A)}{\gamma N_A}. \tag{10.6.6b}$$

The two remaining equations ((10.6.5b) and (10.6.5c)) determine the mean current density j_0 and mean field E_0. Let us assume for simplicity that the photovoltaic contribution j_{ph} is negligible for the material under consideration. The value of E_0 then depends on the properties of any external electric circuit to which the crystal is connected. In the common situation in which no voltage is externally applied to the crystal, E_0 and hence j_0 vanish.

We next consider the equation for the first-order quantities (quantities with the subscript one) by considering the portions of Eqs. (10.5.1) through (10.5.4) with the spatial dependence e^{iqx}. The resulting equations are (we assume that $E_0 = 0$)

$$sI_1(N_D^0 - N_A) - (sI_0 + \beta)N_{D1}^+ = \gamma n_{e0} N_{D1}^+ + \gamma n_{e1} N_A, \tag{10.6.7a}$$

$$j_1 = 0, \tag{10.6.7b}$$

$$-n_{e0} e E_1 = iqk_B T n_{e1}, \tag{10.6.7c}$$

$$iq\epsilon_{dc} E_1 = -4\pi e(n_{e1} - N_{D1}^+). \tag{10.6.7d}$$

We solve these equations algebraically (again assuming that $n_{e0} \ll N_A$) to find that the amplitude of the spatially varying part of the static electric field is given by

$$E_1 = -i\left(\frac{sI_1}{sI_0 + \beta}\right)\left(\frac{E_D}{1 + E_D/E_q}\right), \tag{10.6.8}$$

where we have introduced the characteristic field strengths

$$E_{\mathrm{D}} = \frac{qk_{\mathrm{B}}T}{e}, \qquad E_q = \frac{4\pi e}{\epsilon_{\mathrm{dc}} q} N_{\mathrm{eff}}, \tag{10.6.9}$$

where $N_{\mathrm{eff}} = N_{\mathrm{A}}(N_{\mathrm{D}}^0 - N_{\mathrm{A}})/N_{\mathrm{D}}^0$ can be interpreted as an effective trap density. Note that in the common circumstance where $N_{\mathrm{A}} \ll N_{\mathrm{D}}^0$, N_{eff} is given approximately by $N_{\mathrm{eff}} \simeq N_{\mathrm{A}}$. The quantity E_{D} is called the diffusion field strength and is a measure of the field strength required to inhibit the separation of charge due to thermal agitation. The quantity E_q is called the maximum space charge field and is a measure of the maximum electric field that can be created by redistributing charge of mean density eN_{eff} over the characteristic distance $2\pi/q$. Note from Eq. (10.6.8) that E_1 is shifted in phase with respect to the intensity distribution I_1 and that E_1 is proportional to the depth of modulation m in the common case of $\beta \ll sI_0$.

Recall that the change in the optical frequency dielectric constant is proportional to the amplitude E_1 of the spatially modulated component of the static electric field. For this reason, it is often of interest to maximize the value of E_1. We see from Eq. (10.6.8) that E_1 is proportional to the product of the factor $sI_1/(sI_0 + \beta)$, which can be maximized by increasing the depth of modulation $m = 2|I_1|/I_0$,* with the factor $E_{\mathrm{D}}/(1 + E_{\mathrm{D}}/E_q)$. Since each of the characteristic field strengths E_{D} and E_q depends on the grating wave vector, this second factor can be maximized by using the optimum value of q. To show the dependence of E_1 on q, we can rewrite Eq. (10.6.8) as

$$E_1 = -i\left(\frac{sI_1}{sI_0 + \beta}\right) E_{\mathrm{opt}} \frac{2(q/q_{\mathrm{opt}})}{1 + (q/q_{\mathrm{opt}})^2}, \tag{10.6.10a}$$

where

$$q_{\mathrm{opt}} = \left(\frac{4\pi N_{\mathrm{eff}} e^2}{k_{\mathrm{B}} T \epsilon_{\mathrm{dc}}}\right)^{1/2}, \qquad E_{\mathrm{opt}} = \left(\frac{\pi N_{\mathrm{eff}} k_{\mathrm{B}} T}{\epsilon_{\mathrm{dc}}}\right)^{1/2}. \tag{10.6.10b}$$

Note that q_{opt} is of the order of magnitude of the inverse of the Debye screening length.

The dependence of E_1 on q is shown in Fig. 10.6.2. Note that the grating wave vector q can be varied experimentally by controlling the angle between

* Recall, however, that the present derivation is valid only if $m \ll 1$.

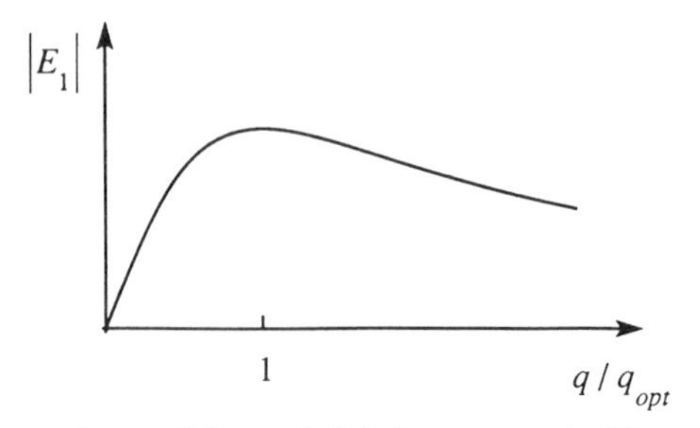

FIGURE 10.6.2 Dependence of the modulated component of the space charge field on the magnitude q of the grating wave vector.

the pump and signal beams, since (see Fig. 10.6.1) q is given by the formula

$$q = 2n\frac{\omega}{c}\sin\theta. \tag{10.6.11}$$

Through an experimental determination of the optimum value of the magnitude of the grating wave vector, the value of the effective trap density N_{eff} can be obtained through use of Eq. (10.6.10b).

Let us next calculate the spatial growth rate that the signal wave experiences as the result of two-beam coupling in photorefractive materials. For simplicity, we assume that the photoionization rate sI_0 is much greater than the thermal ionization rate β (which is the usual case in practice), so that the field amplitude E_1 of Eq. (10.6.8) can be expressed through use of Eq. (10.6.2) as

$$E_1 = -i\frac{A_{\text{p}}A_{\text{s}}^*}{|A_{\text{s}}|^2 + |A_{\text{p}}|^2}(\hat{e}_{\text{p}}\cdot\hat{e}_{\text{s}})E_{\text{m}}, \tag{10.6.12a}$$

where

$$E_{\text{m}} = \frac{E_{\text{D}}}{1 + E_{\text{D}}/E_q}. \tag{10.6.12b}$$

According to Eq. (10.5.5), this field produces a change in the dielectric constant of amplitude $\Delta\epsilon = -\epsilon^2 r_{\text{eff}} E_1$. For the particular geometry of Fig. 10.6.1, the product $\epsilon^2 r_{\text{eff}}$ has the form (Feinberg *et al.*, 1980; see also Feinberg and MacDonald, 1989)

$$\epsilon^2 r_{\text{eff}} = \sum_{ijklm} r_{ijk}(\epsilon_{il}\hat{e}_l^{\text{s}})(\epsilon_{jm}\hat{e}_m^{\text{p}})\hat{q}_k, \tag{10.6.13}$$

where $\hat{e}_l^{\text{s}}$ and $\hat{e}_m^{\text{p}}$ denote that l and m cartesian components of the polarization unit vectors of the signal and pump waves, respectively, and $\hat{q}_k$ denotes the

k cartesian component of a unit vector in the direction of the grating vector. For crystals of point group $4mm$ (such as barium titanate), one finds that for ordinary waves

$$r_{\text{eff}} = r_{13} \sin\left(\frac{\alpha_s + \alpha_p}{2}\right) \tag{10.6.14a}$$

and that for extraordinary waves

$$\begin{aligned} r_{\text{eff}} = n^{-4}\Big[& n_o^4 r_{13} \cos\alpha_s \cos\alpha_p + 2n_e^2 n_o^2 r_{42} \cos\left(\frac{\alpha_s + \alpha_p}{2}\right) \\ & + n_e^4 r_{33} \sin\alpha_s \sin\alpha_p \Big] \sin\left(\frac{\alpha_s + \alpha_p}{2}\right). \end{aligned} \tag{10.6.14b}$$

Here α_s and α_p denote the angles between the propagation vectors of the signal and pump waves and the positive crystalline c axis, respectively, and n is the refractive index experienced by the beam that scatters off the grating.

Note from Table 10.2.1 that for barium titanate the electrooptic coefficient r_{42} is much larger than either r_{13} or r_{33}. We see from Eqs. (10.6.14) that only through the use of light of extraordinary polarization can one utilize this large component of the electrooptic tensor.

The change in the dielectric constant $\Delta\epsilon = -\epsilon^2 r_{\text{eff}} E_1$ produces a nonlinear polarization given by

$$P^{\text{NL}} = \left(\frac{\Delta\epsilon}{4\pi} e^{i\mathbf{q}\cdot\mathbf{r}} + \text{c.c.}\right)(A_s e^{i\mathbf{k}_s\cdot\mathbf{r}} + A_p e^{i\mathbf{k}_p\cdot\mathbf{r}}). \tag{10.6.15}$$

Recall that $\mathbf{q} = \mathbf{k}_p - \mathbf{k}_s$. The part of the nonlinear polarization having the spatial variation $\exp(i\mathbf{k}_s \cdot \mathbf{r})$ can act as a phase-matched source term for the signal wave and is given by

$$P_s^{\text{NL}} = \frac{\Delta\epsilon^*}{4\pi} A_p e^{i\mathbf{k}_s\cdot\mathbf{r}} = \frac{-i\epsilon^2 r_{\text{eff}} E_m}{4\pi} \frac{|A_p|^2 A_s}{|A_p|^2 + |A_s|^2} e^{i\mathbf{k}_s\cdot\mathbf{r}}. \tag{10.6.16a}$$

Likewise, the portion of P^{NL} that can act as a phase-matched source term for the pump wave is given by

$$P_p^{\text{NL}} = \frac{\Delta\epsilon}{4\pi} A_s e^{i\mathbf{k}_p\cdot\mathbf{r}} = \frac{i\epsilon^2 r_{\text{eff}} E_m}{4\pi} \frac{|A_s|^2 A_p}{|A_p|^2 + |A_s|^2} e^{i\mathbf{k}_p\cdot\mathbf{r}}. \tag{10.6.16b}$$

We next derive coupled-amplitude equations for the pump and signal fields using the formalism described in Section 2.1. We define z_s and z_p to be distances measured along the signal and pump propagation directions. We find that in the slowly-varying-amplitude approximation the signal amplitude varies as

$$2ik \frac{dA_s}{dz_s} e^{i\mathbf{k}_s \cdot \mathbf{r}} = -4\pi \frac{\omega^2}{c^2} P_s^{NL}, \tag{10.6.17a}$$

which through use of Eq. (10.6.16a) becomes

$$\frac{dA_s}{dz_s} = \frac{\omega}{2c} n^3 r_{eff} E_m \frac{|A_p|^2 A_s}{|A_p|^2 + |A_s|^2}. \tag{10.6.17b}$$

We find that the intensity $I_s = (nc/2\pi)|A_s|^2$ of the signal wave varies spatially as $dI_s/dz_s = A_s^* \, dA_s/dz_s + \text{c.c.}$, or as

$$\frac{dI_s}{dz_s} = \Gamma \frac{I_s I_p}{I_s + I_p}, \tag{10.6.18a}$$

where

$$\Gamma = \frac{\omega}{c} n^3 r_{eff} E_m. \tag{10.6.18b}$$

A similar derivation shows that the pump intensity varies spatially as

$$\frac{dI_p}{dz_p} = -\Gamma \frac{I_s I_p}{I_s + I_p}. \tag{10.6.18c}$$

Note that Eq. (10.6.18a) predicts that the signal intensity grows exponentially with propagation distance in the common limit $I_s \ll I_p$.*

The treatment of two-beam coupling given above has assumed that the system is in steady state. Two-beam coupling under transient conditions can also be treated using the material equations of Kukhtarev *et al.* It has been shown (Kukhtarev *et al.*, 1977; Refrégier *et al.*, 1985; Valley, 1987) that, under the assumption that $n_e \ll N_D^+$, $N_D^+ \ll N_D^0$, and $\beta \ll sI_0$, the electric field amplitude E_1 obeys the equation

$$\tau \frac{\partial E_1}{\partial t} + E_1 = -iE_m \frac{A_p A_s^*}{|A_p|^2 + |A_s|^2} (\hat{e}_p \cdot \hat{e}_s) \tag{10.6.19}$$

* Here we are implicitly assuming that Γ is a positive quantity. If Γ is negative, the wave that we have been calling the pump wave will be amplified and the wave that we have been calling the signal wave will be attenuated. The sign of Γ depends on the sign of r_{eff}, which can be either positive or negative, and on the sign of E_m. Note that, according to Eqs. (10.6.9) and (10.6.12b), the sign of E_m depends upon the sign of the dominant charge carrier (our derivation has assumed the case of an electron) and upon the sign of q, which is the x component of $\mathbf{k}_p - \mathbf{k}_s$. For the case of barium titanate, the dominant charge carriers are usually holes, and the wave whose wave vector has a positive component along the crystalline c axis is amplified.

with E_m given by Eq. (10.6.12b) and with the response time τ given by

$$\tau = \tau_d \frac{1 + E_D/E_M}{1 + E_D/E_q} \tag{10.6.20a}$$

where

$$\tau_d = \frac{\epsilon_{dc}}{4\pi e \mu n_{e0}}, \qquad E_M = \frac{\gamma N_A}{q\mu}. \tag{10.6.20b}$$

Note that the photorefractive response time τ scales linearly with the dielectric relaxation time τ_d.* Since the mean electron density n_{e0} increases linearly with optical intensity (see Eq. (10.6.6b)), we see that the photorefractive response time becomes faster when the crystal is excited using high optical intensities.

We next write the coupled-amplitude equations for the pump and signal fields in terms of the field amplitude E_1 as

$$\frac{\partial A_p}{\partial x_p} = \frac{-i\omega}{2n_p c} r_{eff} A_s E_1, \tag{10.6.21a}$$

$$\frac{\partial A_s}{\partial x_s} = \frac{-i\omega}{2n_s c} r_{eff} A_p E_1^*. \tag{10.6.21b}$$

Equations (10.6.19) through (10.6.21) describe the transient behavior of two-beam coupling.

10.7. Four-Wave Mixing in Photorefractive Materials

Next we consider the mutual interaction of four beams of light in a photorefractive crystal. We assume the geometry of Fig. 10.7.1. Note that the pump beams 1 and 2 are counterpropagating, as are beams 3 and 4. Thus the interaction shown in the figure can be used to generate beam 4 as the phase conjugate of beam 3.

The general problem of the interaction of four beams of light in a photorefractive material is very complicated, because the material response consists of four distinct gratings, namely, one grating due to the interference of

* The dielectric relaxation time is the characteristic time in which charge imbalances neutralize in a conducting material. The expression for the dielectric relaxation time is derived by combining the equation of continuity $\partial\rho/\partial t = -\nabla \cdot \mathbf{j}$ with Ohm's law in the form $\mathbf{j} = \sigma \mathbf{E}$ to find that $\partial\rho/\partial t = -\sigma\nabla \cdot \mathbf{E} = -(\sigma/\epsilon_{dc})\nabla \cdot \mathbf{D} = -(4\pi\sigma/\epsilon_{dc})\rho \equiv -\rho/\tau_d$. By equating the electrical conductivity σ with the product $n_{e0}e\mu$, we obtain the expression for τ_d quoted in the text.

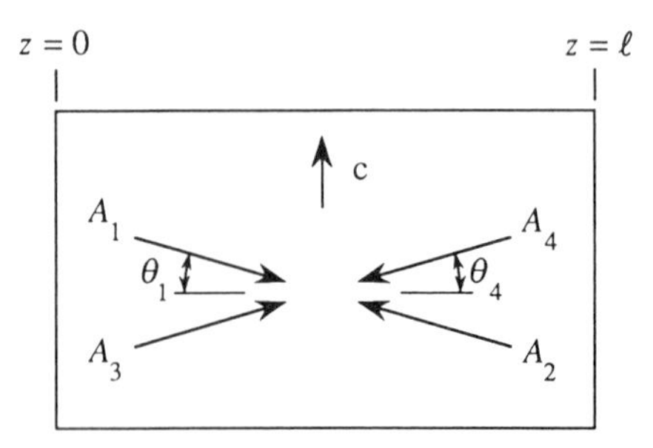

FIGURE 10.7.1 Geometry of four-wave mixing in a photorefractive material.

beams 1 and 3 and of 2 and 4, one grating due to the interference of beams 1 and 4 and of 2 and 3, one grating due to the interference of beams 1 and 2, and one grating due to the interference of beams 3 and 4. However, under certain experimental situations, only one of these gratings leads to appreciable nonlinear coupling among the beams. If one assumes that the polarizations, propagation directions, and coherence properties of the input beams are selected so that only the grating due to the interference of beams 1 and 3 and beams 2 and 4 is important, the coupled-amplitude equations describing the propagation of the four beams become (Cronin-Golomb *et al.*, 1984; see also Fischer *et al.*, 1981)

$$\frac{dA_1}{dz} = -\frac{\gamma}{S_0}(A_1A_3^* + A_2^*A_4)A_3 - \alpha A_1, \tag{10.7.1a}$$

$$\frac{dA_2}{dz} = -\frac{\gamma}{S_0}(A_1^*A_3 + A_2A_4^*)A_4 + \alpha A_2, \tag{10.7.1b}$$

$$\frac{dA_3}{dz} = \frac{\gamma}{S_0}(A_1^*A_3 + A_2A_4^*)A_1 - \alpha A_3, \tag{10.7.1c}$$

$$\frac{dA_4}{dz} = \frac{\gamma}{S_0}(A_1A_3^* + A_2^*A_4)A_2 + \alpha A_4. \tag{10.7.1d}$$

In these equations, we have introduced the following quantities:

$$\gamma = \frac{\omega r_{\text{eff}} n_0^3 E_{\text{m}}}{2c\cos\theta} \tag{10.7.2a}$$

with E_{m} given by Eq. (10.6.12b),

$$S_0 = \sum_{i=1}^{4} |A_i|^2, \tag{10.7.2b}$$

and $\alpha = \frac{1}{2}\alpha_0/\cos\theta$, where α_0 is the intensity absorption coefficient of the material and where for simplicity we have assumed that $\theta = \theta_1 = \theta_4$.

Cronin-Golomb *et al.* (1984) have shown that Eqs. (10.7.1) can be solved for a large number of cases of interest. The solutions show a variety of interesting features, including amplified reflection, self-oscillation, and bistability.

Linear Passive Phase-Conjugate Mirror

One interesting feature of four-wave mixing in photorefractive materials is that it can be used to construct a self-pumped phase-conjugate mirror of the sort illustrated in Fig. 10.7.2. In such a device, only the signal wave A_3 is applied externally. Waves A_1 and A_2 grow from noise within the resonator that surrounds the photorefractive crystal. Oscillation occurs because wave A_1 is amplified at the expense of the signal wave A_3 by the process of two-beam coupling. The ouput wave A_4 is generated by four-wave mixing involving waves A_1, A_2, and A_3. Such a device was constructed by White *et al.* (1982) and is described further by Cronin-Golomb *et al.* (1984).

Double Phase-Conjugate Mirror

Another application of four-wave mixing in photorefractive crystals is the double phase-conjugate mirror of Fig. 10.7.3. In such a device the waves A_2 and A_3 are applied externally; these waves are assumed to be mutually incoherent, so that no gratings are formed by their interference. The nonlinear interaction leads to the generation of the output wave A_1, which is the phase conjugate of A_2, and to the output wave A_4, which is the phase conjugate of A_3. However, A_1 is phase-coherent with A_3, whereas A_4 is phase-coherent with A_2. The double phase-conjugate mirror possesses the remarkable property that one of the output waves can be an amplified phase-conjugate wave, even though the two input waves are mutually incoherent.

The nature of the nonlinear coupling that produces the double phase-conjugate mirror can be understood from the coupled-amplitude equations

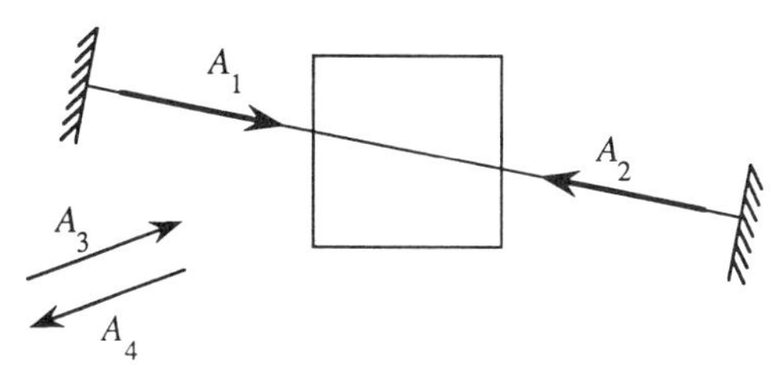

FIGURE 10.7.2 Geometry of the linear passive phase-conjugate mirror. Only the A_3 wave is applied externally; this wave excites the oscillation of the waves A_1 and A_2, which act as pump waves for the four-wave mixing process that generates the conjugate wave A_4.

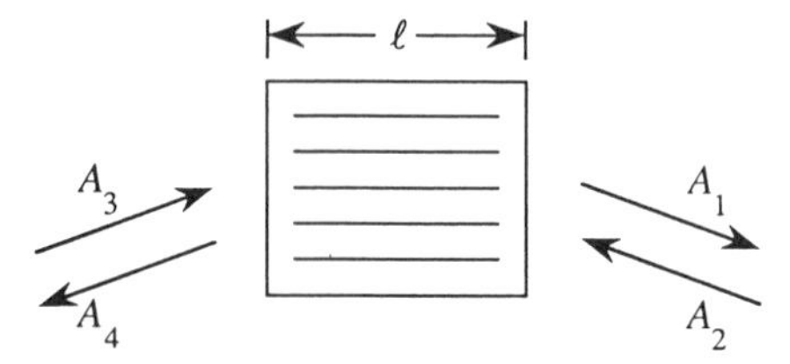

FIGURE 10.7.3 Geometry of the double phase-conjugate mirror. Waves A_2 and A_3 are applied externally and need not be phase-coherent. The generated wave A_1 is the phase conjugate of A_2, and the generated wave A_4 is the phase conjugate of A_3.

(10.7.1). For simplicity, we consider the limit in which α is negligible and in which the input waves A_2 and A_3 are not modified by the nonlinear interaction, so that only equations (10.7.1a) and (10.7.1d) need to be considered. We see that each output wave is driven by two terms, one of which is a two-beam-coupling term that tends to amplify the output wave, and the other of which is a four-wave-mixing term that causes each output to be the phase conjugate of its input wave. It has been shown by Cronin-Golomb *et al.* (1984) and by Weiss *et al.* (1987) that the requirement for the generation of the two output waves is that $|\gamma|l$ be greater than 2. Operation of the double phase-conjugate mirror has been demonstrated experimentally by Weiss *et al.* (1987).

Problems

1. Consider the process of two-beam coupling in barium titanate in the geometry of Fig. 10.6.1. Estimate the numerical values of the physical quantities E_D, E_q, E_{opt}, q_{opt}, E_1, r_{eff}, $\Delta\epsilon$, and Γ. Assume that the effective trap density N_{eff} is equal to 10^{12} cm^{-3}, that the thermal generation rate is negligible, that the modulation index m is 10^{-3}, and that $\theta_s = \theta_p =$ 5 degrees.

2. Verify Eq. (10.6.19).

References

Electrooptic Effect

M. Born and E. Wolf, *Principles of Optics*, Pergamon, London, 1975.

W. R. Cook, Jr., and J. Jaffe, "Electrooptic Coefficients," in *Landolt-Bornstein, New Series*, Vol. II, edited by K.-H. Hellwege, Springer-Verlag, Berlin, 1979, pp. 552–651.

I. P. Kaminow, *An Introduction to Electrooptic Devices*, Academic Press, New York, 1974.

B. J. Thompson and E. Hartfield, in Handbook of Optics, edited by W. G. Driscoll and W. Vaughan, McGraw-Hill, New York 1978.

A. Yariv and P. Yeh, *Optical Waves in Crystals*, Wiley, New York, 1984.

Photorefractive Effect

M. Cronin-Golomb, B. Fischer, J. O. White, and A. Yariv, *IEEE J. Quantum Electron.* **20**, 12 (1984).

J. Feinberg and K. R. MacDonald, in *Photorefractive Materials and their Applications*, Vol. II, edited by P. Günter and J.-P. Huignard, Springer-Verlag, Berlin, 1989.

J. Feinberg, D. Heiman, A. R. Tanguay, Jr., and R. W. Hellwarth, *J. Appl. Phys.* **5**, 1297 (1980).

B. Fischer, M. Cronin-Golomb, J. O. White, and A. Yariv, *Opt. Lett.* **6**, 519 (1981).

A. M. Glass, *Opt. Eng.* **17**, 470 (1978).

A. M. Glass, D. von der Linde, and T. J. Negran, *Appl. Phys. Lett.* **25**, 233 (1974).

A. M. Glass, A. M. Johnson, D. H. Olson, W. Simpson, and A. A. Ballman, *Appl. Phys. Lett.* **44**, 948 (1984).

P. Günter and J.-P. Huignard, editors, *Photorefractive Materials and Their Applications*, Springer-Verlag, Berlin, Part I, 1988; Part II, 1989.

N. Kukhtarev, V. B. Markov, and S. G. Odulov, *Opt. Commun.* **23**, 338 (1977).

N. Kukhtarev, V. B. Markov, S. G. Odulov, M. S. Soskin, and V. L. Vinetskii, *Ferroelectrics* **22**, 949–960, 961–964 (1979).

Ph. Refrégier, L. Solymar, H. Rabjenbach, and J. P. Huignard, *J. Appl. Phys.* **58**, 45 (1985).

G. C. Valley, *J. Opt. Soc. Am.* **4**, 14, 934 (1987).

S. Weiss, S. Sternklar, and B. Fischer, *Opt. Lett.* **12**, 114 (1987).

J. O. White, M. Cronin-Golomb, B. Fischer, and A. Yariv, *Appl. Phys. Lett.* **40**, 450 (1982).

Appendices

Appendix A. Systems of Units in Nonlinear Optics

There are several different systems of units that are commonly used in nonlinear optics. In this appendix we describe these different systems and show how to convert among them. For simplicity, we restrict the discussion to the case of a medium with instantaneous response, so that the nonlinear susceptibilities can be taken to be dispersionless. Clearly, the rules derived here for conversion among the systems of units are the same for a dispersive medium.

In the gaussian system of units, the polarization $\tilde{P}$ is related to the field strength $\tilde{E}$ by the equation

$$\tilde{P}(t) = \chi^{(1)}\tilde{E}(t) + \chi^{(2)}\tilde{E}^2(t) + \chi^{(3)}\tilde{E}^3(t) + \cdots. \tag{A.1}$$

In the gaussian system, all of the fields $\tilde{E}$, $\tilde{P}$, $\tilde{D}$, $\tilde{B}$, $\tilde{H}$, and $\tilde{M}$ have the same units; in particular, the units of $\tilde{P}$ and $\tilde{E}$ are given by

$$[\tilde{P}] = [\tilde{E}] = \frac{\text{statvolt}}{\text{cm}} = \frac{\text{statcoulomb}}{\text{cm}^2} = \left(\frac{\text{erg}}{\text{cm}^3}\right)^{1/2}. \tag{A.2}$$

Consequently, we see from Eq. (A.1) that the dimensions of the susceptibilities are as follows:

$$\chi^{(1)} \text{ is dimensionless,} \tag{A.3a}$$

$$[\chi^{(2)}] = \left[\frac{1}{\tilde{E}}\right] = \frac{\text{cm}}{\text{statvolt}} = \left(\frac{\text{erg}}{\text{cm}^3}\right)^{-1/2}, \tag{A.3b}$$

$$[\chi^{(3)}] = \left[\frac{1}{\tilde{E}^2}\right] = \frac{\text{cm}^2}{\text{statvolt}^2} = \left(\frac{\text{erg}}{\text{cm}^3}\right)^{-1}. \tag{A.3c}$$

The units of nonlinear susceptibilities are not usually stated explicitly in the gaussian system of units; rather one simply states that the value is given in electrostatic units (esu).

There are two different conventions regarding the units of the susceptibility in the MKS system. Some authors replace Eq. (A.1) by

$$\tilde{P}(t) = \epsilon_0[\chi^{(1)}\tilde{E}(t) + \chi^{(2)}\tilde{E}^2(t) + \chi^{(3)}\tilde{E}^3(t)], \tag{A.4}$$

where

$$\epsilon_0 = 8.85 \times 10^{-12}\ \mathrm{F/m} \tag{A.5a}$$

denotes the permittivity of free space. Since the units of $\tilde{P}$ and $\tilde{E}$ in the MKS system are

$$[\tilde{P}] = \frac{\mathrm{C}}{\mathrm{m}^2}, \tag{A.5b}$$

$$[\tilde{E}] = \frac{\mathrm{V}}{\mathrm{m}}, \tag{A.5c}$$

and since one farad is equal to one coulomb per volt (F = C/V), it follows that the units of the susceptibilities are as follows:

$$\chi^{(1)} \text{ is dimensionless,} \tag{A.6a}$$

$$[\chi^{(2)}] = \left[\frac{1}{\tilde{E}}\right] = \frac{\mathrm{m}}{\mathrm{V}}, \tag{A.6b}$$

$$[\chi^{(3)}] = \left[\frac{1}{\tilde{E}}\right]^2 = \frac{\mathrm{m}^2}{\mathrm{V}^2}. \tag{A.6c}$$

The other convention within the MKS system is to replace Eq. (A.1) by

$$\tilde{P}(t) = \epsilon_0\chi^{(1)}\tilde{E}(t) + \chi^{(2)}\tilde{E}(t)^2 + \chi^{(3)}\tilde{E}(t)^3 + \cdots. \tag{A.7}$$

Since the units of $\tilde{P}$, $\tilde{E}$, and ϵ_0 are still given by Eqs. (A.5), it follows that the dimensions of the susceptibilities are as follows:

$$\chi^{(1)} \text{ is dimensionless,} \tag{A.8a}$$

$$[\chi^{(2)}] = \frac{\mathrm{C}}{\mathrm{V}^2}, \tag{A.8b}$$

$$[\chi^{(3)}] = \frac{\mathrm{Cm}}{\mathrm{V}^3}. \tag{A.8c}$$

Conversion among the Systems

In order to facilitate conversion among the three systems introduced above, we express the three defining relations (A.1), (A.4), and (A.7) in the following forms:

$$\tilde{P}(t) = \chi^{(1)}\tilde{E}(t)\left[1 + \frac{\chi^{(2)}\tilde{E}(t)}{\chi^{(1)}} + \frac{\chi^{(3)}\tilde{E}^2(t)}{\chi^{(1)}} + \cdots\right] \quad \text{(gaussian)}, \quad (A.1')$$

$$\tilde{P}(t) = \epsilon_0\chi^{(1)}\tilde{E}(t)\left[1 + \frac{\chi^{(2)}\tilde{E}(t)}{\chi^{(1)}} + \frac{\chi^{(3)}\tilde{E}^2(t)}{\chi^{(1)}} + \cdots\right] \quad \text{(MKS)}, \quad (A.4')$$

$$\tilde{P}(t) = \epsilon_0\chi^{(1)}\tilde{E}(t)\left[1 + \frac{\chi^{(2)}\tilde{E}(t)}{\epsilon_0\chi^{(1)}} + \frac{\chi^{(3)}\tilde{E}^2(t)}{\epsilon_0\chi^{(1)}} + \cdots\right] \quad \text{(MKS)}. \quad (A.7')$$

The power series shown in square brackets must be identical in each of these equations. However, the values of $\tilde{E}$, $\chi^{(1)}$, $\chi^{(2)}$, and $\chi^{(3)}$ are different in different systems. In particular, from Eqs. (A.2) and (A.5) and the fact that 1 statvolt = 300 V, we find that

$$\tilde{E}(\text{MKS}) = 3 \times 10^4\tilde{E}(\text{gaussian}). \quad (A.9)$$

To determine how the linear susceptibilities in the gaussian and MKS systems are related, we make use of the fact that for a linear medium the displacement is given in the gaussian system by

$$\tilde{D} = \tilde{E} + 4\pi\tilde{P} = \tilde{E}(1 + 4\pi\chi^{(1)}), \quad (A.10a)$$

and in the MKS system by

$$\tilde{D} = \epsilon_0\tilde{E} + \tilde{P} = \epsilon_0\tilde{E}(1 + \chi^{(1)}). \quad (A.10b)$$

We thus find that

$$\chi^{(1)}(\text{MKS}) = 4\pi\chi^{(1)}(\text{gaussian}). \quad (A.11)$$

Using Eqs. (1.A.9) and (1.A.10), and requiring that the power series of Eqs. (A.1′), (A.4′) and (A.7′) be identical, we find that the nonlinear susceptibilities in our three systems of units are related by

$$\chi^{(2)}(\text{MKS, Eq. (A4)}) = \frac{4\pi}{3 \times 10^4}\chi^{(2)}(\text{gaussian}) \quad (A.12)$$

$$= 4.189 \times 10^{-4}\chi^{(2)}(\text{gaussian}),$$

$$\chi^{(2)}(\text{MKS, Eq. (A7)}) = \frac{4\pi\epsilon_0}{3 \times 10^4}\chi^{(2)}(\text{gaussian}) \quad (A.13)$$

$$= 3.71 \times 10^{-15}\chi^{(2)}(\text{gaussian}),$$

$$\chi^{(3)}(\text{MKS, Eq. (A4)}) = \frac{4\pi}{(3 \times 10^4)^2}\chi^{(3)}(\text{gaussian}) \quad (A.14)$$

$$= 1.40 \times 10^{-8}\chi^{(3)}(\text{gaussian}),$$

$$\chi^{(3)}(\text{MKS, Eq. (A7)}) = \frac{4\pi\epsilon_0}{(3 \times 10^4)^2}\chi^{(3)}(\text{gaussian}) \qquad \text{(A.15)}$$

$$= 1.24 \times 10^{-19}\chi^{(3)}(\text{gaussian}).$$

Appendix B. Relationship between Intensity and Field Strength

In the gaussian system of units, the intensity associated with the field

$$\tilde{E}(t) = Ee^{-i\omega t} + \text{c.c.} \qquad \text{(B.1)}$$

is

$$I = \frac{nc}{2\pi}|E|^2, \qquad \text{(B.2)}$$

where n is the refractive index, $c = 3 \times 10^{10}$ cm/s is the speed of light in vacuum, I is measured in erg/cm^2 s, and E is measured in statvolts/cm.

In the MKS system, the intensity of the field described by Eq. (B.1) is given by

$$I = 2n\left(\frac{\epsilon_0}{\mu_0}\right)^{1/2}|E|^2 = \frac{2n}{Z_0}|E|^2, \qquad \text{(B.3)}$$

where $\epsilon_0 = 8.85 \times 10^{-12}$ F/m, $\mu_0 = 4\pi \times 10^{-7}$ H/m, and $Z_0 = 377\ \Omega$. I is measured in W/m^2, and E is measured in V/m. Using these relations we can obtain the results shown in Table B.1. As a numerical example, a pulsed laser of modest energy might produce a pulse energy or $Q = 1$ mJ with a pulse duration of $T = 10$ ns. The peak laser power would then be of the order of $P = Q/T = 100$ kW. If this beam is focused to a spot size of $w_0 = 100\ \mu$m, the pulse intensity will be $I = P/\pi w_0^2 \simeq 0.3$ GW/cm^2.

TABLE B.1

Conventional	CGS		MKS	
I	I (erg/cm^2 s)	E (statvolt/cm)	I (W/m^2)	E (V/m)
1 W/cm^2	10^7	0.0458	10^4	1.37×10^3
1 kW/cm^2	10^{10}	1.45	10^7	4.34×10^4
1 MW/cm^2	10^{13}	45.8	10^{10}	1.37×10^6
1 GW/cm^2	10^{16}	1.45×10^3	10^{13}	4.34×10^7
1 TW/cm^2	10^{19}	4.85×10^4	10^{16}	1.37×10^9

Appendix C. Physical Constants

Constant	Symbol	Value	CGS*	MKS*
Speed of light in vacuum	c	2.998	10^{10} cm/s	10^{8} m/s
Elementary charge	e	4.803	10^{-10} esu	
		1.602		10^{-19} C
Avogadro number	N_A	6.023	10^{23} mol	10^{23} mol
Electron rest mass	$m = m_e$	9.109	10^{-28} g	10^{-31} kg
Proton rest mass	m_p	1.673	10^{-24} g	10^{-27} kg
Planck constant	h	6.626	10^{-27} erg s	10^{-34} J s
	$\hbar = h/2\pi$	1.054	10^{-27} erg s	10^{-34} J s
Fine structure constant†	$\alpha = e^2/\hbar c$	7.30	10^{-3}	10^{-3}
	$1/\alpha = \hbar c/e^2$	137	—	—
Compton wavelength of electron	$\lambda_C = h/mc$	2.426	10^{-10} cm	10^{-12} m
Rydberg constant	$R_\infty = me^4/2\hbar^2$	1.09737	10^{5} cm^{-1}	10^{7} m^{-1}
Bohr radius	$a_0 = \hbar^2/me^2$	5.292	10^{-9} cm	10^{-11} m
Electron radius†	$r_e = e^2/mc^2$	2.818	10^{-13} cm	10^{-15} m
Bohr magneton†	$\mu_B = eh/2m_ec$	9.273	10^{-21} erg/G	10^{-24} J/T
		⇒	1.4 MHz/G	
Nuclear magneton†	$\mu_N = e\hbar/2m_pc$	5.051	10^{-24} erg/G	10^{-27} J/T
Gas constant	R	8.314	10^{7} erg/K m	10^{0} J/K mole
Volume, mole of ideal gas	V_0	2.241	10^{4} cm^3	10^{-2} m^3
Boltzmann constant	k_B	1.381	10^{-16} erg/K	10^{-23} J/K
Stefan–Boltzman constant	σ	5.670	10^{-5} erg/cm^2 s K^4	10^{-8} W/m^2 K^4
Gravitational constant	G	6.670	10^{-8} dyne cm^2/g^2	10^{-11} N m^2/kg^2
Electron volt	eV	1.602	10^{-12} erg	10^{-19} J

*Abbreviations: C = coulombs, mol = molecules, g = grams, J = joules, N = newtons, G = gauss, T = teslas.

† Defining equation is shown in the gaussian cgs system of units.

Physical Constants Specific to the MKS System

Constant	Symbol*	Value*
Permittivity of free space	ϵ_0	8.85×10^{-12} F/m
Permeability of free space	μ_0	$4\pi \times 10^{-7}$ H/m
Velocity of light in free space	$(\epsilon_0\mu_0)^{-1/2} = c$	2.997×10^{8} m/sec
Impedance of free space	$(\mu_0/\epsilon_0)^{1/2} = Z_0$	377 Ω

* Abbreviations: F = farad = coulomb/volt, H = henry = weber/ampere.

Conversion between the Systems

$$1 \text{ m} = 100 \text{ cm}$$

$$1 \text{ kg} = 1000 \text{ g}$$

$$1 \text{ newton} = 10^5 \text{ dynes}$$

$$1 \text{ joule} = 10^7 \text{ erg}$$

$$1 \text{ coulomb} = 2.998 \times 10^9 \text{ statcoulomb}$$

$$1 \text{ volt} = 1/299.8 \text{ statvolt}$$

$$1 \text{ ohm} = 1.139 \times 10^{-12} \text{ s/cm}$$

$$1 \text{ tesla} = 10^4 \text{ gauss*}$$

$$1 \text{ farad} = 0.899 \times 10^{12} \text{ cm}$$

$$1 \text{ henry} = 1.113 \times 10^{-12} \text{ s}^2/\text{cm}$$

* Here 1 tesla = 1 weber/m^2; 1 gauss = 1 oersted.

Index

E

F

G

H

I

K

L

M

S

T

ISBN 0-12-121680-2